T0189702

Lecture Notes of the Institute for Computer Sciences, Social Informatics and Telecommunications Engineering 469

The LNICST series publishes ICST's conferences, symposia and workshops. It reports state-of-the-art results in areas related to the scope of the Institute.

LNICST reports state-of-the-art results in areas related to the scope of the Institute. The type of material published includes

- Proceedings (published in time for the respective event)
- Other edited monographs (such as project reports or invited volumes)

LNICST topics span the following areas:

- General Computer Science
- E-Economy
- E-Medicine
- Knowledge Management
- Multimedia
- Operations, Management and Policy
- Social Informatics
- Systems

Weina Fu · Lin Yun

Editors

Advanced Hybrid Information Processing

6th EAI International Conference, ADHIP 2022
Changsha, China, September 29–30, 2022
Proceedings, Part II

Springer

Editors
Weina Fu (iD)
Hunan Normal University
Changsha, China

Lin Yun (iD)
Harbin Engineering University
Harbin, China

ISSN 1867-8211 ISSN 1867-822X (electronic)
Lecture Notes of the Institute for Computer Sciences, Social Informatics
and Telecommunications Engineering
ISBN 978-3-031-28866-1 ISBN 978-3-031-28867-8 (eBook)
https://doi.org/10.1007/978-3-031-28867-8

This Springer imprint is published by the registered company Springer Nature Switzerland AG
The registered company address is: Gewerbestrasse 11, 6330 Cham, Switzerland

Preface

We are delighted to introduce the proceedings of the Sixth European Alliance for Innovation (EAI) International Conference on Advanced Hybrid Information Processing (ADHIP 2022). This conference brought together researchers, developers and practitioners from around the world who are leveraging and developing hybrid information processing technologies as well as related learning, training, and practice methods. The theme of ADHIP 2022 was "Hybrid Information Processing in Meta World".

The technical program of ADHIP 2022 consisted of 109 full papers, which were selected from 276 submissions by at least 3 reviewers each, including 2 invited papers in oral presentation sessions at the main conference tracks. The conference tracks were: Track 1, Information Extraction and Processing in Digital World; Track 2, Education Based Methods in Learning and Teaching; Track 3, Various Systems for Digital World. The technical program also featured two keynote speeches: "Graph Learning for Combinatorial Optimization", by Yun Peng from Guangzhou University, China, which focused on the advantages of using graph learning models for combinatorial optimization, and presented a thorough overview of recent studies of graph learning-based CO methods and several remarks on future research directions; and "Intelligent Fire Scene Analysis Using Efficient Convolutional Neural Networks", by Khan Muhammad from Sungkyunkwan University, Republic of Korea, which presented currently available approaches for early fire detection and highlighted some major drawbacks of current methods, and it also discussed a few representative vision-based fire detection, segmentation, and analysis methods along with the available fire datasets, and the major challenges in this area.

Coordination with the steering chair Imrich Chlamtac was essential for the success of the conference. We sincerely appreciate his constant support and guidance. It was also a great pleasure to work with such an excellent organizing committee team, we appreciate their hard work in organizing and supporting the conference. In particular, the Technical Program Committee, led by our TPC Chair, Khan Muhammad who completed the peer-review process of technical papers and made a high-quality technical program. We are also grateful to the Conference Managers, to Ivana Bujdakova for her support, and to all the authors who submitted their papers to the ADHIP 2022 conference.

We strongly believe that the ADHIP conference provides a good forum for researchers, developers, and practitioners to discuss all the science and technology aspects that are relevant to hybrid information processing. We also expect that future ADHIP conferences will be as successful and stimulating, as indicated by the contributions presented in this volume.

Weina Fu
Lin Yun

Organization

Steering Committee

Chair

Imrich Chlamtac   University of Trento, Italy

Organizing Committee

General Chair

Yun Lin   Harbin Engineering University, China

TPC Chair

Khan Muhammad   Sungkyunkwan University, Republic of Korea

Web Chair

Lei Chen   Georgia Southern University, USA

Publicity and Social Media Chair

Jerry Chun-Wei Lin   Western Norway University of Applied Sciences, Norway

Workshop Chair

Gautam Srivastava   Brandon University, Canada

Sponsorship and Exhibits Chair

Marcin Wozniak   Silesian University of Technology, Poland

Publications Chair

Weina Fu   Hunan Normal University, China

Posters and PhD Track Chair

Peng Gao   Hunan Normal University, China

Local Chair

Cuihong Wen   Hunan Normal University, China

Technical Program Committee

Adam Zielonka	Silesian University of Technology, Poland
Amin Taheri-Garavand	Lorestan University, Iran
Arun Kumar Sangaiah	Vellore Institute of Technology, India
Ashutosh Dhar Dwivedi	Technical University of Denmark, Denmark
Chen Cen	Institute for Infocomm Research, Singapore
Chunli Guo	Inner Mongolia University, China
Dan Sui	Changchun University of Technology, China
Danda Rawat	Howard University, USA
Dang Thanh	Hue Industrial College, Vietnam
Dongye Liu	Inner Mongolia University, China
Fanyi Meng	Harbin Institute of Technology, China
Feng Chen	Xizang Minzu University, China
Fida Hussain Memon	JEJU National University, Republic of Korea
Gautam Srivastava	Brandon University, Canada
Guanglu Sun	Harbin University of Science and Technology, China
Hari Mohan Pandey	Edge Hill University, UK
Heng Li	Henan Finance University, China
Jerry Chun-Wei Lin	Western Norway University of Applied Sciences, Norway
Jianfeng Cui	Xiamen University of Technology, China
Keming Mao	Northeastern University, China
Khan Muhammad	Sungkyunkwan University, Republic of Korea
Lei Ma	Beijing Polytechnic, China
Marcin Woźniak	Silesian University of Technology, Poland
Mu-Yen Chen	National Cheng Kung University, Taiwan
Norbert Herencsar	Brno University of Technology, Czech Republic

Ping Yu	Jilin University, China
Shuai Wang	Hunan Normal University, China
Shuai Yang	Xinyang Vocational and Technical College, China
Shui-Hua Wang	Loughborough University, UK
Tenghui He	Hunan Normal University, China
Thippa Reddy Gadekallu	Vellore Institute of Technology, India
Uttam Ghosh	Vanderbilt University, USA
Wuxue Jiang	The Education University of Hong Kong, China
Xiaochun Cheng	Middlesex University, UK
Xiaogang Zhu	Nanchang University, China
Xinyu Liu	Hunan Normal University, China
Xuanyue Tong	Nanyang Institute of Technology, China
Yanning Zhang	Beijing Polytechnic, China
Yongjun Qin	Guilin Normal College, China
Yun Lin	Harbin Engineering University, China
Zheng Ma	University of Southern Denmark, Denmark
Gabriel Gomes de Oliveira	Public University in Campinas, Brazil

Contents – Part II

Contents – Part I

Composite Fault Signal Detection Method of Electromechanical Equipment Based on Empirical Mode Decomposition

Guolong Fu[1,2], Jintian Yin[1,2(✉)], Shengyi Wu[1,2], Li Liu[1,2], and Zhihua Peng[1,2]

[1] Hunan Provincial Key Laboratory of Grids Operation and Control on Multi-Power Sources Area, Shaoyang University, Shaoyang 422000, China
`yinjintian112@yeah.net`
[2] School of Electrical Engineering, Shaoyang University, Shaoyang 422000, China

Abstract. Aiming at the problem of low detection accuracy when the traditional method is used to detect the composite fault signal of electromechanical equipment, a method for detecting the composite fault signal of electromechanical equipment based on empirical mode decomposition is proposed. In this paper, the operation information of the electromechanical equipment is collected first, and then the complex signal is identified based on the empirical mode decomposition theory, and the location of the complex fault area of the electromechanical equipment is completed to improve the detection accuracy. Finally, experiments are used to prove the advanced nature of the proposed method. The experimental results show that the fault diagnosis accuracy of the proposed method for electromechanical equipment is higher than 75%, the response time is less than 40 ms, and the memory occupation is less than 5500 kB, all of which are superior to the traditional method and have certain application value.

Keywords: Empirical mode decomposition · Electromechanical equipment · Fault signal · Fault detection

1 Introduction

With the rapid development of science and technology, fault diagnosis of electromechanical equipment is a rapidly developing technology all over the world. Fault diagnosis of electromechanical equipment is to master the working state of the equipment in the operation process, find the hidden dangers in the operation process of electromechanical equipment in time, and ensure the operation safety. Although the fault diagnosis technology of electromechanical equipment still uses the traditional diagnosis methods, its technology development has been widely adopted today, but what we see is that the fault diagnosis technology of electromechanical equipment has not formed a complete theoretical system and corresponding effective diagnosis reference technical specifications [1]. The existing technologies are studied one by one for different electromechanical specific faults, which are neither representative nor normative. The diagnosis methods

W. Fu and L. Yun (Eds.): ADHIP 2022, LNICST 469, pp. 1–13, 2023.
https://doi.org/10.1007/978-3-031-28867-8_1

of electromechanical equipment faults are determined according to the type of fault, and the real theories and methods are rarely used in practice, There is no complete system to evaluate this theory and method. The accuracy of fault diagnosis of electromechanical equipment is also a key problem to be solved urgently in this technology. Only by improving the accuracy of fault diagnosis can we reduce the repair time and effectively reduce the economic loss [2]. However, the key to the accuracy of fault diagnosis is to determine the characteristics of the fault, which is a complex problem. Generally speaking, the failure of electromechanical equipment is not a separate failure. The failure may be diversified, including motor manufacturing technology, use of materials, installation, operation, maintenance, etc. In modern mass production, the use of electromechanical equipment is more and more widely, and the operation safety of electromechanical equipment has attracted more and more attention. Based on this, this paper introduces the empirical mode decomposition theory into it, and studies the detection method of the composite fault signal of electromechanical equipment. It first analyzes the fault information extraction structure of electromechanical equipment, completes the collection of composite fault signals, then identifies its signals based on empirical mode decomposition theory, and completes the division and location of its fault areas. Citing the empirical mode decomposition theory can improve the detection quality of composite signals, greatly improve its recognition accuracy, and play an important role in the fault diagnosis and location of electromechanical equipment. It is hoped that through the research in this paper, it can provide literature references for current electromechanical equipment fault identification.

2   Composite Fault Signal Detection of Electromechanical Equipment

2.1   Composite Fault Signal Identification of Electromechanical Equipment

At present, the common monitoring of the health status of electromechanical equipment basically belongs to qualitative analysis and alarm, which can not achieve fault diagnosis and detailed analysis of fault types. Through the in-depth study of vibration signal fault detection algorithm and the research of key technologies such as sensor technology, terminal equipment pair communication and mutual control in industrial Ethernet network structure, a real-time online fault diagnosis model is developed, Realize the full life cycle management and fault diagnosis of electromechanical equipment, so as to facilitate the maintenance personnel to find potential safety hazards as soon as possible and formulate maintenance plans, so as to ensure safe production. The model is composed of computer, power box, vibration monitoring substation, optical transceiver, vibration sensor, shaft temperature sensor and other equipment and host computer software [3]. The composition structure of the model is shown in the Fig. 1 below.

Under the EMD motor equipment architecture, the division of labor for the extraction of electromechanical fault information needs to be determined according to the specific situation and equipment performance [4]. In the extraction of electromechanical fault information, the intelligent terminal needs to work in the field of the device under test, so its performance is limited in many aspects. Considering the limited data processing capability of intelligent terminals, data processing tasks with high computational complexity

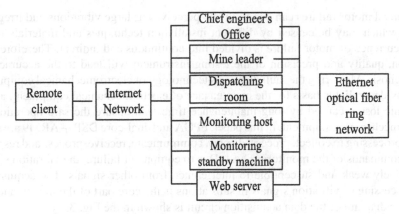

Fig. 1. Schematic diagram of electromechanical fault information extraction structure

require empirical mode decomposition equipment to complete. Empirical Mode Decomposition (EMD) plays an important role in EMD motor device architecture because of its computational power and ease of use [5]. The ultimate purpose of fault diagnosis of electromechanical equipment is fault classification. In this paper, several machine learning methods are used for classification. This relatively complex computational task requires empirical mode decomposition. Within the same inch, some key information obtained through calculation can also feed back some guidance and improvement instructions to the terminal. The operation steps of the equipment fault information detection instruction are as follows (Fig. 2).

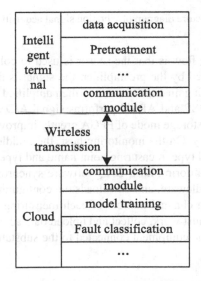

Fig. 2. Operation steps of equipment fault information detection instruction

General motor failures can be intuitively perceived as large vibrations and irregular noises, which may be caused by different installation techniques and materials used. The occurrence of motor failure is divided into continuous and indirect. Therefore, the function, quality and precision of the testing instrument will lead to the accuracy of the diagnosis. Mastering the fault diagnosis technology of electromechanical equipment can provide a reliable basis for the maintenance of electromechanical equipment, avoid economic losses caused by long maintenance time, and ensure the safe operation of electromechanical equipment. In this paper, FPGA and dual-core DSP + ARM9 are used for co-processing to collect, process, display, communicate, receive, process and respond to the commands of the monitoring host. Due to equipment failure, the vibration signal is relatively weak and susceptible to interference from other signals. The acquisition and processing of vibration signals in substations is the core part of the model, and the structure diagram of the data acquisition circuit is shown in the Fig. 3.

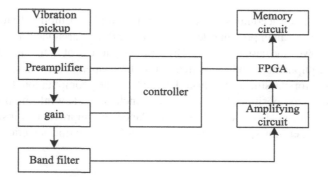

Fig. 3. Structure diagram of vibration signal acquisition circuit

The signal processing flow is that the sensor is used to collect vibration information, the signal is amplified by the preamplifier, the signal is filtered by 4-order low-pass and high pass through gain adjustment, and then amplified twice. The cooperative controller composed of DSP and ARM9 performs signal A/D conversion and storage through FPGA. The hard storage mode of FPGA greatly improves the real-time performance and work efficiency. On the monitoring host, the middleware defines the plant area, equipment name and type, measuring point name and type. This process is called configuration; On the monitoring host, DAQ software synchronizes the configuration information from the middleware, and downloads the configuration information generated by the corresponding channel number of each measuring point to the substation in DAQ. The substation must work, collect and upload data according to the configuration information; When the equipment connected to the substation changes, it must be reconfigured.

2.2 Location Algorithm of Abnormal Area of Electromechanical Equipment Fault Signal

Among the commonly used signals for fault detection at present, noise, current, voltage and vibration signals are particularly favored by scientific researchers. The reason is that these kinds of signals are the most convenient to collect and can reflect the current operation state of electromechanical equipment to a certain extent. The characteristics of these four kinds of signals are shown in the table [6] (Table 1).

Table 1. Performance comparison of different data types

Attribute	Signal type			
	Noise	Electric current	Voltage	Vibration signal
Is it intrusive	Non intrusive	Non intrusive	Intrusive	Non intrusive
Fault type reflection	mechanical failure	Electrical fault	Electrical fault	Mechanical and electrical faults
Fault prediction time in advance	Several weeks before the failure occurs	Several weeks before the failure occurs	Several weeks before the failure occurs	Several weeks before the failure occurs
Scope of application	Leakage of gas, liquid and vacuum equipment	Failure and power consumption of electromechanical equipment	Failure and power consumption of electromechanical equipment	Rotating electromechanical and mechanical equipment failure

Through the experimental comparison of four original data signals, this paper studies whether they are intrusive, reflect the fault type, scope of application and prediction time in advance. At the same time, considering the needs of practical production and application, vibration signal is the most suitable data type for fault diagnosis and prediction of electromechanical equipment. The detection of vibration signal is mainly to detect phase, frequency and amplitude, The most important is fast and accurate fault detection. At present, fault detection is mostly realized by hardware equipment, but the price of high-precision fault detection hardware equipment is very high, which is not suitable for large-scale network application of underground equipment. Therefore, the research on high-precision and fast software fault detection technology is the key to the success of the project [7]. Empirical mode decomposition can transform the original signal between time domain and frequency domain to find a more convenient method of data analysis. In the field of signal processing or data processing, fast empirical mode decomposition is a typical time-frequency domain decomposition method. From the root, HT is a method to quickly calculate discrete empirical mode decomposition. This fast calculation method is more suitable for modern computers to carry out empirical mode decomposition analysis of data. 130fft is not only a fast algorithm of DFT, The definition is also derived from DHT. The definition of FFT is given below. The definition of process DFT is:

$$X_k = \sum x_{N-1} e^{-2\pi \frac{1}{N}} \tag{1}$$

where, $x_0, x_1, \ldots, x_{N-1}$ is a complex number, and its complexity is calculated as e according to its definition. Centralize the data sample $N = \{x_1, x_2, \ldots, x_n\}$, and assume that the new coordinate system after orthogonal decomposition is $\{w_1, w_2, \ldots, w_a\}$, z_i^T is the orthogonal basis vector. If the dimension of the data space after dimensionality reduction is a, the original sample in the new coordinate system is μ, the original data sample point x, and the space between the reconstructed data sample point cage is:

$$\sum_{i=1}^{m} \left\| \sum_{j=1}^{d'} z_{ij} x_i \right\|_2^2 = \mu \sum_{i=1}^{m} z_i^t N - 2 \sum_{i=1}^{m} z_i^t w_a + a \tag{2}$$

If the distance from all sample points to the hyperplane is close enough to meet the best reconstruction and minimize the spacing, then

$$\Delta W = \sum_{i=1}^{m} \left\| \sum_{j=1}^{d'} z_{ij} x_i \right\|_2^2 \min_W -\mathrm{tr}\left(W^T \left(\sum_{i=1}^{m} x_i x_i^T \right) W \right)$$

$$= \sum_{i=1}^{m} \left\| \sum_{j=1}^{d'} z_{ij} x_i \right\|_2^2 \min_W -\mathrm{tr}\left(W^T X X^T W \right) \tag{3}$$

The traditional software fault detection algorithm can not be applied to the single chip microcomputer model because of its complex operation structure, and the operation speed is relatively slow. The project studies a fault detection algorithm based on empirical mode decomposition, which is applied to the embedded model for rapid fault detection, reduces the application cost, and lays the foundation for the wide application of technology. Consider the following sinusoidal signals with amplitude A, period t and initial phase φ:

$$x(t) = A e^{-2\pi \frac{1}{N}} - \Delta W \sin\left(2\pi \frac{1}{N} t + \varphi \right) \tag{4}$$

The standard sinusoidal signal of the same frequency is:

$$y(t) = \sin\left(2\pi \frac{1}{T} - e^{-2\pi \frac{1}{N}} \right) \tag{5}$$

If the period is μ_x and the initial phase is μ_y, the cross covariance function is:

$$C_{xy}(\tau) = E[x(t) - \mu_x] E\left[(y(t) - \mu_y) \right] \tag{6}$$

where, P represents the Fourier transform function of the mean formula as follows:

$$P_{xy}(\omega) = \int_{-}^{+\infty} C_{vy}(\tau) e^{-\omega t} - x(t) y(t) \tag{7}$$

This algorithm has high accuracy after test. The cross power spectrum operation method greatly improves the anti-interference ability. It has significant advantages in the application under the condition of narrow underground space and more interference [8]. In addition, the structure of the algorithm is relatively simple, which greatly reduces the consumption of computing resources, improves the computing speed, and provides conditions for the wide application of low-cost fault detection.

2.3   Realization of Fault Signal Detection of Electromechanical Equipment

According to the characteristics of periodic data and fault signals of electromechanical equipment, a fault signal detection algorithm of motor equipment based on empirical mode decomposition is proposed. The algorithm is divided into two parts: Empirical Mode Decomposition part and intelligent terminal part. The empirical mode decomposition part is mainly responsible for the division of data, the cache of effective data, the calculation of the optimal sliding window size and the judgment of fault signal. The intelligent terminal part is mainly responsible for the data exchange with the empirical mode decomposition end, According to the extraction results of fault signal information by empirical mode decomposition, locate the fault signal and detect the located fault signal [9]. After processing the original data, the extracted and dimensionally reduced feature vectors are transmitted to the. On the one hand, the amount of data transmission is reduced, on the other hand, the reliability of the data to be transmitted is increased. The support vector machine is used to classify the bearing fault feature data. The largest part of the consumption of computing resources is the calculation of the classification hyperplane and the classification process of the data. Putting this part of the calculation on the can greatly improve the accuracy and practical feasibility of classification [10]. And in the construction of fault feature database, in the process of fault diagnosis, constantly improve the feature database, and use new features to replace the old features, not only make the features up-to-date, but also keep the data in the feature database on a certain scale [11]. The establishment of model base makes the algorithm proposed in this paper no longer have one-time learning ability like the traditional sVIM fault feature classification algorithm. The algorithm proposed in this paper can be continuously improved

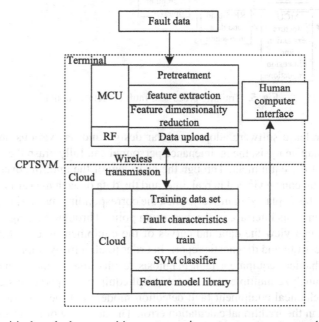

Fig. 4. Empirical mode decomposition motor equipment support vector machine model

and enriched in the process of operation, and has the ability of "lifelong learning". As shown in the figure, the empirical mode decomposition motor equipment support vector machine model used in this paper (Fig. 4).

The empirical mode decomposition online training model adopts the structure of "offline training + online training online classification". In the initial stage, the classified fault data are used to train the empirical mode decomposition model offline and establish the initial empirical mode decomposition model; Then start online fault diagnosis. The diagnosis results of the initial empirical mode decomposition model are notified to the site through the human-computer interface. After the staff check the fault on the spot, they are fed back to the model, and the fault feature vector is saved in the feature model library. The model counts the accuracy of diagnosis. When the accuracy is lower than the specified threshold, call the feature library for secondary training of the empirical mode decomposition model, That is, online training, while avoiding excessive resource occupation. The training process of empirical mode decomposition model is shown in the Fig. 5.

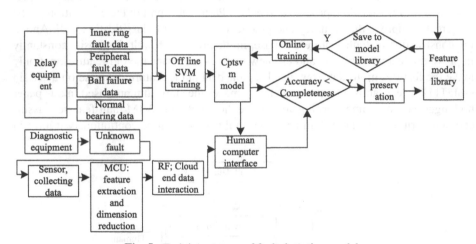

Fig. 5. Training process of fault detection model

The data release software adopts modular design and has various analysis maps. Through the map analysis, the maintenance personnel can fully grasp the operation and health status of the equipment. Through the overview of the unit, the overall operation of the equipment can be viewed in real time, and the data of each measuring point on the equipment can be displayed intuitively. Click the corresponding measuring point to enter the spectrum analysis interface of the measuring point. Through the diagnostic analysis interface, you can view the data and curves of the equipment in each time period, as well as the spectrum and diagnostic report. In order to effectively improve the accuracy of electromechanical equipment fault diagnosis results, the diagnosis model needs to calculate and analyze multiple electromechanical equipment operation signals through the electromechanical equipment fault detection model, and the detection result error will be less than the traditional calculation error. The accuracy of the test results of the electromechanical equipment fault detection model is significantly higher than that of the traditional fault diagnosis methods, which has strong diagnostic value and should be vigorously popularized. With the progress of science and technology, the traditional fault diagnosis technology of electromechanical equipment has a new development trend. The fault diagnosis of electromechanical equipment has integrated sensor technology, artificial intelligence technology, data processing technology and wireless communication technology, and gradually transformed into precision and multidimensional, diversified diagnosis theory and diagnosis model and intelligent diagnosis technology. Mature diagnostic technology will be quickly applied to socialist modernization and national defense security construction.

3   Analysis of Experimental Results

In order to fully verify the practicability of the proposed method, this paper firstly tests the performance of the detection algorithm, and then based on different load conditions and fault changes, the diagnostic accuracy of different equipment states changes, and finally selects the detection accuracy, detection time and operation occupancy. Three indicators of memory are used to test the performance of the proposed method and the comparison method. The experimental equipment uses Dell precision tower5810 server and the terminal equipment uses embedded intelligent terminal. In order to facilitate the experimental comparison and operation, this paper uses the simulated periodic flow data for the test, and adds the periodic gradual fault signal and noise interference signal for the test. The experiment is divided into two parts: the comparison of the parameters of the algorithm itself and the comparison of different algorithms to fully illustrate the effectiveness and superiority of the algorithm. According to the different sampling frequency of the sensor, it is determined that the number of data points added to the fault signal in each cycle is 10–20. The input k value range is determined according to the range of the fault signal. The accuracy of the algorithm with different K values is shown in the Fig. 6.

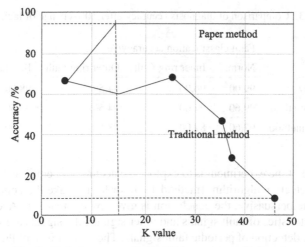

Fig. 6. Variation curve of algorithm accuracy with K value

It can be seen from the figure that the value of K is too large or too small, which will lead to the decline of the accuracy of the algorithm. According to the analysis of the detection results output by the algorithm, if the value of K is too small, it will lead to the algorithm misjudging some fault signals as normal data flow and reducing the detection rate; If the value of K is too large, the normal signal is easy to be misjudged as a fault signal, which increases the misjudgment rate, resulting in a sharp decline in the accuracy of the algorithm. According to the above analysis, the best K value should be between 10 and 15. During the operation of mechanical equipment, the equipment state may change, and the reasons include load change and mechanical damage caused by long-time operation of the equipment. The model solution proposed in this paper is helpful to reduce the diagnosis error caused by this kind of change to the diagnosis model. In this experiment, the data under different load conditions are collected as the change of load state, and the data under different fault diameters are collected as the mechanical wear caused by long-time operation. In order to use a large amount of data for the calculation of diagnosis accuracy, the "manual inspection" in the "algorithm description" part is replaced with a fault tag to judge whether the diagnosis is correct or not. The table shows the comparison between the accuracy of traditional off-line learning and online learning diagnosis proposed in this paper when the load change and fault size change are used to simulate the change of equipment state (Table 2 and table 3).

Table 2. Comparison of diagnostic accuracy after 20% load change

Test method	Fault classification accuracy/%				
	Normal	Inner ring fault	Outer ring fault	Ball failure	Total
Original accuracy	99.00	96.00	94.50	98.00	96.85
Online learning method	98.60	97.00	94.50	98.50	95.95
Offline learning method	93.60	81.50	90.50	95.00	92.00

Table 3. Comparison of diagnosis accuracy after 10% change of fault size

Test method	Fault classification accuracy/%				
	Normal	Inner ring fault	Outer ring fault	Ball failure	Total
Original accuracy	99.00	96.00	94.50	98.00	96.85
Online learning method	99.60	97.00	94.50	96.50	97.85
Offline learning method	97.60	82.00	78.50	90.00	87.89

The purpose of the experiment is to explore the accuracy of empirical mode decomposition fault detection algorithm (method 1). In order to make the conclusion of this experiment more persuasive, the conclusion is given by comparison. According to the different characteristics of fault signals and other signals, the signals are separated, and both include the detection of periodic fault signals. The comparison of the experimental results can better reflect the advantages of this algorithm (Fig. 7).

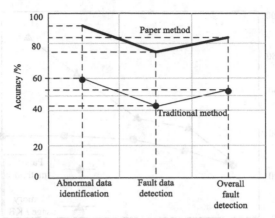

Fig. 7. Comparison of accuracy of different methods

It can be seen from the experimental data that the empirical mode decomposition data fault detection algorithm has the highest accuracy, and the test signal processing method based on cepstrum has the lowest detection accuracy. Although the test signal processing method based on cepstrum can highlight the small period signal through signal weighting, the influence of noise will also increase in the signal with low signal-to-noise ratio. Therefore, although it has good processing effect on the signal with high signal-to-noise ratio, its recognition degree for the signal segment containing a large amount of noise is not very high, resulting in the decline of the overall detection accuracy of this method; The empirical mode decomposition fault detection algorithm proposed in this paper makes full use of the periodic characteristics of fault signals, which can effectively separate noise signals and fault signals, so as to achieve high detection accuracy. In order to compare the execution efficiency of the algorithm, the length of time spent in the operation of the algorithm and the size of memory space are selected for comparison. The experimental data used are divided into five sections, each accounting for 20% of the data set. The average time and average memory space required for the three algorithms to run on the segmented data are counted respectively. The experimental results are shown in the Fig. 8 (Fig. 9).

It can be seen from the experimental data above that the fault signal detection algorithm of empirical mode decomposition motor equipment proposed in this paper is superior to the traditional algorithm in terms of running time and memory. The calculation speed of the proposed algorithm is obviously faster than the other two algorithms, so from the speed analysis, this algorithm is more suitable for the stream data processing with a large amount of data. It can be seen from the figure that the memory consumption of this algorithm is less than that of the other two algorithms, which proves that this method has high practicability and fully meets the research requirements.

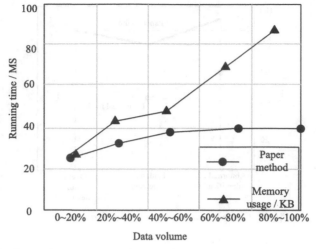

Fig. 8. Comparison of running time of different algorithms in different stages

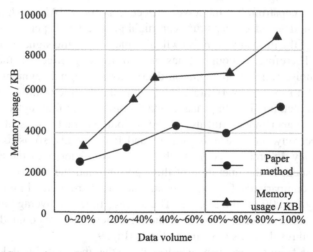

Fig. 9. Comparison of memory usage of different algorithms in different stages

4 Concluding Remarks

With the increasing application of large-scale electromechanical equipment, people's requirements for its fault detection technology are gradually increasing. In view of the existing problems such as difficulty in fault detection and positioning and long mainte-nance cycle, the research on high-efficiency electromechanical equipment fault diagnosis technology is of great significance for reducing maintenance workload, shortening fail-ure time, and improving production efficiency Through the analysis of the research status of fault diagnosis technology, this paper studies and designs the fault detection algorithm based on empirical mode decomposition, and develops the vibration monitoring substation, vibration sensor, temperature sensor and host computer monitoring software based on FPGA and dual core controller, forming a set of electromechanical equipment fault diagnosis model. The model has been tested in Shanxi Tiandi Wangpo Coal Industry Co., Ltd. The operation of the model is stable and reliable. It can judge the operation state of the equipment in advance and give corresponding measures, which greatly reduces the workload of on-site maintenance and maintenance personnel and improves the efficiency of safe production.However, the model in this paper still has shortcomings, that is, it can only identify the faults of some electromechanical equipment, and cannot accurately identify the fault signals of large electromechanical equipment systems, which is also the next research goal.

Fund Project. Hunan Provincial Department of Education Youth Fund Project (21B0690); National College Students Innovation and Entrepreneurship Training Program (202110547057); Hunan University student innovation and entrepreneurship training program (Xiangjiaotong [2021] No. 197, item 3385); Shaoyang City Science and Technology Plan Project (2021GZ039); Hunan Provincial Science and Technology Department Science and Technology Plan Project (2016TP1023).

References

1. Zhao, Y.-B.: Research on multi-sensor fusion multi-fault signal monitoring of Marine turbine equipment. Ship Sci. Technol. **44**(17), 114–117 (2022)
2. Huang, W., Li, S., Mao, L., et al.: Research on multi-source sparse optimization method and its application in compound fault detection of gearbox. J. Mech. Eng. **57**(07), 87–99 (2021)
3. Wang, H., Wang, M., Song, L., et al.: Method of compound fault signal separation using double constraints non-negative matrix factorization. J. Vibr. Eng. **33**(03), 590–596 (2020)
4. Lian, J., Fang, S., Zhou, Y.F.: Model predictive control of the fuel cell cathode system based on state quantity estimation. Comput. Simul. **37**(07), 119–122 (2020)
5. Xiao, Y., Shen, Y., Yang, F., et al.: Fault state variables integral based open-circuit fault detection for power unit of cascaded h-bridge converter. Power Syst. Technol. **45**(11), 4213–4225 (2021)
6. Cui, R.H., Li, Z., Tong, D.-S.: Arc fault detection based on phase space reconstruction and principal component analysis in aviation power system. Proc. CSEE **41**(14), 5054–5065 (2021)
7. Chen, Y., Chen, Y., Liu, Z.-Q., et al.: A gear fault detection method based on a fiber bragg grating sensor. Chin. J. Lasers **47**(03), 232–241 (2020)
8. Liu, H., Wang, Y.-Y., Chen, W.-G., et al.: Fault detection for power transformer based on unsupervised concept drift recognition and dynamic graph embedding. Proc. CSEE **40**(13), 4358–4371 (2020)
9. Wang, B., Cui, X.: Detection method of arc high resistance grounding fault in resonant grounding system based on dynamic trajectory of volt-ampere characteristic. Proc. CSEE **41**(20), 6959–6968 (2021)
10. Yang, S.-Z., Xiang, W., Wen, J.: A fault protection scheme based on the difference of current-limiting reactor voltage for overhead MMC based DC grids. Proc. CSEE, **40**(04), 1196–1211+1411 (2020)

Research on Fast Separation Method of Motor Fault Signal Based on Wavelet Entropy

Jintian Yin[1,2], Li Liu[1,2(✉)], Junfei Nie[1,2], Zhihua Peng[1,2], and Riheng Chen[1,2]

[1] Hunan Provincial Key Laboratory of Grids Operation and Control on Multi-Power Sources Area, Shaoyang University, Shaoyang 422000, China
yinjintian112@yeah.net
[2] School of Electrical Engineering, Shaoyang University, Shaoyang 422000, China

Abstract. The extraction of motor signals by traditional methods will be affected by multi-component signals and non-stationary signals, and the separation effect of motor fault signals is poor. Therefore, a fast separation method of motor fault signals based on wavelet entropy is proposed. Obtain the motor fault vibration signal, convert it to the frequency domain for solution, and denoise the motor fault vibration signal through three-layer wavelet packet decomposition. Based on wavelet entropy, the sliding window is set for simulation, and the optimal features are selected for extraction to quantitatively describe the time-frequency and energy distribution of motor fault transient vibration signal. The second-order VKF filter is selected to extract multiple components at the same time, so as to realize the separation of multi-component signals. Experimental results show that this method can effectively separate and extract motor fault signals, and can achieve good results under high noise intensity.

Keywords: Wavelet entropy · Motor failure · Fault signal · Rapid separation · Signal separation · Fault diagnosis

1 Introduction

In the industrial process, the safety and reliability of mechanical system determine the quality of products. Whether the faults can be identified and classified in time is the key to ensure the safe operation of the system and inhibit the deterioration of faults. As an important power equipment in various production fields, motor has the advantages of low price, relatively simple overall structure and reliability. It undertakes more than 80% of the kinetic energy output in the process of modern industrial and agricultural production. In the modern industrial system, with the rapid development of manufacturing digitization, real-time recording and perception of production and operation status and operation environment have been realized, and a large number of industrial time series data have been accumulated and are being generated. If the motor fails during the operation of the equipment, it will lead to unstable operation and sharp increase in energy consumption. In serious cases, it will even cause damage to the motor and equipment, which will affect

W. Fu and L. Yun (Eds.): ADHIP 2022, LNICST 469, pp. 14–25, 2023.
https://doi.org/10.1007/978-3-031-28867-8_2

the normal operation of the whole equipment. Sudden shutdown and maintenance must be carried out, resulting in problems such as slow production progress and economic loss. In the face of massive data, how to quickly extract the sensitive feature set of motor fault, accurately identify and classify it, and quickly separate the motor fault signal is the key to efficiently find motor fault and avoid serious damage. It is also the key research object of fault diagnosis.

At present, it can be divided into three categories: machine based, model-based and signal-based fault diagnosis. Wang Xin et al. Took a group of air-conditioning fault motors as the experimental object, built a motor fault diagnosis platform, carried out acoustic signal acquisition experiments on air-conditioning motors in four states, and classified the data set by using algorithm [1]. As a new structure depth learning algorithm, algorithm can classify motor fault acoustic signals well. Cai Wenwei and others proposed a fault diagnosis method of micro motor based on sound signal. According to the characteristics of high signal-to-noise ratio and easy to be affected by the environment, the method of maximum correlation kurtosis deconvolution wavelet threshold denoising is used to enhance the periodic impact components in the sound signal and filter out the environmental noise [2]. The envelope and envelope spectrum of the signal are obtained by Hilbert transform. According to the shape of the envelope and the frequency corresponding to the peak value of the envelope spectrum, the fault diagnosis of micro motor is realized. Based on the idea of multi-source information fusion, Zhang Yahui and others adopted the fusion correlation spectrum characteristics of motor stator current and vibration signal as the diagnosis basis of rotor broken bar and stator turn to turn short circuit fault [3]. By fusing the characteristic signals containing the same fault frequency component for correlation analysis and establishing the correlation relationship between different signals, they can effectively suppress the spectrum components not related to fault identification in the single signal spectrum, make the motor fault characteristic frequency component more prominent and reduce the difficulty of fault identification. The fast separation method of motor fault signal can keep the motor itself in good operation condition, but there is still the problem of interaction between multi-component signal separation and feature extraction of non-stationary signal. Motor and its related power equipment are important assets of enterprises. Its reliability and stability in the operation process is the key to ensure the safe and stable operation of mechanical equipment for a long time.

Therefore, it is necessary to study the fast separation method of motor fault signal. Wavelet analysis is a kind of time-frequency analysis, which is developed on the basis of Fourier analysis. It decomposes the signal into the superposition of a series of wavelet functions. Because wavelet transform is a local transformation of signal time-frequency domain, it eliminates the influence of signal redundant information and can extract feature information more effectively. The high-frequency coefficient matrix of each layer of wavelet transform is formed into a sequence related to probability distribution, and its entropy value can be obtained through calculation, which is the representation of the average complexity of wavelet coefficient distribution. Wavelet entropy contains a lot of information that can characterize the fault characteristics. According to the change of wavelet entropy before and after motor fault, the motor state is detected. In this paper, a fast separation method of motor fault signals based on wavelet entropy is proposed. The method first collects the vibration signal of the motor equipment fault, and by estimating the bandwidth of the modal, it is converted into the frequency domain and solved to obtain the motor fault vibration signal to be processed. Then, the noise reduction process is carried out. Based on the feature extraction of the motor fault signal by the wavelet entropy, a rapid separation model of the motor fault signal is established to realize the rapid separation of the fault signal. Finally, experiments are used to prove the advanced nature of this method. The method can improve the accuracy of motor fault diagnosis, has certain practical significance for ensuring the safe production of motors, and can also provide reference for signal separation in related fields.

2   Fast Separation Method of Motor Fault Signal Based on Wavelet Entropy

2.1   Obtain Motor Fault Vibration Signal

Vibration signal is one of the direct feedback of motor dynamic behavior, which contains the characteristics and attributes that can accurately describe the motor fault state. Because the motor vibration signal presents the characteristics of non-stationary and non-linear, and the vibration signal in fault state is quite different from the normal state, it can reflect the characteristics of different motor fault states [4]. In this paper, the vibration signal of motor is obtained by modal decomposition. The flow of modal decomposition is shown in Fig. 1.

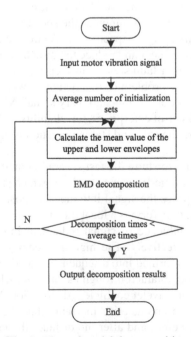

Fig. 1. Flow of modal decomposition

Modal decomposition finds the maximum and minimum values of the motor signal, connects these extreme points in the form of spline interpolation as the upper and lower envelope of the signal, and removes the arithmetic mean of the envelope as the low-pass baseline component. The remaining high-frequency oscillation part is the mode of the motor signal. After modal decomposition, the original motor signal can be expressed as:

$$\beta(t) = \sum_{a=1}^{b} \chi_a(t) + \delta_b(t) \tag{1}$$

In formula (1), t represents time; $\beta(t)$ is the mode of the original motor signal; $\chi_a(t)$ is each IMF component; $\delta_b(t)$ is the residual component; a is the number of components; b is the mean value of the upper and lower envelope. Hilbert transform the mode to obtain its unilateral spectrum. The calculation formula is as follows:

$$f(t) = \left[\alpha(t) + \frac{\lambda}{\pi t} \right] * \beta(t) \tag{2}$$

In formula (2), $f(t)$ represents the single side spectrum of motor fault vibration signal; $\alpha(t)$ is the unit impulse function; λ is imaginary unit; $*$ stands for convolution. Then, the frequency spectrum of the mode is moved to the corresponding fundamental frequency band. By estimating the bandwidth of the mode, the motor fault vibration signal is converted to the frequency domain for solution, and the motor fault vibration signal to be processed is obtained.

2.2 Noise Reduction Processing of Motor Fault Vibration Signal

In addition to rich fault characteristic information, the vibration signal generated by the motor in case of fault usually includes other interference noise. Therefore, it is necessary to denoise the obtained motor fault vibration signal. Wavelet transform can work in time domain or frequency domain, and extract the local information through the analysis of the signal [5]. In this paper, wavelet soft threshold is used to denoise the motor fault vibration signal. A one-dimensional signal model with noise can be expressed as:

$$B(t) = G(t) + \kappa(t) \tag{3}$$

In formula (3), $B(t)$ represents a one-dimensional signal with noise; $G(t)$ is the $\frac{1}{\vartheta}$ noise signal to be extracted, and ϑ is the frequency; $\kappa(t)$ is the background noise signal of the environment and system in the detection process. For the processing of the same signal, the results may vary greatly after using different wavelet functions to transform the signal. Therefore, when processing and analyzing the signal in practical application, we should consider the characteristics of the analyzed signal and the adaptability of the wavelet function, and select the wavelet function that can well reflect the characteristics of the signal. In engineering signal analysis, DB4 shows good performance in frequency domain. Therefore, DB4 wavelet is selected as the wavelet function in this paper. The orthogonal wavelet basis DB4 is selected on MATLAB software, and the wavelet packet coefficients of different frequency bands are obtained by using wpdec function and wpcoef function to decompose and solve the wavelet packet coefficients of each frequency band. After 3-layer wavelet packet decomposition, the signal is also decomposed into the average adjacent subbands by wavelet packet, that is, there will be 8 subbands. After wavelet decomposition, in order to separate the signal from the noise, it is necessary to select an appropriate number as the threshold. When the calculated decomposition coefficient is less than the selected threshold, the decomposition coefficient can be regarded as caused by noise and discarded; On the contrary, it is considered to be mainly caused by signals [6]. According to the principle of wavelet packet energy spectrum analysis, the percentage of each frequency band in the total signal energy can be obtained by dividing the energy of each channel after wavelet packet decomposition by the total signal energy. At this time, the setting method of soft threshold is adopted, which shrinks a fixed value to zero, and then reconstructs the wavelet coefficients after wavelet decomposition through the formula. The soft threshold function mentioned here is:

$$p' = \begin{cases} \operatorname{sgn} p(|p| - \varphi), & |p| \geq \varphi \\ 0, & |p| < \varphi \end{cases} \tag{4}$$

In Eq. (4), p' is the high-frequency wavelet coefficient obtained by operation; sgn is a symbolic function; p the appropriate number selected from the high-frequency wavelet coefficients; φ represents the threshold. Through this capability spectrum analysis, we can roughly know the approximate frequency range of the fault point. After the motor fault vibration signal is decomposed by wavelet packet, because the number and width of samples in each frequency band are equal, the energy spectrum obtained by wavelet packet decomposition can be transformed into histogram, which is convenient and intuitive to observe the feature quantity.

2.3 Feature Extraction of Motor Fault Signal Based on Wavelet Entropy

In the research process, this paper will find out the relevant features in turn, focus on wavelet entropy, set sliding window for simulation, select the best features for extraction, and set parameters for comparison. Because this signal has great difference in frequency distribution, and wavelet analysis has good localization characteristics in frequency domain. Compared with wavelet transform, the high-frequency part of the signal shows better frequency resolution and can show the characteristics contained in the high-frequency part of the signal. Therefore, this paper mainly starts with the two calculation methods of spectrum entropy and energy entropy. The analysis flow of extracting motor fault signal characteristics based on wavelet entropy is shown in Fig. 2.

When the motor fails, the voltage and current on the line will suddenly change. Because the line adopts distributed parameters in this paper, the sudden change signal will make the fault voltage and current signals reflect back and forth continuously, resulting in the waveform distortion of fault voltage or current in the transient process. The calculation of spectral entropy is mainly aimed at the frequency spectrum. After the frequency spectrum of the signal is made by Fourier transform, for the frequency distribution, the amplitude of each frequency point is taken as the sum of squares, and then the square of the amplitude of each point is divided by the total sum of squares to obtain

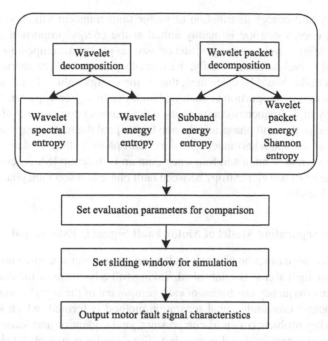

Fig. 2. Analysis flow of extracting motor fault signal characteristics based on wavelet entropy

the probability, that is, the probability of frequency distribution [7]. Relative entropy can be used as a measure of the difference between two signals. When the difference between two signals is large, the relative entropy between signals will show a larger value. When the similarity between two signals is large, the relative entropy between signals will show a smaller value. The calculation formula of frequency distribution probability is as follows:

$$q_x = \frac{A_x^2}{\sum_c A_x^2} \tag{5}$$

In formula (5), q_x represents the probability of frequency distribution; A_x represents the amplitude of each frequency point; x and c represent the serial number and total number of frequency points. The spectrum entropy can be further calculated, and the formula is as follows:

$$H = -\sum_c q_x \log q_x \tag{6}$$

In Eq. (6), H represents the spectral entropy. Therefore, combined with wavelet packet transform and relative entropy, a comprehensive wavelet entropy which can quantify the difference between a fault point and other signal points in motor operation is constructed. If the fault signal is decomposed by wavelet multi-layer decomposition, the order of singular value greater than 0 may be very large after singular value decomposition and transformation. Therefore, wavelet entropy can quantitatively describe the time-frequency and energy distribution of motor fault transient vibration signal. The calculation of energy entropy is mainly aimed at the energy proportion of each frequency band. After wavelet decomposition or wavelet packet decomposition, the signal is decomposed to each subband. Firstly, the sum of squares of all reconstruction coefficients on each node should be calculated, that is, the energy value of each subband. By calculating the energy proportion of each subband, the energy entropy can be obtained. There are many high-frequency components in the transient components of the voltage or current signal of the fault phase, and the non fault signal fluctuates near the fundamental frequency, so the spatial arrangement of the fault phase voltage and non fault phase voltage signal in time domain and frequency domain is different [8]. By comparing the relative difference of wavelet entropy between fault phase and non fault phase, the motor fault signal is identified.

2.4 The Fast Separation Model of Motor Fault Signal is Established

Based on the feature extraction of motor fault signal by wavelet entropy, a fast separation model of motor fault signal is established. Taking full advantage of the characteristics of instantaneous frequency estimation of each component of the signal, a non-stationary multi-component motor fault signal separation method is formed, which successfully solves the two key problems of multi-component signal separation and feature extraction of non-stationary signal in signal processing. The singular points of the signal can be divided into two categories. One is that the amplitude of the signal changes suddenly at a certain time, resulting in the discontinuity of the signal; The other is that the amplitude waveform of the signal is continuous, but its first-order differential is discontinuous. By means of frequency domain filtering, the signal decomposition problem is transformed into a constrained optimization problem. Its ultimate goal is to decompose the real signal with multiple frequency components into a series of discrete quasi orthogonal sub signals, that is, the finite bandwidth eigenmode function, and minimize the sum of the limited bandwidth of all modes [9]. In this paper, if estimation is used as the reference instantaneous frequency parameter of each component, and second-order VKF filter is selected to extract multiple components at the same time, so as to realize the separation of multi-component signals. Each mode has its own center frequency, which can be regarded as tightly supported in the spectrum. The set of all modes is the optimal reconstruction of the original signal. The time-frequency baseline extracted by the instantaneous frequency estimation method can estimate the instantaneous frequency of the signal component of interest to obtain its carrier matrix. Then the reconstruction form of each component signal is:

$$U = WZ \tag{7}$$

In Eq. (7), U represents the reconstruction form of each component signal; W represents the amplitude envelope of the component; Z represents carrier matrix. In case of motor fault, the signal traveling wave generated from the fault will always exist, refraction and reflection will occur at the same time, and the traveling wave will not disappear until the motor fault signal is cut off or restored to normal. On the basis of fully considering the narrowband characteristics of the signal, the optimal Wiener filter is constructed adaptively according to the center frequency of the mode, which makes the frequency band of the mode more concentrated and the signal-to-noise ratio higher. The components realizing signal separation still belong to non-stationary signals, and order analysis can transform the signal from non-stationary in time domain to stationary in angle domain through equal angle resampling [10]. The order ratio (order) is defined as the number of cyclic vibrations of the object to be measured for each revolution. The expression is:

$$\varpi = \frac{\eta \varsigma}{v} \tag{8}$$

In formula (8), ϖ represents the order; η represents vibration frequency; v represents the speed; ς indicates the frequency of motor rotation. Carry out diagonal slice bispectrum analysis on the signal residue. When the input speed is unique, the order diagonal slice bispectrum analysis based on speed can be carried out to extract the amplitude frequency characteristics of the signal residue. Using the interval estimation method, for the collected sample data of similar faults, the value interval of wavelet entropy at each level in the population with sufficient confidence can be estimated from the sample data, that is, the confidence interval. When the wavelet entropy of each scale is within the confidence interval determined by the sample, this kind of fault can be explained. Finally, the instantaneous frequency estimation and order amplitude characteristics and the amplitude frequency characteristics of diagonal slice bispectrum are used to diagnose the vibration signal. Based on the above process, the design of motor fault signal fast separation method based on wavelet entropy is completed.

3 Experimental Analysis

3.1 Experimental Preparation

The motor vibration data used in this experiment are collected by sqi-mfs mechanical fault. In the process of data acquisition, the speed of the motor is controlled by the frequency converter to obtain the vibration data of the motor at different speeds. The frequency conversion meter can display the current speed and corresponding frequency. The acceleration sensor is horizontally installed above the driving end and load end of the motor to collect motor vibration signals in different states. The motor status can be described as normal, broken rotor bar, bearing fault, stator winding and voltage imbalance. 150 samples are collected for each motor state, and the training set and test set are divided according to the ratio of 4:1. The sampling frequency of the experiment is 128 Hz and the motor speed is 1600 rpm. The fast separation methods of motor fault signals based on 1d-cnn algorithm, sound signal and fusion correlation spectrum are selected as the control group, and compared with the design method in this paper. In order to better compare and evaluate the performance of this method and other improved methods for motor fault signal separation and extraction, this paper uses the fault characteristic coefficient as the quantitative index. This index can be expressed as the ratio of the amplitude of the fault characteristic frequency component of the signal to the amplitude of all frequency components. The fault characteristic coefficient is regarded as the ratio of fault component energy to total signal energy in the signal. The larger its value is, the higher the proportion of fault component energy is, indicating that the more obvious the fault characteristic is, and the more effective the fault signal separation is.

3.2 Results and Analysis

In order to fully verify the advanced nature of the method in this paper, Fast separation method of motor fault signal based on 1d-cnn algorithm is selected as comparison method 1, Fast separation method of motor fault signal based on sound signal as comparison method 2 and Fast separation method of motor fault Signal based on fusion correlation spectrum is the comparison method 3, and a comparison experiment is carried out with the method in this paper. The separation effect of the motor fault signal is related to the noise intensity, so this paper conducts experimental tests under different noise intensity conditions, and the results are shown in Figs. 3, 4, 5 and 6.

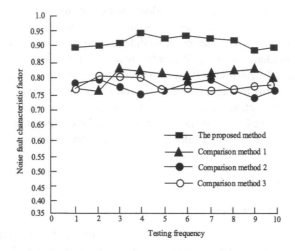

Fig. 3. Fault characteristic coefficient of noise = 0 dB

In the case of noise = 0 dB, after multiple tests, the fault characteristic coefficient obtained by the proposed method is in the range of 0.883−0.942, and the average fault characteristic coefficient is 0.919, which are all better than the comparison method. Compared with method 3, the improvement is 0.150, 0.118 and 0.144.

In the case of noise = 1 dB, after multiple tests, the fault characteristic coefficient obtained by the proposed method is in the range of 0.848−0.886, and the average fault characteristic coefficient is 0.867, which are all better than the comparison method. Compared with method 3, the improvement is 0.149, 0.163 and 0.166.

In the case of noise = 1 dB, after multiple tests, the fault characteristic coefficient obtained by the proposed method is in the range of 0.741−0.786, and the average fault characteristic coefficient is 0.764, which are all better than the comparison method. Compared with method 3, the improvement is 0.143, 0.146 and 0.152.

In the case of noise = 4 dB, after multiple tests, the fault characteristic coefficient obtained by the proposed method is in the range of 0.668 − 0.723, and the average

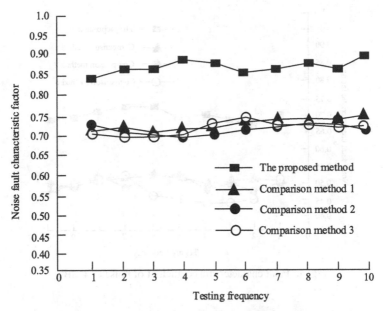

Fig. 4. Fault characteristic coefficient of noise = 1 dB

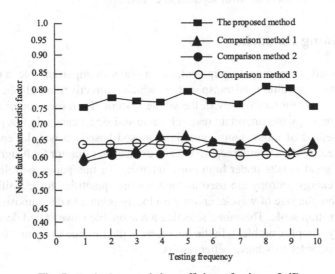

Fig. 5. Fault characteristic coefficient of noise = 2 dB

fault characteristic coefficient is 0.687, which are all better than the comparison method. Compared with method 3, the improvement is 0.191, 0.195 and 0.227.With the increase of noise intensity, the time-frequency baseline of motor signal becomes more and more blurred, and the dispersion of signal peak points increases, which increases the difficulty of fault signal separation. The method based on wavelet entropy can still effectively

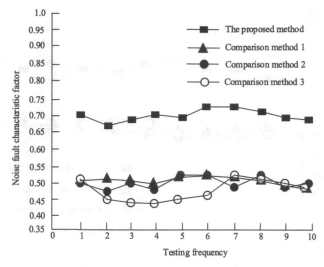

Fig. 6. Fault characteristic coefficient of noise $= 4$ dB

extract motor fault features when the noise intensity increases, realize the rapid separation of fault signals, and show an ideal separation effect.

4   Concluding Remarks

As a widely used power output equipment, motor plays an important role in many fields such as industry, agriculture and transportation, which is directly related to the production efficiency and operation reliability of the whole system. Therefore, motor fault signal diagnosis technology has important research value and significance. In this paper, a fast separation method of motor fault signal is proposed based on wavelet entropy. This method can realize the effective separation and extraction of motor fault signal, and can also achieve good results under high noise intensity. In this paper, wavelet spectrum entropy and energy entropy are used as fault feature quantities for classification and recognition, but the size of wavelet entropy is closely related to decomposition wavelet and decomposition scale. Therefore, selecting a reasonable wavelet and decomposition scale is a very complex problem. In the future, we will continue to try and follow up on this problem in order to achieve better results.

Fund Project. Hunan Provincial Department of Education Youth Fund Project (21B0690); Shaoyang City Science and Technology Plan Project (2021GZ039); Hunan Provincial Science and Technology Department Science and Technology Plan Project (2016TP1023)

References

1. Wang, X., Mao, D.-X., Li, X.-D.: Motor fault diagnosis using microphones and one-dimensional convolutional neural network. Noise Vib. Control 41(2), 125–129 (2021)
2. Cai, W.-W., Huang, J., Li, W.-G., et al.: Research on fault diagnosis method for micro motor based on sound signal. Mach. Tool Hydraulics 48(23), 190–195 (2020)
3. Zhang, Y.-H., Yang, K., Li, T.-L.: Fault diagnosis method of asynchronous motors using fusion correlation spectrum. Electric Mach. Control 25(11), 1–7 (2021)
4. Wang, D.-M.: Sparse decomposition based maintenance simulation of mechanical and electrical faults of aviation equipment. Comput. Simul. 38(1), 37–41 (2021)
5. Shen, H.-F., Shi, J.: Fault feature extraction method of motor rotor vibration signals. Noise Vib. Control, 42(04), 138–143+151 (2022)
6. Wang, C.-L., Lu, X.-J.: A motor bearing fault diagnosis based on deep learning method. J. Lanzhou Jiaotong Univ. 39(2), 43–50 (2020)
7. Zhao, S.T., Wang, E.X., Chen, X.-X., et al.: Fault diagnosis method for large motor based on sound-vibration signal combined with 1D-CNN. J. Harbin Inst. Technol. 52(09), 116–122 (2020)
8. Zhang, Y.H., Wang, Y.-F., Zhao, X.-P., et al.: Motor fault diagnosis based on deep metric learning. Measur. Control Technol. 39(7), 30–37 (2020)
9. Wang, B., Ding, J.-H., Sheng, G.: Design of host computer for vibration signal acquisition of motor fault monitoring system. Autom. Instrum. 36(02), 11–15 (2021)
10. Yan, J.-L.: Motor fault diagnosis method based on vibration analysis. Appl. IC 38(7), 88–89 (2021)

Steering Control Method of Mobile Forklift Based on Sensing Signal

Haochen Zhang$^{(\boxtimes)}$ and Chen Xian

Shaanxi Institute of Technology, Xian 710300, China
ttbm21@163.com

Abstract. Because the steering control method of the mobile forklift has the problem of large steering deviation, a steering control method of the mobile forklift based on sensing signal is designed. Using the yaw rate calculated by the two-degree-of-freedom front wheel steering model as the ideal value, the steering variables of the mobile forklift truck are obtained, the rotation angle relationship between the two steering wheels is coordinated, the types of steering mechanisms are identified, the linear two-degree-of-freedom model is constructed, the desired steering curve corresponding to the steering wheel is written into the compensation controller, and the steering control mode is optimized by using sensing signals. Experimental results show that the steering deviation between the designed steering control method and the other two methods is 3.257°, 5.466° and 5.486° respectively, which indicates that the steering deviation decreases with the combination of sensing signals.

Keywords: Sensing signal · Mobile forklift · Steering control · Motion variable · Sideslip angle of centroid · Yaw rate

1 Introduction

According to the relevant records, the earliest forklift is used for the loading and unloading of military materials at the airport. It is a single-stage gantry forklift modified from a car by Clark Company. Forklifts were put into use during the Second World War because of the need to move military supplies and the high handling efficiency of forklifts that could not be compared with manual work. After the Second World War, with the development of the world automobile industry, the sales and technology of forklifts also developed rapidly. With the development of society and economy, people's pursuit of efficiency is gradually improving, so forklifts are more and more involved in economic development and social construction. Forklifts, also known as forklifts, automatic lifts, forklifts, versatile loaders, etc., is a forklift to take goods to put the mechanical device, is one of China's top 10 construction machinery equipment, is mainly used in freight yards, factories (machinery manufacturers, textile mills, etc.), docks, commercial sites (such as shopping malls, etc.), airports and construction sites, because of its compact body, short axle distance, rotation is flexible, is now widely used in the social national

© ICST Institute for Computer Sciences, Social Informatics and Telecommunications Engineering 2023
Published by Springer Nature Switzerland AG 2023. All Rights Reserved
W. Fu and L. Yun (Eds.): ADHIP 2022, LNICST 469, pp. 26–39, 2023.
https://doi.org/10.1007/978-3-031-28867-8_3

economy vehicles and tools. Since the steering torque of the vehicle can be easily determined by the steering motor, the yaw angle can be predicted by a linear vehicle model of the steering system and an observer after the yaw rate and steering angle are measured.

The main function of a forklift truck is to transfer, stack, store and transport the goods. With the development of social production and science and technology, the performance of forklifts has been gradually improved, and the number and variety of forklifts have increased significantly. Forklifts can also be divided into internal combustion forklifts powered by fossil fuels and mobile forklifts powered by electricity (storage forklifts are also very widely used mobile forklifts) [1, 2] according to the different types of power. On the contrary, the mobile forklift has the advantages of low noise, high reliability, easy operation, environmental protection and energy saving, and has won more and more forklift buyers in the development process. According to the different contact ways between the wheels and the ground, it can be divided into track-type forklift and trackless forklift, which is widely used at present. According to the position of the fork can be divided into straight moving forklift and side moving forklift. Although the proportion of internal combustion forklifts is relatively high and the market share of mobile forklifts is relatively low, the ratio of forklifts to internal combustion forklifts is changing, and the proportion of mobile forklifts is increasing year by year, especially electric storage forklifts, electric storage forklifts have entered an accelerated growth cycle. Then according to the predicted side-slip angle, the system uses a feedback control to carry out effective steering intervention to achieve the correct steering operation.

2 Steering Control Method of Mobile Forklift Based on Sensing Signal

2.1 Obtain Steering Motion Variables for Mobile Forklifts

Yaw rate, sideslip angle of center of mass and steering radius are important factors to evaluate steering performance of forklift. Moving forklift handling stability of the first choice of variables should be considered to select the parameters that can be directly measured or can be estimated. The yaw rate is related to the trajectory retention. When the sideslip angle of the center of mass is close to 0, the yaw rate determines the attitude of the forklift. When the sideslip angle of the center of mass is small, the yaw rate determined by the linear two-degree-of-freedom model is relatively stable for the vehicle motion. The yaw angle and yaw rate of the moving forklift can be regarded as the first target of the main motion variables of the control system, and the yaw rate of the moving forklift can be measured by gyroscope and other instruments, and the yaw angle of the moving forklift can be real-time estimated by the state observer. The yaw rate gained from the two-degree-of-freedom front wheel steering model is taken as an ideal value, and the yaw rate gain of the four-wheel steering vehicle is similar to that of the front wheel steering vehicle, so the driver will not feel great change. The centroid sideslip angle is often used to describe the trajectory keeping problem.

Compared with the traditional steering control system of forklifts, the control target of the stability of the mobile forklift mainly includes two aspects: (1) The high speed operation of the mobile forklift requires the stability and safety of the mobile forklift, that is, the response value of the centroid yaw angle of the mobile forklift should be controlled near zero degree, and the yaw rate of the mobile forklift should be smaller than that of the traditional steering control system. When driving a forklift, the driver expects the forklift to run along the axis, that is, the speed of the vehicle coincides with the longitudinal axis of the vehicle. (2) When the mobile forklift runs at a low speed, it is required that the mobile forklift can improve the mobility of the mobile forklift while maintaining stability, that is, the response value of the centroid angle of the mobile forklift should be as close to zero as possible, and the yaw angle speed of the mobile forklift is larger than the value of the traditional steering control system. Therefore, the formula of ideal yaw rate and pavement adhesion condition is as follows:

$$W = \frac{\phi/H}{\frac{\phi}{H^2}(1 - \gamma)^2} \tag{1}$$

In formula (1), ϕ represents the road adhesion coefficient, H the gravitational acceleration, and γ the lateral acceleration of the vehicle. Therefore, the ideal yaw rate must satisfy the constraint conditions:

$$|\eta| \le \left|\frac{\phi \times H}{\phi}\right| \tag{2}$$

In order to improve the transient response, the ideal yaw rate is modified according to formula (2) by referring to the approximate first-order pure lag system of the sideslip angle of the ideal center of mass in order to improve the transient response, if only taking the constant η as the ideal yaw rate will cause the vehicle yaw rate to oscillate greatly under the condition of the step signal of the yaw angle:

$$\eta' = \min\left(\left|\frac{\phi \times H}{\phi}\right|\right)(\varepsilon) \tag{3}$$

In formula (3), the ε represents the first-order inertia response time constant. To sum up, the sideslip angle and yaw rate of the moving forklift are selected as the main motion variables of the control system. Because of the limitation of working environment and working property, the working speed of forklift truck can not be too large, but the tire sideslip angle is small in this case. The handling stability of a forklift is the ability of the forklift to follow the driver's direction through the steering system and the steering wheels, and to resist interference and keep the forklift running steadily when it encounters external interference. Forklift handling stability, stability of the good or bad will directly affect the handling of the good or bad degree, so collectively known as handling stability. The handling stability of a forklift is closely related to the sideslip angle and yaw rate of the center of mass. When the sideslip angle of the center of mass is small, the yaw rate is usually used to describe the steering characteristics of forklifts, and the yaw rate can reflect the driver's driving intention. When the sideslip angle of the centroid is large, the yaw rate can no longer reflect the steering characteristics of the forklift, because the larger the sideslip angle of the centroid, the lower the control ability of the driver to the lateral motion of the forklift. Therefore, the sideslip angle and yaw rate of the center of mass are important physical parameters to describe the handling stability of forklift.

2.2   Identify the Type of Steering Structure

The steering structure of a forklift truck is a complete set of mechanisms used to control the steering wheel or articulated frame's azimuth and deflection angle, which ensures the forklift truck driver can control the driving direction of the forklift truck. Forklift drivers can make the corresponding steering action according to the operation needs and the driving environment. Forklift truck steering is light and flexible, small turning radius, good maneuverability and rear wheel steering. There is a serious internal leakage in the steering process, which leads to the disorder of the relationship between the steering wheel and the wheel, so the driver can not predict the steering angle of the vehicle according to the position of the steering wheel in the process of driving, so the current position of the wheel can only be determined by turning back repeatedly, which greatly increases the labor intensity of the driver.

Mobile forklift steering system according to the steering mode, there are two kinds of single-axis steering and bi-axis steering. Single-axis steering mode is mainly used in three-wheeled forklifts. The rear wheels are usually connected to the frame by two tires side by side. The single-axis steering mode is simple and does not need any additional steering mechanism to coordinate the steering angle relations between the steering wheels. The steering is flexible and the steering maneuverability is better. In addition, it is found that the leakage operation of the solenoid valve has the same mechanism and effect as the internal leakage of the steering system, and the steering system is out of sync by reducing the flow of oil in the steering cylinder. At present, four-wheel steering forklift belongs to two-axle steering. In this kind of steering, two forklift wheels rotate around different axes. This kind of steering needs to add steering mechanism to coordinate the angle relation between two steering wheels, which is relatively complex.

When the steering mechanism works, the steering wheel drives the valve core of the full hydraulic steering gear to rotate, and the high-pressure oil enters the steering cylinder through the full hydraulic steering gear to push the piston rod to move. Among them, the full hydraulic steering gear is the measuring device of the full hydraulic steering system to ensure that the steering wheel angle input and the wheel angle into a fixed ratio. The double axle steering mode is simple in structure and good in performance. Therefore, the double axle steering mode is mostly used in four-wheeled mobile forklifts. In order to ensure that the two steering wheels of forklifts can rotate at the same time, they need to be connected with certain structure, and the steering mechanism should be the same when the forklifts move left and right. As an open-loop control system, the steering accuracy and stability of the full hydraulic steering system can not be guaranteed. The steering deviation generated by the system cannot be automatically eliminated, resulting in a disorderly relationship between the steering wheel and wheel position [3]. Therefore, it is necessary to analyze and solve the steering deviation. Because the steering angles of the two wheels are different when steering, the steering mechanism generally adopts a trapezoidal structure. The steering structure of a forklift truck is shown in Fig. 1:

As can be seen from Fig. 1, the steering structures of forklifts mainly include single trapezoidal steering structure, double trapezoidal steering structure and crank slider structure. When steering, the axes of all wheels are required to intersect at a point called the steering center, so as to ensure that the forklift wheel makes pure rolling motion when steering. If the wheel slides with the ground, it will increase the resistance of motion,

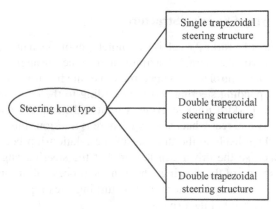

Fig. 1. Steering structure type of forklift truck

thereby speeding up tire wear and reducing the service life of the tire. One of the main features of mobile forklifts is that the working device for loading and unloading goods is placed in the front of the forklift truck and is suspended in front of the wheel supporting the forklift [4]. The main reason of steering deviation of full hydraulic steering system is internal leakage of oil in full hydraulic steering system. At present, the full hydraulic steering gear equipped with cycloidal valve type, this type of full hydraulic steering gear has the advantages of flexible operation, long reliable life, compact structure and easy to arrange, and can realize the manpower steering when the engine is off, so it is widely used in the steering system of forklifts, loaders, cranes and other low-speed and heavy-duty vehicles.

Fork is used to lift and fork cargo parts, through the gantry lifting mechanism to drive the fork to complete the lifting of goods, through the control of hydraulic cylinder mechanism connected with the gantry, fork can also complete the pitching and pitching, relying on these two movements. Plus the forklift truck forward and backward, forklift can easily complete the cargo support, lift, stacking, stacking and other processes, the frame is equipped with a roof rack, the roof rack structure is simple, can provide a good view to the driver, facilitate the driver to observe the surrounding environment [5]. Another feature is that the working device adopts telescopic portal frame and time-saving lifting pulley block, which enables the forklift truck to obtain a larger lifting height under the condition of smaller shape height, improves the passability of the forklift truck and expands the use scope.

2.3 Construction of Linear Two-Degree-of-Freedom Model

Although the L2DOF model only considers the lateral and lateral motion of the forklift and ignores the nonlinear factor of the tire, the simplification of the vehicle model meets the requirements of the vehicle kinematics. In front of the driving axle is the lifting structure of the forklift truck, behind is the frame of the forklift truck. The lifting structure is mainly composed of the door frame, the fork frame and the fork. The tilt of the door frame is controlled by the hydraulic cylinder, and the chain drives the fork frame and the fork to complete the lifting of the goods. The maneuverability and stability of mobile forklift can be divided into two aspects, one is maneuverability, the other is stability. The maneuverability of the forklift means that the forklift can be driven according to the driver's steering intention without deviating from the predetermined track. The dynamic equations of the model are as follows:

$$\begin{cases} (h_1 + h_2)\,\sigma + \frac{1}{\alpha}\|h_1 - h_2\| = \frac{1}{L(\mu)^2} \\ (h_1 - h_2)\,\sigma + \alpha + \frac{1}{g} = L(\mu) - E \end{cases} \tag{4}$$

In formula (4), h_1 stands for centroidal to front wheelbase, h_2 stands for centroidal to rear wheelbase, σ stands for mass of the whole vehicle, α stands for centroidal angle, L stands for steering wheel angle, μ stands for vehicle inertia, g stands for yaw stiffness of front tire, and E stands for yaw stiffness of rear tire. It is equipped with electronic instruments, seats, fans and transmitters for forklifts. The relevant parts of the steering system and the counterweight are installed in the rear part of the forklift. The engine provides power for the forklift, while providing additional stability torque for the longitudinal stability of the whole forklift. Stability is the ability of a forklift to keep steady when disturbed by the outside world, and avoid dangerous situations such as tail wagging and overturning when steering at high speed. The lifting of the forklift can be regarded as a simple lever action, the balance weight above the steering bridge can provide additional stable torque, can increase the quality of the forklift lifting goods.

The steering characteristics of the forklift are mainly characterized by the yaw rate. When the speed of the forklift is constant, the bigger the yaw rate is, the faster the speed is allowed, and the better the turning performance and stability is. In order to ensure the flexibility of forklift, the brake system of forklift is installed on the driving axle. Emergency braking can easily cause the whole forklift to lean forward and longitudinal instability. When synchronous steering control is carried out, the desired steering curve corresponding to the steering wheel and wheel is firstly written into the compensation controller.

When vehicle steering, the steering wheel angle sensor and wheel angle sensor detect the steering wheel and wheel angle respectively, input the detected steering wheel angle into the desired steering curve to calculate the desired wheel angle, and then compare the desired wheel angle with the actual wheel angle. If the steering deviation is greater than zero, the solenoid valve is opened to compensate the steering cylinder with oil, and then the synchronous steering is realized. However, if the wheel angle is too large or the road surface is too complex or the adhesion coefficient is too small, the bigger the sideslip angle of the center of mass of the forklift is, the worse the trajectory keeping ability is, and the driver can not reflect the steering intention of the driver in real time, the smaller the control ability of the driver for the side motion and the yaw motion is, and the dangerous situation such as rollover slip and forklift driving a circle is easy to occur. In this case, the adaptive compensation can be realized by adjusting the opening of the solenoid valve, and the compensation precision and compensation stability are improved.

2.4 Sensor Signal Optimization Steering Control Mode

Sensing signals can be used for gas leak detection, and their prior information plays an important role in signal processing. They provide a systematic research direction for measuring the probability of signal vectors. According to the gas detection technology, the sensing signal carrying gas information can be obtained. Considering the drift of hardware or optical structure, the noise with unknown statistical characteristics in the environment, etc., it is necessary to combine some signal processing algorithms to improve the signal-to-noise ratio and detection precision of the system when detecting the gas concentration or retrieving gas parameters. A lot of research has been done on the closed form of signal apriori.

When gas leakage occurs, according to the detected gas concentration, it is also necessary to combine with some signal processing algorithms to locate the gas leakage source quickly and accurately. For example, Gibbs distribution is to use Laplace matrix to evaluate the probability of image, and the smoothness degree measured by Laplace operator can be used to judge the probability of signal. The role of steering control is to control the direction of forklift, and then achieve steering purposes. The steering process of forklifts can be summarized as steering force through steering transmission and other devices to make the steering wheels deflect at a certain angle to achieve forklifts steering. Through the analysis, for the four-wheel steering mode in line with the existing driving habits, driving vehicles on the driver's requirements are relatively easier. Of course, whether four-wheel steering or four-wheel steering is faced with the need for a four-wheel steering system, the system response and control accuracy can not be verified, the structure is complex, high cost, high risk, very challenging. Considering the working environment of forklifts, most forklifts work in narrow environment such as warehouse, wharf and so on. Loading and transporting heavy goods need very frequent steering. In general, forklift steering system generally has the following steering control requirements: operation to be light, steering action to be sensitive, tracking to be good, steering handling stability to be high.

In addition, the four-wheel conditional steering mode cannot realize the parallel movement of vehicles and the ordinary steering mode (the so-called ordinary mode means that two wheels are fixed in parallel on the coaxial line and the other two wheels are deflected), so in order to realize the parallel movement of vehicles and the traditional ordinary driving mode, the driving mode selection switch needs to be added in the design. For example, when the speed of a forklift is low, the yaw rate of the forklift is usually expected to be increased while the turning radius is reduced, so that understeer can be avoided. After comprehensive consideration, the steering scheme of the vehicle is determined to be: the combination of the four-wheel conditional steering mode and the driving mode. The control circuit makes real-time control and adjustment according to the position of the four wheels, and responds to the driver's steering wheel's control. This model is the "guide point" in the body of the longitudinal axis of the line, the following to the left side forklift as an example to establish mathematical models. On the contrary, when the speed of forklift truck is high, the driver usually hopes not to be too sensitive, because when the steering wheel angle is too sensitive, the driver is prone to nervous state of mind when driving at high speed, easy to misuse the steering wheel and other dangerous behaviors. Therefore, when driving at high speed, the driver usually hopes that the yaw angle speed can be appropriately reduced relative to the steering wheel angle to enhance the driving safety. In the normal driving mode: the driver in the normal driving mode, the side forklift two front wheels according to the steering wheel rotation angle to turn, and two rear wheels keep motionless turn angle is zero. Thus it can be seen that the turning radius and yaw rate can reflect the steering maneuverability of forklift.

The steering flexibility of a forklift usually refers to the turning ability of the forklift, especially in narrow space, which can better reflect the steering flexibility of the forklift. The geometric aspect can be described by turning radius, and the kinematic aspect can be described by vehicle speed and yaw rate. The steering angle of the front two wheels is not limited by mechanical devices such as steering rocker. By using the control of the steering angle controller, the steering stability of forklift truck is improved and the tire friction loss is reduced. The control strategy of isometric reverse rotation of front and rear wheels is a kind of control strategy summed up according to the driver's driving experience of forklift truck. The isometric reverse rotation in the name refers to the control strategy that the virtual front wheel rotation angle and the rear wheel rotation angle are equal and opposite directions. The main idea of the control strategy of isometric reverse rotation of front and rear wheels is to control the center of the instantaneous rotating shaft of forklift truck to the horizontal line of the longitudinal center of the truck body, which is very in line with the driver's driving habit. The lateral force in the coordinate system shall be calculated according to the working conditions of the omnidirectional side forklift under normal driving conditions with uniform speed. The calculation formula is as follows:

$$G = \frac{|k + t|^2}{\|\arctan(\varpi - k)\|} \tag{5}$$

In formula (5), k represents the curve peak factor, t represents the curve shape factor, and ϖ represents the curve curvature factor. On this basis, the right front wheel of a forklift is selected as an example for the calculation of the wheel sideslip angle. The speed of the geometric center of the right front wheel around the center of mass of the whole vehicle is:

$$y = \frac{l\sqrt{p^2/q^2}}{\psi} \tag{6}$$

In formula (6), l represents the distance from the front wheel axle to the center of mass, p represents the horizontal drift of the curve, q represents the vertical drift, and ψ represents the vertical load on the tire. The control of isometric reverse rotation of front and rear wheels can not only reduce the turning radius of the forklift, but also improve the flexibility of the electric forklift. The control of isometric reverse rotation of front and rear wheels also satisfies the Ackermann theorem for three-wheel full steering forklifts. In the mode of in-situ steering, when the driver chooses this mode, the four wheels turn to the angle set by the electronic control unit at the same time, and the forklift turns in any direction under the control of the driver with the center of the whole vehicle as the center. The steering mode provides great convenience for the side forklift to turn around in situ, and improves the mobility of the side forklift in a narrow environment. In order to ensure that the three tires and the road surface of the electric forklift truck are in the state of pure rolling, the virtual front wheel angle and the geometric constraint are used to control the two front wheel angles respectively. The tire under pure rolling condition can exert its mechanical characteristics to the utmost extent, and reduce the wear of electric forklift tire under non-pure rolling condition.

3 Simulation Experiment

3.1 Experimental Preparation

Using MATLAB/SIMULINK platform for simulation analysis. In addition, in order to analyze the effect of joint fuzzy control, the ideal values of yaw rate and sideslip angle of center of mass are obtained by ideal model. ADAMS can create mechanical system simulation model, and MATLAB can build control model. Joint simulation uses the same physical prototype model. If two software are simulated independently and separately, we need to create the system model under two different software environments. Under the high speed simulation of the mobile forklift, the speed of the forklift is 2. Skm/h, the forklift is simulated under the stepping condition. Given the three-wheel angle, the signal jump of the rear wheel is 1.57 rad, the signal jump of the left front wheel is 2.32 rad, the signal jump of the right front wheel is 0.75 rad, and the input angle is given the virtual front wheel angle.

Through the ADAMS own interface and MATLAB/Simulink combined use, to achieve the control of data exchange between variables. In order to establish the joint simulation model, we need to connect ADAMS/View and MATLAB/Simulink through ADAMS/Control tools, create state variables in ADAMS/View, define them as output and input variables, establish the control system model in MATLAB, and exchange the output values of ADAMS as the input values of MATLAB between two software. Especially, the steady state performance of forklift steering is emphasized and the steady state value is observed. The steering performance of forklift truck under sinusoidal condition is studied.

3.2 Experimental Results

In order to verify the feasibility of the control method, the steering control method of mobile forklift based on singular value decomposition and envelope analysis are selected to carry out simulation experiments. The steering deviations of the three methods are tested under different steering angles. The experimental results are shown in Table 1–5.

Table 1. Steering angle 500° steering deviation (°)

Number of experiments	Steering control method of mobile forklift based on singular value decomposition	Steering control method of mobile forklift based on envelope analysis	Steering control method of mobile forklift
1	8.945	8.554	6.548
2	9.036	8.692	5.615
3	9.122	9.307	6.142

(*continued*)

Table 1. (*continued*)

Number of experiments	Steering control method of mobile forklift based on singular value decomposition	Steering control method of mobile forklift based on envelope analysis	Steering control method of mobile forklift
4	8.877	8.546	6.646
5	9.613	9.112	5.849
6	9.250	8.456	6.945
7	9.163	9.001	5.331
8	8.489	8.748	6.128
9	8.331	9.316	6.489
10	9.004	8.549	5.002
11	8.557	9.055	5.221
12	9.161	8.748	5.031
13	8.944	9.163	6.111
14	9.033	9.224	5.499
15	8.247	9.218	5.336

As can be seen from Table 1, the average steering deviation of the steering control method of the designed mobile forklift truck and the other two steering control methods of the mobile forklift truck is 5.860°, 8.918° and 8.913° respectively; as can be seen from Table 2, the average steering deviation of the steering control method of the designed mobile forklift truck and the other two steering control methods of the mobile forklift truck is 4.238°, 7.189° and 7.346° respectively; as can be seen from Table 3, the average steering deviation of the steering control method of the designed mobile forklift truck and the other two steering control methods of the mobile forklift truck is 3.181°, 5.205° and 5.180° respectively; as can be seen from Table 4, the average steering deviation of the steering control method of the designed mobile forklift truck and the other two steering control methods of the mobile forklift truck is 1.892°, 3.710° and 3.889° respectively; as can be seen from Table 5, the average steering deviation of the designed mobile forklift truck and the other two steering control methods is 1.13° and 2.306° respectively. The experimental results show that the bigger the steering angle is, the smaller the steering deviation is, and the experimental results of the steering control method are always lower than the other two methods. Because the method in this paper takes the yaw rate calculated by the two-degree-of-freedom front wheel steering model as the ideal value, the steering variables of the mobile forklift are obtained. A linear two-degree-of-freedom model is constructed, and the expected steering curve corresponding to the steering wheel is written into the compensation controller, and the steering control method is optimized by using the sensor signal to reduce the angle error.

Table 2. Steering angle 750° steering deviation (°)

Number of experiments	Steering control method of mobile forklift based on singular value decomposition	Steering control method of mobile forklift based on envelope analysis	Steering control method of mobile forklift
1	7.364	7.104	4.316
2	7.458	7.649	4.125
3	7.063	7.315	4.691
4	7.442	7.422	4.157
5	6.948	7.244	4.2033
6	7.431	7.106	4.879
7	7.009	7.370	4.221
8	7.066	6.987	4.166
9	6.994	7.451	4.021
10	6.849	6.844	4.334
11	7.124	7.216	4.159
12	6.997	7.469	3.988
13	7.118	7.336	4.173
14	7.645	8.154	4.125
15	7.332	7.522	4.007

Table 3. Steering angle 500° steering deviation (°)

Number of experiments	Steering control method of mobile forklift based on singular value decomposition	Steering control method of mobile forklift based on envelope analysis	Steering control method of mobile forklift
1	6.141	4.316	2.316
2	5.154	6.977	3.011
3	5.216	6.522	5.565
4	5.489	4.167	3.228
5	5.147	5.844	2.974
6	4.889	4.019	3.104
7	5.006	5.490	2.967

(*continued*)

Table 3. (*continued*)

Number of experiments	Steering control method of mobile forklift based on singular value decomposition	Steering control method of mobile forklift based on envelope analysis	Steering control method of mobile forklift
8	4.667	5.337	3.114
9	5.553	5.108	2.879
10	5.022	4.977	3.455
11	4.987	5.313	2.964
12	5.216	4.879	3.117
13	5.337	5.112	3.020
14	4.987	4.415	2.844
15	5.266	5.226	3.161

Table 4. Steering angle 1000° steering deviation (°)

Number of experiments	Steering control method of mobile forklift based on singular value decomposition	Steering control method of mobile forklift based on envelope analysis	Steering control method of mobile forklift
1	3.145	4.030	1.698
2	3.158	3.946	1.596
3	4.005	3.889	2.036
4	3.874	3.849	2.102
5	3.015	3.569	2.137
6	4.112	4.221	1.948
7	3.648	3.878	1.869
8	3.554	4.101	1.774
9	4.022	4.033	2.016
10	3.694	3.948	1.555
11	4.017	3.665	1.644
12	3.898	4.121	2.077

(*continued*)

Table 4. (*continued*)

Number of experiments	Steering control method of mobile forklift based on singular value decomposition	Steering control method of mobile forklift based on envelope analysis	Steering control method of mobile forklift
13	3.121	3.447	1.495
14	4.516	4.119	2.316
15	3.878	3.515	2.117

Table 5. Steering angle 1250° steering deviation (°)

Number of experiments	Steering control method of mobile forklift based on singular value decomposition	Steering control method of mobile forklift based on envelope analysis	Steering control method of mobile forklift
1	2.546	2.005	1.002
2	1.826	1.848	0.645
3	2.331	2.695	1.214
4	2.548	1.463	1.588
5	3.144	2.316	0.948
6	2.008	2.169	1.131
7	2.105	1.944	1.252
8	3.014	2.165	1.055
9	2.642	2.588	1.441
10	2.849	1.996	1.033
11	1.948	2.131	1.145
12	2.113	2.455	1.015
13	1.879	1.878	1.002
14	2.615	2.116	1.143
15	1.022	1.787	1.088

4   Conclusion

Because the forklift needs to turn frequently, the driving state has been in the dynamic process, and the three control methods are derived from the mathematical model, belongs to static control. Therefore, the linear two-degree-of-freedom model of mobile forklift is introduced as the ideal reference model and method verification model. In this paper, the non-step problem of steering system after active steering is studied, and a full hydraulic synchronous steering system is designed to realize steering synchronization by compensating steering oil cylinder. Based on the constant yaw rate gain control and constant lateral acceleration gain control, the variable transmission rate control based on combined gain control and steering efficiency control is designed. It provides a reference for determining the expected steering curve for hydraulic steering vehicles and other hydraulic steering machinery. At the same time, the simulation results show that the steering control method of the mobile forklift is better than that of the control for the forklift. Because of the influence of the research condition, this paper does not consider the influence of eccentric load on the lateral stability of forklifts.

References

1. Fu, H.-P., Liu, Z.-X., Wu, J.-C.: Topology optimization design of forklift steering bridge. Equip. Manufact. Technol. (5), 57–61, 65 (2020)
2. Zhang, Z.-M., Ye, J., Zuo, G.-B., et al.: Dynamic and static characteristics of the steering-bridge for forklift based on the virtual prototype technology. Mech. Res. Appl. 33(1), 69–72 (2020)
3. Wei, L.-B., Wang, J.J.: Aligning torque analysis and both handiness and energy-saving optimization of the forklift hydraulic power steering. Comput. Simul. 37(8), 91–95 (2020)
4. Chen, H., Qing, T., Zhou, Z., et al. Study of multi-peaks F-P interference sensing signal spectrum wavelength peak seeking algorithm. Transducer Microsyst. Technol. 39(6), 30–32, 36 (2020)
5. Cao, Y.: Development of a resonant circuit for micro sensing signal. Telecom Power Technol. 37(13), 14–15, 18 (2020)
6. Liu, H., Wang, B., Lang, D.-Z., et al.: FBG sensing signals process with CS method using improved observation matrix. Softw. Guide 19(8), 202–206 (2020)
7. Shang, Q., Qin, W.: Research progress of signal processing methods of FBG sensing system. Optic. Commun. Technol. 44(5), 5–9 (2020)

Extraction Method of Vibroseis Phase Signal Based on Time-Varying Narrowband Filtering

Haochen Zhang$^{(\boxtimes)}$ and Chen Xian

Shaanxi Institute of Technology, Xian 710300, China
ttbm21@163.com

Abstract. When the phase travel time value increases continuously, the accuracy of the vibroseis phase signal extraction method will be disturbed. Aiming at this problem, a time-varying narrow-band filtering based vibroseis phase signal extraction method is designed. Calculate the linear sine signal expression formula, build a vibroseis simulation model, in the entire relevant time domain, the scanning signal satisfies the set frequency bandwidth, extract the characteristics of the noise generation mechanism, use the time-varying narrowband filter to scan the global seismic phase, and near-field data As a reference model, set the seismic phase signal extraction mode. Experimental results: The average extraction accuracy of the vibroseis phase signal extraction method in this paper and the other two vibroseis phase signal extraction methods are: 76.798%, 66.359%, and 66.694%, respectively, indicating that the band becomes narrower when fully combined. After filtering technology, the accuracy of the designed vibroseis phase signal extraction method has been improved.

Keywords: Time-Varying narrowband filtering · Vibroseis · Seismic phase signal · Extraction method · Sinusoidal signal · Harmonic interference noise

1 Introduction

While the vibroseis is being applied, the disadvantages of long time-consuming and low construction efficiency for the vibroseis acquisition of a single shot greatly limit the scale of its use [1, 2]. Therefore, in order to improve the collection efficiency of vibrators, many companies have developed a variety of high-efficiency operation methods for vibrators and methods to improve the construction quality of vibrators, and have achieved remarkable results. After decades of technological development, vibroseis has become a widely used mainstream vibrator technology. Vibroseis is a large-scale high-precision seismic exploration equipment that excites seismic wave signals through the principle of reaction. It can continuously generate excitation signal gates with constant amplitude and various frequencies during the working process. Since then, various high-efficiency acquisition technologies for vibrators have doubled their construction efficiency into dozens of times of growth, which is far higher than the construction efficiency of explosive sources. The development of high-efficiency acquisition technology of vibroseis has promoted the

W. Fu and L. Yun (Eds.): ADHIP 2022, LNICST 469, pp. 40–54, 2023.
https://doi.org/10.1007/978-3-031-28867-8_4

rapid expansion of the market scope of vibroseis exploration, and the large-scale application of high-efficiency acquisition technology has in turn promoted the improvement and development of this technology. Compared with traditional explosive sources, the advantage of vibroseis is that the operation process is safe. The excitation signal is controllable, does not have a negative impact on the surrounding natural environment, and the work efficiency is significantly improved. At the same time, the cost of exploration operations is significantly reduced. For technicians, the initial high-efficiency acquisition starts with alternate scanning, which has low technical requirements but can only improve construction efficiency by 50–100% [3].

On the basis of alternate scanning, more efficient vibroseis acquisition technology has been developed. Due to the advantages of high efficiency and low cost, vibroseis have been widely used in many countries. KZ series vibroseis is a hydraulic vibrator with independent intellectual property rights independently developed by China National Petroleum Corporation Orient Geophysical Company. KZ-28 type vibrator is the representative work of this series of vibrators. It has simple overall structure, easy operation, convenient maintenance, low technical support, and enhances the operation ability under harsh conditions. At present, in the Middle East, Africa, Central Asia, America and other countries and regions, based on factors such as safety, environmental protection, operation efficiency, and operation cost. Most oil companies, especially well-known international oil companies, use vibrators to reduce project risks when operating conditions permit. In China, vibroseis have also been widely used in exploration areas such as Northern Xinjiang, Tuha, and Inner Mongolia. It is the first time that the redundant structure design is adopted in the key system, which greatly enhances the reliability of the vibrator. Therefore, the KZ-28 vibroseis has received very good response in exploration applications at home and abroad. In addition, the threshold of alternate scanning technology is low, and the equipment requirements are not high, which is the basic technical reserve of vibroseis acquisition technology. At present, various new high-efficiency acquisition technologies are also developed on the basis of alternate scanning technology. Only by mastering alternate scanning exploration technology can they lay the foundation for the research and construction of other high-efficiency scanning exploration technologies.

2 Extraction Method of Vibroseis Phase Signal Based on Time-Varying Narrowband Filtering

2.1 Building a Vibroseis Simulation Model

The seismic data acquisition method of the controllable source can be substantially divided into the following three categories: conventional acquisition method, high efficiency acquisition method, high fidelity acquisition method. The controllable source is a low energy density excitation source, and the excitation energy needs to be accumulated for a long time [4, 5]. During the signal scan, the continuously changing sinusoidal vibration waveform is known that the frequency increases from low by time when the rass scan is scanned. The controllable source car is in traveling, lifting the vibrator and separated from the ground. Before entering the work, the hydraulic cylinder is lifted up, allowing the weight of the source car through 8 air springs for passive vibration isolation on the top plate and the plate of the vibrator to ensure that the vibrator is tightly coupled to the ground. Usually a linear sinusoidal sweep signal is used, and the linear sinusoidal signal expression formula is:

$$D(\delta) = d_0(\phi) \sin \frac{\delta}{d_0} \times \frac{1}{\delta} \tag{1}$$

In formula (1), d_0 represents the expression of amplitude variation with time, ϕ represents the initial phase, and δ represents the sweep length. However, there are usually obvious harmonic interference and adjacent gun interference in the channel concentration. The efficient acquisition methods of vibroseis mainly include alternate scanning and sliding scanning. Vibroseis forward modeling is an important means and method to study and understand the propagation law of vibroseis seismic waves in complex underground formations. It is of great significance to guide the design of field observation systems and optimize inversion algorithms.

When the vibrator is working, the servo valve opens, and the hydraulic system generates high-pressure hydraulic oil, which alternately enters the upper and lower chambers between the weight and the piston rod, thereby driving the weight to move up and down. The reaction force generated by the hydraulic oil is transmitted to the plate through the piston rod, and the signal generated by the vibrator is transmitted to the ground through the plate, thereby exciting the seismic wave signal. Using the excitation method of the vibrator to simulate, not only can accurately simulate the harmonic interference noise of the vibrator and its secondary interference noise. It can also accurately simulate the characteristic noise produced by various efficient acquisition methods. Then on the basis of formula (1), the frequency expression formula of pseudo-random signal is:

$$\gamma = \sum \frac{(\delta - \eta)^2}{2H} \tag{2}$$

In formula (2), η represents the start scanning frequency, and H represents the end scanning frequency. In order to ensure the stability of the model, four uprights are fixed on the lower plate as guide rods for the spring, and four holes are punched in the corresponding positions of the upper plate. The four guide rods respectively pass through the corresponding holes and cannot affect the vertical vibration of the top plate. The four through holes should be as smooth as possible, and the friction damping should be reduced by applying lubricant on the contact surface with the four guide rods. According to the CRF algorithm, at the same time, the orthogonal body-fitted grid layered coordinate transformation technology is proposed to simulate the complex near-surface structure. It can accurately simulate the secondary disturbance and vibroseis caused by the huge undulating height difference of the dunes [6]. The spring is sleeved on the support rod and fixed on the lower plate and the upper plate, so that the direction of the elastic force acting on the two plates is perpendicular to the upper and lower surfaces of the plates. Therefore, it is ensured that the model can only vibrate in a single degree of freedom in the vertical direction, so as to avoid motion in other directions from affecting the accuracy. Vibroseis seismic exploration is to control the plate by generating a continuous vibration signal through an electronic cabinet, generating a scanning signal whose frequency varies with time, and transmitting the energy of the scanning signal to the ground. When the signal generated by the vibration plate propagates downward, because the scanning signal lasts for a long time, each pulse of the underground reflection layer also lasts for a long time. They are combined into overlapping signals, with complex waveforms aliased together that cannot be distinguished and interpreted.

2.2 Extracting Noise Generation Mechanism Features

When vibroseis exploration is carried out in some special areas, the problem of serious degradation of data quality is sometimes encountered. If the on-site geophysicist wants to improve the acquisition effect by increasing the excitation energy, increasing the scan length is the convenient and easiest way. There are several typical characteristic noises in vibroseis, mainly including harmonic interference noise and surface response noise. Harmonic interference noise can be generated for many reasons, due to the nonlinear nature of the mechanical device and the coupled response distortion of the source substrate-ground coupled system. The seismic wave generated by the vibrator propagates into the ground, and there is inevitably harmonic distortion, which greatly reduces the signal-to-noise ratio of the vibrator data. The scanning excitation signal used in vibroseis exploration is a strictly monotonic long-term scanning signal in mathematical expression. There is a strict definition in the physical concept, conversely, not any signal can be used as a signal excitation in seismic exploration. However, with the gradual improvement of production efficiency, the influence of harmonic interference noise on the effective signal also gradually increases. Therefore, the continuous update of harmonic interference noise suppression technology is promoted.

The methods of suppressing harmonic interference noise can be divided into two methods: adjusting outdoor acquisition parameters and suppressing harmonics indoors. In mathematical concepts, a strictly monotonic signal means that the frequency at any point in the entire signal frequency band is unique, and the same frequency does not appear repeatedly at any time. In the vibroseis construction, if it is up-frequency scanning, it means that the frequency of the scanning signal is strictly increased. If it is a downsweep, it means that the frequency of the sweep signal is strictly decreasing. Through the analysis of the mechanism of harmonic interference noise, it can be known that the harmonic interference noise is unavoidable. Especially in the efficient acquisition method of vibroseis data, the generation of harmonic interference noise can be reduced by adjusting the field acquisition parameters. But it cannot eliminate harmonic interference noise. Therefore, thorough harmonic interference noise suppression by indoor means has become a necessary choice.

The above mathematical definition determines that in the correlation process, in the entire correlation time domain, after the scanning signal satisfies the set frequency bandwidth, the correlation wavelet has a unique maximum value. That is to say, in the time domain, when the downlink scanning signal reaches the reflection interface at a certain depth, the propagation path of the ray will change. Correlation maxima and strictly decreasing correlation edge perturbations appear in the correlation records. The vibroseis excites a continuous scanning signal, which is roughly divided into linear signal, nonlinear signal and pseudo-random signal according to the nature of the scanning signal. On the basis of the above description, the scanning signal type is obtained, as shown in Fig. 1:

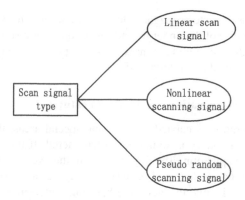

Fig. 1. Schematic diagram of scanning signal tyes

As can be seen from Fig. 1, the types of scanning signals mainly include: linear scanning signals, nonlinear scanning signals, and pseudo-random scanning signals. During vibroseis exploration, due to the influence of various factors, the special noise of vibroseis cannot be avoided. The first type of characteristic noise is generated by the distortion of the mechanical system device or the nonlinearity caused by the seismic plate-ground coupled system. This kind of harmonic noise can be stored by the force signal of the vibration plate, which is called the harmonic noise contained in the force signal, which is the harmonic noise in the narrow sense. In the following text, we call it harmonic interference noise.

The nonlinear scanning signal often sacrifices a certain low-frequency signal-to-noise ratio in exchange for improving the resolution of the geological target. The second type of characteristic noise is due to the coupling difference between the vibrating plate and the surface structure, so that the seismic phase signal stored by the vibrating plate is quite different from the truly propagated vibrator distortion signal. And if the frequency of the vibrator is equal to the natural frequency of the earth, resonances will be generated, which will interfere with the seismic records produced by the vibroseis. That is to say, only on the premise that the linear scanning signal can obtain a better signal-to-noise ratio, it is possible to use the nonlinear scanning signal to further improve the quality of the seismic acquisition data.

This is a typical manifestation of the conflict between signal-to-noise ratio and resolution in seismic exploration work. Such harmonics cannot be stored in the force signal in the vibration plate. Therefore, it is also impossible to use methods such as predictive filtering to separate the shock plate force signal to eliminate the influence of harmonic noise, which is called surface response harmonic noise. If the original excitation is a scan signal with time-invariant amplitude spectrum characteristics, it is easy to adjust by increasing the scan length. However, if the original excitation is a scanning signal with time-varying amplitude spectrum characteristics, it is not easy to adjust by increasing the scanning length. Because increasing the scanning length will lead to the deformation of the signal amplitude spectrum, it is generally necessary to increase the number of vibrators or increase the number of vertical stacking to achieve the purpose of increasing the energy.

2.3 Time-Varying Narrowband Filtering Sweeps Global Seismic Phase

Because the three-phase asynchronous servo motor used in the precision vibrator is limited by the accuracy of the motor servo. Compared with the theoretical value, there is an error between the actual frequency of the source transmitted signal, and the actual transmitted waveform is often difficult to fully satisfy the linear frequency modulation signal. Taking a linear scan signal as an example, when the input of the vibrator control device is a linear scan signal. Due to the distortion of the mechanical system of the vibrator and the distortion of the vibration plate and the earth system, the output signal on the vibration plate of the vibrator is a nonlinear distortion signal [7].

The transmitted frequency often deviates from the theoretical design frequency at certain moments. In order to obtain a more accurate actual transmitted signal, a frequency error model correction method is proposed to complete the correction of the near-field waveform. All nonlinear signals can be extended by Fourier series theory into a superposition of a series of linear signals. Including the frequency equal to the main frequency of the original signal is the fundamental signal. That is, the input linear scanning signal, the frequency equal to twice, three times, and four times the frequency of the original signal is the second harmonic, the third harmonic and the fourth harmonic respectively. We generally call the main frequency more than three times the original signal as high-order harmonics. Force signals containing harmonics include ideal sweep signals and harmonic signals of various orders. Therefore, the vibration signal of harmonic distortion can be expressed by the following equation:

$$\varphi = \sum_{i=1}^{n} \sin \frac{\|\mu + 2\pi(i-1)\|}{\mu_i} \tag{3}$$

In formula (2), μ represents the scanning period, and i represents the n-th harmonic signal. The center frequency of the time-varying narrowband filter is filtered according to the actual frequency sweep curve of the active source. After the sweep frequency signal propagates a certain distance, the delay value of the sweep frequency curve is unknown. Therefore, the premise of filtering the recorded data of the station is that the starting point of the frequency sweep signal needs to be known, that is, the travel time parameter needs to be estimated first. There are many reasons for the nonlinearity of the vibrator mechanism. It mainly includes the nonlinearity of the encoder device, the nonlinearity of the hydraulic system, the nonlinearity of the accumulator, the nonlinearity of the servo valve, the nonlinearity of the accelerometer and the nonlinearity of the coupling between the piston skin and the weight.

In order to detect the first arrival of the vibroseis sweep signal, time-varying narrowband filtering is performed by continuously changing the delay value. The cross-correlation between the filtered far-field station record and the seismic phase signal is used to obtain the maximum amplitude, and the scanning curves of different maximum amplitudes corresponding to different delay values can be obtained. If there is a seismic phase signal, and the frequency corresponds to the passband of the time-varying filter, the maximum amplitude of the cross-correlation obtained after filtering is large. From this, it can be concluded that there is an obvious regularity in the time position of the harmonic interference noise. By cross-correlating the distorted signal expression (2) with the scanning signal expression (3), the excitation signal time using linear upscaling is obtained. The two time intervals of the harmonic interference noise signal are expressed as:

$$R_1 = \frac{(g - 1)}{\Delta h} \times Fh_1 \tag{4}$$

$$R_2 = \frac{(g + 1)}{g \Delta h} \times Fh_2 \tag{5}$$

In formulas (4) and (5), g represents the phase spectrum of the harmonic signal, F represents the bandwidth of the sweep signal, h_1 represents the first-order harmonic energy, and h_2 represents the second-order harmonic energy. Transform domain interference suppression technology has two research priorities, one is the selection of orthogonal transform, and the other is how to effectively identify and suppress interference components in transform domain, that is, the study of transform domain processing algorithms. Simply put, transform domain processing can be divided into two types: trade-off-based and adaptive-based. If there is no vibroseis sweep frequency signal, the maximum amplitude after cross-correlation is small and unstable.

The peaks appearing in the scanning curve are relatively stable and have regular peaks, which indicate that the source waveform and the station waveforms achieve the best match, so this peak can be used as an estimate of the first arrival. As the name implies, the trade-off-based transform domain processing is that when it is determined that a transform domain coefficient contains interference components, the coefficient is completely set to zero, otherwise the value of the transform domain coefficient is not changed. The adaptive-based transform domain processing uses an adaptive algorithm to minimize the statistical error between the output signal and the expected value.

Since the frequency of the frequency sweep curve of the precise control source is determined at a certain moment, it is possible to perform filtering in a very narrow range near the emission frequency corresponding to that moment. It can suppress the influence of noise to a greater extent and obtain data quality with better signal-to-noise ratio. After correcting the frequency deviation of the active phase signal and estimating the first arrival of the vibroseis sweep frequency signal. It can perform time-varying narrow-band filtering and seismic phase separation for far-field station observation data.

2.4 Set the Seismic Phase Signal Extraction Mode

Use the source approach recording signal to correct the sweep signal. Corresponding to a specific moment, the change of the center frequency of the time-varying narrowband filter is consistent with the frequency of the seismic phase signal. Using the time-varying narrowband filter to filter the near-field signal of the source, we can get 61 different filtering results.

When selecting the frequency range for extracting the seismic phase signal, the frequency response of the stratum in the construction area should be considered first. The frequency response of the stratum is different in different regions and strata. Generally speaking, in the same area, the frequency response of the shallow layer is higher, and the response frequency of the deep layer is lower. If the frequency of the seismic phase signal just falls in the filter frequency band with a certain offset center frequency, the output signal energy value at this moment is the largest. That is, the filtering result with the largest energy corresponds to the actual frequency of the frequency sweep signal, so as to determine the offset of the frequency sweep curve of the source to the theoretical value at a certain moment. This is the principle of frequency correction of the seismic phase signal.

If the geological task is to understand shallow structures, the higher frequency extracted seismic signals should be used. If you want to understand the deep structure, you should use the lower frequency extracted seismic signal. If you want to understand the structure and interrelationship of shallow, middle and deep layers. Then the frequency band for extracting the seismic phase signal should be wider to adapt to the shallow, middle and deep frequency responses. The formation frequency response of a region can be obtained from existing geological data and well log tests.

Through a certain step change time point, the actual frequency sweep curve of one cycle can be obtained. in actual data processing. In order to ensure that the time domain signals corresponding to different frequencies can contain a complete periodic signal, the signal points for intercepting and calculating the energy should have a certain number, and the correction frequency points are sampled at 0.3 s. Finally, the frequencies at other times are obtained by linear interpolation. When determining the frequency range for extracting seismic phase signals, attention should also be paid to the characteristics of the surface structure in the construction area. The thickness of the surface layer in different areas is different, and the absorption of different frequencies is also different. When the surface seismic phase signal passes through the low-velocity zone, it is equivalent to passing through a low-pass filter. The realization of the seismic phase separation extraction technique depends on knowing the starting point of the frequency sweep signal in advance. Here, the starting point of each cycle of the sweep frequency signal is set as the same starting point for all vibrations. Using the near-field data as a reference model, the time-varying narrow-band filtering is performed on the data of a certain far-field station at a certain time. And according to the principle of zero-phase filter to eliminate the phase shift of the filtering result.

Use the filtering result to perform cross-correlation with the near-field source data, take the maximum value of the cross-correlation value, change it to the time value in turn, and obtain the maximum cross-correlation value corresponding to different times. If the inner surface of the construction area is loose and dry, and the low-velocity zone is relatively developed, its absorption of high-frequency components is serious. At this time, the frequency range for extracting the seismic phase signal should be lower. The design of the starting frequency of the vibroseis to extract the seismic phase signal is also a very critical selection parameter. At present, most of the seismic phase signals in the excitation process of the vibroseis are extracted by up-frequency. Therefore, the design of the initial extracted seismic phase signal we discussed is usually about the low-frequency initial seismic phase signal. The attenuation of the surface has a great influence on the surface response noise. The relationship between the surface attenuation and the surface response noise is simulated and tested by using vibroseis Ru sound forward modeling. From this, it is concluded that the vibration record is equal to the result of the convolution of the gun force signal and the reflection coefficient:

$$s = \frac{y \otimes \|t - 1\|^2}{2} \tag{6}$$

In formula (6), y indicates that the former term is the fundamental term, and t denotes that the latter term is the harmonic noise term. On this basis, the average value of the amplitude of each channel in each frequency band is calculated. The calculation formula is as follows:

$$V = \frac{\sigma_{mn}}{\sigma} \sum_{n=1}^{y} \frac{1}{r_{mn}} \tag{7}$$

In formula (7), σ represents the average value of the data amplitude of each track, m represents the amplitude value of the y-th track data, and n represents the number of tracks. When there is a stable seismic phase at a certain time, the corresponding mutual maximum value is stable. And using the superposition principle, the scanning processing results of multiple periodic signals are superimposed to obtain the final seismic phase scanning curve. Due to the randomness of noise and the high repeatability of vibrators, the stacking of multiple scans can effectively compress such random fluctuations and highlight the scan peaks of effective signals. For the scanning curves of multiple periods, if the extreme point corresponding to a certain moment can be obtained at the same time, it means that the seismic phase information at that time exists in the record of the station.

Whether it is up-frequency extraction of seismic phase signal or down-frequency extracted seismic phase signal, in the records obtained by excitation, harmonic interference is inevitable, but the degree of interference is different. The harmonic interference produced by the vibrator in the excitation process cannot be eliminated. But not terrible, we have many ways to cut harmonic interference. Furthermore, we can obtain the arrival time information of the seismic phase through the stable peak of the global sweep curve. The repeatability of the scanning results is good, which also shows that the source excitation signal has good repeatability. And the recorded signal of this station has a higher signal-to-noise ratio.

3 System Test

3.1 Test Preparation

The data processing of the vibroseis swept signal extracted based on the time-varying narrowband filtering technology is mainly realized by the following steps. Moreover, in order to improve the computing efficiency, in the process of study and research, the mixed programming method of Matlab and Visual C ++ is used to realize the whole data processing. As a continuous signal, the data processing method of active source signal is very different from that of natural seismic data. At present, vibroseis data processing includes many tedious steps. In order to better analyze and compare the processing results, and can improve the efficiency of data processing. According to the test requirements, connect the FG-506A signal generator to the SA-PA020 power amplifier. The SA-PA020 type power amplifier is connected to the SA-JZ010 vibration exciter that excites the lower board: the PCI-8603 data acquisition card is inserted into the computer containing the PCI slot.

In the learning process, the independently written programs are integrated into the interactive interface operation based on the Matlab GUI, and the software also adds a small software for random signal analysis in the learning process. At the same time, the terminals are extended, and the AIO and AI 1 interfaces are respectively connected to the output ends of the charge amplifiers connected to the piezoelectric acceleration sensors on the upper and lower plates of the vibrator experimental model. The DAO output interface is connected to the input end of the SA-PA010 power amplifier. In order to reduce the computational complexity of the computer and improve the calculation speed, the original recording is first downsampled from 200 Hz to 100 Hz. Both near-field records and far-field observation station records are down-sampled. Initialize the parameters of the calculation program, including the bandwidth of the time-varying narrowband filter, the sampling rate, etc. The output end of the SA-PA010 power amplifier is connected to the JZ-005 exciter that excites the upper plate. After a series of debugging, all the above devices can work normally, thus completing the deployment of all devices. This filter is used to correct the sweep frequency curve, and the variation law of the sweep frequency curve of the actual transmission is calculated.

3.2 Test Results

In order to verify the effectiveness of the designed vibroseis phase signal extraction method, a comparative experiment is carried out. The wavelet transform-based vibroseis phase signal extraction method and the vibroseis phase signal extraction method based on genetic algorithm are selected, and the experimental comparison is made with the vibroseis phase signal extraction method in this paper. The extraction accuracy of the three methods was tested under different seismic phase travel time conditions. The experimental results are shown in Tables 1, 2, 3, 4 and 5:

Table 1. 500T/s2extraction accuracy of seismic phase travel time (%)

Number of experiments	Extraction method of vibroseis phase signal based on wavelet transform	Extraction method of vibroseis phase signal based on genetic algorithm	The method designed in this paper to extract the phase signal of vibroseis
1	85.614	86.949	93.646
2	86.941	85.341	92.878
3	92.313	87.558	93.646
4	84.556	89.616	95.202
5	87.081	86.334	94.779
6	83.445	87.512	96.546

(*continued*)

Table 1. (*continued*)

Number of experiments	Extraction method of vibroseis phase signal based on wavelet transform	Extraction method of vibroseis phase signal based on genetic algorithm	The method designed in this paper to extract the phase signal of vibroseis
7	83.616	86.992	95.223
8	79.345	85.313	96.314
9	87.405	86.077	95.008
10	79.668	86.963	96.317
11	82.057	87.545	97.548
12	83.664	89.011	96.322
13	85.191	85.162	95.814
14	84.206	86.346	96.071
15	86.772	85.009	95.313

It can be seen from Table 1 that the average extraction accuracy of the vibrator phase signal extraction method in this paper and the other two vibrator phase signal extraction methods are: 95.375%, 84.792%, and 86.782%, respectively.

Table 2. 1000T/s2 extraction accuracy of seismic phase travel time (%)

Number of experiments	Extraction method of vibroseis phase signal based on wavelet transform	Extraction method of vibroseis phase signal based on genetic algorithm	The method designed in this paper to extract the phase signal of vibroseis
1	72.131	75.144	83.160
2	75.418	76.319	82.484
3	76.902	75.208	81.619
4	78.994	76.919	83.445
5	79.613	73.218	82.916
6	72.448	74.067	81.447
7	73.106	75.336	83.699
8	75.431	76.912	82.015
9	78.551	77.433	83.479
10	79.055	75.612	82.164
11	76.228	74.338	82.669

(*continued*)

Table 2. (*continued*)

Number of experiments	Extraction method of vibroseis phase signal based on wavelet transform	Extraction method of vibroseis phase signal based on genetic algorithm	The method designed in this paper to extract the phase signal of vibroseis
12	78.606	77.404	85.314
13	79.449	78.994	86.019
14	75.212	76.303	85.443
15	76.303	77.448	86.494

It can be seen from Table 2 that the average extraction accuracy of the vibrator phase signal extraction method in this paper and the other two vibrator phase signal extraction methods are: 83.491%, 76.496%, and 76.044%, respectively.

Table 3. 1500T/s2 extraction accuracy of seismic phase travel time (%)

Number of experiments	Extraction method of vibroseis phase signal based on wavelet transform	Extraction method of vibroseis phase signal based on genetic algorithm	The method designed in this paper to extract the phase signal of vibroseis
1	62.313	62.599	75.664
2	64.588	66.838	76.913
3	63.942	65.087	75.842
4	64.778	64.919	76.319
5	65.209	66.387	77.448
6	66.311	65.338	76.312
7	65.841	66.205	75.814
8	66.933	67.283	76.901
9	65.177	68.966	75.282
10	66.834	65.311	76.316
11	66.009	63.708	75.255
12	65.846	64.118	76.448
13	67.331	63.009	77.911
14	68.955	62.551	78.306
15	65.744	63.058	79.154

It can be seen from Table 3 that the average extraction accuracy of the vibrator phase signal extraction method in this paper and the other two vibroseis phase signal extraction methods are: 76.659%, 65.721%, and 65.025%, respectively.

Table 4. 2000T/s2 extraction accuracy of seismic phase travel time (%)

Number of experiments	Extraction method of vibroseis phase signal based on wavelet transform	Extraction method of vibroseis phase signal based on genetic algorithm	The method designed in this paper to extract the phase signal of vibroseis
1	62.135	55.142	66.544
2	59.847	53.617	65.911
3	61.006	57.915	64.358
4	59.348	56.912	68.122
5	58.616	61.202	69.313
6	57.311	60.78	67.405
7	57.919	58.334	69.212
8	58.206	59.226	68.513
9	53.363	60.447	69.455
10	58.714	59.161	67.506
11	55.221	58.334	68.224
12	54.301	57.105	68.548
13	53.448	58.494	65.314
14	57.616	56.322	66.008
15	55.142	54.770	67.014

It can be seen from Table 4 that the average extraction accuracy of the vibrator phase signal extraction method in this paper and the other two vibrator phase signal extraction methods are: 67.430%, 57.480%, and 57.851%, respectively.

Table 5. 2500T/s2 extraction accuracy of seismic phase travel time (%)

Number of experiments	Extraction method of vibroseis phase signal based on wavelet transform	Extraction method of vibroseis phase signal based on genetic algorithm	The method designed in this paper to extract the phase signal of vibroseis
1	49.815	55.644	58.647
2	48.633	46.778	57.698
3	49.579	44.322	57.848

(*continued*)

Table 5. (*continued*)

Number of experiments	Extraction method of vibroseis phase signal based on wavelet transform	Extraction method of vibroseis phase signal based on genetic algorithm	The method designed in this paper to extract the phase signal of vibroseis
4	52.116	48.944	58.619
5	47.3361	46.311	59.361
6	49.155	45.915	60.248
7	45.202	48.221	61.225
8	46.113	47.050	62.348
9	44.819	48.099	63.776
10	45.225	47.645	62.551
11	46.915	48.317	63.005
12	45.848	46.955	62.144
13	46.913	46.311	63.948
14	45.649	47.088	62.315
15	46.317	48.912	61.778

It can be seen from Table 5 that the average extraction accuracy of the vibrator phase signal extraction method in this paper and the other two vibrator phase signal extraction methods are: 61.034%, 47.309%, and 47.767%, respectively. Due to the limitation of the research conditions, the secondary disturbance caused by the special noise of vibroseis in the desert area caused by the violent undulating surface and loose sand dunes has not been studied. Because the expression formula of linear sinusoidal signal is calculated in this paper, the simulation model of vibration source is established. In the entire correlation time domain, the scanning signal satisfies the set frequency bandwidth, extracts the characteristics of the noise generating mechanism, and improves the extraction accuracy of the signal.

4 Conclusion

The method for extracting the phase signal of the vibroseis in this paper, and a frequency error correction method is proposed, which can obtain the actual frequency value of the vibroseis transmitted signal to the ground at different times. A comprehensive summary of the two characteristic noises of the seismic phase is carried out, and two characteristic noise suppression methods are developed for the characteristic noises of the two vibroseis. Seismic waves in the range contain rich information about the slow structure of the crust and the upper ground. Waveform observations with larger epicentral distances can constrain deeper structures and provide a basis for the operating mechanism of the earth's interior. A frequency division filtering method of artificial sorting gathers is adopted for the surface response noise, and a combined suppression method and process are proposed. At the same time, the frequency tracked when the time-varying narrowband filter works is more accurate, and the center frequency of the filter can be guaranteed to be within the bandwidth tolerance. In this paper, the expression formula of linear sinusoidal signal is calculated, and the simulation model of vibration source is established. In the entire correlation time domain, the swept signal satisfies the set frequency bandwidth, extracting the characteristics of the noise generating mechanism. The extraction of the seismic phase signal is achieved by scanning the global seismic phase and near-field data using a time-varying narrowband filter as a reference model.

References

1. Ma, T., Wang, Y.-C., Liu, X.-G., et al.: Application of deconvolution method with ground force in vibroseis raw shots calculation. Prog. Geophys. **35**(4), 1438–1444 (2020)
2. Wang, Z., Jianjun, X., Li, X., et al.: Influence of vibroseis high-efficiency acquisition on weak signals and corresponding countermeasures. Geophys. Prospect. Petrol. **59**(5), 695–702 (2020)
3. Wang, H.: Vibroseis seismic exploration with customized wavelet. Geophys. Prospect. Petrol. **59**(5), 683–694 (2020)
4. Wang, Y., Li, H., Tuo, X., et al.: Picking the P-phase first arrival of microseismic data with strong noise. Geophys. Prospect. Petrol. **59**(3), 356–365 (2020)
5. Zhang, J., Zhao, G.-Y., Song, N.-N.: Analysis and suppressing of high frequency distortion on loose near surface in vibroseis acquisition. Prog. Geophys. **35**(1), 250–257 (2020)
6. Martuganova, E., Stiller, M., Bauer, K., et al.: Cable reverberations during wireline distributed acoustic sensing measurements: their nature and methods for elimination. Geophys. Prospect. **69**(5), 1034–1054 (2021)
7. Liu, Y., Lu, Y.-H., Liu, M., et al. Information acquisition of earthquake emergency rescue collapse based on lora wireless technology. Comput. Simul. **37**(3), 224–228 (2020)

Research on Intelligent Prediction of Power Transformation Operation Cost Based on Multi-dimensional Mixed Information

Ying Wang[✉], Xuemei Zhu, Ye Ke, Jing Yu, and Yonghong Li

State Grid Fujian Power Economic Research Institute, Fuzhou 350000, China
Yingwang21212@163.com

Abstract. Operation cost is an important link in the operation of power enterprises. In the process of intelligent prediction of power transformation operation cost, there is a problem of low accuracy. Therefore, an intelligent prediction method of power transformation operation cost based on multi-dimensional mixed information is designed. Evaluate the fixed cost of power grid, determine the budget amount in different budget periods, extract the life cycle of power grid substation equipment, establish the cost estimation relationship, use multi-dimensional mixed information to build the cost control model, refine the project category, and optimize the intelligent prediction mode of operation cost according to the different nature of each link cost. Test results: the average prediction accuracy of the intelligent prediction method of power grid substation operation cost in this paper and the other two intelligent prediction methods of power grid substation operation cost are 79.357%, 71.066% and 69.313% respectively, indicating that after using multi-dimensional mixed information, the application effect of the designed intelligent prediction method of power grid substation operation cost is more prominent.

Keywords: Multidimensional mixed information · Power grid · Substation · Operating costs · Fixed costs · Intelligent prediction

1 Introduction

With the accelerated development of power grid, the operation, maintenance, renewal and transformation of power grid assets are becoming more and more arduous. The traditional management method can not meet the needs of development under the new situation. Predicting the operation cost of power transformation has become the top priority of power grid enterprise management [1, 2]. Power industry is the most important basic energy industry in the growth of national economy. It is the first basic industry of national economy and the priority development focus in the economic development strategies of countries all over the world. The characteristics of large initial investment, high operation and maintenance costs and long service time of the main equipment of the power grid make it have broad prospects for the application of multi-dimensional

© ICST Institute for Computer Sciences, Social Informatics and Telecommunications Engineering 2023
Published by Springer Nature Switzerland AG 2023. All Rights Reserved
W. Fu and L. Yun (Eds.): ADHIP 2022, LNICST 469, pp. 55–69, 2023.
https://doi.org/10.1007/978-3-031-28867-8_5

mixed information. With the rapid development of China's economy, the demand for power in the whole society continues to increase, and the expansion of power sales market stimulates the investment in power equipment. Analyzing the cost management of power grid enterprises from the perspective of multi-dimensional mixed information will help power grid enterprises get rid of the shackles of traditional cost concept and build a new cost management system. Power engineering project has the characteristics of large initial investment, relatively long construction time, long operation cycle and high operation and maintenance cost. Because all loads in the power grid system and generator units generate public costs under complex interactive transmission, and the interaction responsibility is difficult to distinguish. Transmission cost is divided into two important components: variable cost and fixed cost. While ensuring the safety and reliability of the power grid, improve the use efficiency of power grid assets, reduce the life cycle cost, and realize the lean, whole process and all-round management of power grid assets. Variable cost mainly includes loss cost, blocking cost and opportunity cost. The fixed cost mainly includes line construction cost, equipment purchase cost, management cost, maintenance cost, depreciation cost, etc. So as to build a new cost prediction system, improve the use efficiency of power grid assets, reduce costs, and realize lean, whole process and all-round management of power grid assets while ensuring the safety and reliability of power grid. Accordingly, the requirements for cost control and management of power engineering projects are also relatively high. Only when the public cost is shared can the stable operation of the power market be guaranteed. There is no direct one-to-one correspondence between the generation of these public costs and the specific load. The main performance is that we should consider not only the purchase and construction cost of the power engineering project, but also the operation and maintenance cost of the project. In other words, we should pay attention to the control of the total cost in the cost management and control of the power engineering project, and the goal should be the optimization of the total cost. The design of electricity price should consider a fair cost sharing, that is, each consumer should pay the corresponding part of the service cost. Moreover, this new type of price must be able to control consumption and power generation patterns at the same time, especially consumption patterns. At present, the traditional intelligent prediction model of power transformation operation cost emphasizes the division of stages and segmented management, that is, the decision-making, design, procurement, construction, installation, operation, decommissioning and disposal of power engineering projects are managed and implemented by different departments, so it is difficult to optimize the total cost of power engineering projects, Then it affects the achievement of performance objectives and sustainable development of power enterprises. Reference [3] proposes a distributed control method based on incremental cost consistency Firstly, based on the consistency theory, the incremental cost consistency algorithm is given to verify the cost optimization problem; Secondly, based on the optimal solution of the incremental cost consistency algorithm, a new distributed droop controller is designed. Under the condition of satisfying the balance of supply and demand, the control strategy makes the operating cost of the microgrid system optimal, the power can be reasonably distributed, and the frequency is stable at the rated value; Finally, the effectiveness of the proposed control strategy and algorithm is verified by simulation Reference [4] proposes a two-level optimal allocation method that takes into account the cost and benefit of energy storage system in the whole life cycle Firstly, the model of life cycle cost and benefit is established to calculate various costs and benefits of energy storage; Then the genetic algorithm combined with simulated annealing algorithm is determined for the inner and outer layers; Finally, benchmark low voltage microgrid is used as an example to simulate the proposed optimization strategy The simulation results show the feasibility and effectiveness of the method. The model can effectively manage the voltage, keep the voltage within a safe range, and reduce the cost through arbitrage However, the above two literature methods have the problem of low prediction accuracy in the process of intelligent prediction of substation operation costs. Therefore, an intelligent forecasting method of substation operation cost based on multi-dimensional mixed information is designed. Evaluate the fixed cost of the power grid, determine the budget amount in different budget periods, extract the life cycle of the power grid substation equipment, establish the cost estimation relationship, establish the cost control model using multi-dimensional mixed information, refine the project category, and optimize the intelligent prediction mode of operation cost according to the different nature of the costs in each link.

2   Research on Intelligent Prediction of Power Transformation Operation Cost Based on Multi-dimensional Mixed Information

2.1   Assess Grid Fixed Costs

The cost of equipment operation and maintenance stage of power grid enterprises refers to the cost of all relevant activities such as operation, maintenance and overhaul in order to maintain its normal operation in the whole life cycle of assets. Most of the fixed costs are "operating capacity costs", which are the costs that must be paid by enterprises in order to achieve production and operating capacity, such as depreciation expenses of fixed assets, house rent, wages and salaries of managers, etc., also known as binding fixed costs. Once the amount of such costs is determined, it is not easy to change and has a considerable degree of binding force. In order to simplify the treatment, only the emergency repair cost is considered in the fault disposal cost, and the condition monitoring cost and material reserve cost are not considered. Each cost item can be further subdivided into direct cost and indirect cost. Compared with the binding cost, discretionary fixed cost refers to the fixed cost formed by determining the budget amount of a plan period according to the enterprise's operating financial resources before the beginning of the accounting year, such as development expenses, advertising expenses, etc. The cost composition is classified by activity type, as shown in Fig. 1:

As can be seen from Fig. 1, the fixed cost of power transformation operation can be divided into: operation and maintenance cost, equipment maintenance cost, fault disposal cost, condition monitoring cost and material reserve cost. The general procedure of power grid fixed cost analysis is to investigate the whole process of product operation process, identify and measure operation, and establish corresponding operation center. Collect the resource cost into the corresponding operation, and evaluate the effectiveness and value-added of the operation. However, in a certain period, it is similar to the binding fixed cost, so it is also a part of the fixed cost. Establish activity-based cost database,

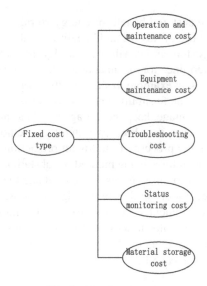

Fig. 1. Fixed cost types

select typical activities, and calculate activity-based cost driver coefficient. Calculate the activity-based cost consumed by products and summarize the output activity-based cost. Since the budget amount is only within the fixed budget period, it can also be determined qualitatively according to the specific budget period. This part of the cost is not binding, and the cost amount can be determined according to different situations. Evaluate the contribution of activity-based cost to output, determine cost efficiency, and collect value-added activities and non value-added activities. From the analysis of cost behavior, the cost is composed of fixed cost, variable cost and mixed cost. Direct cost refers to the material, labor and expense costs directly related to the operation of the cost object during the operation period, including internal and external (outsourcing) direct costs. Within a certain period or relevant scope, the part of the cost whose amount remains unchanged without being affected by the change of business volume is called fixed cost, such as depreciation of fixed assets, salary of managers and house rent. Indirect cost refers to the cost that is indirectly related to the operation period of the cost object, or is directly related but cannot be traced back to the specific object in an economic and reasonable way. It is divided into asset operation indirect cost and asset management indirect cost. However, from the perspective of unit business volume, the increase or decrease of business volume changes in the opposite direction with its fixed cost. After collecting direct costs and allocating indirect costs, they are summarized to form the cost of a certain type of operation of a single asset. The operation cost of a single asset is the total cost of various operations. Therefore, only in a certain period and relevant scope can the cost be fixed. If the business volume within this range changes, the fixed cost will change.

2.2 Extract the Life Cycle of Power Grid Substation Equipment

Equipment life cycle refers to the total length of time from the beginning of putting the equipment into use to the final withdrawal from use due to the complete loss of equipment function. The difference between project life cycle cost analysis and traditional economic evaluation is that it takes life cycle cost as the cost to measure the corresponding utility of the project. For power transformers, the use experience and test results show that after the equipment volume reaches a certain number, the overall statistics of the samples can still calculate the law: within the service life of the equipment, there is a certain law between the number of faults and the service time [5, 6]. The relationship curve between transformer operation time and defect failure rate shows the form of low in the middle and high on both sides. Therefore, the core of life cycle cost analysis is to identify the cost items at each stage of the life cycle, quantitatively estimate and analyze the cost according to a certain cost estimation model and method, and finally obtain the life cycle cost, and then make decisions on this basis. After determining the life cycle cost model of substation project divided by cost generation stage, it is necessary to further structure the cost of each part to the basic cost unit that can be used for specific calculation. The whole life cycle cost of substation project refers to the total cost incurred in the whole life cycle of substation, that is, from the construction of substation project to the scrapping of the project. The cost is divided by stages according to the whole life cycle. The cost is mainly composed of the following parts: as shown in Fig. 2:

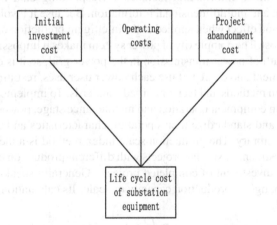

Fig. 2. Phased cost of substation equipment life

It can be seen from Fig. 2 that the life stage costs of substation equipment are: initial investment cost, operation cost and project scrap cost. The purpose of life cycle cost decomposition is to decompose the components of life cycle cost to the required level, so as to establish the cost decomposition structure. The cost breakdown structure is a hierarchical tree structure. It includes all relevant cost units, neither omission nor repetition, from coarse to fine. It is broken down to the basic cost unit that can be estimated. Cost deconstruction shall be consistent with the classification principles of power engineering industry and the financial classification principles of engineering projects. Each basic cost unit must have a precise definition and be independent of each other, so as to obtain cost estimation data from the financial department. In order to compare and make decisions on different schemes and management measures of equipment and system. A very effective method is to quantify them into comparable costs, that is, life cycle cost estimation. The cost estimation is usually carried out before the cost occurs. Therefore, the cost estimation relationship must be established. The basic cost unit obtained through cost deconstruction shall not omit important cost units or double calculate. The cost deconstruction shall follow certain principles: the deconstruction object shall include all relevant expenses incurred in the whole life cycle of the substation project. Solve various algorithm problems. This work is also called cost modeling. The life cycle of equipment and systems is very long, and the estimated costs occur in different years. Due to the different actual values of equal funds at different times, the calculated costs need to be converted or corrected before they can be used for comparison, measurement and analysis. Because the cost composition of substation project is very complex, in order to facilitate calculation and scheme evaluation, the cost structure can be appropriately simplified and the cost units that have a great impact on decision-making can be retained.

2.3 Constructing Cost Control Model with Multi-dimensional Mixed Information

Multidimensional mixed information includes corresponding data information, economic information, and multidimensional information. Because it involves a wide range of levels, it has a good effect when applied to the intelligent prediction of power transformation operation cost. The complexity of power system makes it impossible to accurately judge the actual path of power transmission in the power grid, so it is difficult to determine which equipment and degree of use each power user uses, resulting in a variety of fixed cost allocation methods and lack of unified standards. To implement activity-based cost management in equipment operation and maintenance stage, power grid enterprises first need to unify and standardize their operation characteristics and establish a set of standard operation library. The production scale index method is a method to estimate the equipment investment of similar projects with different production capacity by quoting the equipment investment of completed projects. Generally speaking, the cost will change with the change of production capacity or scale. Its calculation formula is:

$$R_2 = R_1 \left(\frac{L_2}{L_1} \right)^{-1} - \delta \tag{1}$$

In formula (1), R_1 represents the actual engineering cost of similar projects, R_2 represents the engineering cost required by the proposed project, L_2 represents the production scale of the proposed project, L_1 represents the production scale of similar projects, and Q represents the price conversion index. Moreover, due to the particularity of production and operation of power grid enterprises, the principle of fixed cost allocation of power grid enterprises is also different from that of general enterprises. Cost estimation is an important link in life cycle cost evaluation. Under the condition that the information about the project and its cost is very limited. Classify the operations according to the main operation information such as operation equipment, operation parts, operators and operation items. Detailed items are set under the main item category to describe more detailed job information. The reasonable allocation of fixed costs of power grid enterprises should not only be consistent with the necessary production costs of power production, but also its regulation direction and regulation strength should be consistent with the requirements of the optimal allocation of resources [7]. The detailed project categories should include: equipment category, standard work code, operation directory, operation implementation unit, operator, operation content, voltage level, capacity, benchmark price, labor cost, machinery cost, equipment cost, comprehensive work day unit price, comprehensive work day quantity, expense classification, material code, material name Information such as material unit and material quantity. According to the data of similar projects or the limited information of the proposed project, simply and accurately estimate its various expenses or how to obtain the cost to be generated according to the paid cost, these are the problems that must be solved in cost estimation, and the accuracy of cost estimation is inseparable from the establishment of model. The initial investment cost estimation can be expressed as:

$$P = \sum_{\beta=1}^{\alpha} P_{\beta-1} \tag{2}$$

In formula (2), α represents the operation and maintenance cost, β represents the discount rate or basic rate of return, and P represents the initial investment cost. According to the calculation results, in the process of establishing the standard operation database, we should also cooperate with the power grid to develop and establish a unified information system to match it, and formulate relevant management systems and standards. In the operation of equipment operation and maintenance stage, work order is a tool to implement its cost management. The fixed cost allocation of power grid enterprises should first follow the principle that the fixed cost should be reasonably allocated among all kinds of service objects according to the different conditions of service objects, such as different voltage levels and different power consumption characteristics, so that all kinds of service objects can bear the fixed cost fairly. The expression formula of construction project investment cost is:

$$G = \frac{\phi \times \varepsilon}{2} \times \frac{1}{\varepsilon} \tag{3}$$

In formula (3), ε represents the cost of unit construction quantities and ϕ represents the total construction quantities. Work order management process: operators must first select standard operations in the standard operation library according to their operation needs. The system automatically generates operation documents such as standard operation cost, resource consumption and operation instructions. Operators operate according to the work order operation documents. The maximum demand is the maximum power consumption of the user. The variable cost is closely related to the user's power consumption, while the fixed cost is only related to the user's maximum power consumption. This is because although the frequency of the user's maximum demand is low, the system still needs to prepare the corresponding generation capacity and transmission and distribution capacity for its possible maximum power demand. After the operation is completed, the operator performs the work order feedback operation, checks the actual operation amount and resource consumption with the original work order, and the work order system synchronously calculates the actual cost of the operation. Establishing a set of standardized work order format and operation standards is the key link in the implementation of work order management. Therefore, the maximum demand is the main factor affecting the fixed cost. For example, for large industrial users, since the maximum demand is much higher than that of ordinary residential users, power grid enterprises must prepare more installed capacity costs for them in order to meet the requirements of their maximum demand, including the construction of transmission and distribution lines and substations. Activity based cost management in equipment operation and maintenance stage is a system engineering with many projects and huge information content, which needs a perfect information system as a guarantee. A set of standard Activity-Based Cost database module can be established based on ERP system, which includes three sub modules: code table, price database and quota database, so as to realize the informatization of activity-based cost management. In the process of expanding reproduction and capital circulation, the appreciation of capital over time and the interest withdrawn from bank savings are the concrete embodiment of the time value of capital. The equipment purchase cost mainly consists of the original price of the equipment and the transportation and miscellaneous expenses of the equipment. The calculation formula of the equipment purchase cost is:

$$H = \sum \frac{Q \times (1+\gamma)^2}{Q} \tag{4}$$

In formula (4), Q represents the original price of equipment, γ represents the freight and miscellaneous rate, and H represents the equipment purchase cost. For power grid enterprises, the operation and maintenance of equipment is mainly to ensure the safe and stable operation of the power grid, including operation and maintenance, equipment defect elimination, status supervision, temporary accident disposal, spare parts management and other operations. The quality standard is to maintain the safe and reliable operation of the power grid, eliminate all kinds of defects affecting the safe and economic operation in time, improve the economy of power grid equipment, reasonably control equipment maintenance, etc. The time value of funds means that certain funds have different values at different time points, that is, a sum of funds today, even without considering inflation, is more valuable than the same amount of funds obtained in the future. The installation engineering cost can be calculated according to the percentage of the equipment purchase price and the corresponding installation rate, or estimated according to the equipment weight and the corresponding installation cost and other index coefficients:

$$L = \frac{Q \times \eta}{\sum \gamma} \tag{5}$$

In formula (5), η represents the installation rate and L represents the installation cost. In the equipment operation and maintenance stage, a large number of operations can be identified. If the operation center is divided according to the operation activities, the work is relatively too complex. Therefore, the operations can be combined according to the principle of homogeneity, the main operations can be determined, and then the operation center can be established. The equipment and projects we study have a long service life, the cost points are different, and the time value is also very different. If we simply superimpose the costs at these different time points, it will not be comparable.

2.4   Optimize the Intelligent Prediction Mode of Operation Cost

According to the formulation principle of sales price, the operation cost is the basis of formulating sales price. The power supply cost mainly refers to the fixed cost of power grid enterprises. Therefore, it is necessary to analyze the composition of fixed cost of power grid enterprises. The fixed cost of power grid enterprises refers to the cost of providing preparation to ensure power supply to users at any time, while the variable cost refers to the cost of actual power supply. For power grid enterprises, the operation and maintenance of equipment is mainly to ensure the safe and stable operation of the power grid, including operation and maintenance, equipment defect elimination, status supervision, temporary accident disposal, spare parts management and other operations. However, the materials consumed jointly by several operations belong to indirect costs, which should be allocated in a reasonable and simple way and included in the cost of various production, operation and maintenance operations. When the consumption quota is correct, it is usually allocated according to the proportion of material quota consumption or material quota cost of the operation. The calculation formula is as follows:

$$W = \sum \frac{E}{Z} \tag{6}$$

In formula (6), E represents the actual total consumption of materials, Z represents the quota consumption of various operation materials, and W represents the distribution rate. The quality standard is to maintain the safe and reliable operation of the power grid, eliminate all kinds of defects affecting the safe and economic operation in time, improve the economy of power grid equipment, reasonably control equipment maintenance, etc. As the power industry is a capital intensive industry, its investment proportion is quite large, so the fixed cost of power grid enterprises is much larger than the variable cost. In the equipment operation and maintenance stage, a large number of operations can be identified. If the operation center is divided according to the operation activities, the work is relatively too complex. Therefore, the operations can be combined according to the principle of homogeneity, the main operations can be determined, and then the operation center can be established. According to the impact on decision-making, fixed costs can be divided into past capital investment and investment in power grid expansion and transformation prepared for future load growth, as well as debt service, depreciation, taxes, operation management and maintenance fees. Establish a standard operation library for the equipment operation and maintenance stage of power grid enterprises, which can subdivide the above operations, prepare a standard operation directory according to the type of equipment, and formulate detailed standard operation specifications. The contents of the standard operation library must form a work instruction document for operators' reference according to a unified format and standard. At the same time, fixed costs can also be divided according to links. Since fixed cost allocation involves all links of distribution and transmission, according to the different nature of the costs of each link, fixed costs can be divided into fixed access costs, power grid use costs and auxiliary service costs. Fixed access cost is the fixed cost borne by power users, and the power grid company provides power supply services for them. It mainly includes network connection cost and management service cost. The contents of these guidance documents shall include all-round factors involved in on-site operation, such as standardized operation flow chart, preparation work, work requirements, tools and instruments, material list, dangerous point analysis and safety control measures, division of labor of operators, work content, process standard of work steps, acceptance standard, performance evaluation, etc. The cost of network connection mainly includes the cost of lines and equipment connected to the power grid. This part of the cost has one-to-one correspondence, and each user only needs to bear its own part. The calculation is mainly composed of the cost of network connection line and the cost of user network connection facilities in the substation where the connection point is located, which has nothing to do with the amount of transmission [8–10]. The operation cost is generally related to the capacity, scale and equipment performance characteristics of the substation project. The analogy method is usually used to compare the characteristic parameters of the new substation project with the parameters of the existing substation project, and estimate the cost of the new substation project according to the historical cost data of the Operation Substation project. At the same time, the parameter method can also be used for regression estimation. The estimation formula of operating cost can be expressed as:

$$F = \sum_{y}^{x} \sigma_x \times \frac{1}{k_y} \tag{7}$$

In formula (7), σ represents the planned annual consumption, k represents the unit price, x represents the planned annual maintenance, and y represents the cost coefficient of various cost parameters [11, 12]. For routine inspection and maintenance, major and minor repair and planned maintenance, the operation shall be carried out in strict accordance with the operation instruction documents. Unplanned maintenance, fault repair and other operations should be carried out in accordance with the operation instruction documents as far as possible. There is no one-to-one correspondence in the cost of management services, which should be borne by all users who enjoy transmission and distribution services. The use cost of power grid refers to the cost of power system construction, operation and maintenance. The annual operation loss cost is calculated as follows:

$$\varpi = \frac{h+t}{2} \times K \tag{8}$$

In Eq. (8), h represents demolition cost, t represents residual value and K represents electricity price. In view of this situation, we can properly classify and summarize those equipment operation and maintenance operations that are too detailed, select the more typical standard operations for cost calculation, then select the main influencing factors of activity cost for accounting, calculate the adjustment coefficient, and finally calculate the cost of single operation according to the adjustment coefficient [13, 14]. This part of the cost is closely related to the use degree of users, but the cost that users should bear cannot be clearly defined like the cost of connecting to the network, so it can only be borne by all power grid users, resulting in the phenomenon of mutual subsidies [15]. In this way, the workload of formulating standard activity-based costing can be greatly simplified, the single standard cost can be distinguished, and higher flexibility and applicability can be obtained. After the formulation of standard Activity-Based Costing is completed, the standard Activity-Based Costing database is formed through collection and sorting to realize cost intelligent prediction.

3 Application Test

3.1 Testing Environment

China's power system generally has two peak loads in the morning and evening. Therefore, the peak period can be 2 h in the morning and 2 h in the evening, 8 h in the night trough period, and the rest are divided into daytime periods. The specific period division can be adjusted and determined after considering the characteristics of the power system. Operating system platform software configuration: the database server operating system can use Microsoft Windows 2020 server version. If the database server management platform software configures that the database server management platform uses Oracle e database, the optional scheme is: the capacity of a single main transformer is 120 mva and the number is 4. Database server software configuration. Global database name: 1 cam; Tablespace Name: LCAM; Database user name: LCAM_ user. The data comes from a provincial power grid company affiliated to the State Grid. 40 sets of operation cost data are selected. In order to eliminate the impact of different dimensions of substation parameters on the prediction speed and accuracy of the prediction model, it is necessary to normalize all sample data.

3.2 Test Result

The intelligent prediction method of power transformation operation cost based on neural network and the intelligent prediction method of power transformation operation cost based on data mining are selected to compare with the intelligent prediction method of power transformation operation cost in this paper. Test the prediction accuracy of the three methods under different single transformer capacity. The test results are shown in Table 1, 2, 3, 4:

It can be seen from Table 1 that the average prediction accuracy of the intelligent prediction method of power grid substation operation cost in this paper and the other two intelligent prediction methods of power grid substation operation cost are 92.667%, 81.447% and 81.929% respectively. Compared with the other two intelligent prediction methods of power grid substation operation cost, the average prediction accuracy of the intelligent prediction method of power grid substation operation cost proposed in this paper is higher.

It can be seen from Table 2 that the average prediction accuracy of the intelligent prediction method of power grid substation operation cost in this paper and the other two intelligent prediction methods of power grid substation operation cost are 86.401%,

Table 1. Prediction accuracy of transformer capacity 120 mva (%)

Number of tests	Intelligent prediction method of power transformation operation cost based on neural network	Intelligent prediction method of power transformation operation cost based on data mining	Intelligent prediction method of power transformation operation cost designed
1	85.162	79.611	92.346
2	82.164	82.106	91.105
3	81.322	81.915	93.667
4	80.199	83.154	92.548
5	79.826	81.522	91.055
6	78.202	77.848	93.677
7	79.655	79.655	92.154
8	82.348	83.102	93.104
9	82.449	84.254	94.221
10	83.144	86.122	92.788

Table 2. Prediction accuracy of transformer capacity 180 MVA (%)

Number of tests	Intelligent prediction method of power transformation operation cost based on neural network	Intelligent prediction method of power transformation operation cost based on data mining	Intelligent prediction method of power transformation operation cost designed
1	79.648	77.841	85.666
2	76.477	76.912	84.364
3	79.633	76.337	87.223
4	78.548	74.515	86.452
5	78.102	76.902	85.914
6	79.616	77.454	84.906
7	81.217	73.161	85.465
8	82.945	72.955	88.416
9	80.799	73.154	87.149
10	79.540	72.464	88.455

79.653% and 75.170% respectively. Compared with the other two intelligent prediction methods of power grid substation operation cost, the average prediction accuracy of the intelligent prediction method of power grid substation operation cost proposed in this paper is higher.

Table 3. Prediction accuracy of transformer capacity 240 MVA (%)

Number of tests	Intelligent prediction method of power transformation operation cost based on neural network	Intelligent prediction method of power transformation operation cost based on data mining	Intelligent prediction method of power transformation operation cost designed
1	65.316	64.119	76.164
2	64.822	63.784	74.512
3	65.919	62.006	73.604
4	64.378	63.147	72.649
5	63.978	62.922	73.468
6	66.121	63.781	74.915
7	67.106	64.352	74.658
8	68.241	65.464	75.355
9	69.316	66.223	76.461
10	65.752	65.464	74.512

It can be seen from Table 3 that the average prediction accuracy of the intelligent prediction method of power grid substation operation cost in this paper and the other two intelligent prediction methods of power grid substation operation cost are 74.630%, 66.95% and 64.126% respectively. Compared with the other two intelligent prediction methods of power grid substation operation cost, the average prediction accuracy of the intelligent prediction method of power grid substation operation cost proposed in this paper is higher.

It can be seen from Table 4 that the average prediction accuracy of the intelligent prediction method of power grid substation operation cost in this paper and the other two intelligent prediction methods of power grid substation operation cost are 63.732%, 57.070% and 56.028% respectively. Compared with the other two intelligent prediction methods of power grid substation operation cost, the average prediction accuracy of the intelligent prediction method of power grid substation operation cost proposed in this paper is higher.

Table 4. Prediction accuracy of transformer capacity 300 mva (%)

Number of tests	Intelligent prediction method of power transformation operation cost based on neural network	Intelligent prediction method of power transformation operation cost based on data mining	Intelligent prediction method of power transformation operation cost designed
1	61.324	58.641	63.947
2	59.878	59.316	64.518
3	62.316	55.828	63.228
4	58.499	54.161	62.945
5	59.677	55.319	63.701
6	56.322	52.348	62.533
7	52.878	55.227	63.944
8	53.945	54.313	62.586
9	52.146	55.649	64.512
10	53.712	59.481	65.401

4   Conclusion

The method of this design is to propose the selection process of main transformer capacity and number on the basis of considering the economic load rate of main transformer recovery cost. Evaluate the fixed cost of the power grid, determine the budget amount in different budget periods, extract the life cycle of the power grid substation equipment, establish the cost estimation relationship, establish the cost control model using multi-dimensional mixed information, refine the project category, and optimize the intelligent prediction mode of operation cost according to the different nature of the costs in each link. Induce the power users to make rational and correct choices according to their own power consumption, save energy consumption, transfer the peak load, and play a guiding role in the formulation and adjustment of electricity price. Therefore, it not only greatly simplifies the selection process of main transformer, but also ensures the reliability and economy of equipment operation. In the future development, the relevant advanced algorithms of deep learning are introduced to predict the operation cost of power grid substations, which makes the application effect of the intelligent prediction method of power grid substation operation cost more remarkable.

References

1. Wang, Y.l., Wang, S., Zheng, Y., et al.: Calculation and allocation of operation and maintenance cost of power grid project based on elastic net **20** 165–172 (2020)
2. Wang, M.J., Liu, Y.B., Gao, H.J., et al.: A two-stage stochastic model predictive control strategy for active distribution network considering operation cost risk. Adv. Power Syst. Hydroelectr. Eng. **11**, 8–18 (2020)

3. Zhang, Z.X., Wen, C.B., Cai, P.C.: Distributed droop control of islanded microgrid based on incremental cost consistency **4**, 517–523 (2020)
4. Li, T., Xu, Y., Chen, J., et al.: Optimal configuration of energy storage for microgrid considering life cycle cost-benefit **3**, 46–51, 58 (2020)
5. Zhang, Y., Chen, Q.X., Xia, Q., et al.: Active distribution network cost allocation method based on distribution factor method. Electr. Power **4**, 13–21 (2020)
6. Wang, J.F., Kong, L.S., Fan, X.M., et al.: Optimal planning for soft open point integrated with ESS to improve the economy of active distribution network. Electr. Power Constr. **10**, 63–70 (2020)
7. Fang, Y., Chen, J., Tian, X.Z.: Capacity economical optimization of non-grid-connected wind/hydrogen hybrid micro power grid. Comput. Simul. **2**, 110–114 (2020)
8. Wang, S., Liu, X.Y., Liu, S., et al.: Human short-long term cognitive memory mechanism for visual monitoring in IoT-assisted smart cities. IEEE Internet of Things J. **9**, 7128–7139 (2021)
9. Liu, S., He, T.H., Dai, J.H.: A survey of CRF algorithm based knowledge extraction of elementary mathematics in Chinese. Mob. Netw. Appl. **26**, 1891–1903 (2021)
10. Liu, S., Wang, S., Liu, X.Y., et al.: Fuzzy detection aided real-time and robust visual tracking under complex environments. IEEE Trans. Fuzzy Syst. **1**, 90–102 (2021)
11. Qian, J., Wang, P.P., Cheng, G., et al.: Joint application of multi-object beetle antennae search algorithm and BAS-BP fuel cost forecast network on optimal active power dispatch problems. Knowl.-Based Syst. **226**, 107149.1–107149.21 (2021)
12. Li, W.J., Liu, Y.G., Liang, H.J., et al.: A new distributed energy management strategy for smart grid with stochastic wind power. IEEE Trans. Ind. Electron. **2**, 1311–1321 (2021)
13. Huang, H., Jia, R., Shi, X.Y., et al.: Feature selection and hyper parameters optimization for short-term wind power forecast. Appl. Intell.: Int. J. Artif. Intell. Neural Netw. Complex Probl.-Solving Technol. **10**, 6752–6770 (2021)
14. Jakoplic, A., Frankovic, D., Kirincic, V., et al.: Benefits of short-term photovoltaic power production forecasting to the power system. Optim. Eng. **1**, 9–27 (2021)
15. Zhang, Y., Li, Y.T., Zhang, G.Y.: Short-term wind power forecasting approach based on Seq2Seq model using NWP data. Energy **213**, 118371.1–118371.14 (2020)

Research on Automatic Storage of Remote Mobile Surveillance Video Information Based on Cloud Terminal

Zihe Wei[✉] and Zhihui Zou

College of Humanity and Information, Changchun University of Technology,
Changchun 130122, China
weizihe666@163.com

Abstract. Cloud storage technology has become a trend of future storage development. Users can no longer buy hardware storage devices, and fully realize remote networked storage, which greatly reduces the user's use cost. At the same time, it can also provide faster and larger capacity storage and sharing functions. Therefore, in order to achieve the goal of effective storage of massive video, this paper proposes a remote mobile surveillance video information automatic storage method based on cloud terminals, constructs a remote mobile surveillance video information management model, optimizes the remote mobile surveillance video information automatic storage algorithm, and realizes the design goal of remote mobile surveillance video information automatic storage. Finally, the experiment proves that the research method of remote mobile surveillance video information automatic storage based on cloud terminals is highly effective and fully meets the research requirements.

Keywords: Cloud terminal · Remote mobile monitoring · Monitoring video · Information storage

1 Introduction

With the rapid development of mobile Internet, the demand for multimedia applications such as distance education, telemedicine and video conference is increasing greatly, and the demand for video management is also increasing. The progress of high and new technology, such as cloud terminal signal processing and image compression, has greatly promoted the development of multimedia technology. With the development of cloud terminal technology, remote video application has become a research hotspot in the field of computer, including video storage, memory management and scheduling, and video transmission on the network. Cloud terminals are multimedia files that are streamed over a network using streaming technology. They support streaming as they are downloaded and played, making it possible to transmit media data of unknown size [1]. Cloud terminal technology is widely used in video monitoring model. Large video monitoring model is composed of video management server, PTZ controller, video

© ICST Institute for Computer Sciences, Social Informatics and Telecommunications Engineering 2023
Published by Springer Nature Switzerland AG 2023. All Rights Reserved
W. Fu and L. Yun (Eds.): ADHIP 2022, LNICST 469, pp. 70–83, 2023.
https://doi.org/10.1007/978-3-031-28867-8_6

storage server, cloud terminal distributor and so on. Due to the limitation of front-end devices and backbone network bandwidth, when a large number of users access video resources concurrently, the backbone bandwidth is tight and the user access delay is increased [2]. The solution is to avoid the network bottleneck caused by large-scale concurrent traffic by various cloud terminal distribution. Because of the huge amount of video information, the storage space of video is always a problem that people care about. Storing video effectively can save the storage space greatly.

Qin et al. [2] proposed a blockchain based method for coal mine video monitoring data storage and sharing, which uses edge devices to transmit data, uses interstellar file systems to store detection data, and stores all videos in the cloud center. Liu et al. [3] proposed a design method of video monitoring and storage system based on FPGA, which uses flash as the storage unit, reasonably selects the working mode of storage, designs according to the standardized design idea and combines with the requirements of the task book, and realizes the storage of video monitoring data. The above two methods can realize the storage of video data, but the storage capacity is limited.

In order to solve the above problems, a cloud based automatic storage method of remote mobile surveillance video information is proposed. By building a remote mobile surveillance video information management model and optimizing the video storage data management and evaluation algorithm, the remote mobile surveillance video information can be automatically stored.

2 Automatic Storage of Remote Mobile Monitoring Video Information

2.1 Remote Mobile Monitoring Video Information Management Model

Cloud computing is the development of distributed processing, parallel processing and grid computing, in which huge computing processors are automatically split into numerous smaller subroutines over a network and submitted to a large model consisting of multiple servers, the results of which are computed and analyzed and sent back to the user. Through cloud computing, network service providers can process tens of millions or even billions of messages in seconds to achieve a network service as powerful as a "supercomputer". The goal of the cloud computing model is to migrate the individual, personalized operations running on a PC or a single server to a large number of server "clouds", where the cloud terminal is responsible for processing the requests of users and outputting the results. It is a model centered on data computation and processing.

According to the storage resource and storage strategy, the video storage server saves the frame data of the cloud terminal in the storage medium, and sends the saved frame data to the playback terminal to complete the storage and retrieval of the frame data of the cloud terminal. The functional modules are shown in Fig. 1.

The video storage technology not only stores the video data to the hard disk, but also needs to consider the subsequent retrieval function. In video monitoring model, manageability is the requirement of video storage technology, that is, users can retrieve and play back video according to their own needs at any time and anywhere. Therefore, efficient retrieval efficiency is an important goal of video storage technology [4].

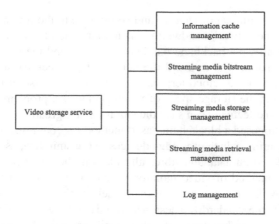

Fig. 1. Video storage service function module

If the stored video can not be retrieved, the video storage technology will lose its meaning of existence. In this paper, the software design of video network storage model is mainly embodied in two aspects: the software design of video network storage technology, and the implementation of iSCS protocol. But the storage design of video data adopts the classified storage strategy of time order, time-sharing, time-sharing and same channel. Video storage will also occur due to the large amount of video data caused by the lack of hard disk space, so in the design of video storage strategy, but also consider the hard disk coverage, video retrieval technology software design. The video retrieval strategy is designed according to the storage algorithm. This paper adopts the retrieval strategy of step by step, mainly through the query index table information to realize the retrieval and playback of data. The software design of video network storage can be divided into three modules, namely video storage module, ISCSI protocol module and embedded database module. These three modules cooperate with each other to complete the network storage of video data. Therefore, the design of the video network storage architecture module based on cloud terminals can be shown in Fig. 2.

Figure 2 deals with the sharing of data between the video capture process and the video storage process after the video capture process has been processed through a ring of shared buffer domains [5]. In this paper, the video data in shared memory is stored in the network storage server by video storage module, embedded database module and ISCSI module. The video stream data generated by the video monitoring model is massive, and these massive video data files pose a severe challenge to the storage model of the model.

2.2 Optimized Video Storage Data Management Evaluation Algorithm

For traditional file models, similar database functions and structures have been developed to meet the requirements of different types of file storage, while for video timeline sequential structures, such complex hierarchical lookups are redundant [6]. However, these two traditional file models are based on the disk allocation strategy of the first

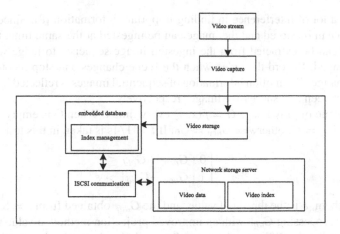

Fig. 2. Video network storage structure module based on cloud terminal

available area, which will cause the large file such as video data to be separated, and after a long time running, the time-related video data will have a high probability to be separated into different parts of the disk, thus resulting in spatial irrelevance, which requires the conversion of multiple pointers to achieve the purpose of lookup and reduce the efficiency of access [7]. Analog signal monitoring model is mainly composed of camera, video matrix, monitor, video recorder and so on. Video transmission line is used to connect video from the camera to the monitor. The host of video matrix is used to switch and control by keyboard. The video is recorded by a long time video recorder using magnetic tape, as shown in the figure. The traditional analog closed-circuit television monitoring model has many limitations (Fig. 3).

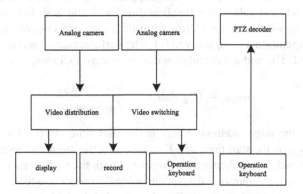

Fig. 3. Composition diagram of analog video monitoring system

Because of the large amount of video information, video monitoring model often needs a lot of hard disk storage space. Although it is possible to compress the video, if the monitoring time is too long, it still requires a lot of storage space, and most of the video recording is the same scene without change. This not only wastes storage space, but also has a lot of interference in finding important information [8]. Since the video compression can be stored and the image can be ingested at the same time, the motion information can be extracted from the ingested image sequence to judge whether the scene is changed. Record the video when the scene changes and stop recording when there is no change. The motion information of sequential images is reflected in the degree of change between two successive images respectively.

Take a frame of gray image $G = f(x, y)$ every time t. When G is empty, that is, the first frame, $G_{obl} = G$, otherwise $G_{nev} = G_0$. If $t = 1/4s$ is taken in this test:

$$G_{diff} = \begin{cases} 0 \mid G_{nev} - G_{old} \leq \varepsilon \\ 1 \mid G_{nev} - G_{old}| > \varepsilon \end{cases} \tag{1}$$

Among them, ε is the threshold value, and the G_{diff} obtained from $\varepsilon = 8$ is a binary image. After processing G_{diff} with an improved projection method, a value that can be used to determine the difference between G_{nev} and G_{old} is obtained, assuming that the head is at a particular location, and if the probabilities of the location of the next data to be accessed occur at all locations of the disk, the expectation of the seek time required for the head to move to the next data access is defined as the average seek time at that particular location, that is:

$$T_{avgi} = G_{diff} \varepsilon_T - D \sum P_i T_i \tag{2}$$

where T_{avgi} is the average seek time at the i-th data block, ε_T is the expected value of a single seek time, D is the total number of data blocks, T_i is the probability of the ith data block, and P_i is the time required for the head to turn to the i-th data block. Due to the use of a fixed size data area to store video data, there is a strict one-to-one linear correspondence between the time stamp of the stored video frame and the disk space address, and the retrieved keyword happens to be the time stamp itself. Then we can use the address calculation function directly as the hash function to calculate the corresponding video storage address from the time stamp keyword, which further reduces the computational complexity to 0.1, The performance of search and location is greatly improved. The address calculation during storage includes:

$$A_{target} = T_{avgi} A_{offset} + \frac{t_{now} - t_{begin}}{T_{timeperblock}} \tag{3}$$

where A_{target} is the target address, A_{offset} is the start time offset address, t_{now} is the current time, t_{begin} is the start time, and $T_{timeperblock}$ is the duration represented by each BK. it is determined during formatting. Compared with the traditional file model based on file name search, the number of 10 is undoubtedly reduced, which greatly changes the performance of video data access. Based on the disk logical storage structure in this scheme, the minimum unit of data organization is block. Therefore, the size of V DB cache built in memory is an integer multiple of the size of block. Its calculation method is as follows:

$$VDBSize = \left\lceil \frac{BR/8 + (ML + HS) \times FR}{BS} \right\rceil \times A_{target} \tag{4}$$

where, FR is the video code rate, BS is the length, BR is the size of the video frame header, HS is the frame rate, and ML is the size of the data unit in the disk structure. Data storage scheme provides data storage and video management interface for network video monitoring model, which determines the storage efficiency, management performance and reliability of video data. This chapter designs the data storage scheme from the disk logical storage structure, video index information management, data cache mechanism and data reconstruction method, and realizes the storage operation interface of disk formatting, data reading and writing, video retrieval and data reconstruction [9]. In order to improve the efficiency of video storage, optimize the performance of video retrieval and enhance the security of video data, a video data caching mechanism and data reconstruction method based on cloud terminal are designed. According to the characteristics of data storage in network video monitoring model, a logical storage structure of disks is designed, based on which all data are organized in format and stored directly in bare disks. The disk logical storage structure is shown in Fig. 4.

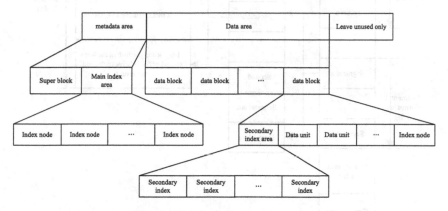

Fig. 4. Disk logical storage structure

Disks are logically divided into metadata area and data area. The metadata area is composed of two parts: super block and main index area, which is used to record the structure information of disk and video segment index information. The data is divided into equal sized data blocks for storing video data, picture data and sub- index information, and video and picture mixed storage. The remaining disk space is reserved unused. In order to realize the optimization of remote mobile surveillance video information automatic storage algorithm.

2.3 Realization of Automatic Storage of Surveillance Video Information

The new storage scheme is no longer considered from the point of view of optimizing the timing long storage mode, but adopts the fixed space storage mode. That is, the total space is initialized at one time into a number of fixed-size data files, each of which may be stored for a different length of time in the cloud terminal, and the data file is not deleted during cycle coverage, but is rewritten and the index is updated [10]. The new scheme effectively avoids the generation of disk fragmentation, but at the same time brings the following problems: maintenance of the cloud terminal storage time period index becomes very complicated; when the remaining space of the data file is not enough to store the cloud terminal of one second, this remaining space will be idle, resulting in a loss of space utilization, this situation is called "one-second tail"; the length of the alarm cloud terminal varies greatly, and the storage is bounded by events, if each alarm is stored in a different data file, a large amount of remaining space will be idle, this situation is called "event tail". The data deployment for the storage unit is shown in Fig. 5.

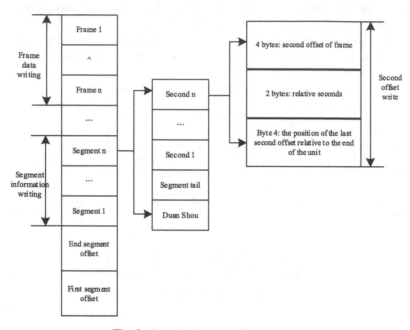

Fig. 5. Data deployment of storage unit

The video network retrieval strategy is designed according to the storage strategy, and adopts the method of progressive retrieval (hard disk index, file index, frame index) to realize the video retrieval of video channel, time and type. When the user needs to replay a certain video, the ARM target board will query the index information according to the user's retrieval interface, that is, search conditions to find eligible video data in turn. The module of video retrieval is similar to the storage design. The structure design of the concrete video retrieval module is shown in Fig. 6.

The storage and transcoding of massive video data based on cloud terminals, and the solution of management model are proposed. It not only improves the performance of video transcoding, but also provides convenience for users. Users can view video through the ordinary player, and also realize the management of video data based on cloud terminals. As shown in Fig. 7, the management model of the entire monitoring video function is presented.

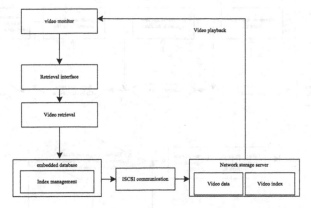

Fig. 6. Composition of video retrieval module

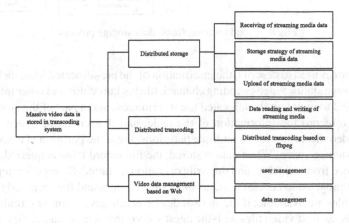

Fig. 7. Management model of various functions of monitoring video

Set the frame cache and segment index cache when the storage unit is written, and the cache time is set to 5 s to avoid frequent writes. If the current storage location of the channel is full, set the pre-allocation location to the current location and set the status of the corresponding location in the storage location index to used. The frame data is written backwards, and the index is written backwards. When the index is written, the end time of the segment is written even if the segment does not actually end, and the end time of that cell in the channel storage unit index is updated. Based on the above design, spatial strategy and index management are designed to solve the problem of "1 s tail" and "event tail" effectively. Although the complexity of the index is increased, the cost is worth it. Under the new storage management scheme, the cloud terminal frame data storage process is shown in Fig. 8.

The device does not store other types of files, and is dedicated to storing video files. This condition is used as a precondition for the policy. Prior to storage, the storage device is pre-initialized, that is, several empty files are created according to the specified size,

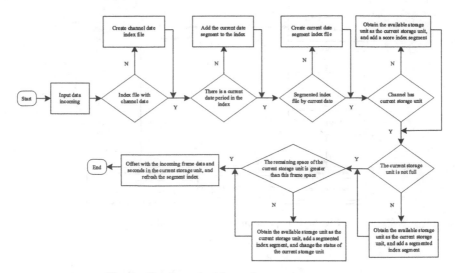

Fig. 8. Cloud terminal frame data storage process

and a structure is used to describe the information of the pre-allocated files, including file name, file usage status, video encoding channel, file backup status and other information, a table is used to maintain the allocated file information, and a part of the storage space is divided to record the information of the table, and the remaining space is used to store the video files. If there are multiple partitions, the same partition is made for each partition's storage space. When data is stored, the file record table is queried, an empty file is obtained from the table, and the write operation is started. Storage Strategy Design an assistant program to check the status of storage devices and files regularly to ensure that empty files are available; if the storage device is full, give a hint or circular storage, overwrite the earliest video files with the latest video stream in 68 cases. Figure 9 shows the video storage strategy.

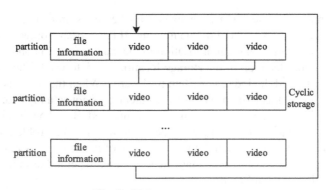

Fig. 9. Video storage strategy

During the operation of the model, the model accesses the storage device, the model detects the storage device, acquires the capacity of the storage device, calculates the maximum number of files that can be stored according to the capacity and the size of the pre-allocated empty files, and creates a continuous file of fixed size. When storing video data, the file information table is read first, and the available empty files are determined by looking up the status information of the files in the table; without the empty files, the earliest created files can be found again to store the video information, so the latest files can be used to overwrite the oldest files for circular storage; or the storage space is full and the files still need to be saved, the user is prompted to change the storage device. Because of the use of tables to maintain device file information, a secondary storage program is needed to manage the files, provide available files for the application to store video data, and update the latest information of the files in the table. In the process of storing information on the video server, the available empty files are obtained first, and then the video data is written. Helper programs need to periodically detect the condition of storage devices and files to ensure that empty files are available; if storage devices are full, prompt or circulate storage to overwrite the oldest video files with the latest video streams.

3   Analysis of Experimental Results

Apache-tomcat60 36 as the running web server, MyEclipse 90 is used as the development tool of the whole software. The versions of the three frameworks of struts 2 + Spring + Hibernate. At the same time, the cloud terminal version used is the same as the version deployed in the cluster in the previous section. Cloud terminal-0.2. In order to develop a rational storage scheme, this paper compares the monitoring storage scheme with the previous storage scheme. The results are shown in Table 1:

As can be seen from Table 1, although the expression form of monitoring storage data is the same as that of previous image storage, reflecting the irregular and unstructured characteristics, it is structured in the organization mode and has strong regularity, which is its biggest feature and the difference from the traditional storage model.

By further comparing the data upload rate, it can be concluded that with the increase of the number of redundant backups, the file upload rate will be reduced due to the time overhead of the cloud terminal pipeline. Based on this, the data storage time consumption with different number of redundant backups is compared, as follows:

As can be seen from Fig. 10, for the same total amount of video data with different slice sizes, the required download time is basically unchanged. Because the overhead of cloud terminal pipeline is not required in the download process, the time required for downloading is less than that for uploading the same data. When the number of redundant backups changes, the result is the same. Use the fast multimedia storage information model on the disk volume to test the same items as the above cloud terminal monitoring video information storage information model. Since there is no concept of file fragments, there is only one case. The test data are shown in Table 2:

Further analysis of the total data throughput and read-write ratio of storage information can be obtained:

The results in Fig. 11 show that the performance of video storage of cloud terminal monitoring video information storage model is lower than that of fast multimedia storage

Table 1. Comparison of characteristics between monitoring storage and traditional storage files

Project	Past storage		Monitoring storage
	Database table	Text/image	Image/audio
Data representation	Structured	Unstructured	Unstructured
Data organization	Unstructured	Unstructured	Structured
Data saving time	Disordered, indefinite duration	Disordered, indefinite duration	Delete and update regularly
Data update method	Disorderly, uncertain way	Disorderly, uncertain way	Orderly, increasing from the tail
Data reading and writing mode	Disorderly, repeatedly read and write	Disorderly, repeatedly read and write	Orderly, write a small amount of read/no read at a time
Storage block size	9-65KB	512B-1MB	64KB-1MB
performance requirement	10PS	10PS	Broadband
Storage hotspot	2/9 principle, 20% of data carries 80% of access	2/9 principle, 20% of data carries 80% of access	Equal access to data
Data importance	Important, high data value	Uncertain	Mostly useless data

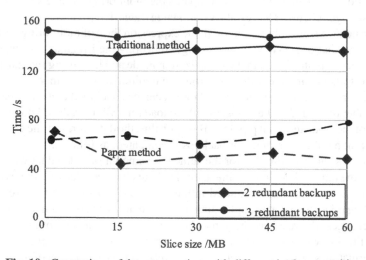

Fig. 10. Comparison of data storage time with different backup quantities

information model on the whole. When the cloud terminal monitoring video information storage is initially written into a large file, because there is no file fragment, the efficiency

Table 2. Performance test results of fast storage files

Parameter model	Data throughput (MB/s)	
	Paper method	Traditional method
Sequential storage	81.65	59.62
Sequential readout	69.58	33.62
100% random write	9.52	2.4
100% random readout	7.95	3.1

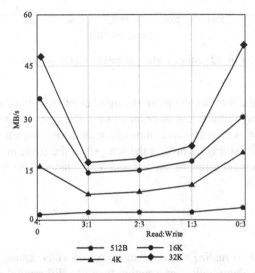

Fig. 11. Relationship between total data throughput of storage information and read-write ratio

is very high, which is close to the storage limit value. Because each frame needs to be written to the index, the rate is lower than the limit value. However, in the long run, the cloud terminal monitoring video information storage will operate with low efficiency, so the efficiency of cloud terminal monitoring video information storage is still higher in the long run. Select video streams with different bit rates as the data input source to test the storage efficiency. The video stream data is written to the disk device according to the writing mode, and the average storage bandwidth is calculated by recording the time spent continuously writing video data. In order to compare the storage efficiency with the traditional file model storage scheme, the storage efficiency of the traditional file model storage scheme is tested in the same environment and in the same method. The test result curve is shown in Fig. 12.

The test results show that the data storage scheme in this paper has high data storage efficiency and can fully meet the storage requirements of various common video bit rates. At the same time, compared with the file model, this scheme can effectively reduce the number of data network transmission and disk consumption. Under the same amount

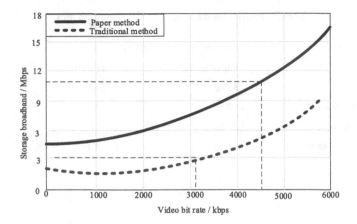

Fig. 12. Storage effect comparison test results

of data, the smaller the bit rate, the more the number of reduction, and the greater the improvement of storage efficiency. In order to save the cost of storage resources, the 512 Kbps low bit rate video stream commonly used in the current video monitoring model is used for video storage. Through the analysis of the curve in the figure, it can be seen that the improvement effect of storage efficiency under these two bit rates is very significant.

4   Conclusion

Using cloud terminal to realize video acquisition and video storage, using improved projection method to obtain video information from the differential image of sequence image, and then decide whether to record video according to the size of video, which can greatly save hard disk storage space and improve the storage proportion of useful video information. The method proposed in this paper has fast speed and good effect. It is a good video storage method, which fully meets the research requirements.

5   Fund Project

Scientific Research Project of Jilin Provincial Education Department: Research on key technology of industrial Internet data information security (JJKH20221280KJ).

References

1. Mao, Q., Jia, R., Zuo, L., et al.: A traffic surveillance video vehicle detection method based on deep learning. Comput. Appl. Softw. **37**(09), 111–117+164 (2020)
2. Qin, X., Zhang, J., Wang, B., et al.: Research on storage and sharing of coal mine video surveillance data based on blockchain. Metal Mine (06), 138–143 (2022)

3. Liu, H., Cai, J., Wang, X., et al.: Design of video monitoring and storage system based on FPGA. Instrum. Tech. Sens. (04), 64–68+100 (2022)
4. Liu, X., Nan, Y., Xie, R., et al.: DDPG optimization based on dynamic inverse of aircraft attitude control. Comput. Simul. **37**(07), 37–43 (2020)
5. Liu, S., Liu, D., Khan, M., Weiping, D.: Effective template update mechanism in visual tracking with background clutter. Neurocomputing **458**, 615–625 (2021)
6. Liu, S., et al.: Human memory update strategy: a multi-layer template update mechanism for remote visual monitoring. IEEE Trans. Multimed. **23**, 2188–2198 (2021)
7. Lian, J., Fang, S., Zhou, Y.: Model predictive control of the fuel cell cathode system based on state quantity estimation. Comput. Simul. **37**(07), 119–122 (2020)
8. Gao, Y., Yan, T., Man, C., et al.: Design of substation intelligent video monitoring system based on monitoring and control configuration software. Mod. Electron. Tech. **43**(16), 18–20+25 (2020)
9. Wang, B., Liu, Y., Sun, Q., et al.: Design of mobile video surveillance system based on GB/T 28181 and WebRTC. Electron. Meas. Technol. **43**(18), 112–116 (2020)
10. Xia, Z., Xiang, M., Huang, C.: Hierarchical management mechanism of P2P video surveillance network based on CHBL. Comput. Sci. **48**(09), 278–285 (2021)

Recognition of Self-organized Aggregation Behavior in Social Networks Based on Ant Colony Algorithm

Nan Hu[✉] and Hongjian Li

China University of Labor Relations, Beijing 100048, China
Yingsongvip@163.com

Abstract. In order to effectively detect the real network community structure and improve the accuracy of user stage partitioning to the corresponding self-organized community, a self-organized clustering behavior recognition method based on ant colony algorithm is proposed. According to user's individual attribute and collaborative attribute, the node with high aggregation coefficient under user's knowledge quality scale is chosen as the core to construct social network aggregation behavior community. The evolutionary types of group trajectory are divided into seven types. Ant colony algorithm is used to track the group trajectory. Abstract tagged basic events from user attributes, establish recognition model to identify abnormal behavior, and realize self-organized aggregation behavior recognition in social network. Experimental results show that the self-organized aggregation recognition method based on ant colony algorithm can get more reasonable group structure, better quality of community partition, and improve the accuracy of user stage partition to the corresponding self-organized community.

Keywords: Ant colony algorithm · Social network · Self-organization · Aggregation behavior · Behavior recognition · User characteristics

1 Introduction

With the vigorous development of information technology and computer technology, the world is developing e-commerce industry, electronic payment in China has also been widely popularized. The communication between people has become simpler and more convenient, and the spread of information has broken through the geographical restrictions to become faster and more convenient, so that people's daily lives are more and more colorful, but also to further promote people-to-people exchanges and information sharing. Various kinds of social media have sprung up, changing the way people live, and the circle of people's lives is becoming more and more complex. Microblogs, short videos, live broadcasts and other forms of information are spreading at an alarming rate. Because of the increasing number and complexity of these connections, the society of life is becoming more and more "networked", that is to say, these "invisible" connections make the individual, the collective, the city, the country and even the whole world, these

W. Fu and L. Yun (Eds.): ADHIP 2022, LNICST 469, pp. 84–98, 2023.
https://doi.org/10.1007/978-3-031-28867-8_7

human societies, large and small, closely linked together, making the society a network covering every participant. With the development of specialization, longitudinalization and intellectualization of information media, the platform of information communication is gradually moving towards mobile Internet. Countless "networks" aimed at connecting everyone are being created and constructed so that everyone around the world can participate and benefit from them, and the development of networks is becoming more and more "social." A network community structure with the characteristics of high cohesion and low coupling is formed between the changes and interactions of network members, which is composed of user member nodes [1]. Mining the characteristics of these community structures is of great significance to network research.

There are some common characteristics among complex networks, such as small world, scale-free, high aggregation and strong community structure. The research on complex networks has been a hot topic in many fields. Nodes in the community usually have some common attributes, which reflect the local regularity and global order of the social network to some extent. Community structure refers to the cluster of similar nodes in a complex network, that is, there are frequent links between nodes in the community and scattered links between nodes outside the community. Analyzing the community structure of complex networks is not only helpful to discover the potential relationship and function of complex networks, but also has great theoretical research value and practical application. A social network is a complex network, which is based on human society. Man is the main participant, but it is not an individual, or a group of people, or even a combination of people and things. For example, the participants in a cooperative network can be researchers, research institutes, or enterprises. It is not only the basis of the evolution of social network, but also can promote the development of related applications, such as recommendation system, privacy protection, network marketing and so on. It is of great significance to study the fission and reconstruction of network community system for the development of its information ecology.

At present, scholars in related fields have carried out research on the identification of relevant behaviors in social networks. Reference [2] proposes a dynamic behavior analysis and online detection method for navy users in social networks. Construct the dynamic behavior characteristics of social network users, and analyze the differences between normal users and navy users. Based on the semi-supervised model, combined with dynamic and static behavior characteristics, an online detection model is constructed. Through static behavior feature clustering and dynamic behavior feature filtering, the semi-supervised model uses the most valuable unlabeled user data to perform incremental learning to detect navy users. The average training time of this method is shorter, but the recognition accuracy of this method is lower. Reference [3] proposed an accurate identification method of inappropriate behaviors of naval user teams in social networks. Describe the social network and extract the dynamic characteristics of user behavior. The problem of identifying inappropriate behaviors of naval user teams in social networks is regarded as a binary classification problem, and the corresponding samples of dynamic features are extracted as inputs to build a decision tree. The decision tree is used to identify the misconduct of the naval user team on the new social network data set. This method has certain feasibility, but its recognition effect is poor.

Therefore, this paper proposes an ant colony algorithm based recognition method of self-organized aggregation behavior in social networks, according to the user's individual attribute and collaborative attribute characteristics, build a social network aggregation behavior community, track the group trajectory using ant colony algorithm, establish a recognition model to identify abnormal behaviors, and realize self-organized aggregation behavior recognition in social networks. The group structure obtained by the proposed method is more reasonable, the quality of community division is higher, the community structure of the real network can be effectively detected, and the accuracy of user stage division into the corresponding self-organized community can be improved.

2 Ant Colony Algorithm Based Recognition Method for Self Organized Aggregation Behavior in Social Networks

2.1 Analysis of User Attribute Characteristics in Social Networks

When users choose to participate in self-organizing groups in social networks, they should consider not only the individual attributes of users, but also the cooperative attributes of members. The user individual attributes and collaboration attributes considered in this article are shown in Fig. 1.

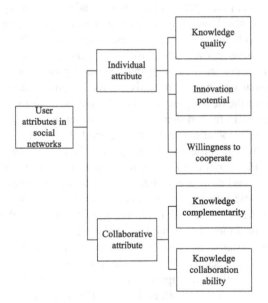

Fig. 1. Characteristics of user attributes in social networks

Individuals in social networks form a "relational structure" through various connections and produce a lot of information dissemination. This "relational structure" is the self-organized agglomeration behavior community. Self-organized agglomeration behavioural communities are an important structure of complex networks, which are composed of multiple communities, where nodes are very similar and compact, and where the connectivity between communities is fragmented [4]. Self-organizing aggregation in social networks requires a range of knowledge. User's knowledge quality includes situational knowledge quality, intrinsic knowledge quality and accessible knowledge quality. Since the target knowledge domain is not specified, only the intrinsic knowledge quality and accessible knowledge quality of user content are measured here. The knowledge quality of users can be quantified as follows:

$$A = \sum \alpha \ln \beta e^{-t/w} \tag{1}$$

In formula (1), A represents the user's knowledge quality; α represents the value coefficient; β represents the post over weight; t represents time; w represents the decay coefficient of knowledge; and e is a natural constant. Users have different knowledge potentials because of their knowledge structure. Some users have a deep understanding of a specific knowledge domain, while some users have only a little or no knowledge of that domain. Community structure can also be represented by user knowledge structure, and a matrix is constructed by the membership coefficient between each node and each community. In complex networks, communities can be divided by simply representing nodes of the same community with the same knowledge tag. Through the community structure, we can more intuitively reflect the relationship between the two sides of the network, and at the same time we can find the relationship between the community. The different structure of a user's knowledge determines its innovation potential. The innovation potential of users can be quantified from knowledge breadth and knowledge depth. When the user's interest in knowledge participation is high, the task of self-organization aggregation will advance more smoothly. The user engagement motivation indicator is used to quantify the user's willingness to collaborate, with the following expression:

$$B = \ln c \ln h \tag{2}$$

In formula (2), B represents the user's willingness to cooperate; c represents the number of replies of the user within the time t; and h represents the points of the user within the time t. In order to balance the distribution of data, both of them are processed logarithmically. The quality of users' knowledge in a social network has different impacts on the network depending on their location, and the closer the user is to the network center, the higher the importance of the user, the greater the impact on other nodes of the network, and the greater the value generated [5]. If the knowledge situation between users is too different, it will affect the knowledge communication between users, and even increase the cost of negotiation between users. Therefore, the complementarity of user knowledge will be considered. The knowledge complementarity coefficients of user s_1 and user s_2 can be expressed as follows:

$$\begin{cases} r = \sum g e^{-t/w} \\ g = |z(s_1) - z(s_2)| \\ z = \beta p \end{cases} \tag{3}$$

In formula (3), r represents the coefficient of knowledge complementarity between user s_1 and user; g represents the comparative advantage of knowledge; z represents the value of knowledge capability of user; and p represents the knowledge stock of post text, $z(s_1), z(s_2)$ represents the knowledge ability value of user s_1 and user s_2, respectively. Using the replies of users to measure the degree of common interest between users. The knowledge interaction ability between users is quantified by the comment relation between users.

2.2 Building Social Network Aggregation Behavior Communities

Community can be regarded as the local epitome of the network, especially in the online social network with a large scale of nodes. Because of the large amount of data, it is difficult to study the whole network directly. Proximity is a measure of a node's ability to occupy the center of a network. It is defined as the length of the shortest path to any other node in the network. The higher the value near centrality, the higher the importance of the node in the whole network. However, it takes a lot of time to get the close-centricity by getting the structure information of the whole network. According to user attributes, the nodes with local maximal characteristics are selected as the core of each community to form the initial structure of each community. Then the external nodes are divided into more attractive communities by comparing the attractiveness of the internal nodes of each community to their connected external nodes, and the process is repeated until all nodes are divided into corresponding communities. The topology of the network is sensitive to proximity to the center. If there are many edge connections in a local area of the network, it can find the center node well, but it can't detect the result well in the sparse area. In this paper, the nodes with higher aggregation coefficient under certain user knowledge quality scale are selected as the core of community, and such nodes are called local maximal aggregation nodes. For a user node, if the product of its aggregation coefficient and its own knowledge quality is greater than or equal to all its neighbor nodes, the node is called local maximum aggregation node. In the random walk process of a node, the probability of jump will change with the change of the node. The importance of nodes in the set is sorted by descending order and the first node is selected as the initial core node. The whole self-organizing aggregation behavior community is regarded as a whole, and the ownership of nodes is determined by the attraction of the community to nodes [6]. The attraction between two user nodes in a social network can be expressed as:

$$u_n(s_1, s_2) = x(s_1, s_2) \, o \, (s_1) \, o \, (s_2) \tag{4}$$

In formula (4), u_n represents the attractiveness between the user s_1 and the user s_2; $x(s_1, s_2)$ represents the elements of the adjacency matrix of the social network in row s_1 and column; and $o(s_1), o(s_2)$ represents the degrees of user s_1 and user s_2 nodes, respectively. When an unupdated node conflicts during the update phase, a label is chosen arbitrarily. The core node is partially expanded by random walk, so that the initial core community is constructed. Considering the closeness of the nodes in the community, the nodes which are three to four sides away from the core nodes are taken as the nodes of the random walk. This strategy of random label selection results in non- unique algorithm results. In the network, there is a two-way interaction between nodes. In the case of label update conflict, the label is selected according to the label weight. The label weight can be expressed as the quotient of attraction between nodes and length of neighbor nodes, and the final node is updated according to the label with the highest weight. At the end of the tagging process, the social network can be divided into two aggregation communities through an iteration, and the aggregation community structure is the same as the real community structure. In the process of traversing the non-community core nodes in the network, some nodes can be regarded as overlapping nodes because they are attracted by multiple communities simultaneously. The membership of overlapping nodes to each community is calculated to measure the degree of subordination of the node to each community. The formula for calculating community membership of overlapping nodes is as follows:

$$\chi(s_1) = \frac{u_n(s_1, s_2)}{\sum\limits_n u_n(s_1, s_2)} \tag{5}$$

In formula (5), χ represents the membership of overlapping nodes to the community; n represents the number of self-organizing aggregation communities in a social network. The whole network is traversed many times, each traversal calculates the attraction between nodes and communities, divides nodes into corresponding communities until all nodes enter into the communities. Based on this, the self-organized aggregation behavior community structure is defined.

2.3 Tracking Self-organizing Group Trajectories Based on Ant Colony Algorithm

Individual tracking in self-organized agglomeration community in social networks involves multiple tracking objects, which is difficult to track. In this paper, the ant colony algorithm is used to match and track the trajectories of self-organized groups. If the core node gravity chain changes, the dissimilarity between nodes in two neighboring time sliced communities satisfies a certain threshold condition, then a certain evolutionary relationship is defined [7]. Generally speaking, the changes of social networks over time mainly include: exiting original nodes, joining new nodes, breaking old connections, and new links between nodes. The change of network structure will also make the internal community structure change correspondingly. In this paper, the evolution of population trajectory is divided into seven types: persistent, shrinking, growing, splitting, merging, disappearing and forming. The selection of data is also more general, which can well mine the types of evolution events under different time slices. Community formation is a process in which some unconnected nodes in the network form a community structure at the due to increasing contact with each other. Observing the changes of the core node chain of each community to judge the evolutionary behavior patterns of community groups. For example, if the core chain of a community breaks, the community splits; if the core chain of multiple communities is connected, the communities merge; and if a new core chain appears, a new community emerges. The way the community is structured is shown in Fig. 2.

The first stage is to partition the community structure of the network snapshot at the time of t using the static social network discovery algorithm based on the attraction

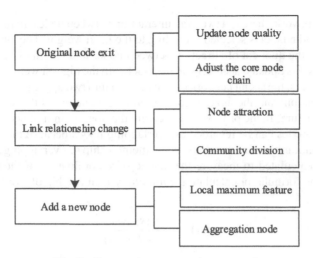

Fig. 2. Community restructuring approach

between nodes. In the second stage, the impact of network changes on social networks is fully considered, including the exit of original nodes, the change of link relationships and the addition of new nodes. In the original ant colony algorithm, the ant mainly relies on the probability function to select the next node. But in the initial stage of the algorithm, the difference of the pheromone concentration is very small, which can not guide the ant to choose the route effectively. Hence the pseudo-random state transition function shown in formula (6).

$$y = \arg \max\{v_1, v_2\} \tag{6}$$

In formula (6), y represents the selection function of candidate nodes; v_1 and v_2 represent the relative importance of pheromone and distance expectations, respectively. Candidate nodes will be selected based on the highest pheromone concentration in the selection function and the highest expected value of the path node. Because the pheromone is updated in the global direction in the ant cycle model, this paper uses the ant cycle model as the updating model, and all ants will update the pheromone content after the iteration. Then, the probability of any candidate node is calculated according to the probability formula, and the next node is determined by roulette. In different iteration periods, the requirements for pheromone increment are different. Therefore, in order to accelerate the convergence rate of ants and avoid local optimization, dynamic pheromone updating method is used here. The calculation formula is as follows:

$$P(t) = P_0 + \eta t \tag{7}$$

In formula (7), P represents the pheromone update value; P_0 is the initial value of the pheromone; and η is the pheromone update mechanism. The volatilization coefficient of pheromone also affects the global convergence and the optimal solution, and it also affects the pheromone content of the path to some extent. The size of pheromone volatilization factor indicates the degree of pheromone change with time for ants. In general, the setting of this factor can not be too large or too small, if set too large, although it can speed up the convergence of the algorithm but will also make the ant on the path that has not yet walked in the volatilization of pheromone faster, if set too small, the ant will fall into the current local optimal state when searching for the path so that the ant can not jump out to find more solutions. In addition, if the volatility factor is set to a constant amount during the search, the search will be reduced. By reserving the pheromone on each path, the global searching ability of the algorithm is ensured in the early stage, and the local optimization caused by excessive pheromone volatilization is avoided [8]. In each part, the quality (degree) of nodes and the attractiveness among nodes of the network should be updated, the core node chain of each community should be adjusted, and the evolutionary behavior of the community should be judged according to the changes of the core node chain of each community. Then, the structure of the self-organized aggregation network should be redivided or adjusted according to the attractiveness among the updated nodes. In the middle and later stages of the algorithm, a larger pheromone volatilization coefficient (i.e. a smaller pheromone residue coefficient) should be used, because there is a certain gap in pheromone on the path after the previous operation, and by using a larger pheromone volatilization coefficient, the difference in pheromone on each path should be increased to ensure the convergence speed of the algorithm. Finally, we consider the new nodes in the network, and add these nodes to the algorithm and related links. The addition of new nodes will make some nodes in the network more difficult. After adjusting the core node chain of each community, the rest nodes in the network should be redivided according to the attractiveness of the updated nodes.

2.4 Establishment of Self-organizing Aggregation Behavior Recognition Model

According to the tag of user attribute data, the self-organized aggregation behavior is transformed into a binary classification problem to judge whether the event is normal or abnormal. When there are only normal events in the detection model, the model belongs to the normal event detection model of clustering, and the abnormal events are far away from the center of clustering. The structure of the self-organizing aggregation behavior recognition model is established as shown in Fig. 3.

First, the tagged basic events are extracted from the user attributes, and then the events are learned to build a classifier. Finally, the events are classified by a classifier. Nodes will select the largest number of neighbor tags in the label update stage, and will select a tag randomly when unupdated nodes collide in the update stage. The occurrence of anomalous events does not exist independently at a certain point in time. Events may change slightly over time, and the whole event cannot be viewed from a single point in time. Therefore, when the events occurring at all points in time are combined together, a complete event can be seen [9]. In the network, there is a two-way interaction between nodes, so this paper considers the interaction between nodes on the impact of label selection, according to the interaction force index to improve the update strategy. Select the node j to be updated from the node update sequence. During propagation, the node j to be updated will select the label corresponding to the maximum force to update its own label. The formula is as follows:

$$f(j) = \arg\max \vartheta(\varphi) \tag{8}$$

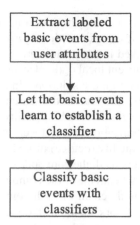

Fig. 3. Establishing the structure of the self-organizing aggregation behavior recognition model

In formula (8), f stands for node update function; ϑ stands for node force value; and φ stands for label. The node force value shall be comprehensively measured by the node's own properties and the relationship between nodes. The calculation formula is as follows:

$$\vartheta = \varepsilon\kappa + (1 - \varepsilon)\theta \qquad (9)$$

In the formula (9), ε represents the damping coefficient, which is used to balance the influence of a node j on its neighbors and the influence of a neighbor on the node j; κ represents the similarity of a node j to its neighbors; θ represents the force of a neighbor on a node j, and the larger the value is, the greater the influence of a neighbor on a node is and the deeper the influence is. Based on the objective knowledge system, the key elements existing in the system are analyzed, including user element, text element and knowledge element. According to the mapping relationship between elements, the integrated model of user knowledge hyper-network model is established. The method based on feature reconstruction considers that there are correlations between normal events, and they are similar to a great extent. It is usually possible to represent a normal event linearly with other normal events, and to reconstruct its main features. Its process is to train the normal events, get the dictionary after learning, and calculate its reconstruction error for the sample to be tested. The normal events can be reconstructed with very small error, but the abnormal events are relatively large. Through the full expression of heterogeneity elements in the target knowledge system, it provides material and framework for the analysis of the elements such as users and knowledge. It is worth noting that most of the current research into user mining, often based on the user's contribution to the community of all knowledge modeling, ignoring the changes in user behavior online. It is worth noting that the earlier the historical behavior of the user occurs, the less its reference value. Therefore, the time dimension is added, that is, the panel data in the target knowledge system is used for dynamic modeling of user knowledge super-network. So we can judge whether the event is abnormal or not according to the result that the reconstruction error is larger than the set threshold. At present, the usual method is sparse representation, which makes a dictionary of the normal events, and then uses the sparse representation of the test events by the dictionary atoms. Thus, the self-organized aggregation behavior recognition method based on ant colony algorithm is designed.

3 Experimental Study

3.1 Experimental Settings

In this paper, three networks are used to verify the effectiveness of the proposed ant colony algorithm for self-organized clustering behavior recognition in social networks. The three representative datasets selected were the Karate Club Social Network (Karate), the Dolphin Social Network (Dolphin) and the Enron Social Network. This paper needs to analyze the structure and characteristics of complex network communities on these datasets. Each of their snapshots is treated as a static network and is identified as a self-organizing aggregation behavior. The Karate network consists of 34 nodes and 78 edges; the Dolphin network consists of 62 nodes and 159 edges. The nodes represent the members of the club, and the edges represent the social relationships among the members. The Enron Social Network Dataset describes the data that Enron employees exchange email. The Enron Social Network contains information on 56 employee nodes and 854 e-mails. Through preprocessing, it was divided into 12 time slice snapshots by month. This method not only ensures the high similarity of the original topology of the network community, but also effectively increases the number of evolutionary events to provide sample support for the subsequent experimental analysis. The self-organized aggregation behavior recognition method based on GN algorithm and Fast algorithm in social networks is used as a comparison method to carry out experimental tests. In this paper, modularity and Normalized mutual information are used to test the effectiveness of the proposed self-organized aggregation behavior recognition method in social networks.

3.2 Experimental Results and Analysis

Modularity, also known as modularity measure, is a commonly used method to measure the structural strength of a network community. The value of modularity represents the result of community partition of self-organized aggregation behavior, and the larger the value is, the better the result of community partition is. The modular results of the respective organization aggregation behavior recognition methods in the Karate, Dolphin, and Enron networks are shown in Tables 1, 2 and 3.

In Karate network dataset, the average modularity of self-organizing aggregation behavior recognition method based on ant colony algorithm is 0.383, which is 0.146 and 0.180 higher than that based on GN algorithm and Fast algorithm. It can be seen that the results of community division of self-organized aggregation behavior of the self-organized aggregation behavior identification method in social network based on ant colony algorithm are better.

In Dolphin network dataset, the average modularity of self-organizing aggregation behavior recognition method based on ant colony algorithm is 0.362, which is 0.153

Table 1. Comparison of modularity results for Karate networks

Number of tests	Recognition of self-organized aggregation behavior in social networks based on ant colony algorithm	Recognition of self-organized aggregation behavior in social networks based on GN algorithm	Fast algorithm based recognition method for self-organizing aggregation behavior in social networks
1	0.386	0.251	0.206
2	0.395	0.224	0.218
3	0.378	0.238	0.197
4	0.382	0.245	0.204
5	0.393	0.222	0.205
6	0.386	0.233	0.212
7	0.375	0.242	0.196
8	0.382	0.255	0.182
9	0.394	0.229	0.203
10	0.361	0.226	0.204

Table 2. Comparison of modularity results for Dolphin networks

Number of tests	Recognition of self-organized aggregation behavior in social networks based on ant colony algorithm	Recognition of self-organized aggregation behavior in social networks based on GN algorithm	Fast algorithm based recognition method for self-organizing aggregation behavior in social networks
1	0.374	0.206	0.208
2	0.342	0.188	0.217
3	0.358	0.185	0.234
4	0.366	0.193	0.226
5	0.373	0.186	0.202
6	0.385	0.222	0.213
7	0.352	0.235	0.225
8	0.361	0.219	0.202
9	0.344	0.232	0.191
10	0.369	0.224	0.194

and 0.151 higher than that based on GN algorithm and Fast algorithm. It can be seen that the results of community division of self-organized aggregation behavior of the

self-organized aggregation behavior identification method in social network based on ant colony algorithm are better.

Table 3. Comparison of modularity results for Enron networks

Number of tests	Recognition of self-organized aggregation behavior in social networks based on ant colony algorithm	Recognition of self-organized aggregation behavior in social networks based on GN Algorithm	Fast algorithm based recognition method for self-organizing aggregation behavior in social networks
1	0.401	0.218	0.219
2	0.389	0.209	0.232
3	0.416	0.225	0.226
4	0.407	0.236	0.203
5	0.404	0.218	0.245
6	0.391	0.247	0.267
7	0.388	0.252	0.234
8	0.392	0.234	0.181
9	0.386	0.242	0.195
10	0.373	0.223	0.228

In Enron network dataset, the average modularity of self-organizing aggregation behavior recognition method based on ant colony algorithm is 0.395, which is 0.165 and 0.172 higher than that based on GN algorithm and Fast algorithm. From the perspective of modularity, the self-organized aggregation behavior recognition method proposed in this paper is more sensitive to community structure and has achieved good results. The method of this paper can correctly divide the user stage into the corresponding self-organization community.

Normalized mutual information is used to measure the similarity of clustering results. Normalized mutual information is an index used to evaluate the similarity between the real network structure and the community structure generated by the algorithm partition. The higher the value is, the higher the similarity between real network and partitioned community structure is. The Normalized mutual information results of the aggregation behavior identification methods in the Karate, Dolphin, and Enron networks are shown in Tables 4, 5 and 6.

In Karate network dataset, the mean of Normalized mutual information is 0.745, which is 0.107 and 0.175 higher than the comparison method based on GN algorithm and Fast algorithm. It can be seen that the more similar the real network and the divided community structure of the self-organized aggregation behavior identification method in the social network based on the ant colony algorithm, the higher the similarity.

In Dolphin network dataset, the mean of Normalized mutual information is 0.701, which is 0.160 and 0.190 higher than the comparison between GN algorithm and Fast

Table 4. Comparison results of normalized mutual information for Karate networks

Number of tests	Recognition of self-organized aggregation behavior in social networks based on ant colony algorithm	Recognition of self-organized aggregation behavior in social networks based on GN algorithm	Fast algorithm based recognition method for self-organizing aggregation behavior in social networks
1	0.726	0.603	0.582
2	0.784	0.629	0.597
3	0.749	0.645	0.604
4	0.750	0.658	0.571
5	0.762	0.626	0.552
6	0.725	0.662	0.563
7	0.717	0.635	0.542
8	0.728	0.627	0.555
9	0.736	0.654	0.578
10	0.772	0.641	0.560

Table 5. Comparison results of normalized mutual information for Dolphin networks

Number of tests	Recognition of self-organized aggregation behavior in social networks based on ant colony algorithm	Recognition of self-organized aggregation behavior in social networks based on GN algorithm	Fast algorithm based recognition method for self-organizing aggregation behavior in social networks
1	0.687	0.559	0.518
2	0.688	0.525	0.497
3	0.695	0.536	0.483
4	0.716	0.542	0.526
5	0.705	0.553	0.519
6	0.722	0.525	0.505
7	0.694	0.531	0.512
8	0.681	0.540	0.523
9	0.703	0.552	0.510
10	0.722	0.545	0.521

algorithm. It can be seen that the more similar the real network and the divided community structure of the self-organized aggregation behavior identification method in the social network based on the ant colony algorithm, the higher the similarity.

Table 6. Comparison results of normalized mutual information for Enron networks

Number of tests	Recognition of self-organized aggregation behavior in social networks based on ant colony algorithm	Recognition of self-organized aggregation behavior in social networks based on GN algorithm	Fast algorithm based recognition method for self-organizing aggregation behavior in social networks
1	0.684	0.516	0.564
2	0.691	0.538	0.577
3	0.686	0.545	0.551
4	0.723	0.521	0.548
5	0.755	0.505	0.555
6	0.762	0.539	0.536
7	0.714	0.512	0.523
8	0.827	0.546	0.545
9	0.796	0.553	0.554
10	0.805	0.526	0.529

In Enron network data set, the mean of Normalized mutual information is 0.744, which is 0.214 and 0.196 higher than the comparison between GN algorithm and Fast algorithm. From the standardised mutual information point of view, the self-organized aggregation recognition method presented in this paper is stable and can effectively detect the real network community structure.

4 Conclusion

Based on ant colony algorithm, a recognition method of self-organized aggregation behavior in social networks is proposed. By analyzing the characteristics of user attributes in social networks, the nodes with high aggregation coefficient under the scale of user knowledge quality are selected as the core to build a social network aggregation behavior community. Ant colony algorithm is used to track the group trajectory, extract labeled basic events from user attributes, establish a recognition model to identify abnormal behaviors, and realize self-organized aggregation behavior recognition in social networks. This method can improve the accuracy of community partition and dynamically adjust the group structure in the network. With the coming of big data era, how to choose a better time window or automatically divide the time window according to the dataset is a problem worthy of study.

Fund Project. This paper is the phased achievement of Motivation Attribution and system solution (19ZYJS021) of collective action in the process of China University of Labor relations of the first-class program of characteristic disciplines of China Institute of labor relations (basic scientific research business expenses of Central Universities); General project of Beijing Social

Science Foundation: phased achievements of employment structure change and staff training in the development of digital economy (20JJB008).

References

1. Zhang, K., Sun, Y.J., Han, H.: Customer credit modeling and credit granting method based on hybrid algorithm of collaborative filtering and social network. Telecommun. Sci. **36**(2), 52–60 (2020)
2. Li, Y., Deng, S., Lin, J.: Dynamic behavior analysis and online detection of spammer user in social network. Comput. Eng. **45**(08), 287–295 (2019)
3. Qiu, G., Li, X., Cheng, X., et al.: Accurate identification of misconduct of water user team in social network. Sci. Technol. Eng. **19**(07), 177–182 (2019)
4. Yuan, L., Gu, Y., Zhao, D.: Research on abnormal user detection technology in social network based on XGBoost method. Appl. Res. Comput. **37**(3), 814–817 (2020)
5. Peng, Y., Jiang, R., Xu, L.: An algorithm for identifying multi-class abnormal behavior of population based on rough set model. Sci. Technol. Eng. **21**(11), 4524–4533 (2021)
6. Xie, Q., Li, Z.: Network malicious behavior identification and detection based on big data association rules. J. Hefei Univ. **38**(2), 85–91 (2021)
7. Rong, W., Jiang, Z., Xie, Z., et al.: Clustering relational network for group activity recognition. J. Comput. Appl. **40**(9), 2507–2513 (2020)
8. Jin, X.: Deep mining simulation of unstructured big data based on ant colony algorithm. Comput. Simul. **37**(11), 329–333 (2020)
9. Yang, X., Sun, Y.: A survey on user behavior of social network based on knowledge graph. J. Hebei Univ. (Nat. Sci. Ed.) **41**(1), 77–86 (2021)

Data Clustering Mining Method of Social Network Talent Recruitment Stream Based on MST Algorithm

Hongjian Li[✉] and Nan Hu

China University of Labor Relations, Beijing 100048, China
abrajim@sohu.com

Abstract. In order to solve the problem that the data clustering mining method of social network talent recruitment stream is affected by the score of graph area and has a long time of index updating, a data clustering mining method of social network talent recruitment stream based on MST algorithm is designed. Based on the six-degree segmentation theory, the features of social network talent recruitment are extracted, the flow computation framework is established, the recruitment data processing process is optimized, and the similarity coefficient is used as similarity measure to construct the flow data clustering model and the mining pattern is designed by using the MST algorithm. Experimental results show that the maximum update time of the proposed method is 16.638 ms, which shows that the proposed method can shorten the update time of the index and is of high value.

Keywords: MST algorithm · Social network · Talent recruitment · Stream data · Clustering · Data processing

1 Introduction

In the early research of stream data clustering, the problem of stream data clustering was regarded as a special case of "large database" clustering problem. This kind of early algorithm belongs to single-layer algorithm structure in frame structure. There are many examples of streaming data, such as: in the business world, large warehousing supermarket transaction data. The supermarket chain's data center collects a large number of transactions from each store every day, each of which includes attributes such as customer purchases and consumption amounts, in chronological order. Single-layer algorithm structure tries to transform the dynamic characteristics of data flow into the traditional static mode, so that it can apply more mature traditional methods to solve the problem.

If each transaction completes and records can be collected immediately by the data center, the recorded data will flow to the data center continuously in the time dimension from the data abstraction point of view. At this stage, the focus of the algorithm research is to improve the performance of the traditional algorithm to adapt to the dynamic characteristics of the stream data.

© ICST Institute for Computer Sciences, Social Informatics and Telecommunications Engineering 2023
Published by Springer Nature Switzerland AG 2023. All Rights Reserved
W. Fu and L. Yun (Eds.): ADHIP 2022, LNICST 469, pp. 99–111, 2023.
https://doi.org/10.1007/978-3-031-28867-8_8

Small space, the algorithm is the representative of this kind of algorithm, it uses the improved K-center point algorithm to make it can be applied in the new problem domain. In the telecom industry, the mobile company will collect the user's call record every time. These records also include a number of attributes, such as calling number, called number, call time, the amount of money and so on. This paper analyzes the requirements of stream clustering model, and summarizes some clustering models that may be suitable for data stream. It is considered that there are three requirements to be satisfied in data stream clustering: (1) compressed expression, (2) rapid and growing processing of new data points, and (3) rapid and clear determination of outliers. In this paper, a two-layer algorithm framework for stream data clustering is proposed. The algorithm is divided into two parts: online layer and offline layer. Large numbers of call logs are arranged in chronological order, pooled in a mobile company's data center, and can be abstracted as a "stream." The online layer algorithm is responsible for fast and simple processing of the stream data and generating the profile data structure. In addition, a series of physiological signals, such as heartbeat, pulse and blood pressure, are transmitted to the analysis module in real time to predict the patient's health status at each time. Line layer algorithm makes use of these summary data information for more complex analysis, and generates more accurate clustering results. In addition to its theoretical value, the study of data stream clustering is of great practical significance. Many practical problems can be solved by clustering the data stream [1, 2]. In the industrial production, some large equipment safety testing instruments will collect the operation parameters of the equipment at this time at every moment, as a data signal for analysis and processing. And so on, these are examples of streaming data. For example, in network monitoring, the network state is analyzed by clustering the network flow, which provides a reliable basis for improving the network performance. It is not hard to see that these data patterns all have some common features. First of all, the volume of data is very large, the number of these data over time to rise sharply, such as Wal-Mart supermarket import and export transactions can reach hundreds of thousands of times. Dynamic analysis of user call records recorded on telecom switches using data flow clustering technology can help balance the load. Clustering analysis of the visit records of large websites can provide a basis for the decision-making of websites.

However, in the process of data processing, the above data stream clustering mining method is affected by the score of graph area, which has the defect of a long index update time. In order to simplify the recruitment data processing process and reduce the index update time, this paper proposes a social network talent recruitment flow data clustering mining method based on the MST algorithm. Innovatively based on the six degree segmentation theory, extract the characteristics of social network talent recruitment, establish a process calculation framework, and optimize the recruitment data processing process; The streaming data clustering model is constructed, and the MST algorithm is used to design the mining pattern, which improves the locality of calculation and realizes efficient data mining. Experimental results show that this method can reduce the update time.

2 Data Clustering Mining Method of Social Network Talent Recruitment Stream Based on MST Algorithm

2.1 Extracting the Recruitment Characteristics of Social Network Talents

Due to the characteristics of no time and space restrictions, enterprises and recruiters can browse recruitment information on electronic devices anytime, anywhere, communicate with each other, and strengthen the information exchange between the recruitment and the applicants. For employers and job seekers, online recruitment is conducive to both sides to use the least time in the broadest scope to find the right talent or position.

Social network recruitment is based on social network, is the application of social network in the field of human resources recruitment, refers to the enterprise recruitment activities based on social network platform. The breakthrough time limit and geographical scope of the Internet can quickly connect users around the world, so that job seekers can see the recruitment information of enterprises on the Internet platform, and can carry out rapid and two-way interaction between users in different regions and countries at any time. This feature of the Internet is more conducive to the recruitment units to find their potential suitable candidates. The enterprise establishes the enterprise talent database based on the social network platform, as well as the enterprise carries on the staff recruitment based on the social network platform release recruitment information. However, the traditional recruitment channel is restricted by time and region, and the network recruitment is not limited by time and space to make it possible to achieve a reasonable flow of talent. Meanwhile, social network recruitment refers to the process of employing social network service platforms such as Microblog and WeChat as recruitment channels.

Social networking is based on the "six-degree segmentation theory," which assumes that the world is unfamiliar to everyone and that it takes just six people to make a connection. The "six-degree separation" emphasizes the important role of "weak relationship" in interpersonal communication. The characteristics of the Internet, such as "no time and space limitation, interactive and real-time", are favored by enterprises. They will also use the official website of enterprises to show the corporate image to job seekers in an all-round and multi-level way, so as to provide job seekers with a true way to understand the enterprises and help job seekers choose suitable recruitment positions. Social networks move offline relationships into the network, widening the range of social interactions and creating a large network of relationships. But social network recruitment relies on the unique advantage of social network to play a strong role of weak relationship. Most enterprises can select suitable talents through online recruitment, the openness and rapid proliferation of online recruitment information for enterprises and candidates to provide more options.

The network regarding the present society, may say is the content rich, the function is formidable, makes the contribution for the enterprise employment advertise effective promotion. Social network recruitment is mainly divided into three parts, namely, personal management, recruitment information and networking. Social networking sites target job seekers through the management of resume information on personal platforms. Network information system, when dealing with a large number of resume information of job seekers, uses its own powerful function of data processing to sort out, screen and analyze more quickly and accurately, thus establishing a powerful talent database for enterprises and providing a basis for the future talent vacancy. Based on the social network, the candidate can know the relevant information of the enterprise in real time, and match the information published by the enterprise with his ability and intention to find a satisfactory job. Candidates can also learn about the culture and values of the company through information posted on social networking sites. In the recruitment website, the enterprise recruitment information and talent supply and demand information is updated in time to facilitate enterprises and recruiters choose at any time. Enterprises can select job candidates according to search engines, and the function of automatic classification helps to find suitable job candidates quickly and effectively, and feed back to job candidates in time, thus improving the efficiency of recruitment. The recommendation function of the social network recruitment platform, based on the matching of the personal information of the candidates and the recruitment information of the enterprises, can recommend the highly matched candidates to the recruitment enterprises. The internet recruitment can improve the ability of information collection, analysis and processing, reduce the time and manpower cost of manual information collection, so as to reduce unnecessary capital waste and make the recruitment more effective.

Social networks are also able to recommend well-matched companies to job candidates, thereby enabling efficient communication and matching of information among candidates, social networks and recruitment companies, laying a solid foundation for person- and organization-matching, as shown in Fig. 1:

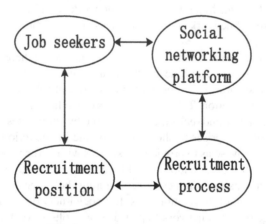

Fig. 1. Recruitment match diagram

As can be seen from Fig. 1, social network recruitment uses social network platform to realize the understanding, communication and interaction of the information of both sides, thus accomplishing the recruitment goal. Online recruitment is based on the initiative of both employers and job seekers to communicate on the Internet, which can not only improve the timeliness of online recruitment, but also provide good opportunities for job seekers to understand the enterprises and positions. Therefore, this kind of online recruitment is being accepted and used by more and more enterprises and job seekers.

Specifically, enterprises use social platforms to implement recruitment, big data technology as a support to analyze whether the characteristics and capabilities of candidates match the organization, thus enhancing the effectiveness of recruitment. Different from the traditional way of recruitment, it does not demand the absolute consistency of time and space, which facilitates the choice of time between the two sides, and saves a lot of time.

2.2 Optimize Recruitment Data Processing Process

The prominent feature of social network is that the network coverage is divergent, and the network technology makes the network between people become larger and larger, so as to enlarge the user's network. High throughput, low latency, good scalability and stable running system is an ideal system for stream data mining, which can not be realized without reasonable design and planning of port architecture, programming interface and high available technology.

Social network recruitment based on social network platform is to make use of the wide network of social network platform, which can make the information spread sufficiently and make the best use of social resources, and provide continuous and sustainable impetus for the career development of job seekers and the long-term development of enterprises. The so-called system architecture is the combination of various subsystems in the computing platform. The mining of stream data is inseparable from the specific streaming computing architecture. There are one master node and several slave nodes in the master-slave computing architecture. The master node is responsible for task allocation, resource scheduling, load balancing and fault tolerance. For enterprises, in the network of social networks with the social platform to find the right person can achieve twice the result with half the effort. The scheduling of the whole port depends entirely on the master node, and the slave node receives the tasks assigned by the master node and completes them respectively. In the symmetrical architecture, there is no master node for scheduling, and each node has the same function and good scalability. Resource scheduling and load balancing need to be coordinated by related distributed protocols. For job seekers, can make up for their own narrow circle of the shortcomings of the full received a large number of external recruitment information, in order to find suitable positions to pave the way. Job seekers in the production of resumes and interviews, the presentation is not necessarily the real daily state, there will be deliberate performance of the phenomenon.

Generally speaking, in the streaming data environment, the programming interface that uses directed task topology to represent the relationship between computing tasks is designed to facilitate programmers to implement the functions in the task topology. But by browsing the homepage space of the job seeker on the social networking platform, through a large number of data analysis to obtain the inner information of the job seeker's friends and personality traits. Unlike batch computing, where data is stored on a persistent device in advance, large data streaming does not allow data to be similarly persisted. The main processing flow of the recruitment data is shown in Fig. 2:

As can be seen from Fig. 2, the main processing flow of recruitment data includes three steps: data selection, data preprocessing and data transformation. Using MST algorithm, we can analyze the interpersonal communication of job seekers and understand

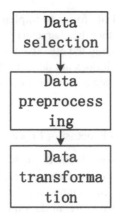

Fig. 2. Main processing flow of recruitment data

the quality and word-of-mouth image of job seekers indirectly through the circle of friends. Social networking platforms provide job seekers with a freer and fuller space to express their opinions and choices than a stylized resume. Therefore, it is not easy to replay data after node failure in the big data streaming environment, and the high availability technology of batch computing can not be fully applied to the streaming environment. That is to say, social network recruitment can make the human resource managers know more about the personality, interest and experience of the candidates, and help them to match the positions offered by the enterprises, which improves the success rate of recruitment, at the same time, reduces the turnover rate to a greater extent. In data mining, the main task of classification mining is to discover value more accurately through learning. An algorithm of classification mining can be divided into two main parts: building model and using model to predict. The classification algorithm for stream data can build decision tree and mine incremental data, and the incremental data mining is also dynamic.

2.3   Build Streaming Data Clustering Model

Using the existing initial data to build a clustering model, the data to be classified will continue to arrive, the classification model can be used to classify these objects. Clustering is the process of dividing data objects into different subsets. In the application of data mining and data analysis, it is often necessary to quantitatively express the differences between data points or between individuals, so as to evaluate the similarity of data points or individuals and their categories [3–5]. There will also be an endless stream of data to update the decision tree. The data to be predicted in the future will be predicted by the latest decision tree.

Using the existing initial data to build a clustering model, the data to be classified will continue to arrive, the classification model can be used to classify these objects. Clustering is the process of dividing data objects into different subsets. In the application of data mining and data analysis, it is often necessary to quantitatively express the differences between data points or between individuals, so as to evaluate the similarity of data points or individuals and their categories [3–5]. There will also be an endless stream of data to update the decision tree. The data to be predicted in the future will be predicted by the latest decision tree.

$$L(p, q) = \sum_{\beta=1} |p - q_\beta|^{\alpha-1} \tag{1}$$

In formula (1), p represents the data set, q the data space, β the data distance difference, and α the data dimension.

Therefore, the calculation of similarity is an important index in the research field of data mining. Since the properties of clustering model are divided according to similarity, it is very important to calculate the accuracy of similarity. This type of data, produced with the growing reach of the Internet and the widespread use of automatic digital devices, goes beyond the limits of existing information systems that can be persisted, processed accurately, and accessed repeatedly. Set the fixed clustering parameters in advance, and then get the final result according to the parameters. In addition, we can set the threshold to distinguish the strength of similarity through the threshold, and then carry out the subsequent clustering mining process [6, 7].

In clustering model, the selection of similarity measure is often represented by distance measure, which measures the difference and similarity between data points. Data matrix can be used to represent the characteristics of the corresponding data, through the matrix list to declare the corresponding attribute structure, each row in the matrix represents a data. The structural characteristics of t data can be represented by the following matrices T:

$$T = \begin{bmatrix} t_{11} & \cdots & t_{1x} \\ \vdots & \ddots & \vdots \\ t_{y1} & \cdots & t_{xy} \end{bmatrix} \tag{2}$$

In formula (2), t represents the number of clustering objects, x represents the attributes of data, and y represents the structural characteristics of data. Stream data has many characteristics: the data arrives in real time, the data flow speed is unpredictable, and it requires fast and instant response. The data changes over time, on a vast and unbounded scale. Among many clustering models, Euclidean distance measure is usually used to measure the attribution degree of the target point and the cluster. However, large datasets are often accompanied by higher dimensions.

On the basis of formula (2), if distance is used as the similarity, then the value on the diagonal line is 0, if similarity coefficient is used as the similarity measure, then the value on the diagonal line is 1, and the formula of the transformation matrix T' of the matrices T is as follows:

$$T' = \begin{bmatrix} r_{11} & \\ \vdots & \ddots \\ r_{x1} & \cdots & r_{xx} \end{bmatrix} \tag{3}$$

In formula (3), r represents the horizontal columns of the matrix that represent the corresponding data points. Data processing is basically single-pass scanning algorithm, once the data is difficult to be taken out again. The degree of data structuring is low, which needs multilevel and multidimensional processing. As the dimension of the set of numbers to be clustered increases, the distance between two points will become smaller and smaller, and the farthest neighbor and the nearest neighbor will be almost the same. Therefore, Euclidean distances are no longer suitable for measuring large, high-dimensional data sets.

Based on the above principle, assuming that the h central point is known, the $h + 1$ central point can be selected by the following principle $G_{(h+1)}$. Can be expressed as:

$$G_{(h+1)} = \frac{1}{\max\left[g(d, h)^2\right]} \tag{4}$$

In formula (4), g represents the shortest distance between the candidate center point and the former h center point, and d represents the maximum value of the shortest distance. The results show that the similarity of objects in clusters and the difference of objects in clusters are evaluated according to the attribute values of the objects, usually involving distance vectors.

Clustering may find previously unknown groups in the data object. On the other hand, the length of the data stream is theoretically considered to be infinitely long, and its range is also considered to be infinite. For example, if a router only routes 10,000 different IP address pairs, then despite the large number of IP packets, there is no challenge for querying IP address pairs whose traffic is greater than a certain threshold. By comparison, Manhattan distance is a better representation of the actual similarity between data than Euclidean distance in high dimensional space. But when the IP address pair space is far beyond the real storage, the small number of IP packets also makes many queries difficult.

On the same dataset, different clustering methods may produce different clustering. Clustering analysis is not divided through the people, but through the algorithm automatically. Finally, the infinity of stream data makes it impossible for stream data mining to retain the original data, but can only maintain a series of profiles in memory, and generate the final results based on the profiles.

2.4 MST Algorithm Design Mining Patterns

The MST is constructed in each grid, and then the MST of each grid is connected to obtain the MST algorithm of the original dataset. Then the clustering is completed. The process of data mining is a process of discovering various models, summaries and derived values from a given set of data. The main stages involved include: data preparation, data mining, result representation and interpretation. The mining method based on MST is a significant research field in the mining method based on graph theory. MST is an important research direction of connectedness in graph theory. It has many excellent structural properties in the representation of set data, and has been widely used in many fields. Data selection mainly refers to extracting the relevant data from the existing database or data warehouse to form the target data. Data preprocessing is to merge the data, to solve the semantic fuzziness, data processing omission and cleaning dirty data, etc.

In the MST-based clustering algorithm, firstly, the relationship among all the samples in the data set should be represented as the graph structure of MST. Among them, the node of graph structure is all the sample points in the data set, and the edge of graph structure is the similarity measure between the two connected sample points. The aim of da transformation is to eliminate the dimension of data, that is to find out the really useful features from the initial features, narrow the processing range and improve the quality of data mining. Then, according to the given weight closed value or the number of clusters, delete the edges with the largest weight in turn, and obtain the connected subgraphs of the graph structure, each of which represents a cluster.

Data preparation is an important step of data mining. Whether the data preparation is good or not will affect the efficiency and accuracy of data mining and the effectiveness of the final pattern. This stage is the actual work of mining, first of all to determine the task of mining or what is the goal, such as data summary, classification, clustering, association rules or sequential pattern discovery. Among them, the nodes belonging to the same subgraph have the greatest similarity, and the similarity with other subgraph nodes is greater than the nodes within the subgraph. In general, for a given dataset, the MST generation algorithm is not unique for each execution because there may be two or more edges of the same length. According to the task of mining, the choice of algorithm is the most important step, and the choice of algorithm directly affects the quality of mining.

The extracted information is analyzed according to the end user's decision goal, and the most valuable information is distinguished and submitted to the decision maker by the decision support tool. The social network talent recruitment stream data clustering mining process is shown in Fig. 3:

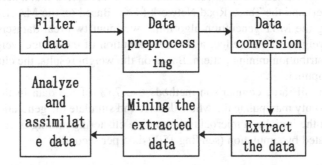

Fig. 3. Social Network Talent Recruitment Flow Data Clustering Mining Flowchart

As can be seen from Fig. 3, the process of data clustering and mining in social network recruitment flow mainly includes: data selection mainly refers to extracting relevant data from existing databases or data warehouses to form target data. Data preprocessing is to merge data, to solve semantic fuzziness, to deal with the missing data and dirty data in the data.

However, the non- uniqueness of MST constructs on the same dataset does not affect the clustering result based on the minimum generated cluster, because the generated cluster is independent of the specific MST structure. The clustering algorithm based on

MST has its own unique advantages. The task of this step is not only to present the results (e.g., using information visualization), but also to filter the information. If the decision makers can not be satisfied, we need to repeat the above data mining process. MST is an important data structure in graph theory, and it has good geometric properties.

The essence of MST is the minimal cost spanning tree, i.e. the weight of all the edges and the minimal connected subgraph in the graph structure. Mining the web data after data preprocessing, using various mining technologies to mine the rules and patterns hidden behind the data, and according to the specific application, filter out the rules or patterns that are not used in the pattern mining stage, and transform the useful rules and patterns into knowledge and apply them to the specific field. So that the structural characteristics of the data set is more prominent, segmentation MST clustering results produced more scientific, rational and easy to operate. The performance of the algorithm is not limited by the shape of clusters, and irregular classes can be identified.

3 Experimental Study

3.1 Experimental Preparation

In order to test the effectiveness of the proposed method, we design a comparative experiment in the environment of Eclipse.

The experimental environment is a cluster environment composed of 16 computing nodes. Among them, the resource configuration of each machine is as follows: CPU: Intel (R) Core (TM) i 3-2100 @ 3.10 GHz (dual core), memory: Apacer 4G DDR3 × 2, hard disk: Hitachi SOOG/7200 RPM operating system: Red Hat Linux 6.1 JDK version: jdk -1.6.0-30. The experiment used two real datasets, the Live Journal social network (SOC) (default data set) and the Texas Road Network (TX). Based on the MATLAB R2015a program, using the MST generation algorithm, we input two real datasets, randomly generate the weights of each edge, and the distribution of each edge weight satisfies the uniform distribution mining pattern. Based on the weight results, the cluster mining results are computed.

In addition, all three comparison methods use 75% of the data in the dataset to initialize, randomly manipulate the whole dataset and simulate the generation of stream data. Based on the initialized microclusters, online clustering mining is carried out for the data generated by simulation (sending 1500 data per second).

3.2 Experimental Results

For the same dataset, we use the clustering mining method based on genetic algorithm and the clustering mining method based on deep learning to compare the two methods. As the number of graph partitions is larger, the index updating process is more complex and the updating time is longer. Therefore, taking the number of graph partitions as the test index, the maximum update time of the three methods is tested when the number of graph partitions is different. The experimental results are shown in Tables 1, 2 and 3:

According to Table 1, for area score index of 4 maps, the maximum update time of the social network talent recruitment flow data clustering mining method in this paper is

Table 1. Maximum update time for area score index of 4 maps (ms)

Number of experiments	Data clustering mining method of social network talent recruitment stream based on genetic algorithm	Data clustering mining method of social network talent recruitment stream based on deep learning	Data clustering mining method of social network talent recruitment stream
1	16.514	15.699	12.141
2	15.334	16.347	11.547
3	16.254	16.925	10.698
4	17.121	16.548	12.146
5	16.992	15.377	11.214
6	15.847	16.201	12.096
7	16.224	15.649	13.147
8	17.131	17.003	12.588
9	16.255	16.528	13.164
10	15.288	16.945	12.410

Table 2. Maximum update time for area score index of 8 maps (ms)

Number of experiments	Data Clustering Mining Method of Social Network Talent Recruitment Stream Based on Genetic Algorithm	Data Clustering Mining Method of Social Network Talent Recruitment Stream Based on Deep Learning	Data Clustering Mining Method of Social Network Talent Recruitment Stream
1	19.488	21.004	14.512
2	18.554	19.822	16.528
3	23.516	21.225	15.494
4	21.334	22.313	15.461
5	23.818	24.164	14.332
6	22.156	23.255	13.520
7	24.549	24.818	14.511
8	22.616	25.191	15.260
9	19.477	24.462	14.220
10	18.463	23.646	13.744

13.164 ms, the maximum update time of the genetic algorithm based social network talent recruitment flow data clustering mining method is 17.131 ms, and the maximum update

Table 3. Maximum update time for 16 chart area score indexes (ms)

Number of experiments	Data clustering mining method of social network talent recruitment stream based on genetic algorithm	Data clustering mining method of social network talent recruitment stream based on deep learning	Data clustering mining method of social network talent recruitment stream
1	36.154	35.164	22.314
2	32.166	34.917	23.487
3	31.255	33.626	22.619
4	32.848	32.588	24.466
5	33.649	34.649	23.718
6	32.717	34.718	24.513
7	32.478	33.644	22.462
8	33.915	32.502	23.647
9	32.477	33.784	22.164
10	33.025	34.519	21.005

time of the deep learning based social network talent recruitment flow data clustering mining method is 17.003 ms; According to Table 2, for area score index of 8 maps, the maximum update time of the social network talent recruitment flow data clustering mining method in this paper is 16.528 ms, the maximum update time of the genetic algorithm based social network talent recruitment flow data clustering mining method is 24.549 ms, and the maximum update time of the deep learning based social network talent recruitment flow data clustering mining method is 25.191 ms; According to Table 3, for 16 chart area score indexes, the maximum update time of the social network talent recruitment flow data clustering method in this paper is 24.513 ms, the maximum update time of the genetic algorithm based social network talent recruitment flow data clustering mining method is 36.154 ms, and the maximum update time of the deep learning based social network talent recruitment flow data clustering mining method is 35.164 ms. This is because this method is based on the six degree segmentation theory, extracts the characteristics of social network talent recruitment, establishes a process calculation framework, optimizes the recruitment data processing process, and improves the mining efficiency; At the same time, the similarity coefficient is used as the similarity measure to construct the stream data clustering model, and the MST algorithm is used to design the mining pattern, which improves the locality of calculation and eliminates unnecessary time overhead.

4 Conclusion

By using the distributed index technology, the proposed method can make sure that two vertices belong to the same connected component, and only need several communication operations, not need to detect. At the same time, this paper analyzes the data mining technology, the related concepts and characteristics of stream data, and the key technologies of stream data mining. Aiming at the cluster mining in data mining, this paper analyzes the related concepts and challenges of cluster and stream clustering. In addition, the use of zoning acceleration technique effectively improves the locality of the computation and eliminates the cost of unnecessary communication. Future research will focus on improving the accuracy of mining methods on the basis of ensuring low mining time.

References

1. Cheng, H., Liao, Z., Wang, S.: Simulation of mixed attribute data clustering mining based on feature selection. Comput. Simul. **37**(7), 399–403 (2020)
2. Zheng, L., Zhang, H.: Big data clustering mining technology based on swarm intelligence algorithm in cloud environment. Mod. Electron. Tech. **43**(15), 115–118 (2020)
3. Gu, D.: Large data clustering mining based on P-WAP in Hadoop cloud platform. J. Changchun Normal Univ. (Nat. Sci.) **39**(5), 29–35 (2020)
4. Hua, T., Yi, H.: Big data mining based on around-centroid clustering algorithm. Appl. Res. Comput. **37**(12), 3586–3589 (2020)
5. Zang, Y., Xie, L., Zhang, Y., et al.: Data mining algorithm based on power marketing clustering analysis. Inf. Technol. **44**(4), 56–59, 64 (2020)
6. Li, X., Wu, X., Tong, B.: The research on data mining based on dynamic fuzzy clustering—taking the comprehensive strength analysis of Anhui city as an example. J. Guiyang Coll. (Nat. Sci.) **15**(1), 52–57 (2020)
7. Jin, H.: Research on artificial bee colony clustering data mining algorithm for accurate prediction. Digit. Technol. Appl. **38**(10), 95–97 (2020)

Risk Control Method of Enterprise Social Network Marketing Based on Big Data Fusion Technology

Zhiyi Deng[1] and Jin Li[2,3]([✉])

[1] School of Labor Relations and Human Resource, China University of Labor Relations, Beijing 100048, China
zydeng01@126.com
[2] School of Journalism and Communication, University of Chinese Academy of Social Sciences, Beijing 102401, China
idsf54854@yeah.net
[3] Miyun Branch, Beijing Municipal Tax Service, State Taxation Administration, Beijing 101599, China

Abstract. The popularization and application of social network provides a broad platform and space for enterprises to carry out marketing activities, but at the same time, the virtualization of marketing environment increases the difficulty of marketing activities. Based on this, the enterprise social network marketing risk control method based on big data fusion technology is discussed, the enterprise social network marketing risk identification method is optimized, the enterprise social network marketing risk control evaluation index and evaluation algorithm are constructed, and the enterprise social network marketing risk control process is simplified. Finally, it is confirmed by experiments, The enterprise social network marketing risk control method based on big data fusion technology has high practicability in the process of practical application and fully meets the research requirements.

Keywords: Big data fusion · Social networks · Enterprise marketing · Risk control

1 Introduction

Marketing risk refers to the various risks that enterprises bear in the marketing process due to the incompatibility between their marketing strategies and strategies and the market due to the complexity, variability and uncertainty of the enterprise environment and the limited cognitive ability of the enterprise to the environment, which may lead to the obstruction, failure or failure of marketing activities or failure to achieve the expected marketing objectives. Social network marketing risk has the following characteristics: diversified participants, means and strategies required for operation, which makes the sources of risk diversified. Social networks expand the market risks faced by enterprises.

© ICST Institute for Computer Sciences, Social Informatics and Telecommunications Engineering 2023
Published by Springer Nature Switzerland AG 2023. All Rights Reserved
W. Fu and L. Yun (Eds.): ADHIP 2022, LNICST 469, pp. 112–126, 2023.
https://doi.org/10.1007/978-3-031-28867-8_9

Because after enterprises connect to social networks, their business scope involves more problems [1].

The concept of data fusion originates from the needs of war and depends on military applications. However, with the development of data fusion, it has become an independent discipline, which is not obviously affected by a certain application. Instead, it puts forward its own common problem by means of generalization and specialization of concepts with the help of reasoning. Data fusion is a concept with a wide range of applications. It is difficult to give a unified definition. Data fusion is a research direction aiming at the specific problem of using multiple sensors in a system. Its definition can be summarized as: the information processing process of automatically analyzing and synthesizing the observation information of several sensors obtained according to time sequence under certain criteria by using computer technology, so as to complete the required decision-making and estimation tasks. According to this definition, multi-sensor system is the hardware basis of data fusion, multi-source information is the processing object of data fusion, and coordinated optimization and comprehensive processing are the core of data fusion. The movement of information flow, logistics and capital flow at each stage in time and space will be accompanied by risks of different nature. The functional structure of the information platform and the risks generated at each level of the marketing decision-making process bring different losses to the enterprise. The dependence on technology is stronger, and technology affects marketing to a considerable extent. It is often the innovation of new technology that will bring changes in marketing mode and business environment. Sixth, uncertainty. It needs both technical support and the participation of managers, which makes the risk faced by the enterprise's marketing activities very uncertain. Marketing risk early warning uses identification model to identify risk factors in the risk identification stage.

Reference [2] proposes a multi-layer template update mechanism to achieve effective monitoring in multimedia environments. In this strategy, the weighted template of the high confidence matching memory is used as the confidence memory, and the unweighted template of the low confidence matching memory is used as the cognitive memory. By alternately using confidence memory, matching memory and cognitive memory, it is ensured that the target will not be lost in the monitoring process. Reference [3] proposes a fuzzy detection strategy to pre judge the tracking results. If the pre judgment process determines that the tracking result in the current frame is not good enough, the stored target template is used for subsequent tracking to avoid template pollution. The test results on otb100 data set show that the proposed auxiliary detection strategy improves the tracking robustness in complex environments by ensuring the tracking speed.

On this basis, a risk control method of enterprise social network marketing based on big data fusion technology is proposed.Identify the characteristics of enterprise social network marketing risk, and carry out early warning management for the above three different dimensions of marketing risk sources. AHP divides the risk system into several main levels (three levels in this paper), and then uses expert meeting method or Delphi method to score the established risk index factors within the given scoring range. Compare the scores of each risk factor, establish the corresponding judgment matrix, determine the risk weight, and determine its importance. The two-dimensional analysis method based on the identification of marketing risk factors can observe the marketing risks of enterprises that cannot be investigated alone from the stage characteristics or risk characteristics of the marketing process, and organically combine the stage characteristics and risk characteristics of marketing risks.

2   Enterprise Social Network Marketing Risk Control

2.1   Identification of Risk Characteristics of Enterprise Social Network Marketing

The economic impact is whether the products and services operated by enterprises meet the economic characteristics of social network consumers, and also involves the identification and control of enterprise social network marketing risk characteristics. The imperfection of e-commerce marketing law brings very uncertain impact risk to enterprise management. The mode of social network operation is closely related to the development trend of social culture [3]. While the development of science and technology is diversified, it is very important for enterprises to grasp and apply new technologies, which should be synchronized with the development of new technologies. A simple enterprise social network marketing risk feature identification model can be shown in Fig. 1.

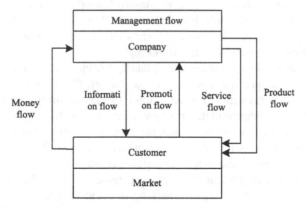

Fig. 1. Enterprise social network marketing risk characteristic identification model

As can be seen from Fig. 1, The first mock exam consists of two main elements: sales organization and market. Connecting these elements are three information communication processes and three entity exchange processes [4]. The company transmits information to the market through various marketing means. The response of customers in the market to the company's marketing activities is reflected in the sales performance of products and the return of payment for goods. Then the company collects these response information and makes plans for future behavior based on current and past information. Entity process refers to the movement that customers buy products or services and obtain benefits at the same time. The process of entity exchange has the characteristics of all commodity transactions [5]. The process of information communication is the remarkable feature of modern marketing model. From the analysis of enterprise operation process, marketing risk is also accompanied by each process of operation process. The poor operation of any process in information flow, capital flow, business flow and logistics will affect the operation efficiency of the enterprise. See Fig. 2 for details.

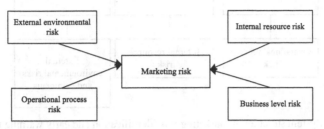

Fig. 2. Determination of enterprise social network marketing risk information identification

Enterprise social network marketing risk information identification refers to a management model that makes the enterprise management model have the ability of "alarm" and "immunity" in preventing, correcting or avoiding marketing errors and management fluctuations [6]. Early warning management carries out detection, identification, judgment, evaluation and pre control of management errors, management fluctuations and management adversity. The identification of enterprise social network marketing risk information takes the adversity phenomenon in enterprise business activities as the research object, so as to achieve the purpose of studying the essential characteristics, cause background and development law of adversity phenomenon and constructing enterprise early warning management mechanism [7]. Enterprise social network marketing risk information identification management is to monitor, identify, evaluate, warn and pre control the marketing risk based on the object of enterprise marketing risk. This paper will explain how to carry out early warning management according to the three different dimensions of the above marketing risk sources. Figure 3 shows the structure of marketing risk identification and early warning model.

As shown in the figure, it is an inclusive relationship from top to bottom. The highest level is the business level, including the following three risk levels: decision-making level, function level and operation level, and then detailed to the specific level. The operation level is generally the use of internal resources; The functional level usually manages the operation process; The decision-making level is a comprehensive evaluation of the statistical conclusions of the external environment and functional level, and then make decisions [8]. At the same time, there are risks at every level and stage. Therefore, to determine the marketing risk early warning in the social network environment. The advantage of this model is that once the risk occurs, it can quickly understand the source of the risk, and the enterprise can respond quickly. Due to different understanding angles of marketing risk, the performance of marketing risk is also different. From the perspective of the consequences caused by marketing risk, it can be divided into pure marketing risk and speculative marketing risk. Pure marketing risk refers to the uncertain state in which there are only loss opportunities but no profit opportunities. There are only

Fig. 3. System structure of marketing risk identification and early warning model

two consequences caused by pure marketing risk, either bring losses to the enterprise or no losses, and it has no possibility of profit [9]. Such as damage and deterioration of goods, bad loans, etc., are pure risks. Speculative marketing risk refers to those uncertain states with both loss possibility and profit possibility. It leads to three possibilities: no change in loss and profit. Speculative marketing risk mainly depends on people's risk handling skills to prevent. It is a complex, changeable and very difficult risk. It often puts enterprises in a dilemma. The handling of speculative marketing risk is not only a science, but also an art, but also the focus of marketing risk discussion (Fig. 4).

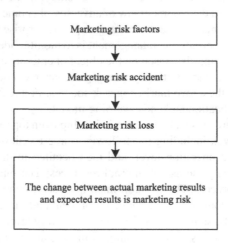

Fig. 4. Risk factors of enterprise social network marketing

Marketing activities are complex business activities. There are many uncertain factors in every link, from the formulation of market strategy to the establishment of marketing mix, from the issuance of products to the settlement of payment for goods. In a complex environment, due to the limitations of the experience and ability of marketing subjects, they do not fully understand and grasp the generation, development and consequences of risks, or fail to take timely and effective measures to prevent them, which will lead to various losses. The most common is the mistakes of marketing managers themselves, that is, the risks and losses caused by their sense of responsibility and work quality [10]. The actual loss of the risk caused by this management failure will be much greater than expected. A new risk will be caused by a risk, and a big risk will be caused by a small risk.

2.2 Social Network Marketing Risk Control Evaluation Algorithm

Determine the weight of each expert. The weight of each expert can be obtained in this way. Given the score range, the larger the score, the better you know about the risk system. Give yourself a score within the score range [11, 12]. Let x represent the score given by the first expert, and then calculate the average score, expressed by X. Then the expert weight vector is y_{in}^j, in which the weight of the expert judge is $i = 1, 2, ..., n$ and the weight constitutes the weight vector $j = 1, 2, ..., n$. . The analytic hierarchy process divides the risk system into several main levels (mainly divided into three levels in this paper), and then uses the expert meeting or Delphi method to score the established risk index factors respectively in the given score range. The greater the score, the more important this kind of risk is, Then compare the scores of each risk factor and establish the corresponding judgment matrix to determine the risk weight, and then determine its importance. Matrix s represents the evaluation matrix of the i-th layer from the risk perspective of the expert.

$$
s_{i\alpha}^j = \begin{bmatrix}
\dfrac{y_{i1}^j}{y_{i1}^j} & \dfrac{y_{i2}^j}{y_{i1}^j} & \cdots & \dfrac{y_{in}^j}{y_{i1}^j} \\
\dfrac{y_{i1}^j}{y_{i2}^j} & \dfrac{y_{i2}^j}{y_{i2}^j} & \cdots & \dfrac{y_{in}^j}{y_{i2}^j} \\
\vdots & \vdots & \ddots & \vdots \\
\dfrac{y_{i1}^j}{y_{in}^j} & \dfrac{y_{i2}^j}{y_{in}^j} & \cdots & \dfrac{y_{in}^j}{y_{in}^j}
\end{bmatrix}
\tag{1}
$$

In the judgment matrix s (indicates the target number of criteria in the previous layer, and Y indicates the score given by the expert to the k-th risk factor of / layer. The weight value of each expert's index factor for each layer can be obtained by applying the principle of analytic hierarchy process. The size of the risk factor can be obtained from the vector]. Therefore, in the result of artificial subjective judgment, we can judge and prevent those risk factors from light and heavy, and then take relative measures to prevent them from happening. According to the above-mentioned operation For the specific characterization and cause analysis of marketing risk, according to the ideas, principles and methods of the above-mentioned early warning index design, after the preliminary selection of indicators and the improvement of the system, we established the enterprise internal marketing risk early warning index system, as shown in Table 1.

The so-called marketing reliability refers to the probability that the elements of marketing environment E_t, marketing model S_t and marketing management M_t interact

Table 1. Enterprise internal marketing risk early warning index system diagram

Internal cause marketing risk early warning evaluation index system	Primary index	Secondary index	Auxiliary indicators / data points
	Organizational strategic risk evaluation index	Feasibility of marketing strategy	Accuracy of business direction; The scientificity of formulating strategy; Rationality of marketing objectives
		Accuracy of marketing strategy	The number of marketing strategies used accurately; Total number of marketing strategies developed
	Organization risk evaluation index	Self organizing ability	Coordination and adaptability of the organization
		Tissue activity	Organizational learning and innovation ability
		Rationality of management level	Actual and required management levels
		Rationality of management range	Actual leadership and number of managers who should be able to manage
	Organization function risk evaluation index	Proportion of marketing function	Many functions of marketing activities; Realization degree of comprehensive marketing indicators

to achieve the predetermined marketing function in a specific period of time. Multiple forces are interrelated and interact with each other, which can be measured by formula.

$$K_t = s^j_{i\alpha} E_t \cdot S_t \cdot M_t \tag{2}$$

The opposite of reliability is unreliability, that is, marketing accidents. Unreliability = 1 - Marketing reliability. Marketing accident (Marketing unreliability) does not represent marketing risk, but it is closely related to marketing risk and is the proximate cause of marketing risk. High marketing accident rate means high possibility of marketing risk loss. We use the marketing risk coefficient to represent the degree of marketing risk, and the marketing risk degree is the product of marketing risk coefficient and marketing unreliability. Marketing risk management is to improve marketing reliability and reduce marketing risk. It can be seen from the above formula that there are many factors affecting the market share, one of which is the price factor, and the price of products depends on the price of raw materials, equipment efficiency, wages and management level. The other is non price factor, which depends on technology development ability, market development ability, enterprise reputation and customers' preference for its products. The overall effect of these factors is different from that of time. The following table lists the threats of the decline of enterprise market share. In fact, the complete marketing threat analysis should uniformly list and analyze the possible marketing accidents, as shown in Table 2.

Table 2. Marketing threat analysis based on big data fusion

Threaten	Reason	Result	Risk mitigating factors	Loss estimation	
				economic loss	Market loss
Market share decreased by 10%	The price is on the high side	Lose some customers	cost reduction	550000 yuan	3%
	Backward technology development	Lose some customers	Increase technological innovation	250000 yuan	1%
	Weak market development	Can't attract new customers	Increase market investment	350000 yuan	2%
	Low enterprise popularity	Can't attract new customers	Enhance corporate image	250000 yuan	2%
	Customers do not form preferences	Lose some customers	Strengthen brand building	250000 yuan	1%
	Poor sales service	Lose some customers	Implement satisfactory service	150000 yuan	1%

It should be noted that in the macro environment, although political and legal environmental risks, macroeconomic risks and social and cultural risks all have an important impact on enterprise marketing activities, most enterprises can only adapt to these risks, and their changes are beyond the ability of enterprises. Moreover, their influence on enterprise marketing is very complex, so they are not included in the key object of this paper; The influence and significance of science and technology on enterprise marketing activities are significant and far-reaching, which is reflected in the reduction of supplier costs, the enhancement of competitors' technological innovation, the emergence and competition of alternative products and so on. Other influencing factors will be reflected in the follow-up study of the evaluation index system. Finally, a two-dimensional table as shown in Table 3 is established.

Table 3. Two dimensional analysis of marketing risk factors based on big data fusion

	Type	Internal marketing risk		Exogenous marketing risk			
Stage		Marketing organization operation risk	Marketing actor risk	Competitive risk	Customer risk	Supply risk	Third party risk
Value choice	Customer segmentation	K11	K12	K13	K14	K15	K16
	Market segmentation	K21	K22	K23	K24	K25	K26
	Value orientation	K31	K32	K33	K34	K35	K36
Value creation	product development	K41	K42	K43	K44	K45	K46
	Service development	K51	K52	K53	K54	K55	K56
	Product manufacturing	K61	K62	K63	K64	K65	K66
Value communication	distribution service	K71	K72	K73	K74	K75	K76
	Personnel promotion	K81	K82	K83	K84	K85	K86
	Sales promotion	K91	K92	K93	K94	K95	K96

Horizontal lines represent all stages of the marketing process, and vertical lines represent risks with certain characteristics. Where k represents the characteristic risk of the second stage of the marketing process. Based on the two-dimensional analysis method of marketing risk factor identification, we can observe the enterprise marketing risk that can not be investigated separately from the stage characteristics or risk characteristics of marketing process, and organically combine the stage characteristics and risk characteristics of marketing risk. This helps to clarify the relationship and characteristics between marketing risk factors and each stage of the marketing process. In the marketing process, with the advancement of the process, the change of the marketing environment and the increase of relevant participants, the uncertainty factors gradually increase, and the probability of risk also gradually increases. Due to the accumulation of investment, the possible loss caused by risk is increasing. As for each stage and its related risks, it will be discussed in detail in the construction of follow-up index system.

2.3 Vision of Social Network Marketing Risk Control

Enterprises must establish corresponding organizations to command and coordinate marketing risk control. Enterprise marketing risk control organization is the center of enterprise marketing risk control. In the process of daily production and operation activities, the enterprise marketing risk control organization is responsible for the knowledge training and education of marketing risk control, formulating corresponding rules and regulations, analyzing and evaluating possible marketing risks, putting forward response plans, and organizing simulation exercises of marketing risk disposal; After the occurrence of marketing risk, the company will uniformly command and coordinate the forces of all aspects of the enterprise, and be fully responsible for the handling of marketing risk. Depending on the enterprise scale and business complexity, the establishment scale of enterprise marketing risk control organization is also different. For small enterprises, their marketing risk control organization sets up a marketing risk control group under the marketing manager and is equipped with several corresponding professionals to specialize in marketing risk control. Its organization is shown in Fig. 5 below.

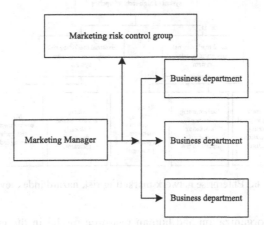

Fig. 5. Organizational structure of enterprise marketing risk based on big data integration

Generally speaking, the index system is a collection of a series of interrelated indicators that reflect the basic situation of a social and economic phenomenon. It reflects the quantitative performance and quantitative relationship of social and economic phenomena from multiple perspectives and levels. Similarly, to build the index system of enterprise marketing risk early warning model is to analyze all kinds of risks reflected in the index system according to the method of model analysis, and divide it into thousands of parts. Construct the framework of the index system from different aspects, and then subdivide each part and side until the evaluation can be described by statistical indicators. Marketing risk has different forms in different marketing environments, marketing stages and marketing subjects. From different angles and different reasons for the occurrence of risks, we can divide enterprise marketing risks into many types. Marked by the category of risk loss, marketing risk can be divided into natural risk, social risk, economic risk, political risk and technical risk; Marked by the controllable degree of risk, marketing risk can be divided into controllable risk and uncontrollable risk; With the level and scope of risk involved as the symbol, marketing risk can be divided into macro risk and micro risk, with the object of risk as the symbol, marketing risk can be divided into property risk, personal risk and responsibility risk, with the nature of risk as the symbol, marketing risk can be divided into dynamic risk and static risk, with the cause of risk as the symbol, marketing risk can be divided into subjective risk and objective risk, with the source of risk as the symbol, Marketing risk can be divided into internal risk and external risk. In short, the types of enterprise marketing risk are diverse, and the formation of risk is the result of the joint action of many factors. When the effect reaches a certain degree, the occurrence of an event plays a leading role, resulting in the occurrence of marketing risk. According to the different causes of enterprise marketing risk, it is divided into internal cause risk and external cause risk, that is, marketing organization risk and external environment risk. The evaluation index system structure is shown in Fig. 6.

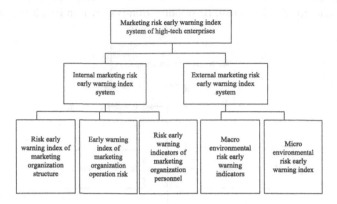

Fig. 6. Enterprise network marketing risk hazard index level

The marketing organization and human resource model in the enterprise network marketing risk hazard index model have many indicators that are related. They can be combined into a marketing organization model to retain more objective indicators as much as possible. The fund control model can achieve good control in financial risk early warning; The sub model of marketing strategy control model can be indirectly reflected by product control sub model and organization personnel control model. In order to eliminate the redundancy of indicators, they are also excluded from the control model. According to the above selection and selection of marketing control models, what remains are product control model, competition control model, public control model, collaborator control model and organization control model. Therefore, in this paper, the enterprise marketing control model is composed of these four sub models, as shown in Fig. 7.

Common big data fusion measurement methods include modeling method, variable control chart and set discrimination, but these methods determine the risk based on qualitative analysis or uniform value, propose to use the industry risk level as the central value, and then use the big data fusion analysis model to determine the risk grade method, which can flexibly determine the early warning grade according to the situation of the industry, the acceptance level and management ability of enterprise marketing risk. Based on this, the scale of marketing risk classification is determined, and the early warning level is determined according to the marketing risk level, as shown in Table 4.

Generally speaking, in the changing environment, the key to determining the position of enterprises in the market is the effectiveness and creativity of marketing organizations. Therefore, the internal marketing risk is mainly caused by the low efficiency of enterprise marketing organization. The purpose of analyzing the internal marketing risk factors

Fig. 7. Adjusted enterprise marketing risk level early warning model

Table 4. Marketing risk warning level

	Issue 13	Issue 14	Issue 15	Issue 16	Issue 17	Issue 18
BP marketing risk evaluation index	63	58	69	83	73	78
Risk alarm level	Central police	Central police	Central police	Light police	Light police	Light police
BP marketing risk evaluation index	19	18	10	48	39	48
Risk alarm level	Heavy police	Heavy police	Heavy police	Central police	Central police	Central police

of enterprises is to find and master the key factors affecting the marketing results of enterprises in the marketing organization, so as to create good conditions for the smooth implementation of early warning and pre control. For large enterprises, risk control is more complex. Its organizational structure is shown in Fig. 8 below.

The establishment of a special marketing risk management organization is conducive to the coordination and disposal of marketing risk management internally and externally, and provides an organizational guarantee for the effective prevention of marketing risks and the rapid and effective disposal of existing risks.

3 Analysis of Experimental Results

When measuring the effect of enterprise social network marketing risk control based on big data fusion technology, we rely on the normal distribution curve, which is the characteristic of variance covariance method and Monte Carlo simulation method. However, for the marketing risk of an industry or enterprise, the normal distribution curve may

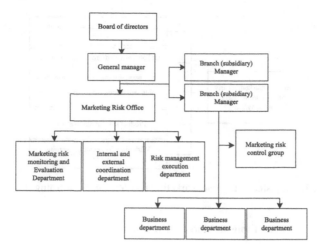

Fig. 8. Enterprise marketing risk management organization

not be the most appropriate model. Take the actual value of enterprise capital stock as a reasonable deflator, fully consider different systems, and add the deflator of capital marketing in income and product accounting, but the system does not take the adjustment of product quality into account. The system mainly includes the regression quality adjustment of characteristic information and the estimation of "near" vector autoregressive function:

$$R_t^q = \kappa_0 + \kappa_1 R_{t-1}^q + \kappa_2 R_{t-1}^n \tag{3}$$

In the formula: R_t^q is the risk factor obtained from the data set; R_t^n is the official enterprise capital marketing index; κ is the coefficient of each interaction item of the enterprise. To verify the practical value of the new enterprise social network marketing risk data behavior characteristic analysis method, the following comparative experiment is designed. Although the risk distribution does not completely obey the normal distribution, as shown in the figure below, the normal distribution can be used for simulation and prediction as a reference for risk management and marketing decision-making (Fig. 9).

In the above two curves, the normal distribution curve is symmetrical. It means that the opportunity for the return of the enterprise market to rise and fall is equal. The return of product sales in the industry is asymmetric. We know that Literature [2] method underestimate the possibility of huge losses because they rely too much on the normal distribution curve. In comparison, Monte Carlo simulation method overcomes this deficiency. Further, taking the once financing marketing behavior of an enterprise as the experimental object, record the changes of the negative impact of data behavior on the marketing revenue of enterprise social network before and after the application of this method in the experimental time of 5 months. The following figure reflects the change of the negative impact of horizontal data behavior on enterprise marketing revenue in the experimental time of 5 months (Fig. 10).

The analysis shows that with the increase of experimental time, the control effect of enterprise marketing risk under the guidance of this method is obviously better.

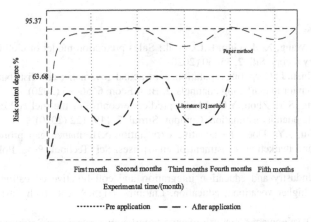

Fig. 9. Normal distribution of enterprise marketing price risk and market risk

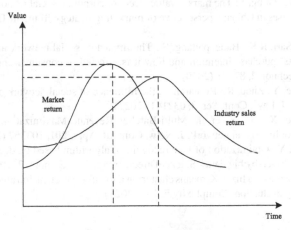

Fig. 10. Comparison of control degree of negative impact of enterprise marketing

4   Conclusion

By analyzing the source of marketing risk in the social network environment, a set of marketing early warning model with hierarchical and inclusive relationship is obtained, and the risk identification and control are carried out in combination with big data fusion technology to appropriately improve the early warning of marketing risk in the social network environment. This paper discusses the enterprise social network marketing risk control method based on big data fusion technology, optimizes the enterprise social network marketing risk identification method, constructs the enterprise social network marketing risk control evaluation index and evaluation algorithm, and simplifies the enterprise social network marketing risk control process.Put forward early warning countermeasures according to marketing risks, hoping to provide reference for operating enterprises in the social network environment.

References

1. Wang, X.T., Wang, X.M., Guo, R.L., et al.: Sales prediction model of clothing enterprises based on grey theory. Silk **2**, 55–60 (2020)
2. Song, F.S., Chen, J.: Corporate reputation, contractual governance and distributor's extra-role altruistic behavior in marketing channels. J. Bus. Econ. **6**, 56–65 (2020)
3. Lian, J., Fang, S.Y., Zhou, Y.U.: Model predictive control of the fuel cell cathode system based on state quantity estimation. Comput. Simul. **7**, 119–122 (2020)
4. Guan, J., Yin, J.Y.: Does the negative expectation-performance gap promote the R&D investment and marketing investment of enterprises. Sci. Technol. Prog. Policy **11**, 79–88 (2020)
5. Zhang, Y.: industry and education integration and collaborative education in marketing specialty of higher vocational education. Liaoning High. Vocat. Tech. Inst. J. **10**, 29–32 (2020)
6. Wang, Y.l.: Development & analysis of export marketing strategies of Baijiu enterprises under the background of big data. Liquor-Making Sci. Technol. **1**, 127–130 (2020)
7. Hu, B., Shen, L., Gong, J.: The market value effect of corporate social responsibility of listed tourism enterprises in China: a perspective of marketing strategy. Tourism Tribune **10**, 15–27 (2020)
8. Prabowo, H., Sari, R.K., Bangapadang, S.: The impact of social network marketing on university students' purchase intention and how it is affected of consumer engagement. Int. J. New Media Technol. **2**, 87–91 (2020)
9. Xiong, C., Xue, Y., Zhou, R.: Research on the influence of social network on risk control of New Ventures. J. Phys. Conf. Ser. **3**, 032141 (2021)
10. Liang, G., Yao, X., Gu, Y., et al.: Multi-Batches Revenue Maximization for competitive products over online social network. J. Netw. Comput. Appl. **201**, 103357 (2022)
11. Zhu, Y., Jiang, Y.: Optimization of face recognition algorithm based on deep learning multi feature fusion driven by big data - ScienceDirect. Image Vis. Comput. **2**, 104 (2020)
12. Chen, Z.: Using big data fuzzy K-means clustering and information fusion algorithm in english teaching ability evaluation. Complexity **5**, 1–9 (2021)

Intelligent Push Method of Human Resources Big Data Based on Wireless Social Network

Xiaoyi Wen[1] and Jin Li[2,3]([⊠])

[1] School of Labor Relations and Human Resource, China University of Labor Relations, Beijing 100048, China
Wenxiaoyi008@sina.com

[2] School of Journalism and Communication, University of Chinese Academy of Social Sciences, Beijing 102401, China
idsf54854@yeah.net

[3] Miyun Branch, Beijing Municipal Tax Service, State Taxation Administration, Beijing 101599, China

Abstract. Human resources big data has a wide distribution range, a large amount of data and a variety of data types. Aiming at the problem of low integration of human resources raw data, an intelligent push method of human resources big data based on wireless social network is proposed. Combined with wireless social network, the human resources data is integrated and mined, and the human resources data is preprocessed to build an OAP data warehouse; then a human resources recommendation algorithm combined with the wireless social network latent semantic model is proposed. Behavior, mining the potential job characteristics of job seekers, and then realize the intelligent push and matching of human resources big data. The test results show that the intelligent push method of human resources big data based on wireless social network proposed in this study has a significantly better recall rate than the traditional single latent semantic model and deep forest algorithm, and effectively improves the integration degree and push efficiency of human resources raw data.

Keywords: Wireless social network · Human resources · Data intelligent push

1 Introduction

With the continuous deepening of the application of electronic human resource management, online career recruitment platforms, social network applications, online labor market and other technologies in organizational human resource management, as well as the maturity of big data technology, organizational collection, storage and use are closely related to human resources. Mass data related to resource management will become more convenient and systematic. The quantification of these data in structured, semi-structured and unstructured forms will also be possible. In this context, human resource big data that can be used for macro and micro human resource management research came into being. Compared with the previous sample survey data, human resources big data has the

W. Fu and L. Yun (Eds.): ADHIP 2022, LNICST 469, pp. 127–141, 2023.
https://doi.org/10.1007/978-3-031-28867-8_10

characteristics of large sample size, real-time, dynamic, and valuable, which can more truly and deeply reflect the activities and status of individuals and organizations, and maximize the control of quantitative research. The measurement bias and statistical bias, as well as the exploration of more detailed research questions, make human resource management research usher in new opportunities.

Job hunting and recruiting have always been hot topics that people care about. With the advancement of modern information technology, job hunting and job application through intelligent data push has become a mainstream [1]. At the same time, considering the cost, more and more employers publish recruitment information through the Internet. Due to the gradual increase in the number of positions and job seekers, the corresponding job information and personal information of job seekers have increased, so there is a problem of information overload in the field of Internet recruitment. A large number of positions and user information make the management of human resources more cumbersome, and job seekers are also very easy to fall into the trap of "information trek". However, at present, most recruitment websites mostly push popular jobs indiscriminately and users complete job matching by searching keywords, and do not carry out personalized job matching according to changes in user interests. In response to this problem, many companies comprehensively use the Simrank algorithm and hierarchy. The intelligent recommendation model is built by the analysis method, and the matching of personnel and jobs is realized [2]. Some enterprises have used Naive Bayes classification, case reasoning and context retrieval to complete job matching and recommendation, and achieved certain results. On the basis of the above research, in order to improve the accuracy of job recommendation in human resources recruitment, an intelligent push method of human resources big data based on wireless social network is proposed.

2   Wireless Social Network Human Resources Big Data Intelligent Push

2.1   Human Resource Management Data Information Management System

Human resource management refers to the rational allocation, organization and deployment of human resources according to the requirements of enterprise development strategy, using modern scientific methods and certain material resources, through the recruitment, training, use and assessment of employees in the enterprise. Through a series of processes such as planning, organization, command and control and coordination of human resources, a series of activities such as planning, organization, command and control and coordination of human resources can mobilize the enthusiasm of employees and give full play to the potential of employees, so as to achieve the goals of the enterprise, create value for the enterprise, bring benefit [3]. The specific content system of human resource management is shown in Fig. 1.

As shown in Fig. 1, human resource management activities mainly include human resource planning, recruitment and deployment, training and development, performance appraisal, salary and benefits and labor relations, etc. The focus of human resource management is on selecting, educating, employing and retaining people, Make the best use of talents, so that employees and enterprises can grow together. Human resource management mainly includes six modules, including human resource planning, recruitment

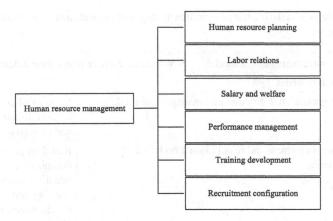

Fig. 1. Human resource management content system

and allocation, training and development, performance management, salary and benefits, and labor relationship management.

In order to ensure that enterprises can use faster and more effective management to respond to changes in the market environment. Human resource management is a very important part of enterprise management. Therefore, new practices of human resource management such as future organization, comprehensive talent supply chain, employee career development, and employer brand building are crucial for enterprises to respond to changes. To this end, a comparative analysis is made between the old human resource management model and the new human resource management demands, as shown in Table 1.

Organizational management is the foundation of enterprise human resource management, and also the foundation of human resource management software design. First, the model will design basic organizational element information such as companies, business units, and departments according to enterprise management demands [4]. These organizational element information can retrospectively modify the change history. Build associations on elements, and then form an organizational chart. Based on organizational element dimensions and organizational management, set the organization's preparation management and generate related management reports based on the organization. The design frame diagram of the organization management module and sub-module is shown in Fig. 2.

The human resources data integration recommendation model includes five parts: wireless social network application server, cache model, recommendation algorithm server, data warehouse and business database. The basic processing flow is shown in Fig. 3.

Based on the data warehouse data, the attribute features of users and positions are extracted to form a user-post feature vector; the speech is analyzed by the latent semantic model to obtain implicit features, and then spliced to form fusion features; the above features are used as input, input In the deep forest algorithm, the user post matching results are obtained [5]. Finally, under the condition of comprehensive consideration of job matching degree and user interest characteristics, and based on the results of

Table 1. Comparison of the old human resource management model and the new human resource management appeal

Old human resource management model	New human resource management demands
HR management business	
Hierarchical organizational structure and management mode	Networked and modular organizational structure and management mode
Focus on work arrangement and result evaluation based on management process	Based on project-based organization, pay attention to employees' learning and development, innovative achievements and influence
Pay attention to the performance results, and the evaluation process is evaluated according to a fixed cycle	Focus on the process of goal setting, decomposition, tracking and improvement
The assessment results will be directly applied to cadre selection and evaluation, employee family planning or salary results	Transparent assessment process. Employee performance can be assessed in multiple dimensions such as performance achievement
Only pay attention to the working experience in the enterprise	Pay attention to the current employment experience of employees and label their abilities
Focus only on the leader's evaluation and performance to achieve results	Pay attention to employee evaluation. In addition to the official assessment process, employees are encouraged to socialize and evaluate each other
Sanctify "leadership"	Everyone can be a leader and try the "maker" organization

deep forest prediction, the classification results are linearly fused to establish a job recommendation part.

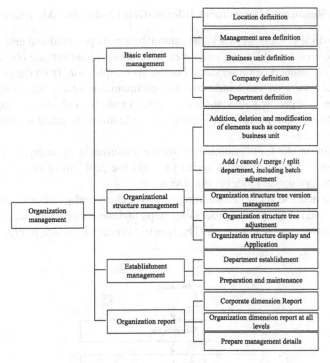

Fig. 2. Design framework of the organization management module and sub-modules

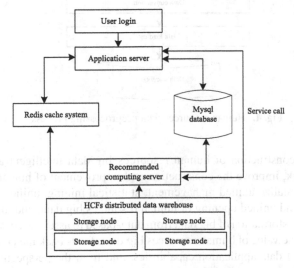

Fig. 3. Data collection model process based on wireless social network

2.2   Human Resource Management Information Evaluation Algorithm

Through the wireless social network, the establishment of personal and enterprise information characteristic profiles can more clearly understand the service objects and provide accurate recommendation services. For the above question, from the perspective of wireless social networks, which industries and companies have this person worked for in history, where have people with similar experiences gone to, and which companies have the need to absorb such people [6]. Through these features, targeted recommendations can be made.

A prediction model for the number of trainee positions in an enterprise is proposed. In this process, a penalty factor is added to avoid the problem of overfitting, and the accuracy of the prediction model is improved.

In order to achieve accurate matching of job seekers and positions, raw data need to be preprocessed. The specific processing steps include data extraction, data cleaning and transformation, and data loading. The specific process is shown in Fig. 4.

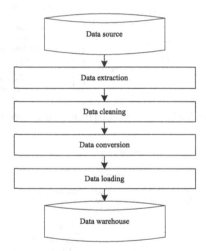

Fig. 4. Human resources data preprocessing process

Through the construction of human resources big data intelligent analysis wireless social network, improve the construction and improvement of human resources in data capabilities, realize unified management, statistical mining, unified analysis, unified application, and unified opening of human resources big data, and improve human resources. The transformation of big data from data-based to asset-based, while continuously improving the value of human resources big data, feeds back the business through the construction of data application capabilities, and from the perspective of wireless social networks, lays a good data foundation for the improvement of human resources business capabilities. HR big data value. The data used comes from a human resources platform [7]. It contains structured data in each sub-model of the platform and the model log of the platform, so as to obtain job information, job seeker information and user behavior data information. There are three matrices R, P, and Q in the latent semantic model. Matrix R is the score of job seeker's interest in recruitment positions, matrix P_{uk} is the interest of job seekers in the hidden class, and matrix Q_{ki} is the job seeker's interest in the hidden class. Weights. It can be seen that the hidden class has the function of linking job seekers and recruiting positions in the implicit semantic model, and matrix R is decomposed into the product of matrix F and matrix K. . According to the Latent Semantic Model (LFM), the number of hidden classes is 6, and the calculation formula of the prediction score of job-seeker 7 for the recruitment position i is as follows

$$r_{ui} = u/R \sum_{k=1}^{F} P_{uk} Q_{ki} - F_{KI} \tag{1}$$

Human resource data mining is not a one-shot solution, it needs repeated iterative experiments, and the optimization model is adjusted according to data changes, which are inseparable from the effective evaluation method of the model. Support offline evaluation and online evaluation of two model evaluation methods, simplify model evaluation work, support the visual display of evaluation results, make the evaluation results easier to understand, and make the trained model more suitable for business problems. According to the optimization loss function, the matrix p_u and Matrix p_i is solved, and a penalty factor r_{ui} is added to avoid the overfitting problem in this process, satisfying:

$$\lambda = r_{ui} \sum_u |p_u|^2 + r_{ui} \sum_i |p_i|^2 \tag{2}$$

The apprenticeship employment rate x can better reflect the contribution of an enterprise to the local employment situation, and is an important indicator to measure whether an enterprise can become a trainee base. Apprenticeship employment rate—the number of large human resources in the company's apprenticeship who stay in the local employment/the number of apprentices in the enterprise Through preliminary analysis, the estimation method of the apprenticeship employment rate of the enterprise is obtained, the relevant characteristics of the enterprise are extracted, and the characteristic factors that affect the apprenticeship employment rate are identified. And the linear regression method is used to predict the trainee employment rate. See the trainee employment rate and prediction model.

$$\begin{cases} A = \lambda u_N / v \\ B = \eta \lambda / a_1 x_1 + a_2 x_2 + a_3 x_3 \end{cases} \tag{3}$$

Among them, a_1, a_2, a_3 represents the weight value of the apprentice employment rate in the past three years, x_1, x_2, x_3 represents the apprentice employment rate of the company in the past three years, u_N represents the number of trainees employed locally in the apprenticeship human resources, v represents the number of trainees, and η represents the apprentice employment rate in the next year. Predicted value. According to the trainee data and employment data of some enterprises, the least squares method is used to fit a linear regression model. Then the loss function calculation formula is:

$$c = A \sum_{u,i} \left(r_{ui} - \sum_{k=1}^{F} P_{ub} q_{ki} \right)^2 + B \sum_u \eta |p_u p_i|^2 \tag{4}$$

According to the loss function, the stochastic gradient descent method is used to minimize the root mean square error between the job seeker's actual rating and the predicted rating, and realize the optimization of the objective function. To this end, we put forward a prediction model for the number of probationary positions in the enterprise. The model comprehensively considers the number of newly increased human resources, the number of newly increased human resources, and the number of existing employees in the enterprise in the past few years. The linear regression model is used to predict The optimal number of apprenticeship positions in an enterprise is the optimal number of apprenticeship positions. This model can be used for recommendation by companies that have not yet become apprenticeship bases, and can also be used to evaluate the scope of mid-term apprenticeship jobs for companies that have become apprenticeship bases.

2.3 Realization of Intelligent Push of Human Resources Data

Informatization can closely gather and analyze the scattered information about human resource management, automate and optimize the business process of human resource management, so that the information flow is accelerated and more comfortable. The functions are more tightly integrated. The human resource management information model can not only completely cover and clearly divide the work functions of the human resource management department, but also reflect the optimized business process in the model to standardize and improve the business process of human resource information management, which can make human resource managers Get away from it, and thus have more time to think about strategic human resource management. From the above general recommendation model analysis, it can be concluded that the data acquisition and processing of the input module, the user preference matching of the recommendation module and the model selection of the output module will affect the final recommendation result of a model, but the biggest factor that determines the accuracy of the model is the recommendation. Algorithm.

In order to complete the design requirements of the human resources data integration model, it is necessary to complete the design of the data integration and recommendation model based on the construction of the data warehouse. The construction frame diagram is shown in Fig. 5.

The recommendation model is divided into presentation layer, application layer, recommendation algorithm layer and storage layer from bottom to top. And the recommendation calculation method layer is divided into two types: online calculation method and offline calculation method. The presentation layer embodies the recommendation model and the user's interactive interface in the form of web pages, and provides users with functions such as registration, login and job personality recommendation. The application layer is mainly composed of user management section, job browsing section, query section and recommendation section. This layer is mainly responsible for receiving and processing user requests. The recommendation algorithm layer is the core part of the human resources recommendation model, which consists of two parts: online calculation method and offline calculation method. Among them, the online calculation method mainly uses the scoring or classification results to obtain the candidate's interest in the position and the matching degree of the position. Offline calculation method: mainly based on human data, construct user behavior, and establish recommendation

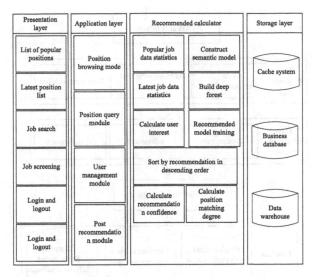

Fig. 5. Human resources data recommendation model structure

algorithm. The human resource intelligent recommendation model relies on the data mining algorithm based on clustering, so the effective data acquisition and processing is of great significance to the accuracy of the experimental results. Mainly describe the database establishment and data anonymization scheme and implementation process. Specifically, the source data acquisition and preprocessing process of the data are introduced, and then a data standardization and vectorization method based on the value weighting method is proposed. Secondly, a clustering effect is used to evaluate the DB index, and through the good convergence of the data in the graph The feasibility of this method is proved. At the same time, a joint anonymity hierarchical protection model suitable for data release is introduced, and an appropriate level of anonymity model can be selected according to the actual situation. Finally, the related models are compared and analyzed through experimental verification. Based on this, the process of database establishment and anonymization scheme is constructed, as shown in Fig. 6.

The real-time message push management module consists of two sub-modules, the task storage module and the task sending module. Among them, the task storage sub-module uses the API organization as the core to build the equipment. It can select the push instructions that the model needs to execute according to the emergency running rules in the background of the wireless social network environment, and temporarily store the information data in the IO chip in the form of a packet structure.. This sub-module performs connection judgment on the data structure to be stored in the form of WEB push to ensure that the API organization can obtain a push message queue with strong scalability. The task sending sub-module is the lower-level execution unit of the task storage sub-module. It cannot modify the existing information instructions in the API organization. It can only mark and forward the instructions to be pushed according to their storage sequence until the task storage sub-module. Existing information in

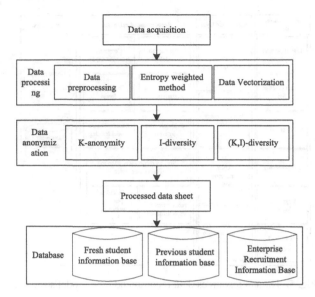

Fig. 6. Process flow of database establishment and anonymization scheme

is completely consumed. The detailed real-time message push management module structure is shown in Fig. 7.

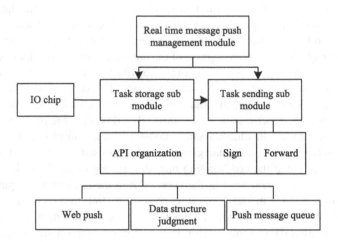

Fig. 7. Structure diagram of real-time message push management module

The business process of human resources information management should take into account the recruitment process, performance management process, employee training and development process, employee career plan, and resignation process. The corresponding templates can be designed according to the standard work process to complete the HC Group can sort out the business processes involved in the human resources management system one by one, and analyze the feasibility of establishing business flow informatization, and gradually establish and improve human resources informatization The business process, the designed human resources intelligent recommendation model for the field of college recruitment and employment can not only be used by colleges and universities to accurately recommend graduates to improve the employment rate during the recruitment and employment of fresh graduates. Emergency push human resource information distribution is based on the wireless social network bit sequence. Coded modulation operation performed on the information to be transmitted. Human resource information is a set of specific coded strings including head pointer, intermediate encryption algorithm and tail pointer, which can be freely communicated between the model center computer and various hardware execution devices. It realizes the organic integration of human resource information and communication nodes, and can accelerate the physical response speed of wireless social networks on the premise of ensuring accurate model push transmission. The intermediate encryption algorithm and the tail pointer are the main structure of the human resource information distribution operation of emergency push. On the one hand, it should comply with the communication operation requirements of hardware devices at all levels. Reasonably connect components to ensure that model push operations get good execution results. So far, the construction of the model software operating environment has been completed to realize the smooth operation of the emergency wireless social network large-capacity information real-time push model. The specific human resource information allocation principle is shown in Fig. 8.

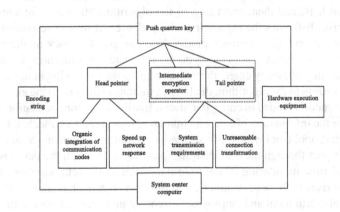

Fig. 8. Schematic diagram of emergency push data distribution

It can also be used by recruiting companies to effectively screen advantageous resources. Due to the relationship between the original data and the display effect, this chapter takes recruitment and employment as the starting point to conduct experiments, and this model is referred to as the intelligent recommendation model. The specific process is shown in Fig. 9.

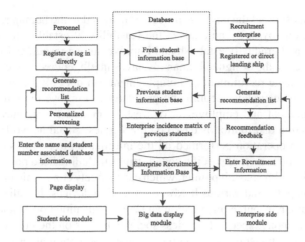

Fig. 9. Flow chart of intelligent recommendation model operation

The main user interface has three main ports, namely: human resources login terminal, enterprise login terminal and big data display terminal. The specific execution process of each port is as follows: Human resources login terminal: First, fresh graduates enter the account password (password) to log in to their own account. If there is no account, you can register an account (monitored by the background administrator, and illegal users cannot register); secondly, human resources Entering your own student number can automatically retrieve its school information; again, select the number of recommended lists, and then, according to the algorithm, the model recommends the top N enterprise lists with the highest matching degree of human resources to human resources users from high to bottom; finally, You can give feedback on the recommendation results or bookmark a company you are interested in; Enterprise login terminal: First, the recruiting company enters the account and password to log in to the enterprise account. According to the algorithm, the list of graduates who match the enterprise from high to low is recommended to enterprise users; finally, the recommendation results can be fed back or the information of a certain human resource can be collected; big data display terminal: school employment office or government management Organization, you can enter this port through the administrator password. Through the data visualization interface, real-time monitoring of multiple data such as the total number of contracts signed, enterprises with more contracts, and the flow of human resources employment to the national recruitment and employment trends can be carried out intuitively.

3   Analysis of Experimental Results

The software development environment of the entire experimental program is Windows, the test environment is macOS, the development language used is Python 3.7, and the packages called by the program include Sklearn, Django, Pandas, Pymysql, Spyder, etc. The initial data collection format is Microsoft Office Excel, and the later import database format is MYSQL. The specific development and test environment is shown in Table 2.

Table 2. Development and testing software environment table

	Operating system	Windows XP
Development environment	Development platform	Python 3.9.5
	Database	Oracle Enterprise 11G
	Data visualization	Navicat 12.2.22
	Necessary Kit	Django 2.1.7
Testing environment	Operating system	Windows 10
	Test platform	Python 5.8.9
	Database	SQL Server2008
	Interface browser	Google Chrome 80.0

Taking the Simrank algorithm and the naive Bayesian classification algorithm as the experimental comparison methods, the performance of the three algorithms for intelligent push of human resources big data is compared. The results are shown in Fig. 10.

From the above experimental results, it is not difficult to see that the recall rate and F1 value of the single Simrank algorithm and the naive Bayesian classification algorithm are significantly lower than those of the recommended algorithm in this paper. This improves the recommendation performance of the algorithm. The single model has the problem of overemphasizing user behavior data and pushing job positions only from the matching value of users and positions, ignoring the one-sidedness of users and positions' own content and user interests. A hybrid recommendation algorithm based on deep forest and latent semantic model, the algorithm fully considers the matching value of the job and the user's interest bias, that is, the user's interest bias and the job matching probability are linearly integrated to achieve the final recommendation.

On this basis, the time-consuming of the push method of human resources big data intelligent push method based on wireless social network is analyzed, and the comparison results are obtained as shown in Fig. 11.

Analysis of the results in Fig. 11 shows that the push time of the algorithm in this paper is less than 2 s, while the classification performance of the Simrank algorithm and the Naive Bayes classification algorithm are 6.2 s and 4.8 s respectively. It can be seen that the push method of human resources big data intelligent push method based on wireless social network proposed in this paper takes less time to push and has high push efficiency.

4   Conclusion

The research shows that the recall rate of human resources big data intelligent push method based on wireless social network is significantly higher than that of a single model or algorithm, and fully reflects the advantages of deep forest and latent semantic model. The experimental results also show that the recall rate of the recommendation

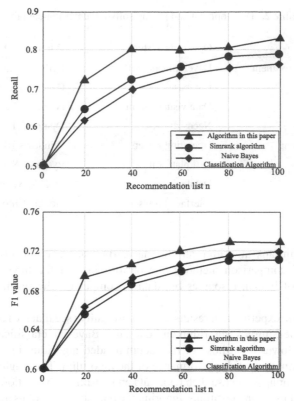

Fig. 10. Performance comparison test results of hybrid recommendation

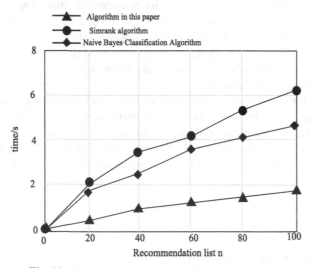

Fig. 11. Time-consuming comparison results of push

algorithm in the human resource big data intelligent push method based on wireless social network is significantly better than the recommendation performance of traditional collaborative filtering and content-based recommendation algorithms.

References

1. Zhang, M., Zhao, L., Zhao, S.: The influence mechanism of HR attributions on voice behavior: mediating role of psychological contract. Res. Econ. Manage. **41**(04), 120–131 (2020)
2. Zhou, X., Chen, S., Li, L.: Research on the relationship among big data capacity, technological innovation and competitiveness of human resource service enterprises. Manage. Rev. **33**(07), 81–91 (2021)
3. Dong, L.: Co-governance of integrity in human resource service industry: the logic of framework and path of realization. Chin. Public Adm. **04**, 46–51 (2021)
4. Chen, X., Xiao, M.: Control and commitment HR systems and firm performance—voluntary turnover rate mediation model based on moderated ownership. Res. Econ. Manage. **41**(06), 131–144 (2020)
5. Liu, X., Nan, Y., Xie, R., et al.: DDPG optimization based on dynamic inverse of aircraft attitude control. Comput. Simul. **37**(07), 37–43 (2020)
6. Zhang, B., Xu, S.: Research on the influence of high-involvement human resource practices on employee innovation: the role of thriving at work and humble leadership. Sci. Technol. Prog. Policy **38**(07), 141–150 (2021)
7. Wei, D., Zhao, Y., Zhao, S.: The development trends in human resource management in Chinese enterprises—from a perspective of human capital. Chin. J. Manage. **18**(02), 171–179 (2021)

Situational Simulation Teaching System of Information Literacy Education Based on Mobile Terminal

Xin Wang[1(✉)] and Fan Lv[2]

[1] Shandong Institute of Commerce and Technology, Jinan 250103, China
wangxin220106@163.com
[2] Sanmenxia Polytechnic, Sanmexia 472000, China

Abstract. The traditional situation simulation teaching system has the problems of high space limitation and low response rate, which affect the teaching effect of information literacy education. Therefore, this paper proposes a scenario simulation teaching system for information literacy education based on mobile terminals. Analyze the teaching needs of mobile terminal applications, achieve information compression of information literacy education through DSP chips, and design the microprocessor core module to achieve the scene simulation teaching environment of information literacy education; Adjust the teaching content by calculating the feedback index of teaching effect. After setting the data table format of the system storage simulation scene, the software part is transplanted to complete the system design. The experimental results show that the response time of the designed teaching system is less than 2S, and the information literacy ability of students can be improved. The scene simulation teaching effect of information literacy education is good.

Keywords: Mobile terminal · Information literacy education · Scenario simulation · Teaching system · Teaching feedback

1 Introduction

The information society constantly puts forward new and higher requirements for everyone's information literacy ability. Information literacy has become a necessary basic literacy and key variable for people's survival, learning and development. With the development of educational informatization and the rapid development of mobile terminal technology, many schools have applied mobile terminals in Classroom Teaching [1]. Schools in many areas have achieved certain results in using mobile terminals in classroom teaching, which not only improves the efficiency of classroom teaching, but also greatly mobilizes students' learning enthusiasm. Mobile learning is carried out with the help of mobile devices and mobile communication, which is not limited by place and time. It combines mobile computing and digital learning. It is a new digital learning mode. Using mobile terminals for learning has the characteristics of learning materials

W. Fu and L. Yun (Eds.): ADHIP 2022, LNICST 469, pp. 142–155, 2023.
https://doi.org/10.1007/978-3-031-28867-8_11

that can be obtained anytime and anywhere, efficient resource search ability and flexible interaction. In recent years, teaching methods or teaching models have become a research hotspot in the field of information literacy education. Many novel and effective teaching methods have attracted the attention of researchers. The first threshold that information literacy teaching should lead learners through is to establish this belief: the mastery of knowledge is a process of constant conviction. Whether it is relatively stable or violently changing knowledge, its degree of certainty should be expressed as a subjective probability distribution, rather than all right or all wrong, so as to leave room for the creation of knowledge. In view of the complex picture of the interaction between information, the main content of the information literacy teaching actually carried out at present is to train students how to effectively reveal the interrelations created by predecessors under specific problem situations [2]. The research objects of information literacy education abroad mainly include teachers, subject librarians and students. The objects of information literacy education in Colleges and universities are not limited to college students, but also teachers engaged in professional knowledge and quality education. Teachers' information literacy has a very important impact on students' education.

Jia Xiaoting and others designed a multimedia based mobile terminal remote education security monitoring system [3]. The hardware design of the system was realized through Mina network framework equipment, usbssc32 channel steering gear control board, Ethernet dfrduino w 2100 expansion board and remote monitoring camera; The software part of the system is composed of webcam XP software and Dr positioning and tracking software. Use multimedia technology to realize the safe interaction of mobile terminal education monitoring system. This method can improve the accuracy of the monitoring system, but the monitoring response delay is too long.

In view of the above problems, this paper designs a situational simulation teaching system for information literacy education based on mobile terminals, which can effectively improve the teaching effect of the situational simulation teaching system for information literacy education by calculating the feedback index of teaching effect and adjusting the teaching content.

2 Hardware Design of Situational Simulation Teaching System for Information Literacy Education Based on Mobile Terminal

The hardware of situational simulation teaching system for information literacy education based on mobile terminal is mainly composed of network server, single chip microcomputer, teacher end, network camera and student mobile phone. The hardware framework of the teaching system is shown in Fig. 1 below.

It can be seen from the analysis of Fig. 1 that the hardware of the scene simulation teaching system is composed of wireless network communication interface, Ethernet interface, mobile terminal, camera and power module. The camera collects the interactive scene information in the scene simulation teaching, supplies power to the system through the power module, and connects the mobile terminal through the wireless network communication interface and Ethernet interface to ensure the transmission and feedback of information and the stable operation of the packaging system [4].

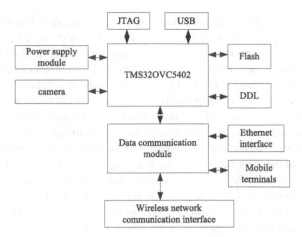

Fig. 1. Hardware framework of scenario simulation teaching system

The hardware design of situational simulation teaching system mainly includes clock power module design, A/D and D/A conversion module of voice signal, McBSP interface design of multi-channel buffer serial port between PCM CODEC and DSP, extended memory design, interface design of network adapter and DSP, etc. The mobile terminal interacts with the server through the mobile Internet, which can be composed of WIFI, 4G, 3G or 2G networks, or directly connected to the network teaching system server group in the school through the switch and wireless routing in the school. The transmission speed, quality and delay of different networks are also different [5]. Therefore, in order to reduce the impact of different networks on student client programs, this paper adopts thread pool technology, which minimizes the overhead of resources through the use of multithreading, so as to reduce the adverse impact of network congestion on the graphical interface.

2.1 Microprocessor Core Module Design

According to the performance analysis of digital multimedia terminal system and considering the cost performance of the system, the microprocessor selects DSP high-speed digital signal processing chip TMS320VC5402. The serial processing between the DSP chip and the DSP chip contains a multi-channel buffer to realize the communication between the DSP and the DSP decoder. As an external UO device of DSP, network adapter realizes the design of network architecture. The processing chip includes 8/16 bit SRAM/NOR flash interface, 16 bit SDRAM interface, hardware NAND flash controller, support NAND flash self starting 10 m/100 M adaptive Ethernet MAC, support RMII interface 64 K byte high-speed on-chip SRAM USB1.1 Device, and the full speed can reach 12 mbps. Its on-chip DPLL supports a variety of power consumption modes, such as idle, slow, normal and sleep.

Most peripherals in the system are connected to the system by I/O port, so it is necessary to allocate the I/O end of each module reasonably. The function of UART (serial port) is to realize asynchronous transmission, that is, data transmission with another asynchronous communication transceiver according to the set baud rate without clock synchronization. UART shall send or receive signals that meet RS-232 protocol bit by bit according to user configuration information. The system has four serial ports, in which serial ports 0 and L are level converted through MAX232, serial ports 2 and 3 are directly led out through TTL, and the level conversion chip max3221 is used. The definition of serial port pin is shown in Table 1 [6].

Table 1. Definition of max3221 chip serial port pin

Serial number	Definition	Working level
RXD	Receive data	1
TXD	Send data	1
GND	Logically	0

In order to meet the real-time requirements of audio and video data transmission used in the teaching process, the transmission layer protocol of the equipment selects UDP protocol, that is, connectionless data transmission, while the application layer protocol adopts Real-time Transmission Protocol RTP protocol. After the audio and video encoded data is generated (TS stream), the data type is analyzed by the data type parser, the serial number generator generates the serial number of the packet data, generates the time stamp according to the system time, and then generates the synchronization source identifier according to the identification ID of the sender, then inserts the information obtained in the above process into the frame header of the RTP message, and fills the data part into the payload of the RTP message, Finally, it is encapsulated by UDP and transmitted to the network. The microprocessor must select the appropriate packet length according to the actual network situation to control the transmission of teaching data resources.

2.2 Communication Interface Design

The mobile terminal connects to the Internet through the wireless network provided by the operator, interacts with the server of the system, and performs operations such as login or file download; At the same time, the administrator can connect to the Internet through the personal computer and manage the files stored on the server through the browser. In consideration of compatibility and universality, this design uses HTTP protocol to realize the communication between mobile client program and server. This is because HTTP protocol is a protocol with perfect performance and function, and has been widely used; Secondly, the mobile terminal device can well support the protocol; Here, the use of HTTP protocol makes the upper program not care what kind of mobile Internet the lower layer uses to communicate, that is, for the upper program, HTTP protocol makes the lower communication technology transparent; Finally, HTTP protocol supports firewall traversal, which makes HTTP protocol applicable to more complex network environments. In HTTP communication, the communication between the client and the server adopts the mode of request response, that is, after the client establishes a connection with the server, if the client needs a service, it will first send the service request to the server, and then the server will process the request according to the content of the request, and send the processing result to the client after processing [7].

The network interface adopts the chip DM9161E of DAVICOM Company, which is a high-performance network physical layer transceiver. It adopts RMII interface and realizes 10 M/100 Mbps adaptive network port with the MAC controller inside the processor. The main clock required by the system is determined by Provided by an external 50 MHz crystal oscillator. As a device (USB DEVICE) control module defined in the USB protocol, the USBD module is responsible for communicating with the host defined in the USB protocol, completing the processing process specified in the protocol, and completing the data transmission between the chip and the USB host controller through the USB protocol.

In the teaching system, the pins related to the processor: VDD_USB needs to be connected to the 3.3 V logic power supply, VSS_USB needs to be connected to the logic ground, and the other data differential transmission signals D+ and D− are directly connected to the connector of the USB DEVICE. At the same time, in order to realize the stable transmission of data, it is necessary to add an appropriate matching resistance near the USB DEVICE interface. In addition, in order to realize the full-speed operation of USB DEVICE, the D+ pin needs to add a 1.5 K pull-up resistor to 3.3 V.

In this design, a complete information transfer process is divided into four stages: connection establishment, service request sending, response information return and connection closing. For the server, there is an HTTP resident program running on it, which is responsible for responding to client service requests and feeding back the processing results to the client. On the client side, it can send service requests to the server at any time, and these requests can be authentication requests, query requests, function requests, and so on.

On the basis of the hardware framework designed above, the software part of the teaching system is designed to realize the function of situational simulation teaching.

3 Software Part Design of Information Literacy Education Scenario Simulation Teaching System Based on Mobile Terminal

3.1 The Setting of Teaching Content of Information Literacy Education Simulation Scenarios

Different from traditional classroom teaching and online online education, mobile terminals are lighter and more portable, and mobile services are more personalized, so that mobile learning further extends the time dimension of traditional teaching. Students can use fragmented time to conduct MOOC and micro-lectures. Independent study, even scattered time can "turn waste into treasure". In the environment of mobile learning, the content and form of learning have also changed. The information literacy education that students need to learn is no longer just the basic knowledge and skills of information literacy such as basic information awareness, information processing, and information utilization, but adapting to modern information. Information identification ability, information retrieval ability, information utilization ability and information sharing ability required by technology [8].

As the object of simulation, information literacy application scenarios are relatively limited, and a scenario and problem background can be described in different scenarios. When conducting information literacy education simulation scenarios, the following points need to be observed:

First, the main body is the student, and the teacher is the organizer and guide. Second, the specific teaching content, the actual situation of the students, and the social reality are the basis for creating situational simulation activities. Third, simulation activities simulate the environment, process and factors of the occurrence and development of things in life or work scenarios. Fourth, students experience, solve problems, and complete tasks to recognize, experience and perceive in the scenario simulation activities; and in the discussion and analysis after the activity, they jointly explore their emotions, attitudes, value orientations and attitudes towards people, things and things. Problem-solving strategies, which enable students to understand knowledge, touch emotions, expand their thinking, and explore appropriate methods and techniques for dealing with problems.

Figure 2 shows the process of students using mobile terminals to conduct information literacy education scenario simulation teaching.

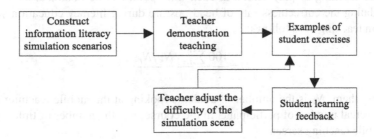

Fig. 2. The situational simulation teaching process of information literacy education

Mobile search is the first and most important part of acquiring mobile information. Training and teaching to improve the mobile search ability of college students should become the primary task of college students' mobile information education. In the teaching process, mobile search should be included in the focus of mobile information literacy education, so as to cultivate students' mobile search knowledge and improve mobile search ability. With the rapid development of information technology in the mobile information environment, the application of voice search makes the search no longer a simple text input of keywords or topics, but a simpler, faster and more direct voice input. By setting information retrieval requirements in different information search scenarios. Students complete the learning of the corresponding teaching content in a given scenario.

3.2 Calculation of Feedback Indicators for Teaching Effect in Simulated Scenarios

By strengthening the evaluation, sharing and feedback mechanism of students, it can improve the enthusiasm and participation of students to join mobile learning, so as to improve the teaching quality of college students' mobile information literacy education. In the process of using the teaching system for teaching, in addition to the statistics of the students' attendance rate, task completion rate and completion quality through the system background, the camera of the mobile terminal device can also be used to count the status of the students experiencing the situational simulation teaching, so as to facilitate timely adjustment of teaching plans and teaching content [9]. The feedback indicators of simulated situational teaching effect studied in this paper are mainly composed of the following contents:

3.2.1 Head Posture Acquisition

Read the head gesture recognition statistical result file, you can get the number of students looking at the teaching surface of the mobile terminal at each time point. Comparing the number of students looking at the mobile teaching display screen at each time point with the total number of attendance, the ratio of the number of students looking at the mobile teaching display screen at each time point can be obtained. The formula for calculating the correctness rate of head posture during literacy education scenario simulation teaching is as follows:

$$D = \frac{100 \sum_{i=1}^{n} N_k / N_r}{n} \tag{1}$$

Among them, N_k is the number of students looking at the mobile teaching screen; N_r is the actual number of participants in the course; n is the number of times looking at the mobile teaching screen.

3.2.2 Expression Recognition

Read the expression recognition statistical result file, you can get the number of students who are recognized as listening expressions at each time point. Comparing the number of students who were identified as attentively listening to the total number of attendance, the ratio of the number of students who were attentively listening at each time point was obtained. Assuming that the number of serious listening times is measured, take the average value of all the listening time points in the class, and calculate it on a 100-point scale. The formula for calculating the expression seriousness rate is as follows:

$$R = \frac{100 \sum_{i=1}^{n} N_t / N_r}{n} \tag{2}$$

Among them, N_t is the number of people who listened carefully.

3.2.3 Head Up Rate Identification

Read the header recognition statistical result file, and compare the number of headers at each time point with the total number of attendees to obtain the header rate at each time point. Assuming that the class is raised n times and the number of heads is N_h, the average of all head-up time points in the class is taken and calculated on a 100-point scale. The head-up rate indicator is:

$$H = \frac{100 \sum_{i=1}^{n} N_h/N_r}{n} \tag{3}$$

3.2.4 Calculation of Hand Raising Rate

Read the number of people who have raised their hands each time in the human body gesture recognition statistical result file, and compare the number of people who raised their hands each time with the total number of attendance to get the j hand-raising rate U_j. Assuming that m times of hand-raising in class are measured, the average of all hand-raising time points in the class is taken and calculated on a 100-point scale. The score of the hand-raising rate indicator is:

$$U = \frac{100 \sum_{j=1}^{m} U_j}{m} \tag{4}$$

According to the above indicators, in the current teaching process, the learning status of students using mobile terminals to learn the content of information literacy education can be obtained. According to the comprehensive average of the scores of each index, evaluate whether the current teaching content and simulated teaching scenarios can mobilize students' enthusiasm for learning, so as to obtain teaching feedback information.

3.3 Database Design

Each functional module of the teaching system involves a large number of data processing, and data processing is an essential part of any management information system. The design of the system data architecture is actually the design of the data table according to the business requirements. According to the analysis of the main business process, the main data tables involved in the system include: teaching resource information data table, simulated scene information data table, interactive information data table, and teaching feedback information data table. The following Table 2 is the main content of each data table of the system database [10].

The teaching scenarios and teaching contents used in the simulated scenario teaching of information literacy education are stored in the database according to the data table. During teaching, the corresponding simulated scene information can be extracted according to the design of the teaching content for teaching assistance.

Table 2. Database part data table

Field name	Types of	Whether the primary key (foreign key)	Illustrate
ID	int	Primary key	Simulation teaching resource ID number
Name	nchar		name
MakerID	int	Foreign key	Resource producer ID
ResTypeID	int	Foreign key	Resource Type ID
SubjectID	int	Foreign key	Scene type
Abstract	text		Introduction to Simulation Scenarios
QuestionTag	nchar		Question label
StudentID	int		Student ID number
IfAnswer	bool		Have you answered
If Response	bool		Whether to give feedback to students
HardLevel	int		Content difficulty

After the software part designed above is loaded on the system hardware framework, the design of the mobile terminal-based information literacy education scenario simulation teaching system is completed.

4   Test Experiment

The realization of the system is based on the design scheme of the system. Before the actual application of the system, it is necessary to conduct all aspects of performance testing and research on the system.

4.1   Experimental Content

Functional testing is to verify whether the user requirements are realized and whether the system functions conform to the functions of the design content. Therefore, the response rate and packet loss rate of the system in this paper are tested. At the same time, in order to test the actual application of the system, students of different grades in a university were selected to use the system to study, and the students' information literacy ability was improved before and after using the system. Under the guidance of experts in relevant fields, this evaluation adopts a scoring system to obtain corresponding data.

4.2 Experimental Results

4.2.1 Response Rate Comparison

Table 3 below shows the comparison of the response rates of the augmented reality system, the multimedia system and the method in this paper under different response requests.

Table 3. Comparison of system response rates

System requests/103 times	System response time/s		
	Design system	Augmented reality system	Multimedia system
100	1.23	6.3	33.9
150	1.41	8.9	36.2
200	1.49	12.3	39.0
250	1.53	16.8	42.8
300	1.55	22.9	48.3
350	1.58	32.0	52.3
400	1.64	38.0	58.1
500	1.70	59.7	69.2

By analyzing the data in Table 3, it can be seen that when the number of system requests is 300×103 times, the system response time of the design system is 1.55 s, the system response time of the augmented reality system is 22.9 s, and the system response time of the multimedia system is 48.3 s; The overall analysis shows that the response time of the teaching system designed in this paper is less than 2.0 s under different service requests, which meets the technical index requirements of the current teaching activities.

4.2.2 Comparison of Packet Loss Rate

Table 4 below shows the comparison of data transmission packet loss rate of augmented reality system, multimedia system and the method in this paper under different response requests.

By analyzing the data in Table 4, it can be seen that when the number of system requests is 100×103 times, the data transmission packet loss rate of the designed system is 0.20%, the data transmission packet loss rate of the augmented reality system is 8.2%, and the data transmission packet loss rate of the multimedia system is 6.9%; The overall analysis shows that the packet loss rate of data transmission of the teaching system designed in this paper is less than 0.55% under different service requests, which meets the technical index requirements of the current teaching activities.

Table 4. Comparison of packet loss rate of system data transmission

System requests/103 times	System data transmission packet loss rate/%		
	Design system	Augmented reality system	Multimedia system
100	0.20	8.2	6.9
150	0.25	12.5	8.2
200	0.32	16.2	18.3
250	0.39	18.9	22.7
300	0.43	20.0	18.0
350	0.47	26.5	28.2
400	0.48	29.3	26.0
500	0.55	32.1	32.3

4.2.3 Comparison of Students' Information Literacy Ability

Figure 3 below shows the comparison of the improvement of students' information literacy ability after using the system to teach information literacy-related content.

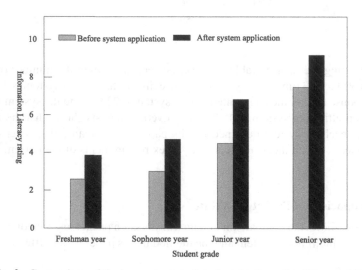

Fig. 3. Comparison of the improvement of students' information literacy ability

Analysis of the information in Fig. 3 shows that after using the system in this paper to study, the students' information literacy ability has been improved to varying degrees, with a minimum improvement of 17.5%, which has a good practical application effect.

4.2.4   Evaluation Results of Students' Information Literacy

The teaching system designed in this paper is compared with the actual application effect of the traditional information literacy education teaching system. 200 students are randomly selected from each grade in colleges and universities, and 800 students are divided into experimental group and control group according to grade. The students in the experimental group used the teaching system designed in this paper when receiving information literacy education, while the students in the control group used the traditional information literacy education teaching system. In addition to the differences in the teaching system used by all students participating in this system application effect comparison test, the factors that interfere with the assessment of students' information literacy level, such as the composition of the teaching staff, teaching content, and plans, remain the same. Before and after the students in the experimental group and the control group applied the teaching system, the distribution of the number of students in different levels of the information literacy assessment in each group was used as an analysis index to compare the actual effects of the two teaching systems applied to teaching. The information literacy of the students in the experimental group and the control group was evaluated by professional experts by scoring and formulating test papers. Among them, the full score of the scoring system is 100 points, and the full score of the test paper is 100 points. The students' information literacy assessment level scores are calculated by weighting. 70 points are the qualified line for information literacy, 85 points are the good line for information literacy, and students higher than 85 points are rated as excellent in information literacy. Tables 5 and 6 below show the comparison results of the actual effects of the systems when different teaching systems are used for information literacy education.

Table 5. Information literacy assessment results of students in the control group

Grade	Forward			Back		
	Excellent	Good	Qualified	Excellent	Good	Qualified
1	8	21	71	10	27	63
2	10	16	74	15	20	65
3	7	18	75	16	24	60
4	10	27	63	17	31	52

Analyzing the data in Table 5 and Table 6 respectively, it can be seen that the use of the teaching system in information literacy education can improve the information literacy level of students of different grades to a certain extent, that is, in the process of information literacy education, the use of the teaching system to assist It can improve the effectiveness of information literacy education. Comparing and analyzing the data in Tables 5 and 6, it can be seen that in the process of information literacy education, using the teaching system designed in this paper, the number of students whose information literacy assessment level has reached excellent and good is significantly more than the results of students' information literacy assessment after using the traditional teaching

Table 6. The results of the information literacy assessment of the students in the experimental group

Grade	Forward			Back		
	Excellent	Good	Qualified	Excellent	Good	Qualified
1	9	19	72	20	54	26
2	11	18	71	24	56	20
3	10	22	68	22	48	30
4	12	28	60	17	31	52

system.. It shows that in the teaching process, using the teaching system designed in this paper can obtain better information literacy education effect, that is, the practical application effect of the system in this paper.

In summary, the mobile terminal-based information literacy education scenario simulation teaching system designed in this paper is suitable for schools to carry out teaching activities. Can meet the teaching needs of the school.

5 Concluding Remarks

With the development of big data, blockchain, artificial intelligence and other fields and their application in real life, the society's demand for high-level talents with information literacy has surged. Information literacy education is the best way to achieve lifelong education and create a learning society. The importance cannot be ignored. With the development of computer-assisted teaching, information literacy has gradually become the focus of user education. Strengthening the education of students' information literacy and improving their information literacy ability is of great practical significance for the future career development of students. It is also the realistic requirement of the information society for education and meets the needs of national development for high-quality and high-quality talents. At present, many regions are also actively promoting the use of mobile terminals, and using mobile terminals to carry out teaching activities can reduce the time and space limitations of traditional teaching systems. Modern mobile terminal technology has made rapid progress. With the reduction of terminal cost and the improvement of technology, the functions are more perfect, and various products are flooding the market. This paper designs a situational simulation teaching system for information literacy education based on mobile terminals, and verifies the feasibility of practical application of the designed system by means of system testing.

References

1. Xu, N., Fan, W.: Research on interactive augmented reality teaching system for numerical optimization teaching. Comput. Simul. **37**(11), 203–206+298 (2020)
2. Yu, L., Wang, J.: The foundational structure of information literacy as its empowerment mechanism. J. Libr. Sci. China **47**(05), 4–19 (2021)

3. Jia, X., Li, M.: Design of security monitoring system for mobile terminal distance education based on multimedia technology. Mod. Electron. Tech. **42**(18), 77–80 (2019)
4. Liu, H.: Optimization of school education and teaching system based on reducing burden and improving quality. Prim. Secondary Sch. Manage. **18**(02), 57–59 (2022)
5. Xiao, X.: On the theoretical origin, connotation and components of information literacy—also on international experience of information literacy education. E-educ. Res. **42**(08), 116–121+128 (2021)
6. Song, Y., Song, S.: Design and development of a personal smart-device based distributed virtual reality field teaching system. Res. Explor. Lab. **39**(02), 227–232 (2020)
7. Li, J.: Design of the timeliness analysis system of college education and teaching based on big data. Digital Commun. World **10**, 87–88 (2021)
8. Zheng, W., Zheng, L., Zhang, L., et al.: Application of scenario simulated flip class in nursing practice teaching of nursing students in higher vocational education. Chin. Nurs. Res. **33**(04), 676–678 (2019)
9. Li, S., Fang, D.: Research and implementation of intelligent computer network teaching system. Sci. Educ. Guide (Mid term) **16**(11), 54–55 (2020)
10. Ji, X., Chen, X., Yu, Y.: Unity3D based real-time transesophageal echocardiography simulation system and its key techniques. J. Comput. Appl. **35**(S1), 235–238 (2015)

Dynamic Integration Method of Economic Teaching Resources Based on Information Fusion

Fan Lv[1]($\boxtimes$), Ying Chen[1], and Xin Wang[2]

[1] Sanmenxia Polytechnic, Sanmenxia 472000, China
lvfan88889@126.com
[2] Shandong Institute of Commerce and Technology, Jinan 250103, China

Abstract. In order to manage massive teaching resources more effectively, taking the major of economic education as an example, this study proposes a dynamic integration method of economic teaching resources based on information fusion. Firstly, a scientific and reasonable collection system of economic teaching resources is constructed combined with information fusion technology, then the evaluation criteria of dynamic integration of economic teaching resources are divided, and the steps of dynamic integration of economic teaching resources are simplified. Finally, experiments show that this method has high practicability in the process of practical application.

Keywords: Information fusion · Economic teaching · Resource integration · Evaluation criterion

1 Introduction

Teaching resources refer to the elements that can form teaching ability, have value and can be applied in teaching in the process of education and teaching. The coverage of teaching resources mainly includes information fusion and electronic teaching resources and information. Multimedia materials, teaching plans, courseware, test questions, homework exercises, educational papers, e-books, network courses, etc. required in the process of information technology application in education and teaching activities can be called teaching resources [1].

In the context of information explosion, effective management and scientific screening of massive teaching resources is one of the main research contents of current teaching resource management. Due to the different emphases of resource management and application in the teaching process, and the characteristics of the pattern of teaching resources to be collected and identified, it is very easy to have the problems of repeatability and error of data extraction [2].

In the process of Information Fusion Teaching Resource Construction in Colleges and universities, the current teaching resource management effect is still difficult to meet

W. Fu and L. Yun (Eds.): ADHIP 2022, LNICST 469, pp. 156–169, 2023.
https://doi.org/10.1007/978-3-031-28867-8_12

the current research requirements. Take the major of economic education as an example. This major focuses on theoretical economics and has the attributes of Applied Economics. It aims to train high-quality economic professionals with solid professional basic knowledge and theory, as well as international vision and innovation and entrepreneurship ability. For the teaching of economics majors, the dynamic integration of economic teaching resources is more conducive to helping students get the latest economic development trends and better improve the teaching quality. Therefore, this study proposes a dynamic integration method of economic teaching resources based on information fusion. The specific research ideas are as follows:

(1) Use information fusion technology to collect economic teaching resources.
(2) Mark the integration characteristics of economic teaching resources.
(3) Classify the dynamic characteristics of economic teaching resources.
(4) Establish the management model of college teaching resources, and design the dynamic integration framework.
(5) We will focus on five aspects of teaching resources and content, teaching plans and design, teaching organization, teaching effectiveness, and reform and innovation, and build an evaluation standard system for the integrated application of economic education resources.
(6) The integration and sharing of resources are completed through the process of downloading, storing, sharing, regrouping and combining resources.

This method improves the goal of storage and management of teaching resources from different channels through information fusion from the technical level, and realizes the fusion processing of teaching resources.

2   Dynamic Integration Method of Economic Teaching Resources

2.1 Information Fusion Collection and Feature Marking

"Teaching resource integration" is an existing state of optimized combination of digital resources. It is to integrate, cluster and reorganize the data objects, functional structures and their interactive relationships in each relatively independent digital resource system according to actual needs, so as to form a new, more effective and efficient digital resource system.

Information fusion storage architecture is the data integration platform of information fusion resource integration system, occupying the bottom layer of three-tier structure [3]. Information fusion storage system is a storage system that can share large capacity and high transmission rate. It does not occupy campus network (LAN) resources, but also can realize remote disaster recovery, data backup and high scalability. It is becoming an ideal solution for integrating resources such as digital economy teaching in colleges and universities.

Since the SAN storage area system separates the storage device from the server in the network environment, the information fusion storage system connects the RAID disk array with the teaching resource server through FC optical fiber switch. When there is a demand for resource data access, the resource data is transmitted at high speed between the relevant server and the background RAID disk array through information fusion, Moreover, the resource data server can access any storage device on the San, which improves the availability of the resource data system. The logical topology of a typical information fusion storage system is shown in Fig. 1.

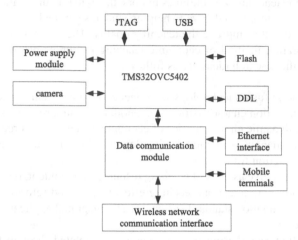

Fig. 1. Logical topology diagram of information fusion collection and management

The learner centered information fusion learning resource integration model mainly adopts metadata and web2 0 and other technologies, provide relevant resources to learners according to the teaching plan and taking the learning process as the main line. The formal structure is mainly characterized by the logical continuity of knowledge, and the substantive structure forms a collection of relevant resources according to the requirements of teaching design [4].

On this basis, the integration mode of information fusion learning resources is studied, as shown in Table 1.

Table 1. Characteristics of information fusion learning resource integration mode

Type	Resource centered integration model	Learner centered integration model
Subject	Resource construction and organizer	Integrate designers and teachers
Basis	Resource attribute characteristics	Learner centered
Technology	Computer integration, etc	Semantics, metadata technology, etc
Form	Subject classification, subject navigation, learning resources, etc	Online courses, learning platforms, etc
Characteristic	Integrating resources based on attribute characteristics	Integrate resources according to the learning process

Based on the current problems faced by the construction and management of information fusion teaching resources, as well as the advantages and value of feature marker oriented digital resource construction, this paper constructs a feature marker model according to the characteristics of information fusion technology and teaching resources, and carries out the classified management of teaching resources with the help of information fusion teaching resources feature marker related technology. The feature marking system model of information fusion teaching resources is shown in Fig. 2.

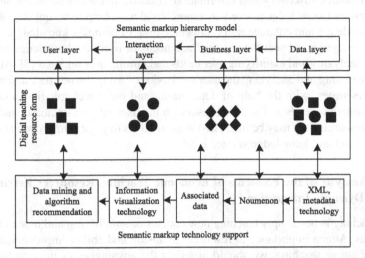

Fig. 2. Feature marking system model of Information Fusion Teaching Resources

In Fig. 2, the data layer is used to store various types of data. In feature tags, there are rich data types, including a large number of semi-structured and unstructured data in addition to the traditional structured data. The data layer is used to store different types of data. It is the basis of feature marking system [5].

The data layer is not only the traditional teaching data resource library with a single resource as the storage unit, but also a large number of databases with fine-grained "knowledge unit" as the independent storage unit, such as XML database, relational database and so on. The establishment of these databases meets the information needs of users to realize the rapid acquisition and efficient utilization of knowledge, deepens the information service from the resource itself to the knowledge unit within the resource, and makes the feature mark have the function of knowledge service.

On the basis of storing all kinds of data and establishing an independent knowledge base of knowledge units in the data layer, the business layer identifies the knowledge units and realizes the automatic association between the knowledge units, so as to build a factory wide knowledge network full of characteristic connections and realize the automatic discovery of knowledge [6]. The business layer can structurally process the data, and carry out fine-grained identification and feature annotation on the knowledge units within the resources, connect many resources, and establish multi-level and rich knowledge associations to meet the needs of users for information and knowledge integration and centralized acquisition in the current environment. The association established by the business layer not only includes the correlation between traditional teaching resources, such as establishing the connection between a resource and related resources, but also includes the connection between various parts of the resource and the link between the internal knowledge unit of teaching resources and external resources, which greatly enriches the knowledge content of teaching resources.

The interaction layer is directly user-oriented and guides users to discover, acquire and utilize knowledge by providing user interaction interface [7]. The navigation design based on resource discovery and information visualization window in the interaction layer can present knowledge to users in a more vivid way, help users quickly obtain the desired knowledge and enhance users' understanding of complex knowledge.

The user layer is to meet the personalized needs of users for knowledge and information by analyzing and identifying user needs and mining user interests [8]. At the user level, by recording and analyzing the user's behavior of using the network and browsing teaching resources, with the help of data mining and other technologies, analyze the user's information needs and interests, establish the user interest database, and recommend the resources that may be interested to users, so as to greatly improve the resource utilization level and knowledge service level.

2.2 An Analysis of the Elements of Economic Teaching Resources Retrieval and Dynamic Integration

Online teaching is not simply teaching new courses or simplifying and repeating previous courses. Attach importance to resource expansion and ability improvement. In the process of online teaching, we should highlight the advantages of information fusion resources of online teaching, which are convenient for transmission, diverse types, function oriented to support students' online learning, and highlight the cultivation and improvement of students' online learning ability. Focus on discussion and interaction, answer questions and solve doubts [9, 10].

To this end, information fusion technology and content management system are used to integrate heterogeneous data, integrate distributed data resources of different sources, types and formats, and build the middle layer of the three-tier model structure of resource integration. The data flow of the economic teaching resource retrieval system is shown in Fig. 3.

For the scientific management of teaching resources in colleges and universities, based on the existing teaching resources information, this paper analyzes, counts, stores, queries and backs up all the information, and establishes a multi-functional resource management model on this basis. So as to realize the informatization of teaching resources, scientific management, office automation and networking of communication among colleges and universities. According to different functions, the information fusion center can be divided into infrastructure service layer, virtualization layer, management middleware layer and application service layer. The center is a resource pool composed of infrastructure, software and application platform. In this resource pool, the underlying hardware platform with blade server as the core is responsible for data processing, and users obtain services on demand through the upper software and application platform. The information fusion platform realizes the integrated management of server teaching resources, and can expand or shrink resources according to the actual needs of users. It has the characteristics of high resource utilization, good scalability and low complexity of equipment management.

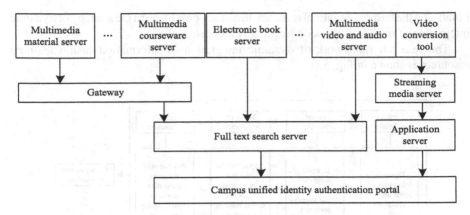

Fig. 3. Data flow diagram of economic teaching resource retrieval system

The management model of teaching resources in colleges and universities is shown in Fig. 4.

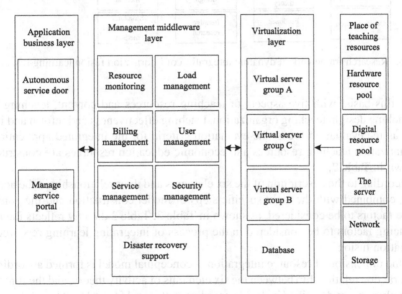

Fig. 4. Management model of teaching resources in Colleges and Universities

In the research on the evaluation of the dynamic integration model of information fusion learning resources, this research adopts the research strategy of combining qualitative and quantitative research. Based on the Expectation Confirmation Theory of information system, through the practical application of the dynamic integration model of information fusion learning resources in teaching, this paper carries out the research on the influencing factors of learners' willingness to continue to use, so as to verify the effectiveness of the dynamic integration model of information fusion learning resources.

Finally, on the basis of empirical research, this study puts forward the strategy of dynamic integration of information fusion learning resources,

The research framework of dynamic integration of information fusion learning resources is shown in Fig. 5.

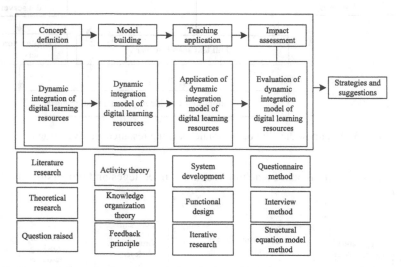

Fig. 5. Research framework of dynamic integration of information fusion learning resources

On this basis, with five aspects of teaching resources and content, teaching plan and teaching design, teaching organization, teaching effectiveness and reform and innovation as the evaluation focus, the evaluation criteria for the integrated application of informatization teaching resources and economic education resources are constructed, as shown in Table 2.

According to the description of the six elements and their relationship in the activity theory, combined with the eight specific steps of model construction and the comprehensive factors to be considered, as shown in Table 3. Table 3 directly reflects the main influencing factors to be considered in the process of integrating learning resources of information fusion.

In the dimension of resource integration, a conceptual model is formed according to the interaction relationship between the six elements of activity theory, and the functions of each element are described in detail according to the goal formed by "demand action", which can realize the two-way interaction of effective information.

The dynamic integration elements of information fusion learning resources are shown in Table 4.

In the learning process, learners participate through the functional design of dynamic integration of information fusion learning resources, complete learning activities by using information fusion learning resources, participate in feedback and sharing according to their experience in the learning process, and feed back effective information to implementers, so as to realize two-way interaction of effective information.

Table 2. Evaluation criteria for integration of teaching resources

No	Dimension	Index	Content decomposition
A	Technical	Teaching resources and contents	Curriculum resources can meet the curriculum standards and enrich the content; Teaching resources can better develop new ideas and serve the curriculum objectives and academic requirements; Rich resources, able to meet the needs of ordinary assessment
B	Instructional design	Reorganization and reconstruction of granular resources	The teaching objectives are clear, the teaching arrangement and design are reasonable, the teaching content is properly designed, and the key points and difficulties of the course can be accurately grasped
C	Usability	Vocational skills training	The process of vocational skills training is arranged reasonably, which can reflect the characteristics of online courses, the arrangement of teaching interaction is reasonable, the focus of online teaching process is prominent and well-organized
D	Resource integration	Teaching effectiveness	Teachers' advanced teaching concept and good teaching effect; Students actively participate, have strong interaction, and can achieve learning goals
		Reform and innovation	Be able to continuously improve the effect of online teaching according to the teaching characteristics of online courses

2.3 Dynamic Integration of Economic Teaching Resources and Realization of Relationship

The process of resource classification, integration and extraction is shown in Fig. 6.

The teaching resource integration platform supports users to select required resources from material library, courseware library and case library. For the selected results, you can browse the details of resources or open the resource file to view. When you decide to select a resource, click the "join" link on the right side of the resource title. The system will automatically add the selected resources to the defined consolidated resource file

Table 3. Element analysis of dynamic integration model of information fusion learning resources

Modeling steps	Essential factor	Influence factor
Clarify the objectives of the activity	Object	Social environment
Identify the executor of the activity	Subject	Ability
Clarify the tools used to support the main body to complete the activities	Tool	Technology and method
Clarify the planning and specifications in the implementation of activities	Rule	Task type and characteristics
Clarify the responsibilities of different subjects	Community	Task allocation
Clarify the role division of different subjects	Division of labor	Situational factors
Specify the specific environment in which the activity takes place	–	Situational factors
Clarify the expected results	–	Achievement degree of mission objectives

and store them in the SQL Server database. The sharing model of economic teaching resources is shown in Fig. 7.

The co construction and sharing of information resources is the starting point of the integration and utilization of teaching resources in Colleges and universities. It is an effective measure to integrate teaching resources by regrouping and combining teaching resources according to relevant disciplines or themes and certain principles, and then integrating teaching resource databases in the network environment.

3   Analysis of Experimental Results

The following experiments are designed to verify the effectiveness of this method.

The computer configuration required for the experiment is as follows: the experimental environment is composed of five machines, one is the test machine (2-core processor above 2GHz, 4 GB DRAM and SATA hard disk at 250B) as the test machine, and the test program 1 Roadrunner 1L is installed: one server is the main controller of cloud cluster, and the other three (2-core processor above 2GHz, 2 GB Drau and SATA hard disk at 250GB) as the node controller, Installation of node RPM software package: equipped with San switches and routers above 100bps: the installation system includes Linux, eucalyptus, EC2, what software and MYSQL to build a test system for university teaching resource management model under the information fusion environment. In this test requirements, the response time of the page for registering and querying student information shall not exceed 8 s.

The experiment first shows the statistical results of teaching resource management, as shown in Table 5.

Table 4. Analysis of dynamic integration elements of information fusion learning resources

Activity system	Activity elements	Activity content
Target	Activity objectives	Realize the two-way interaction of effective information
Subject	Learner	Carry out the learning process according to the learning objectives and participate in dynamic integration activities
Object	Digital learning resources	Open learning resources, commercial sex education resources, etc
Tool	Dynamic integration of instructional learning resources	Design corresponding functions to achieve activity objectives
Rule	Feedback principle	Use resources and participate in feedback and sharing
Community	Teachers and other learners	Have the same learning objectives and learning situations
Division of labor	Subject	Learners are not only the main body of the learning process, but also the participants of dynamic integration
Performance	Specific activities	Through the corresponding functions and modules, the two-way interaction of effective information is realized

The test values in Table 5 reflect that the design of university teaching resource management model is feasible and has good stability.

On this basis, the traditional method of dynamic integration of teaching resources based on information navigation is compared with the method in this paper, and the results are shown in Table 6.

Based on the comparative analysis of the above experimental results, compared with the traditional methods, the dynamic integration effect of teaching resources of this method is significantly better, which can fully meet the research requirements.

Finally, the application performance of different methods is verified by taking the resource response amount in the same time as an index, which can reflect the integration response efficiency of different methods. In order to avoid the singleness of the experimental results, two comparison methods were selected, namely, traditional method 1 based on information navigation and traditional method 2 based on B / S architecture. The comparison results are shown in Fig. 8.

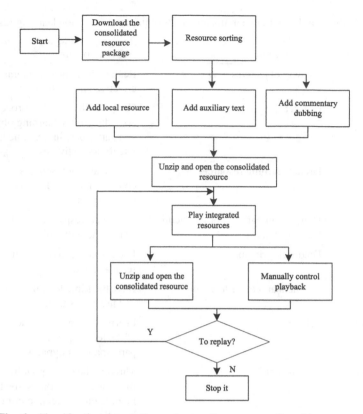

Fig. 6. Classification, integration and extraction process of teaching resources

From the analysis of Fig. 8, it can be seen that under the same environmental impact, compared with the two traditional methods, after the integration and extraction of teaching resources by using this method, the response volume of resources is more, indicating that the amount of data that can be integrated and processed by this method is also significantly larger, which meets the design requirements.

4 Conclusion

The integration of information fusion and teaching resources is an important problem in the construction of information resources in local colleges and universities. Its information integration and service should be people-oriented, and pay attention to the diversity, comprehensiveness, personalization and specialization of users' information needs. Because users are the utilization and judge of the results of information integration. Therefore, local colleges and universities should pay attention to strengthening the research on the utilization efficiency of information fusion teaching resources on campus, assign full-time personnel in economic teaching in colleges and universities to be responsible, feed back opinions and utilization at any time, make statistical analysis and evaluation on the utilization of integrated resources, and explain the problems with

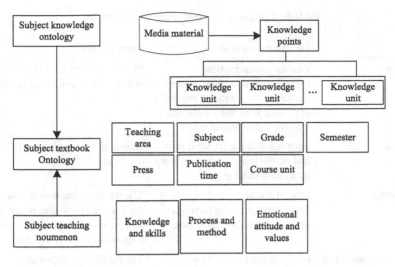

Fig. 7. Economic teaching resource sharing model

Table 5. Statistical results of teaching resource management

Test item	Target value	Practical value	Pass no
Login corresponding time	< = 2.9	2.653	adopt
Query response time	< = 2.9	1.652	adopt
Login transaction success rate	100%	2165	adopt
Query transaction success rate	100%	2165	adopt
Total number of completed transactions	2200 in 30 min	2165	adopt
CPU utilization	75	53.58	adopt
Memory usage	70	52.16	adopt

statistical data, It can be provided to the competent department as the reference basis for the planning and construction of information fusion teaching resources in Colleges and universities, so as to realize the value-added of the development and utilization of information fusion teaching resources.

At present, the online courses of colleges and universities established in China have achieved initial results in the collection of resources, but in terms of application, they pay insufficient attention to the application and application effect after the release of online course construction. The utilization rate of these high-quality teaching resources is not high, they have not played their due role in the improvement of teaching quality and school running efficiency, large investment has no large output, and high investment has not doubled efficiently. The research on how to use online courses to restructure and reengineer high-quality resources and how to organically combine this high-quality

Table 6. Comparison of two integration methods

	Traditional method	Paper method
Function	Resource construction, organization and classification, sequence integration, using links to provide retrieval access and quickly locate resources	Provide references, information summary and automatic retrieval
Retrieval	Integrate into the retrieval portal and establish a one-stop resource navigation database	One stop information query and obtain results
Basic feature	Ease of use and effectiveness of the system; Friendly retrieval interface and reasonable link	The retrieval function is powerful, the knowledge system is complete, and the acquisition and document transmission are convenient
Resource organization	Integrate the information resources through the organization system and deeply reveal the resources	The distributed information resources are seamlessly connected in a conventional way to facilitate access
Integration technology	Use navigation to process the search results and select reasonable results to pass to users	Cross database integration, merging multilingual search results

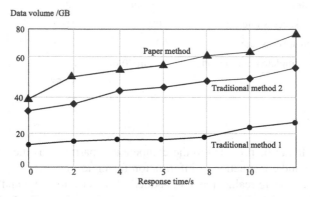

Fig. 8. Comparison of resource search response and detection results

resources with the teaching of the course itself is not in-depth, and there are no effective measures and countermeasures, which need to be further studied.

References

1. Li, X.: Design of university distance education system based on ASP technology under background of the internet. Mod. Electron. Tech. **43**(23), 178–181+186 (2020)
2. Zhang, B., Ge, S., Wang, Q.: Multi-order information fusion method for human action recognition. Acta Automatica Sinica **47**(03), 609–619 (2021)
3. Mo, M., Sun, Z., Zou, S.: A global information fusion method based on least square applied to distributed POS. Journal of Chinese Inertial Technology **28**(01), 35–40+34 (2020)
4. Xu, N., Fan, W.: Research on interactive augmented reality teaching system for numerical optimization teaching. Comput. Simul. **37**(11), 203–206+298 (2020)
5. Liu, H., Shen, J., Wang, G., et al.: Construction of virtual simulation experimental teaching resource platform based on engineering education. Exp. Technol. Manage. **36**(12), 19–22+35 (2019)
6. Sun, Y.: Design of practice teaching network platform for ideological and political teaching resources in colleges and universities. Microcomput. Appl. **36**(02), 152–155 (2020)
7. Liu, Y., Liu, Y.: Online and offline mixed teaching reform of computer public training course in higher vocational education. Exp. Technol. Manage. **37**(06), 243–245 (2020)
8. Shi, Y., Zhang, J.: Design of information? Based teaching resources sharing system based on multimedia technology. Mod. Electron. Tech. **44**(20), 32–36 (2021)
9. Kong, L., Ma, Y.: Big data adaptive migration and fusion simulation based on fuzzy matrix. Comput. Simul. **37**(03), 389–392 (2020)
10. Yao, W.: Discussion on the construction of professional teaching resources based on design-oriented thought. Liaoning High. Vocat. Tech. Inst. J. **22**(05), 22–26 (2020)

Business Information Mining Technology of Social Media Platform Based on PageRank Algorithm

Ying Liu(✉)

School of Business, Jinggangshan University, Ji'an 343009, China
liuyingzhende@126.com

Abstract. The existing methods of business information mining are flexible and cannot effectively mine the business information of media platform. In order to summarize and manage the mass business information effectively, a research method of business information mining based on PageRank algorithm for social media platform is proposed. Different from the existing methods, it innovatively optimizes the business information evaluation algorithm of social media platforms, increases the flexibility of information mining, and realizes the business information mining of social media platforms. The experiment proves that the technology of business information mining based on social media platforms can effectively summarize and manage a large amount of business information.

Keywords: PageRank algorithm · Social media · Business information · Data mining

1 Introduction

The commercial information platform is a special type of management information platform, which provides support for market-oriented operation activities in enterprise activities. Since market competition requires constant innovation in a mature economy and society, and innovation requires scientific decision-making methods and an adequate information base, the value of such platforms as ancillary tools continues to increase, and the modalities for providing assistance continue to evolve and evolve, integrating more and more advanced theories and technologies with the times [1]. Typical areas of business activities in which ancillary support is provided include:

Market operations, specifically including macro-market analysis, analysis and evaluation of product sales strategies, evaluation of business value of research and development plans for new products, analysis and evaluation of market trends of products, evaluation and optimization of product portfolios and pricing schemes, etc.

Customer Management, such as Customer Relationship Management, New Customer Marketing Strategy Analysis and Evaluation, Customer Channel Performance Analysis and Contribution Analysis [2].

© ICST Institute for Computer Sciences, Social Informatics and Telecommunications Engineering 2023
Published by Springer Nature Switzerland AG 2023. All Rights Reserved
W. Fu and L. Yun (Eds.): ADHIP 2022, LNICST 469, pp. 170–186, 2023.
https://doi.org/10.1007/978-3-031-28867-8_13

Product R&D management, such as product line portfolio evaluation and optimization, product line profitability calculation and analysis, product modification program income evaluation, new product risk and value comprehensive calculation and analysis.

Field of financial investment management of enterprises, such as long-term investment planning and optimization of production resources, risk analysis of investment projects of enterprises, analysis of optimal allocation of short-term production resources, value assessment and operation analysis of intangible assets, analysis of enterprise operation cost structure and elements, etc. [3].

In the modern business information service environment, information is becoming more and more intensive and information-based. If enterprises want to develop steadily in the increasingly fierce market competition, they must extract the information which is beneficial to the business operation after analyzing a large amount of business data deeply, so as to improve their own decision-making ability.

Madhumathi S and the others introduced that the work flow model can be implemented with the data mining in the E-commerce platforms. It helps the product/project manager in several ways. The multiple queries have figured out and those are solved here. The data and reviews are generated automatically. The text are generated with web crawler and stored in database as a raw data. The data are cleaned with Natural Language Processing methods and algorithms [4]. Fauzan A and the others use the association rules with rapid miner software, data mining approach, and predictive analysis that contains various data exploration scenarios. The study provides important evidence for adopting data mining methods in the industrial sector and their advantages and disadvantages. The business information mining of social media platform is realized based on chevron Pacific Indonesia platform [5]. The above methods do not evaluate the business information according to the web page ranking, and the flexibility of the business information mining is poor, and the effective media platform business information mining results cannot be obtained.

Data mining, as a tool that can analyze and check a large amount of business data, can help the business enterprise to extract valuable information from the updated and accumulated data. It can collect and analyze a large amount of data in the business environment based on the established business objectives of the business enterprise, and then extract the key data and information from it, so as to provide effective information services for the scientific business decision-making activities. Different from existing methods, this paper innovatively introduces data mining technology to conduct business information mining on social media platforms. Design the business information management system of social media platform; PageRank-based data mining algorithm evaluates business information according to web page ranking to realize business information mining of media platform.

2 Social Media Platform Business Information Mining

2.1 Business Information Management System for Social Media Platforms

Data mining based on PageRank, as a professional data processing technology, has been widely used in commercial information services, among which Data mining is also widely used in the information services of commercial media platforms, mainly in the risk assessment of media platforms, product supervision and the establishment of commercial competitive intelligence platforms. Some experts have a popular explanation for data mining, that is, to explore knowledge in databases. But there are also some experts who regard it as only one of the most basic and vital links in the process of knowledge discovery. The process of knowledge discovery is shown in Fig. 1.

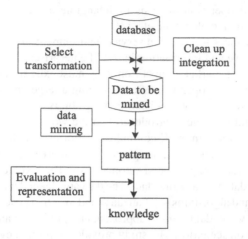

Fig. 1. Steps to turn mining data into feature information

As shown in Fig. 1, the result of data mining is the ability to interact with users using a knowledge base. Through the intuitive model to the user, or stored in the knowledge base as a new knowledge for the user to save. But in today's big data era, "data mining" than "knowledge discovery in the database" as a new concept is more popular. Therefore, this paper chooses terminology for data mining. In addition, the data mining process of the customer marketing response rate prediction model in this paper is basically in accordance with the above process and method steps to study the operation [6]. If we can realize the marketization and normalization of the "mortgage of control" mechanism and reduce the difficulty of its application, we can greatly reduce the cost of the debtor's recourse to funds and at the same time provide a mechanism for high-quality enterprises to send their own quality signals. The direction of business information business data management is shown in Fig. 2.

E-commerce can form cross-validation with other information of enterprises according to the historical data mastered, which is a powerful supplement to the scientific nature of product information management. When you do not have a social media business of your own, mine the available energy data into energy information to provide information services to businesses or social media organizations [7]. For example, when a company wishes to obtain credit from a social media platform, it can provide a report on whether the company's historical product data and historical information match for a fee. Similarly, social media institutions can be provided with services to monitor and analyse the electricity consumption of their leaders [8]. In this model, one is only responsible for the data he provides, and the information he analyzes, and is paid accordingly, without

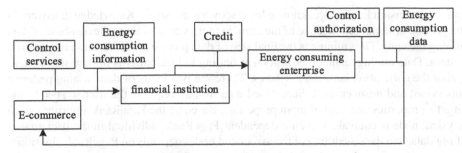

Fig. 2. Business information business data management direction

the social media risk itself. The overall architecture of the business model is shown in Fig. 3.

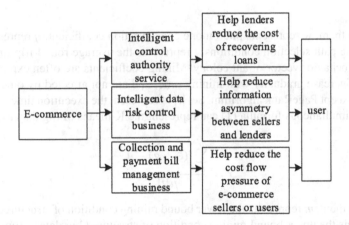

Fig. 3. Overall architecture of the business model

In this architecture, the main business relationships with users are divided into three parts: energy flow, information flow and cost flow. The information flow refers to the relationship between the user and the user due to the information generated by the transaction between the user and the user, and the energy consumption behavior of the user. Energy-use data risk control business shall be carried out around information flow, and help narrow the information asymmetry between debtors and credits. Receipt and payment billing management business shall be carried out around expense flow, and the optimized allocation of cash resources may be realized.

2.2 PageRank Based Business Information Evaluation Algorithm

Data mining is a process of extracting the unknown data and information which may exist in the data by some methods in the irregular, large quantity, unclear and random data. Data mining also has many other terms, different fields are different, for example, data analysis, decision knowledge, knowledge discovery and so on. Knowledge discovery in the general artificial intelligence of the term, in fact, is equivalent to the database of data mining meaning. Data mining is the final step of data processing in business intelligence system. Data mining technology is good at finding hidden patterns of data sets by analyzing the correlation between attributes. There are two kinds of data mining patterns: supervised and unsupervised. Supervised data mining requires the user to specify the target's properties and a set of input properties. Based on the PageRank algorithm, each physical node is equivalent to an independent PageRank individual in the background of big data, and the scheduling of the associated nodes depends on PageRank algorithm. Set α_0 to represent the position information of the PageRank and α_1 to represent the position information of the information source. Using the above physical quantities, the selected parameters of the PageRank algorithm can be represented as follows:

$$i = \int\limits_{0}^{\infty} \frac{w^2 |\alpha_1 - \alpha_0| \times q}{\lambda \dot{u}} dw \tag{1}$$

Among them, w represents pheromone concentration coefficient, q represents optimal foraging path selection conditions, t represents the average round-trip time, $\dot{u}$ represents the behavior vector of ant colony. Mining coefficients are often expressed as $\dot{\xi}$, which can increase gradually with mining time, and are not affected by other physical quantities except PageRank algorithm. Set $|t|$ to represent the execution time of a mining cycle, the simultaneous formula (1) can represent the K-means of big data as:

$$x = \frac{\sum\limits_{m}^{m'} i(\dot{\xi} + g) \cdot |j|}{\mu |t|} \tag{2}$$

Among them, m represents the lower bound mining condition of structured big data, m' represents the upper bound mining condition of structured big data, g represents the basic application privilege of PageRank algorithm, j represents the mining execution vector of large data parameters, and μ represents the execution load constant term of node organization. At this point, the calculation of all parameter vectors is completed.

Further exploiting the support and hyper-chaining relationship, the tradeoff between the two is that the final result will directly affect the result of the algorithm by giving the user a step-size parameter of the index page, which is more acceptable, even if the user's search efficiency loss is not significant. In this case, the definition of the distance with the step-size parameter needs to be compared:

$$d = \frac{p(b|a)}{D(x \cap y / a \cap b) \times ||T - p(a \cup b)||} \tag{3}$$

Expressed as item sets x and y, the two item sets are independent of each other and do not have duplicate attributes, which can be represented as: $x \cap y = \emptyset$. The personalization rules between set X and set Y must meet the following conditions: the personalization rules of two sets have certain universality and interesting. For example: $T = \{T_1, T_2, \cdots T_m\}$ is a collection of items, related data D is a collection of database information, where each information d is a collection of items, making $d \subseteq 1$. Each message has a glyph, called an d_1. Let a be an itemset, $a \in d$. Personalized association rules have $a \Rightarrow b$ implication, in which $a \subset 1$, $b \supset 1$, and $a \cap b = \emptyset$.

Based on the above association rules, the support and confidence of personalized characteristic data are calculated. If the $a \Rightarrow b$ is established in the information set D, it has the support of characteristic attribute. The percentage of the information set D that contains $a \cup b$ is also known as support, or $p(a \cup b)$. If the personalized association rule $a \Rightarrow b$ is established in the information set D and has confidence c, the information set D contains information a and all the information ratios of b, which is also called confidence $p(b|a)$.

There are three main algorithms in the standard data mining model: Bayesian algorithm, PageRank algorithm and PageRank algorithm. Unsupervised data mining techniques do not need to have predictable attributes. Before the research of data source, it is necessary to ensure the content of user's demand for data, which requires the clear characteristics of operational data and analytical data. Because the traditional business platform is mainly used in the daily transaction processing of enterprises, the data stored in the database accord with the characteristics of operational data. The data stored in order to meet the requirements of business intelligence analysis and mining, basically accord with the characteristics of analytical data. The difference between the two is shown in Table 1.

Table 1. Characteristics of operational data and analytical data

Operational data features	Characteristics of analytical data
Indicates the dynamic status of business processing	Indicates the static state of business processing
It is correct at the moment of access	Represents past data
It can be updated. It is updated by the entry personnel entering transactions	It is not updatable, and the access rights of end users are often read-only
Handle details	More attention is paid to the conclusive data, which is comprehensive or refined
The operation requirements can be known in advance, and the system can be optimized according to the expected workload	The operation requirements are not known in advance, and Yongyuan does not know what the user will do next
There are many transactions, each affecting a small part of the data	There are a small number of queries, and each query can access a large amount of data
Application oriented, supporting daily operation	Analysis oriented, support management requirements
Users don't have to understand the database, they just input data	Users need to understand the database to draw meaningful conclusions from the data

The basic logical structure of data warehouse includes star structure and snowflake structure. Compared with the two structures, the star structure contains fewer layers of links for information retrieval than the snowflake structure, is easier to manage, and queries are more efficient; moreover, the star structure based on the star model facilitates the definition of data entities from the perspective of supporting decision makers, and the entities just reflect the core content (theme-related content), and the star model diagram is easy to understand, read and query. The data warehouse star structure diagram is shown in Fig. 4.

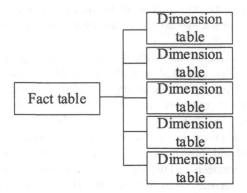

Fig. 4. Data warehouse star structure diagram

Based on the collected and accumulated customer information, the commodity database is used to analyze and mine the customer data, and to predict the customer's trust and desire to buy the product. Database is the process of information leading marketing, the process of information driving sales, and the process of information promoting efficiency. Database is a normal trend, is the trend of the customer as the center of the enterprise, is the trend of the data platform as the business philosophy, is to maintain customer resources to achieve sales goals to expand the trend of sales. Compared with the traditional marketing, the database has significant differences in marketing model, marketing channels, determining goals and many other aspects, as shown in Table 2.

From the macro point of view, the database can accurately predict and feedback the trading market in real time, from the micro point of view, the database can accurately plan the inventory customer's contribution to the enterprise. Therefore, an excellent database can promote the enterprise to understand the real needs of customers, and then interpret the ultimate customer goal in an all-round way, provide a platform for enterprises, humanization, in light of the marketing plan, reduce the operating costs of enterprises, accelerate the marketing efficiency of enterprises, and increase the adhesion degree of customers.

In addition, PageRank algorithm is increasing exponentially in recent years. Using database can effectively help enterprises to intercept effective new information, provide professional theoretical basis for follow-up strategy planning and new product R&D, and improve the competitive advantage of their own products.

Business intelligence refers to that an enterprise makes use of the existing internal and external database data as the source data, through a series of standardized operations such as cleaning, extracting, converting and loading, and then summarizes and

Table 2. Comparative analysis of traditional marketing and database

Measurement standard	Traditional marketing	Database marketing
Marketing model	One way communication marketing mode, passive acceptance of products and advertising information by the audience	Interactive marketing attracts the audience to actively participate in the whole process by designing various scenes
Delivery channel	Television, newspapers, advertisements and other mass media	Personal letters, e-mail, mobile phones, etc.
Target audience	—Communicate to many and face the public	One to one communication, accurate face to each customer and meet customer needs
Deliver content	It is strongly restricted by time, layout and cost	Convey rich advertising information, with less time and space constraints
Customer relationship	Unable to grasp every piece of feedback information in hand, the controllability is low	Through reliable feedback analysis, we can understand the psychology of the audience and establish a customer relationship database to maintain close contact with the audience
Cost	Facing the public, the general price is relatively expensive, especially the price of TV media is higher	Every advertising fee is spent on potential customers to avoid waste of supply and high cost performance
Sustainability	The repeatability is low, and the process is still complex	Develop new service items and simplify the purchase process, bringing the possibility of repeated purchase

stores them to the data warehouse, and then analyzes and processes the data in the data warehouse by adopting online analytical processing, data mining and other information technologies so as to convert them into the information and knowledge that users need to be able to understand, and transmit the potential characteristics and forecast results to the decision-makers so as to assist decision-making and improve performance. Therefore, business intelligence is not a new information technology, it is just a combination of data warehouse, online analytical processing and data mining. The business information feature mining management architecture is shown in Fig. 5.

To sum up, the ultimate goal of business information intelligence is to transform the predicted data and decision information into knowledge successfully by cleaning, processing and analyzing the information flow of all kinds of data sources.

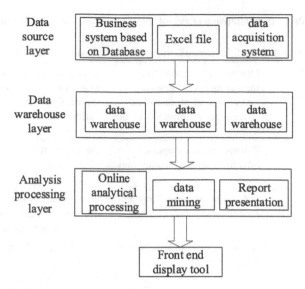

Fig. 5. Business information feature mining management architecture

2.3 Design the Realization of Business Information Mining in Media Platform

In the process of data mining, it is necessary to determine the mining object, prepare the data, build the model, analyze the results of data mining and apply knowledge. Data mining is very useful for finding and describing hidden patterns in a particular cube. Because the data in a cube grows rapidly, it is very difficult to find information manually. SQL Server2015's Data mining algorithm allows automatic pattern finding and interactive analysis During interactive analysis mining, the specific data mining processing flow is shown in Fig. 6.

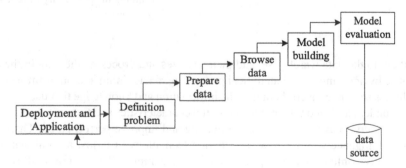

Fig. 6. Business information platform data collection and processing process

In terms of information security, the framework mainly integrates the existing operation platform and the security control mechanism on the database platform to guarantee the data security and operation security, protects the product sales business data by encryption at the database level, and invokes the identity authentication service to confirm that each business has the correct principal user when the user enters the business view or functional navigation unit for the first time. The list of internal variables in the metrics calculation is also restricted to the subject matter of the business leader or the principal user of the calculation through security rules. The architecture of the marketing support platform is shown in Fig. 7.

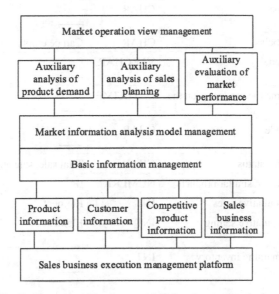

Fig. 7. Structure of business-to-information media platform

By assigning explicit fields to entities in the overall database scheme and implementing reference relationships between entities through external attributes, the logical design in the previous section is transformed into a relational data model, so that the database can be deployed on the current mainstream database management platform. The following are some examples of data tables that describe the field attributes. The product list is shown in Table 3, and each item represents one model of product, including basic information about the competitive products of the Company and other manufacturers. The sales record table is shown in Table 4.

This feature has performance advantages in some measurement tasks that require processing a large number of data sets to determine more reliable results. In the design scheme, the parallel model objects access different input data, and the output results are stored in different temporary data tables, so there is no interaction between them. This simple running logic is good for performance and also good for proper programming. Calls to the following algorithms can be executed as a combination of parameters, each of which is executed as a separate input item, or executed in parallel based on a multithreaded mode, to speed things up, as shown in Fig. 8.

Therefore, the enterprise should improve its information ability and perfect its internal information technology system. Now we are in an information age with rapid expansion and development. Overwhelming information is either actively or passively accessed

Table 3. Product information management table

Field	Field description	Field type	remarks
UQN	Product identification	CHAR	Primary key
DIN	Manufacturer identification	CHAR	Primary key and all zeros indicate the company
PQ	Product type	CHAR	____
PSQ	Product subclass	CHAR	Can be empty
PME	Product model	VARCHAR	____
PTK	The first time the product is on the market	DATETIME	____
QOD	Quality grade	CHAR	____
SOU	Current sales status	CHAR	Trial sale, sale and shutdown
CUG	Market share of similar products	NUMERIC	%
AOI	Current cumulative sales	INT	Element
AIU	Current cumulative selling expenses	INT	Element
AYU	Current remaining inventory	INT	____

Table 4. Sales record table

Field	Field description	Field type	Remarks
OIN	Record number	INT	Primary key
CIN	Product identification	CHAR	Primary key
CPU	Time record	DATETIME	____
CY	Sales batch	INT	____
XIN	Selling price	INT	Element
CI	Customer number	INT	Foreign key
XQYM	Batch sales expenses	INT	Promotion expense items
CIYK	Contract amount	INT	Element
CIYX	Contract No	CHAR	Foreign key

by people. Therefore, selective access to information is very necessary. A successful

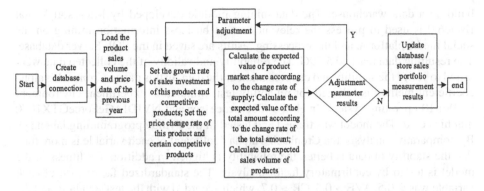

Fig. 8. Business platform information mining management process

analysis begins with the collection and screening of information, begins with the establishment of a pagerank network, and then calculates the probability of a conclusion in each pagerank network based on the quantized information. Finally, the correct conclusion can be reached only after repeated deliberation. Therefore, enterprises should filter purposefully, try their best to quantize every information, and put the quantized information into the PAGERANK network for analysis and collation, and summarize the possible logical relations. Only by combining the pagerank network with quantitative information can a variety of possibilities be analysed to arrive at a more factual conclusion; secondly, with the development of science and technology and economic integration, the increasing interdependence of production technology and economy among enterprises, the trend of resource exchange and complementarity is becoming more and more evident. Therefore, enterprises shall seek truly influential partners on the basis of resource sharing, mutual benefit and reciprocity, and through the combination of the strong and the strong, learn from each other to the maximum extent possible to make use of data and resources, improve their own products, promote the development of technologies, realize the upgrading of key components and technologies, effectively avoid homogenized competition, and enhance core competitiveness. Finally, enterprises must have a deep understanding of user needs, research appropriate products and services with users as the center, and form differentiated competition. Therefore, enterprises need to establish management process and decision-making system based on customer demand management and product planning, transform demand into products with differentiated competitive advantages, and quickly respond to the explicit and implicit demands of users, so as to grasp the market development trend and enterprise rhythm and maintain high profitability.

3 Analysis of Experimental Results

Collect the information, information level, cooperative innovation ability and enterprise performance of a national chain retail enterprise from a social media network platform, and evaluate it with a five level evaluation scale. When developing the website ranking evaluation data mining module, Microsoft SQL Server 200 is selected as a tool for building a data warehouse; The data mining module developed by Microsoft Visual Basic6.0 is used to process the relevant data of business information mining on the social media platform, and the processing results are saved in the SQL Server database. The resulting data reached 5.29G. The reliability and validity of the collected data were tested to verify the stability and centralization of the data. In this paper, Cronbach's alpha coefficient method is selected for reliability test. Simulation experiment environment is: Win10 operating system, Intel i5-9400F processor and NVIDIAGeForceGTX1070 graphics card. The model was trained and tested in the Python programming language. By comparative analysis, the Cronbach's alpha coefficient of each variable is more than 0.7, the stability of data is better, the reliability is higher. In addition, the fitness of the model is tested by confirmatory factor analysis. The standardized factor load of each variable was > 0.5, AVE > 0.5, CR > 0.7, which accorded with the test standard, and the polymerization validity of the model was high. The AVE of each variable was higher than the value of correlation coefficient of this factor and other factors by the discernibility validity test, and the discernibility validity was good. The correlation test results and model fit indices are shown in Tables 5 and 6.

Table 5. Correlation test

Variable	Information acquisition	Information mining	Information creation	Technical cooperation	Organization cooperation	Strategic cooperation	Enterprise performance
Information acquisition	0.828	0.552**	0.482**	0.352**	0.342**	0.358**	0.335**
Information mining	0.552**	0.742	0.741**	0.582**	0.625**	0.575**	0.595**
Information creation	0.482**	0.742**	0.765	0.602**	0.633**	0.599**	0.628**
Technical cooperation	0.358**	0.582**	0.602**	0.825	0.509**	0.565**	0.671**
Organization cooperation	0.342**	0.623**	0.635**	0.509**	0.752	0.606**	0.652**
Strategic cooperation	0.381**	0.576**	0.599**	0.565**	0.618**	0.755	0.655**
Enterprise performance	0.335**	0.595**	0.625**	0.671**	0.652**	0.655**	-

Table 6. Model fitting index

Fitting index	Absolute fitting index			Relative fitting index			Parsimony fit index	
	X2/df	GFI	RMSEA	NFI	TLI	CFI	PGFI	PNFI
Standard value	<4	>0.9	<0.09	>0.9	>0.9	>0.9	>0.6	>0.6
Actual value	1.513	0.832	0.052	0.838	0.952	0.963	0.723	0.771

The attributes of the indicators shown in Table 6 are explained as follows: the absolute fitting index is the measure to which a model interprets the various relationships found by the sample data. The relative fitting index is a measure of how well a model fits the data. The parsimony fit index refers to the overall fit of the proposed model to the sample data.

In the phase of test implementation, professional testers verify the correctness of function implementation item by item based on the pagerank method according to the planned test cases and sample data, as well as the confirmed function requirement use cases. Using a sensitivity curve, the larger the area under the curve, the higher the accuracy, and the smoother the curve, the better. Business information mining flexibility detection results are shown in Fig. 9.

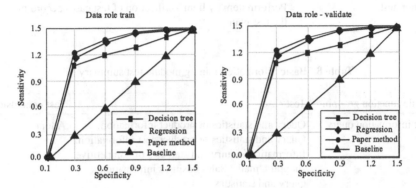

Fig. 9. Business information mining flexibility detection results

As a whole, the curve smoothness of the traditional method is not good, which shows that the stability of the traditional method is worse than the other two models. The PageRank algorithm is better than regression, but the two algorithms are very close and have no obvious advantages or disadvantages. Testing is done according to the planned priority. In addition to verifying that the expected response exists under the correct input conditions, it also verifies the stability in the case of exceptions such as permissions and incomplete data. According to Table 7, the phased testing tasks are summarized.

Applying the above method to specific functions, the test cases and test contents of each module are determined and implemented. Tables 8 and 9 are the module test items and results summary in the implementation phase.

Based on the above research results, the proposed PageRank -based business information mining technology for social media platform has high practicability and accuracy in the practical application.

Table 7. Phased data mining test tasks

Development stage	Main tasks related to testing
Requirement analysis	Determine the test cases according to the functional requirements analysis report, and preliminarily determine the test conditions, priorities and contents of each case
Software design	Determine the implementation difficulty of each function according to the design scheme, and then determine the test conditions, main test tasks, data samples and test plan of each test case
Programming implementation	Complete the source program audit and program debugging of the program unit
Program test	Perform item by Item Verification of test cases according to the test plan

Table 8. Basic information mining module test summary

Module function grouping	Test case	Test conclusion
Basic information query	Query and statistics of product information; Query and statistics of product cost information; Information inquiry and statistics of competitive shopping malls; Sales business information query and statistics	Adopt
Basic information analysis	User defined indicator statistical analysis; Multidimensional view generation; User defined index correlation analysis	Adopt
Business data import	Import actual business data from the company's sales business processing platform and generate records according to the database structure of the system	Adopt

Table 9. Product requirement information mining module test summary

Module function grouping	Test case	Test conclusion
Parameter estimation of measurement model	Estimation and setting of market aggregate change rate, price change rate and other parameters	Adopt

(continued)

Table 9. (*continued*)

Module function grouping	Test case	Test conclusion
Analysis and calculation of product sales and short-term demand for sales	Demand calculation based on total amount analysis method and demand calculation based on sales factor combination method; Demand calculation based on modified sales factor combination algorithm	Adopt
Analysis and calculation of product sales volume and short-term demand	Decomposition calculation of trend component, volatility component and periodicity component of sales volume and sales volume time series; Sales volume and sales volume time series extrapolation prediction calculation; Simple estimation calculation	Adopt
Other auxiliary functions	Query of calculation results; View generation; Report export	Adopt
	Report format conversion	Adopt

4   Conclusion

This paper proposes the business information mining technology based on PageRank, and creatively introduces the data mining technology to design the business information management system of the social media platform; Based on PageRank data mining algorithm, the business information is evaluated according to the ranking of the web page, and the business information mining of the media platform is completed. The experimental results show that this method has high practicability and accuracy in the practical application of social media platform. With the continuous application of data mining in business information services, I believe that the potential of data mining will be greatly stimulated, and the application of data mining in business information services will be further promoted. However, due to the limited conditions, the data mining efficiency of this method has some shortcomings. Future research will continue to use data mining technology to improve the efficiency of business information mining.

References

1. Madhumathi, S., Gomathi, R.: Data mining in Ecommerce platforms for product managers. Res. J. Eng. Technol. Int. Peer-reviewed J. Eng. Technol. **12**(1), 1–7 (2021)
2. Asrin, F., Saide, S., Ratna, S.: Data to knowledge-based transformation: the association rules with rapid miner approach and predictive analysis in evergreen IT-business routines of PT Chevron Pacific Indonesia. Int. J. Sociotechnol. Knowl. Dev. **13**(4), 141–152 (2021)
3. Su, J., Xu, R., Yu, S., Wang, B., Wang, J.: Idle slots skipped mechanism based tag identification algorithm with enhanced collision detection. KSII Trans. Internet Inf. Syst. **14**(5), 2294–2309 (2020)

4. Su, J., Xu, R., Yu, S., Wang, B., Wang, J.: Redundant rule detection for software-defined networking. KSII Trans. Internet Inf. Syst. **14**(6), 2735–2751 (2020)
5. Un, N.: On the tradeoffs of internet personal information and its governance mechanism. Hebei Law Sci. **38**(07), 96–113 (2020)
6. Zhang, Z., Shang, Y., Xu, Y.: Impact of IT-business alignment on business model innovation: based on data empowerment perspective. Sci. Technol. Manage. Res. **41**(12), 1–8 (2021)
7. Wang, C., Yin, S., Liu, W., et al.: High utility itemset mining algorithm based on improved particle swarm optimization. J. Chin. Comput. Syst. **41**(12), 1–8 (2021)
8. Ang, Z., Chen, J.: Sustainable business model of platform context: logic and implementation. Sci. Sci. Manage. S. & T. (Mon.) **42**(02), 59–76 (2021)

Intelligent Mining Algorithm of Macroeconomic Information for Social Network

Ying Liu[✉]

School of Business, Jinggangshan University, Ji'an 343009, China
liuyingzhende@126.com

Abstract. In the operation process of macroeconomic information intelligent mining algorithm, in the face of different data set types, there is the problem of long running time. In order to solve the above phenomenon, an intelligent macroeconomic information mining algorithm for social network is designed. Extract environmental factors, adjust the diversification of service content, describe the length of economic cycle, identify the performance characteristics of macroeconomic cycle, obtain the degree of distance between nodes, use social network to build mathematical model and design intelligent mining algorithm. Experimental results: the running time of the macroeconomic information intelligent mining algorithm in this paper and the other two algorithms are 22.210 s, 33.433 s and 33.082 s respectively, which proves that the designed macroeconomic information intelligent mining algorithm has better performance after making full use of social networks.

Keywords: Social network · Macroeconomic · Economic information · Intelligent mining · Economic model · Information data

1 Introduction

From the perspective of business domain division, economic information business can be attributed to information mining industry. As an intelligence intensive knowledge service industry, it takes professional knowledge, data information and industry experience as resources to provide solutions or decision-making suggestions to solve specific problems according to the needs of different users. No matter how the macroeconomic situation changes, the market will immediately respond to new market information through price adjustment. According to the openness of market information, the efficient market hypothesis divides the market into three forms: weak efficient market, semi strong efficient market and strong efficient market. At the initial stage of putting into use, the society as a whole lacks sufficient understanding of the function and function of information intelligent mining, and many enterprises and managers are skeptical about information intelligent mining. Enterprises rely on experience and managers' ability to make decisions, and have low demand for information mining. These three forms of markets are based on the assumption that the market can operate effectively and make corrections

W. Fu and L. Yun (Eds.): ADHIP 2022, LNICST 469, pp. 187–200, 2023.
https://doi.org/10.1007/978-3-031-28867-8_14

immediately, so the possibility of market prediction model is excluded. In addition, the overall industry volume is small, there is a lack of industry norms and supervision, the quality of personnel is uneven, and few people in the professional field participate.

Based on the theory of EMH, it is different from the short-term market. The randomness of this market movement makes the stock price unpredictable and there is no effective trading strategy to continuously beat the market. The market capacity of domestic information mining is small, and there is a big gap compared with the same industry abroad. Therefore, intelligent information mining technology once developed slowly and was in an embarrassing situation. Due to the uncertainty of characterization parameters and continuous transformation, it is difficult to achieve the ideal mining effect. Financial related market information is an important reference basis for stock market prediction, which includes all kinds of information sources, from financial related forums to financial media portals and then to the portals of relevant government departments, which provide a large amount of information supply for market investors. With the acceleration of economic globalization and the development of market economy and competition, the government and enterprises gradually realize the role of information mining in decision-making, especially the mining of economic information business has an important impact on the future economic development of the country and enterprises. The dependence of the government and enterprises on it has gradually deepened, and the domestic information mining market has gradually developed.

These different types of information represent the market sentiment from different aspects. Especially with the development of the information age, information data has a more and more profound impact on human society, especially in the field with high sensitivity to information such as macroeconomic market. Information will not only cause market fluctuations, but also affect the subsequent trend of stock price. In reference [1], in order to reduce the risk of economic management benefit loss and improve the risk early warning ability, an automatic early warning system of economic management benefit loss risk based on data mining is proposed The panel data statistical regression analysis technology is used to detect the characteristic information of economic management benefit loss risk. According to the detection results, the big data information sampling and fusion processing of the economic management benefit loss risk early warning system are carried out. The characteristic information of large data related to economic management benefit loss risk is mined. The maximum likelihood estimation method is used to automatically evaluate the risk of economic management benefit loss, The association rule set and frequent item set of economic management benefit loss risk big data are extracted, and the robustness test and automatic prediction of economic management benefit loss risk assessment are carried out in combination with the piecewise regression analysis method, so as to realize the automatic early warning of economic management benefit loss risk. The software development and design of economic management benefit loss risk early warning system is carried out under the embedded Linux environment Reference [2] discusses the application advantages of data mining technology in economic statistical surveys, analyzes the application of data mining technology in economic statistical surveys, and puts forward suggestions to promote the in-depth application of data mining technology in economic statistical surveys for reference However, because macroeconomic information is greatly affected by the environment, its performance in mining macroeconomic information needs to be further optimized.

This paper proposes an intelligent mining algorithm for macroeconomic information based on social network. On the basis of extracting environmental impact factors, macroeconomic information is mined to further optimize the mining performance.

2 Intelligent Mining Algorithm of Macroeconomic Information for Social Network

2.1 Extraction of Environmental Factors

As an external environment, market economy is the basis for the survival and development of consulting industry. With the economic globalization and the rapid development of information and communication technology, the scope and methods of information mining research have been further expanded and upgraded. The development of the Internet is one of the factors affecting economic growth, and economic growth is also the result of the joint action of many factors. The analysis of economic growth naturally needs to consider the influence of many factors. Information mining is the product of the development of market economy. With the continuous improvement of the requirements for enterprise management level, based on the theories, methods, experience and consulting tools of the same industry abroad, China's consulting industry has gradually developed. With the change of economic environment and the deepening of economic cognition, and then to the research process of total factor productivity - especially in recent years, many studies use the method of total factor productivity to analyze economic growth.

As an emerging industry, information mining has attracted the attention of all countries and developed rapidly. After the end of the "cold war", with the strengthening of the trend of global economic and trade integration, the center of national strength competition has shifted from the political and military fields to the scientific, technological and economic fields. It is generally believed that the improvement of total factor productivity mainly depends on two ways. One is micro technological progress, which mainly depends on strengthening technology R & D investment and high-end technology introduction, so as to improve production efficiency. Second, rely on various ways to improve the efficiency of resource allocation. Macroeconomic needs developed information mining to provide information guarantee, and needs the full development and utilization of information resources [3, 4]. China's economic model has changed from planned economy to market economy. Fierce market competition has promoted the close combination of information and economy. The main environmental impact factors of macro economy are shown in Fig. 1:

As can be seen from Fig. 1, the main environmental factors affecting macro-economy are: political factors, private economic factors, social factors and technical factors. Throughout many economic information theories and related literature, material capital, labor force and technological progress are recognized as the driving forces of economic growth. At the same time, studies on the contribution of other factors to economic growth are also emerging. The amount of various information data is huge, and the demanders often have no way to start. The way of sorting, processing, refining, refining and deep

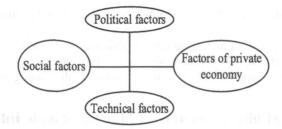

Fig. 1. Main environmental factors affecting macro economy

excavation of information can meet the demands of customers. Practice has proved that the results of information mining can effectively avoid decision-making mistakes and economic losses caused by poor information. The so-called expansion of scale mainly refers to the expansion of Internet related infrastructure, the increase of influence scope, the accumulation of resources, the diversification of service content and so on. Market economy is inseparable from scientific decision-making and information mining.

In economically developed countries and regions, the demand for information is distributed in various fields and industries, including politics, economy, military, culture, science and technology. Specifically, infrastructure expansion includes the laying of necessary equipment for fixed Internet access such as telephone lines, cable TV optical cables and optical fiber lines, as well as the coverage of mobile internet signal range. This is a basic work in the development of the Internet. If these works are not improved and implemented, the development of the Internet will not be possible. The huge market demand promotes the rapid development of information mining. Information mining methods help customers improve management and operation methods, improve operation efficiency and promote scientific decision-making. The Chinese market has a huge demand for information mining, and the potential demand is even greater. The expansion of the scope of influence mainly refers to the increase in the number of Internet users, including the increase in the popularity of Internet users. Of course, it also refers to the improvement of the impact of the Internet in some specific groups, such as the expansion of the scale of rural Internet users and the growth of the number of mobile Internet users. The growth of the number of all kinds of Internet users is the key to the deepening impact of the Internet. For example, in the past, many economic development in China relied on the consumption of natural resources to stimulate economic growth. Resource consumption and waste are very common. Although the economic level has been improved, the problems of resource waste and environmental degradation also follow.

At present, the Chinese government has taken relevant measures to alleviate the problems of development and environment, such as controlling haze and water pollution. These are not short-term actions, but long-term comprehensive treatment. In addition, the Internet access price can also affect the number of Internet users. Reducing the Internet access price is to reduce the entry threshold of the Internet and reduce the network access cost, which is conducive to the promotion of the Internet. How to ensure economic development without damaging the ecological environment, how to deal with the imbalance of economic development, and how to solve the gap between the rich and the poor in the process of urbanization, so as to ensure the steady development of China's economy. This requires think tanks to put forward feasible strategies and solutions from the perspective of information mining, pay attention to social and environmental benefits while emphasizing economic benefits, and take the road of sustainable development.

2.2 Identify the Performance Characteristics of Macroeconomic Cycle

The economic cycle consists of four stages: expansion in most economic activities almost at the same time, followed by the same general recession, contraction and recovery connected with the expansion stage of the next cycle. The effectiveness of economic information mining is also an important symbol of overall comprehensive strength. Economic information is the business pillar and strategic focus of business. The goal of economic information business development strategy is to serve the national economic and social development in the field of economic information and occupy the leading position in the field of economic information. The order of this change occurs repeatedly, but the time is not fixed. Its duration ranges from 1 to 10 or 20 years, and it is no longer divided into shorter periods with similar characteristics close to its own amplitude. The overall business strength has been improved. The voice and influence in the field of macroeconomic information match China's comprehensive national strength and international status, meet the needs of national strategy and serve the national economic information security.

In economic analysis, the economic cycle is often divided into two periods of contraction and expansion, four stages of prosperity, recession, depression and recovery, and two turning points of peak and valley. In the boom stage, investment demand rose, money and credit increased, production and employment increased, corporate profits increased and expanded to the peak. After the peak of the economic cycle passed, the economy began to decline gradually, which is the recession stage. Focus on building the core business platform, change the development mode, optimize the product structure, and improve the information service level, ability, strength and influence. According to the situation of the United States, gross domestic product (GDP) fell for two consecutive quarters, that is, it entered a recession. In the recession stage, effective demand decreases and production investment decreases, resulting in an increase in unemployment. The increase of unemployment will reduce the income of residents, which will further lead to the contraction of demand, the further decline of enterprise profits and the deterioration of enterprise operation. Concentrate resources and efforts to promote the tackling of the four key economic information business products of Xinhua Finance and economics, Xinhua Silk Road, Xinhua credit and Xinhua index, continuously improve the business quality and service level, drive the overall business development with the four pillar economic information business platforms, and improve the market competitiveness and market share.

When the recession is serious, there is a large amount of idle production capacity, and if the machinery and equipment that are worn and repaired are not supplemented and replaced temporarily, it can also meet the needs of production. The part of the economic cycle close to the trough belongs to the depression stage. Its characteristics are: high unemployment rate, lower consumption level of residents, a large number of idle enterprise production capacity, serious inventory backlog, low profit or even loss, and enterprises are unwilling to add new investment. The business implements high-quality projects, innovation projects, demonstration projects and rooting projects, transforms the growth momentum, innovates the models of product development, market promotion and user service, promotes the marketization and specialization of business, pays equal attention to both inside and outside, and promotes the international development of economic information business. Many banks and industrial and commercial enterprises have declared bankruptcy due to poor turnover. When the economic cycle is at its lowest point, recovery means the beginning of the recovery stage. The factors driving the recovery are diverse. For example, a large number of machines need to be replaced after years of wear and tear, inventories need to be replenished, and enterprise orders increase. The employment rate is gradually rising, and the income and consumption expenditure are increasing. With the increase of production and sales, the profits of enterprises are also increasing. Enterprises are optimistic about the investment prospect.

The types of economic cycles are usually divided according to the standard of cycle length. In terms of macroeconomic background, it has the advantages of strategic positioning and global vision that other information mining methods do not have, so that the station height and goal significance of economic information business development strategy are different. In the process of mining business, we can coordinate multiple departments to work together and obtain first-hand data and materials from relevant departments and institutions in the economic field for basic research and analysis of macroeconomic information business [5, 6]. Although the length of the economic cycle is not described in detail in the definition of the economic cycle, scholars finally made a formal division of the economic cycle according to the standard of cycle length in combination with the phenomenon of the economic cycle. According to this division method, there are generally four types of economic cycles shown in Fig. 2:

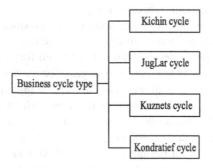

Fig. 2. Types of business cycle

As can be seen from Fig. 2, the economic cycle mainly includes four types: kichin cycle, Jugra cycle and Kuznets cycle. Kichin cycle is summarized by American economist kichin when inspecting bank clearing and wholesale price data in Britain and the United States. The length of this type of economic cycle is about 2–4 years. Generally speaking, it is mainly believed that the cycle of enterprise inventory investment leads to the production of kichin cycle. Therefore, this economic cycle is also called inventory cycle, which is a short cycle. Transfer the backbone of the collection and editing business, reorganize the integrated information collection and editing center, realize the overall collection, editing and distribution of macroeconomic information, and ensure the overall supply of self collected original information of key business products [7, 8].

Based on the statistics of bank loans, interest rates and prices, this type of economic cycle is found in the process of studying the changes of industrial equipment investment in developed countries such as Britain, the United States and France. The duration of this cycle is 7–11 years, and the cycle fluctuation corresponds to a life cycle of investment products that cause fluctuations in GDP, inflation rate and employment. According to the actual needs of business development, complete and strengthen full-time information collectors step by step and with emphasis, and strengthen the overall command and unified dispatching of information collection. In addition, the Jugra cycle belongs to the economic cycle in commercial and trade activities, so it is also regarded as a "real business cycle". At present, most foreign scholars have reached the following conclusions through research: it is precisely because of the fluctuation of unemployment and price with the investment in machinery and equipment that the zugra cycle is produced, which leads to the periodic fluctuation of about 10 years. Now this conclusion has been widely accepted, so the zugra cycle is also known as the "equipment investment cycle". Develop information collectors outside the society in key regions and industries, and expand information collection sources by means of market-oriented remuneration. Strengthen the selection and employment of chief economic analysts in subdivided fields, establish an effective assessment and incentive mechanism, and give full play to the advantages of information analysis. The duration of Kuznets cycle is about 15–20 years. If the average duration of Kuznets cycle is 20 years, this situation is usually considered to be caused by the cyclic change of construction activities. Therefore, Kuznets cycle is also known as "construction cycle". Strengthen the collection and compilation of original information, exclusive information and authoritative information, and improve the quality of economic information collection and compilation. Build an intelligent technical data system for economic information business. Establish a technical committee to coordinate technology construction and projects. In the process of studying the iron consumption of Britain and Germany, kantraev found that this type of economic cycle grows for about 40–60 years. Actively participate in the construction of the company's big data center, accelerate the implementation of the data Lake project, realize the interconnection and sharing of data resources, and improve the efficiency of resource use. Build an integrated information collection, compilation and distribution system to provide technical platform support for the overall collection, compilation, distribution and sharing of information. There are many reasons for Kant raev cycle, but it is generally believed that the main reason is the change of economic structure caused by technological change. After investigating the economic development process of nearly 200 years, kantraev concluded that three epoch-making technological revolutions led to three long cycles.

2.3 Mathematical Model of Social Network Construction

The research of social network began from anthropology. Anthropology is a discipline that makes a comprehensive study of human beings. As individual members of society, each human gradually forms a relatively stable relationship system through various ways of communication and interaction. These relationship systems are social networks. Social networks provide users with various services. Users can enjoy various services brought by social networks, such as making friends, obtaining information, games, instant messaging and so on. The basic fact of social network is that each member has more or less relationship with other members.

A social network is actually a collection of multiple nodes and connections between nodes. Each node represents a member, and each connection represents the relationship between the two members connected. In the era of Web 2.0, the value of users has been gradually released. Users will generate a large amount of data through the network, especially social networks, every day, such as microblogging, commenting, scoring goods, praising articles or news, checking in at places of interest, clicking/browsing favorite goods or news, collecting and purchasing products of interest Upload pictures on the journey to social networks, upload videos taken by yourself to video websites, etc. By calculating the shortest distance between nodes, the degree of distance between nodes is obtained. The higher the compactness and centrality of a node, the shorter the distance between the node and other nodes. The calculation method of the compactness centrality of a node is to calculate the sum of the distances between a node and other nodes, and then take the reciprocal of the calculation result, which is the compactness centrality score of the node. The specific calculation process is shown in formula (1):

$$T(\varepsilon) = \frac{\sum_{i=1}^{n} p + \varepsilon_i}{L} \tag{1}$$

In formula (1), p represents the total number of nodes, ε represents any node, and L represents the shortest distance. With the rapid development of the Internet, contacting and sending information on the network has become an important way for people to communicate. Social network has been widely used in various fields, such as mining network social relations, exploring key factors and tracking information flow. There are also many practices to judge, speculate and explain information behavior and attitude through network analysis. The concept of user-centered has been continuously penetrated in various industries and fields. Enterprises and governments have realized the importance of analyzing and grasping user behavior and preferences. Social network analysis involves various industries and fields. As an interdisciplinary method, it has become an important research paradigm in practical application. In the social network diagram, nodes are used to represent the disseminator and receiver of information, the direction of information transmission is indicated by the direction of arrows, and the thickness of arrows is used to represent the frequency and amount of information transmission. The intermediary centrality of a node is judged by measuring the number of shortest paths passing through the node. The more shortest paths a node passes through, the higher the mediation centrality of the node. Specific calculation process (such as formula 2):

$$G = \sum_{i=1}^{n} \frac{t \times \varepsilon_i}{\sqrt{t+1}} \tag{2}$$

In formula (2), t represents the number of shortest paths. At present, the network has become a feature of the information age. The distributed storage of data on the network is a common way of data storage. There are generally two methods for data crawling.

One is to obtain data directly according to the API interface provided by the platform, and the other is to develop crawler programs according to the crawling objects, simulate user behavior, analyze web pages, and finally get the required information. Although there are a large number of existing methods based on rough set theory, most of them always assume that the analyzed data is stored centrally, and can not directly deal with the distributed stored data. In most cases, websites do not want data to be obtained by third parties, so most websites do not provide open interfaces. Therefore, crawler programs need to be developed to crawl data.

In the distributed data environment, it is generally necessary to store the distributed data in a unified format through the data integration process, so as to make use of the existing rough set methods. Raw data integration requires a lot of space overhead. At the same time, considering the possible heterogeneity of distributed data and the need to maintain and process data changes to maintain data consistency, raw data integration is an expensive work. Crawler scheduler is the main control program of the whole crawler. This program is mainly responsible for scheduling each module in the whole crawler program and controlling the execution and data flow of each module. In addition, it also needs to simulate the behavior of the browser and control the access frequency in the crawling process.

As the preprocessing part of these existing methods, data integration and data maintenance affect the performance of these methods. The URL manager is responsible for managing the URL information in the whole crawling process, including the crawled URL and the URL information to be crawled. When a new URL is found, the manager needs to first judge whether the URL has been crawled. If it has been crawled, it will be discarded, otherwise it will be added to the crawling queue. Moreover, this integration and maintenance of raw data is only a simple integration and maintenance, and does not provide more direct and effective help for rough set methods, such as improving performance. The social network will be divided into modules, and the distinction between each module is very high. The specific calculation process is shown in formula (3):

$$W = \frac{\left[\frac{1}{\eta} - \frac{2}{\varphi}\right]}{2} \times R \qquad (3)$$

In formula (3), η represents the adjacency matrix of the network, φ represents the total number of connections, and R represents the total number of degrees. Therefore, it is necessary to study the less expensive distributed data processing mechanism. The page loader is mainly responsible for requesting the page in the crawling process. The module needs to request according to the request information assigned by the crawler scheduler and the URL given by the URL manager, and obtain the request result. HTML parser mainly realizes the parsing of web pages and parses them into objects convenient for program operation.

2.4 Design Intelligent Mining Algorithm

This section focuses on the main steps of designing macroeconomic information intelligent mining algorithm. Step (1) problem definition: first analyze the needs of the project, clarify the problems to be solved, whether they belong to classification problems, regression problems or other problems, determine the goal of information intelligent mining, and formulate information intelligent mining plan. Attribute reduction and kernel are the two most important concepts in rough set theory. Attribute reduction is a set of conditional attributes obtained by deleting unnecessary and redundant attributes while maintaining the classification ability of the system. It can simplify information and form knowledge without losing the basic information of the system. Step (2–3): data collection and data preprocessing: collect the data to be collected through the information collection tool and save it to the system. Information mining assumes that the data has been mined and stored in the relational database, and information mining mainly focuses on the mining of natural language documents rather than structured data [9]. The comparison results of two attributes are defined by similarity. However, this paper defines two types of attribute values, unique value attribute and list value attribute. The calculation methods of attribute value similarity of these two types are different. The calculation formula of attribute similarity of unique value is:

$$Y(i,j) = \frac{|i,j|}{D_{\max}} \times \frac{1}{D_{i,j}} \qquad (4)$$

In formula (4), D represents the number of the same characters between i and j. . Because there are often many errors in real data, such as the loss of some key data, the collection of wrong data, or the different display of data corresponding to specific problems in different places, these data may cause the distortion of the results. According to whether to distinguish between decision attributes and conditional attributes, attribute reduction can be divided into absolute attribute reduction (not distinguishing between conditional attributes and decision attributes) and relative attribute reduction (distinguishing between conditional attributes and decision attributes). Therefore, the questionable data will be cleaned, the meaningless or valuable data will be filtered out, the missing information will be filled, and the data will be processed into the data form in information intelligent mining, that is, data preprocessing. Step (4) data modeling: select the appropriate model from the clustering model, regression model, classification model, etc., and use specific tools such as statistics, swarm intelligence algorithm, connection model, prediction model and rule-based reasoning to process the data and find the information of interest to users. For the attribute similarity calculation of list value, if the calculation method of table value attribute similarity, such as formula (4) i and j, is the value of list value attribute, the attribute similarity of list value is defined as:

$$S(i,j) = \frac{1}{Q_{\max}} \sum_{i=1}^{n} \max \frac{D}{\delta_{i,j}} \qquad (5)$$

In formula (5), δ represents the maximum list length and Q represents the minimum list length. According to these problems, this paper puts forward the basic architecture of text information mining system. The text information mining system architecture proposed in this paper, in which the bottom layer is the information extraction layer, which is the main data source of the whole macroeconomic information mining algorithm. Step (5) interpretation and evaluation: analyze and evaluate the quality of the results obtained by information intelligent mining, and explain the reasons for such results by analyzing the models and tools used. The top layer is the data processing layer, which mainly cleans the data obtained from information extraction, calculates the cleaned data using the data processing model, and finally obtains the information needed in this paper. Step (6) implement and deploy the model: use computer graphics or image processing technology to transform the results of intelligent information mining into a model that is easy to accept or understand by users. The middle layer is the data storage layer. Due to crawling a lot of data, the data needs to be reasonably stored for subsequent data processing. The structured data in this paper is mainly stored in mysql, and the data is added, deleted, modified and queried through a unified interface.

To sum up, the algorithm flow of intelligent mining of macroeconomic information in this paper is summarized as follows:

Step 1: problem definition
Step 2: data collection
Step 3: Data Preprocessing
Step 4: Data Modeling
Step 5: Interpretation and evaluation
Step 6: implement and deploy the model

Intelligent information mining can infer the general direction of the development of some things or situations in the future, so that enterprises can actively make judgments. Using intelligent information mining to find patterns and relationships in data can help enterprises make better business decisions. Therefore, mining useful information from unstructured and semi-structured text data is an important task in text mining.

3 Experimental Study

3.1 Setting up Experimental Environment

Experimental running environment: 4-core 3.20 ghz CPU, 4 GB memory computer, operating system is Ubuntu 14.04 64bit, IDE is eclipse, programming language is lava, CPU: Intel Core i5 CPU (2-core, 2.70 ghz), memory: DDR3 8 GB, hard disk: 100 GB. The termination condition is set to 10000 iterations and the population size is set to 20. Under this condition, the experimental test is carried out.

3.2 Experimental Process

In the above environment, the macroeconomic information in a website is mined. The mining results can not be directly used for subsequent analysis and processing, which is mainly caused by the difference in the quality of the original data of each data source, specifically: 1) the electronic data is subject to the management level and execution, which makes it difficult to unify the format and standard; 2) It is inevitable to make mistakes in data entry, which is prone to the problem of missing or inconsistent information. Therefore, it is necessary to clean the integrated data so as to eliminate noise and facilitate pattern recognition in the later stage. For the preprocessed data, the method in this paper, the macro-economic information intelligent mining algorithm based on big data and the macro-economic information intelligent mining algorithm based on association rules are used to mine and test.

3.3 Experimental Results

Select the macroeconomic information intelligent mining algorithm based on big data and the macroeconomic information intelligent mining algorithm based on association rules to compare with the macroeconomic information intelligent mining algorithm in this paper. According to part 3.2, after data cleaning, the disk related attributes are recorded in the data. The disk information of each virtual tape library can be recorded by name and target in the information table_ The values of the three attributes ID and entity are merged to build a new attribute. The running time of the three algorithms is shown in Tables 1, 2 and 3.

Table 1. Data set chess running time (s)

Number of experiments	Big data	Macroeconomic information mining	In this paper
1	23.146	22.088	15.602
2	22.581	23.646	14.313
3	21.622	22.584	15.611
4	23.602	21.336	14.902
5	24.158	22.502	13.722
6	23.645	23.691	13.844
7	22.135	24.188	14.601
8	20.018	25.316	15.007
9	22.607	24.101	14.355
10	21.313	22.133	15.206

It can be seen from Table 1 that the running time of the macroeconomic information intelligent mining algorithm in this paper and the other two algorithms are 14.716 s, 22.483 s and 23.159 s respectively; It can be seen from Table 2 that the running time of the macroeconomic information intelligent mining algorithm in this paper and the other two algorithms are 24.909 s, 41.082 s and 41.413 s respectively; It can be seen from Table 3 that the running time of the macroeconomic information intelligent mining algorithm in this paper and the other two algorithms are 27.006 s, 36.733 s and 34.673 s respectively. According to the above data, the running time of the proposed algorithm is shorter. The main reason is that the proposed algorithm extracts environmental factors and processes incremental data, which improves the data mining effect.

Table 2. Running time of dataset mushroom (s)

Number of experiments	Big data	Macroeconomic information mining	In this paper
1	42.615	43.612	26.485
2	39.348	42.154	27.111
3	41.202	41.202	25.199
4	40.165	39.774	24.612
5	41.302	41.225	25.747
6	40.225	38.633	26.551
7	39.688	41.203	25.442
8	41.206	42.371	23.669
9	43.522	41.648	22.501
10	41.549	42.309	21.774

Table 3. Data set connect running time (s)

Number of experiments	Big data	Macroeconomic information mining	In this paper
1	35.211	35.481	25.613
2	36.821	32.164	26.548
3	37.125	33.588	27.136
4	38.914	35.156	28.414
5	37.205	32.011	27.120
6	36.442	34.642	25.842
7	35.194	36.158	26.311
8	36.253	38.146	27.166
9	38.015	35.161	28.545
10	36.146	34.225	27.363

4 Conclusion

The macro-economic information intelligent mining algorithm designed this time sub-divides the types of incremental data according to the characteristics of incremental data and some information saved in the original calculation process. Data cleaning is divided into two parts: attribute value cleaning and duplicate record cleaning. At the same time, the impact of incremental data on the original maximum distribution reduction and rule set is effectively reduced. In the future, further research will be carried out on incremental learning in the algorithm.

References

1. Wang, Y., Wang, N.: Automatic early warning system of economic management benefit loss risk based on data mining. Autom. Instrum. (4), 133–136 (2020)
2. Hu, J.: Analysis on the application of data mining technology in economic statistical investigation. China Sci. Technol. Overv. (16), 20–21 (2021)
3. Gao, M.: National accounts in view of macroeconomic supply side. Stat. Res. **37**(2), 15–25 (2020)
4. Yin, D., Zhang, W.: A review of current macroeconomic nowcasting research. Stat. Inf. Forum **35**(1), 121–128 (2020)
5. Ma, Y., Zhang, Z.: Fixed assets information and macroeconomic growth forecasting. Account. Res. (10), 50–65 (2020)
6. Xiao, Q., Chen, H., Zhang, Y., et al.: The impacts of carbon tax on China's macro-economy and the development of renewable energy power generation technology: based on CGE model with disaggregation in the electric power sector. China Environ. Sci. **40**(8), 3672–3682 (2020)
7. Li, W., Jiao, J.: From the vacant, should-be to the reality: on the formation of macroeconomic regulation and control in China. J. Xiamen Univ. (Arts Soc. Sci.) (4), 22–38 (2020)
8. Wang, S., Fan, C.: Heterogeneity analysis of the impact of financial cycle on macroeconomic stability. Econ. Theory Bus. Manag. (10), 4–20 (2020)
9. Lu, J., Fan, Z., Liu, C.: Supply chain information oriented mining model based on TF-IDF algorithm. Comput. Simul. **38**(7), 153–156, 349 (2021)

Nonlinear Time-Varying Weak Signal Enhancement Method Based on Particle Filter

Zhiming Li[✉]

Nanjing Vocational Institute of Railway Technology, Nanjing 210031, China
lizhiming_zhiming@163.com

Abstract. Traditional weak signal enhancement methods have low signal-to-noise ratio, which affects the accuracy of weak signal recognition. Therefore, this paper proposes a nonlinear time-varying weak signal enhancement method based on particle filter. Collect nonlinear time-varying weak signals, extract nonlinear time-varying diffusion coefficients using particle filter, simulate the real distributed sampling process, build an adaptive neural network model, use subtractive clustering algorithm to determine the selection and number of hidden layer neuron centers, and improve the weak signal enhancement mode. The experimental results show that the signal-to-noise ratio of this method is up to 37.303 db, which shows that the nonlinear time-varying weak signal enhancement method designed by combining particle filter algorithm is more effective.

Keywords: Particle filter · Nonlinear time-varying · Weak signal · Signal enhancement · Signal-to-noise ratio · Background noise

1 Introduction

As an important part of the channel, the main purpose is to transmit the signal sent by the sending device to the receiving device. Coherent detection technology utilizes two opposite characteristics of coherent signal and incoherent noise, and will remove the noise part that is different from the signal phase. However, the channel gives the signal both the chance to successfully reach its destination and the risk that it will be riddled with all kinds of noise or interference. In the narrowband detection technology, based on the known signal frequency as a fixed value, a narrowband filter that limits the system bandwidth is designed to filter out the noise outside the bandwidth, so as to achieve the effect of suppressing noise and enhancing weak signals. Usually, noise is random and comes in a variety of forms. A common problem with signals passing through any communication channel is the presence of additive noise. Generally, additive noise is caused by the internal components of the communication system, such as resistors and solid-state devices. This noise is also known as thermal noise. Coherent and narrowband detection techniques are commonly used to process weak periodic signals in the frequency domain. When the weak signal is a pulse wave and the spectrum is a very wide segment, the coherent and narrowband detection technology for

W. Fu and L. Yun (Eds.): ADHIP 2022, LNICST 469, pp. 201–216, 2023.
https://doi.org/10.1007/978-3-031-28867-8_15

weak signal enhancement will fail [1, 2]. Other sources of noise and interference are caused by outside the system, such as harsh transmission environments. Under normal circumstances, the communication system has a certain anti-noise performance, but if the noise intensity is too large, it will interfere with normal communication, resulting in the inability to effectively and completely transmit and receive information. Since the randomness of noise is manifested in the positive, negative and magnitude of the amplitude, the noise interference can be eliminated to a certain extent by sampling the noisy signal multiple times point by point and averaging, thereby realizing the extraction and detection of weak signals. There are two ways to understand the "weakness" of a signal: one is that the amplitude or intensity of the target signal is too weak relative to the noise background, that is, the signal-to-noise ratio (SNR) is low. The other is that the amplitude of the signal itself is small. When the measured signal is irrelevant to noise, it can adaptively adjust the filter to the optimal state through successive iterations, so as to improve the extraction and detection performance of weak signals under low signal-to-noise ratio. Adaptive noise cancellation technology is suitable for the cancellation of both broadband and narrowband types of noise. Therefore, it has been widely used in signal detection, radar, sonar, array processing, oceanography and other fields. In a communication system, the signal-to-noise ratio is an important indicator to measure the reliability of communication quality. Since there is always noise in the actual received or measured signal, in the actual signal, the expected power of the useful signal component and the objectively existing noise power The ratio is always a finite value, which is the signal-to-noise ratio. The higher the signal-to-noise ratio, the stronger the anti-noise ability of the signal during the transmission process, and the higher the "fidelity".

Relevant scholars have studied this. For example, reference [3] proposes a method to enhance the characteristics of weak current signals of power distribution devices based on vector control, constructs a resonance model of weak current signals of power distribution devices through nonlinear coupling, obtains the characteristics of weak current signals of power distribution devices according to wavelet transform, realizes signal reconstruction using reversible transformation, and completes the enhancement of weak current signal characteristics through vector control. This method can reduce the line distance, but the signal-to-noise ratio is low.

In view of the above problems, this paper designs a nonlinear time-varying weak signal enhancement method based on particle filter.

2   Nonlinear Time-Varying Weak Signal Enhancement Method Based on Particle Filter

2.1   Analysis of Weak Signal Detection Technology

Weak signal detection technology is not only a subject but also a comprehensive technical means. With the deepening of research, it is gradually applied in many fields, including biomedicine, astronomy, seismology, mechanical fault detection, etc. Weak signal detection is a multi-disciplinary comprehensive application of detection methods. By using different methods to study and analyze the statistical characteristics of signals and noise, and use a variety of signal processing methods to analyze and process the input signal, the weak signal can be changed from strong noise to strong noise. In order to meet the requirements of high-precision detection technology required by modern scientific research and technical applications. Weak signal detection is to use the correlation principle to calculate the autocorrelation and cross-correlation of the signal containing noise, so as to extract the useful signal submerged in the noise [4]. When there is only noise and the signal to be measured, the noise is not correlated before and after the time axis and is also not correlated with the signal to be measured, and the signal to be measured has periodicity. After the autocorrelation calculation, the frequency value can be retained to filter out the noise. The characteristics of weak signal detection are as follows: the signal-to-noise ratio of the detected weak signal is relatively low. On the one hand, the measured characteristic signal itself is very weak, and on the other hand, it is disturbed by strong noise. If the failure of mechanical equipment is in the early stage, various characteristic signals caused by the failure are often mixed with the signals of other characteristic sources in a special way, which causes the characteristic signals to become relatively weak. When the two periodic signals have the same fundamental frequency, the useful signal in the extracted noise can be processed by cross-correlation. During cross-correlation, the fundamental frequency can be retained and the common harmonics can be left to remove the noise uncorrelated with the signal.

2.2 Detecting Weak Signals

The ratio of the power of the system output signal to the average power of the noise under the same frequency background noise is the output signal-to-noise ratio of the stochastic resonance system. The signal-to-noise ratio can be expressed by formula (1):

$$G \approx \frac{\pi}{2} \sum \left| \frac{\alpha - 1}{E} \right|^2 \tag{1}$$

In formula (1), α represents the average transition frequency, and E represents the number of times the potential barrier is flipped. It can be seen from the above formula that there is a nonlinear relationship between the output signal-to-noise ratio and the noise intensity, that is, it increases and then decreases with the increase of the noise intensity. When the noise intensity is close to 0, the signal-to-noise ratio should be 0. When the noise disappears, the output signal-to-noise ratio of the system should be infinite, which shows that the assumptions of the adiabatic approximation theory have certain limitations. If the device is running, the signal is also mixed with strong noise interference. The detection of signals has rapidity and real-time requirements. In engineering practice, it is necessary to collect and analyze signals, and the length of data or the duration of the acquisition process is limited, such as the collection of weak signals in the fields of pipeline leakage, communication facilities, earthquake detection, industrial measurement, and real-time monitoring of mechanical systems., the length of the collected data is short, so there are certain requirements for the rapidity and real-time detection. The relevant detection principle is simple, easy to implement, and has high detection efficiency, and can be widely used in actual engineering signal detection. Although this method has the above advantages, most of the noises are assumed to be uncorrelated in the detection, and the noise with small time interval in the actual detection may be correlated. Phenomenon. The characteristics of adaptive filtering are very prominent: in order to adjust the filter to the best filtering state, the parameters of the system can be adaptively adjusted according to some optimal criterion. The input and output power spectrum amplification factor is defined as follows:

$$H = \frac{\left|\frac{\alpha-1}{E}\right|^2}{\phi} \times \frac{1}{\alpha} \tag{2}$$

In formula (2), ϕ represents the response function. This can be achieved even without prior statistical knowledge of the signal and noise, and even when the statistical properties of the input signal change, the filter can satisfy the adaptive "learning process" by adjusting its own parameters. When the statistical characteristics of the input signal change, the process that the filter can adjust its own parameters to the optimal value is called the "tracking process", which reflects the ability of the system to learn and track. By constantly "screening" the signal, the different frequency components in the signal containing noise are decomposed into different modes, and several modes and residuals are formed. The EMD time-frequency analysis method is superior to the classical method in processing capability and effect in many application fields.

2.3 Extraction of Nonlinear Time-Varying Diffusion Coefficients by Particle Filter

Particle filter algorithm is the application of Monte Carlo method in Bayesian estimation. The algorithm uses the random state set of system state variables to obtain statistical indicators such as expected value and variance corresponding to the dynamic process. It is an algorithm based on statistical principles. The elements in these sets of random states are called "particles". Particle filtering can be used for systems described by any state space. Its core lies in constructing a posterior probability density function. The constructed posterior probability density function needs to reflect the real probability distribution. The constructed posterior probability function is sampled to simulate the real probability distribution. Distribution sampling process. The working process of nonlinear time-varying diffusion coefficient extraction is extremely complex, including the selection of the number of network layers, the selection of the number of nodes in each layer, and the determination of the connection mode between the transfer function of each node and the nodes. The PF algorithm based on the SMC structure recursively estimates the nonlinear system by continuously generating a series of particles with weights (weights). The weight of the particle is used to measure the conformity of the particle to the measured value after the transformation of the measurement equation. After the model structure of the extracted system is determined, the parameters in the model results must be extracted. In general, the extraction process is that the particle filter can learn and adjust the parameters to minimize the objective function. Among them, the particle filter parameter extraction is the most commonly used. The objective function of is the mean-squared error function. When the system input signal and noise exist, the bistable can be described by its Langevin equation. The specific expression formula is:

$$\varepsilon = \frac{\phi + \lambda + \eta}{2} \tag{3}$$

In formula (3), ϕ represents the periodic driving force, λ represents the potential barrier height, and η represents the system parameter. Then the algorithm calculates the required posterior distribution through this series of particles and their weights. Considering that the recursive estimation is convenient for computer processing, the nonlinear time-varying diffusion coefficient obtained by sampling must have a sequential relationship. Since Bayesian importance sampling requires the data of the state of the particle at all times when estimating the state, when the latest observation data comes, the weight value of the entire state sequence needs to be recalculated. In each recursive calculation step of the algorithm, more and more particles will deviate from the estimated true value, and their weights tend to 0. In order to avoid this situation, the algorithm needs to constantly eliminate particles with small weights, keep particles with large weights and multiply them to ensure a sufficient number of particles. With the increase of time, the amount of computation tends to be complex, so sequential importance sampling is introduced to solve it. When the latest observation data comes, it is only necessary to add the particles sampled at this moment to the existing particle set. On the basis of formula (3), the average mutual information of input and output is obtained, and the specific expression formula is:

$$D = \sum \frac{d(\varpi - 1)^2}{\|R\|} \tag{4}$$

In formula (4), d represents the information entropy of the output, ϖ represents the conditional entropy, and R represents the conditional probability. Online extraction is completed in the actual operation process of the extraction system, so this extraction process has the characteristics of real-time. The offline extraction can complete the relevant learning and training in advance before the particle filter works on the extracted system. The "array enhancement effect" formed by the combination of multiple parallel nonlinear units is found in weak signal processing devices. The parallel formation of a stochastic resonance array network for the same nonlinear stochastic resonance subsystem can amplify the output response of the system. Extended to array stochastic resonance theory. However, because the input and output training sets of its network are difficult to cover all possible working ranges of the system, and it is difficult to adapt to the parameter changes of the system during the working process, in order to overcome the shortcomings of these two extraction methods, two methods can be used. Extraction methods are used in combination. In an uncoupled parallel array network composed of arbitrary static nonlinear sub-modules, the SNR gain expression for weak periodic signals under strong background white Gaussian noise is derived, and it is demonstrated that the SNR gain is very important for a given stochastic resonance sub-module type. And the noise intensity increases monotonically with the increase of the number of arrays. Firstly, discrete extraction is performed, the weight matrix of particle filter is obtained through discrete training, and then online extraction is performed, and the obtained weight matrix is used as its initial weight through online learning, which is conducive to speeding up the learning process of online extraction. Due to the self-learning and self-adaptive characteristics of particle filter itself, when the characteristics of the extracted system change to different degrees, particle filter can adaptively track the extraction system by continuously adjusting the network connection weights or thresholds. Trend. It is worth noting that when the noise intensity in the array network is optimal, the system SNR gain cannot be further enhanced. Since the optimal array performance in reality is difficult to achieve, people can often generate more superior system response by increasing the number of array elements and adding array noise to the parallel array network composed of sub-optimal nonlinear submodules. The series-parallel model is realized by time-delay particle filter, and the parallel model is realized by internal time-delay feedback particle filter and output feedback network. Since the structure of the series-parallel model uses the input and output signals of the extracted system as the extraction information, and its network training can ensure the convergence and stability of the extracted model, the series-parallel model structure is used in particle filter extraction. Widely used. With a sufficiently large number of array elements and different types of noise, the input-output signal-to-noise ratio gain of the threshold nonlinear array proved to be able to be greater than 1.

2.4 Building an Adaptive Neural Network Model

The feedback particle filter is used for the identification and control of nonlinear time-varying systems, which can fully reflect the time-varying characteristics of nonlinear systems. Therefore, the research on the identification and control of nonlinear time-varying systems based on feedback particle filtering not only has important theoretical innovation value, but also has great guiding significance for practical engineering applications. In practical applications, the dynamics of physical systems change continuously with running time. In order to better describe the time-varying characteristics of nonlinear time-varying systems, time-varying neural networks have become an important choice for system identification. Then, the conventional RBF network and the time-varying RBF network are respectively used for the multi-input multi-output nonlinear time-varying system identification. After the network structure, input and state dimensions are determined, the network weights are adjusted by an iterative learning algorithm with the help of the repeated operation process to conduct network training. Then, the effectiveness of the proposed enhancement algorithm is verified by real-time simulation of the specific nonlinear system. After the network structure is determined, weight adjustment becomes the key to time-varying network applications. After the network structure, input and state dimensions are determined, the subtraction clustering algorithm is used to realize the selection of the hidden layer neuron center and the determination of the number. Considering the approximation error, the weights of the conventional RBF network are adjusted according to the integral learning law with dead zone correction. Considering the coexistence of periodic signals and random excitations, the overdamped motion of Brownian particles in three potential asymmetry cases can be described as the following Langevin equation:

$$\frac{h(\mu)}{g} = -\sum \frac{1}{\sqrt{|\mu - g|^2}} \tag{5}$$

In formula (5), h represents the amplitude of the external periodic modulation signal, μ represents the frequency of the external periodic modulation signal, and g represents the additive white Gaussian noise. And, formula (5) satisfies the following conditions:

$$\begin{cases} \frac{\mu}{g} = \frac{h^2}{2} \\ \sigma \langle h + g \rangle = \sum 2\frac{\mu}{T} \end{cases} \tag{6}$$

In formula (6), T represents the intensity of additive Gaussian white noise. Finally, the particle filter based enhancement algorithm is applied to the modeling of temperature and humidity in the process of wood drying. The effectiveness of the proposed enhancement algorithm can be further verified by comparing the real value of temperature and humidity in the process of wood drying with the simulation results. For time-varying RBF networks, learning from the idea of iterative learning, a semi saturated iterative learning law with dead zone correction is used to complete the training of network weights. Then, the convergence performance of identification error in each case is analyzed by Lyapunov like method. The theoretical analysis shows that with the increase of the number of iterations, the system identification error gradually converges to the given boundary value (the boundary value depends on the dead band range). As a kind of dynamic system model, nonlinear time-varying system is widely used in practical engineering field. However, due to the nonlinear dynamic characteristics of nonlinear time-varying systems, people can not obtain the dynamic model of the system in most cases. Therefore, when using control theory to solve practical problems, establishing the dynamic model of the system is the key to the successful application of control theory in production practice, and the task of system enhancement is to establish the dynamic model of the system. Finally, the conventional RBF network and the proposed time-varying RBF network are used to simulate the nonlinear time-varying system, which further shows the effectiveness of the proposed identification algorithm. Nonlinear time-varying systems have complex nonlinear dynamic time-varying characteristics. How to realize the effective modeling and identification of such systems has been widely studied. With its own strong approximation ability and generalization ability, neural network can complete the training of network weight by measuring the input/output of the system without predicting the system model, so as to realize the effective enhancement of nonlinear system. Therefore, particle filter has been widely used in the enhancement and control of nonlinear system [5]. Dynamic neural network has become an important choice. As a typical structure of neural network, the structure of dynamic neural network is different from feedforward neural network. Dynamic neural network introduces feedback mechanism to train network weights with system input and output data obtained by on-line measurement. Dynamic neural network has been successfully used in nonlinear system identification. The essence of particle filter enhancement is to transform the system enhancement problem into the approximation problem of specific nonlinear function, and the function approximation problem is one of the most basic problems in the research of particle filter. Particle filter trains the network weights by collecting the system input/output data, and realizes the effective approximation of the system output by minimizing the system output error function, so as to complete the whole signal enhancement process.

2.5 Improved Weak Signal Enhancement Mode

For approaching any complex nonlinear function, through theoretical analysis and practical verification, it is concluded that when the network has about 10 middle layer components, after "sufficient" learning, it can meet the identification accuracy requirements in most cases. The core of weak signal enhancement is that the nonlinear stochastic resonance system driven by weak signal and noise can achieve the best synergy among the system input signal, noise and nonlinear stochastic resonance system by adjusting and optimizing the system parameters, so as to make the stochastic resonance system at the resonance point, so as to enhance the weak input signal of the system [6, 7]. At the same time, when the network has the simplest structure, it is also conducive to practical application and meet the high real-time requirements of online identification. Therefore, the simplification of network scale is very necessary. This theory avoids the defect of traditional stochastic resonance theory that excites stochastic resonance by adding noise, and the way of adjusting system parameters is easier to operate and realize. SPN is defined as a network with multiple sensor nodes. The weight attenuation method is a common pruning method and belongs to the regularization method. Its working mechanism can be explained from the perspective of a priori distribution, that is, the minimization of loss function is equivalent to the maximization of a posteriori probability of weight parameters, which confirms the simplest principle of network structure design: for the network that has reached a given training accuracy, the fewer effective parameters, the better the generalization ability, Thus, it provides a theoretical basis for the rationality of designing the minimum structure network. It should be noted that the signal-to-noise ratio mentioned here is different from the concept of signal-to-noise ratio in the field of communication. It is for single frequency signals, that is, the signal is in the form of single peak impact in the frequency domain. Therefore, its definition formula is:

$$Y = \lim \int_{n=1}^{m+\Delta k} \sum \frac{km}{V(n)} \tag{7}$$

In formula (7), k represents the power spectral density, m represents the noise intensity near the signal frequency, n represents the cross-correlation coefficient, and V represents the signal amplitude at a specific target frequency. The input of each node is a common input information source of the network and their independent noise, and all these node outputs are fused to the sink Center for processing to obtain the compressed network output, which does not reduce (or slightly reduce) the amount of system mutual trust while compressing the data. Due to the unique properties of sink pool network, SPN shows the output response of redundancy compression. From the relationship between network structure and generalization ability, it can be seen that after simplifying the structure, the generalization ability of the network will be improved, which is also the direct reason why the regularization method can improve the generalization ability. Therefore, this part uses the particle filter algorithm to achieve the goal of reducing the structural complexity of the network. The algorithm introduces the regularization term representing the structural complexity into the objective function of identifying the network. The concept of SPN is applicable to system simulation in biological neural coding, Nano Electronics, distributed sensor networks, digital beamforming arrays, image processing, multiple access communication networks and social networks [8, 9]. Array stochastic resonance system is composed of nonlinear stochastic resonance subsystems of parallel array. Each subsystem is driven by a common input signal and independent and identically distributed noise. The output processed by the subsystem is merged at the fusion center and calculated, and then the output response of array stochastic resonance system is obtained. In order to avoid over fitting, a multidimensional Taylor network with the smallest structure should be designed, that is, if the fitting effect on learning samples is the same, the generalization ability in the average sense with the simplest structure is the best. The signal-to-noise ratio of the input and output signals of the array stochastic resonance system increases first and then decreases with the increase of the array noise intensity. The non-zero noise intensity corresponding to the peak signal-to-noise ratio is the optimal noise intensity of the stochastic resonance system [10]. In particle filter, the shortest description length is used to represent the complexity of machine learning, that is, given the learning data, the optimal model should have the shortest total description length. The lock-in amplifier can detect the amplitude and phase of the signal submerged in the noise. The weak signal enhancement method is composed of four parts, as shown in Fig. 1:

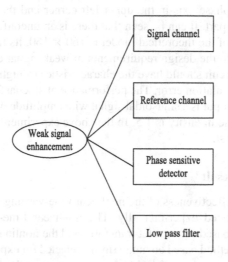

Fig. 1. Weak signal enhancement mode

It can be seen from Fig. 1 that the weak signal enhancement methods include: signal channel, reference channel, phase sensitive detector and low-pass filter. The main idea is to use the phase characteristics of the signal to select the signal with the same frequency and phase as the measured signal as the reference signal, According to the coherence characteristics of the phase sensitive detector, only the signals with the same frequency and phase as the reference signal respond during detection, while other frequency and phase signals are suppressed, so as to achieve the purpose of extracting useful signals. When the number of arrays is large enough, the output signal-to-noise ratio of the stochastic resonance system can be greater than the input signal-to-noise ratio of the system in a certain range of noise intensity range, that is, the signal-to-noise ratio gain of the parallel array stochastic resonance system can be greater than 1. Using this phenomenon can further improve the extraction and detection performance of weak signals in the field of signal processing. The total description length is the sum of description length (data model) and description length (model). The former is the residual of the model, and the latter is used to measure the complexity of the model. Therefore, the shortest description length is an integrated measure to comprehensively evaluate the residual error and model complexity. Its goal is to find an identification network that meets the target accuracy and has the best generalization ability.

3 Experimental Test

3.1 Experimental Preparation

According to the needs of experimental test, ad734 chip of ad company is selected in the circuit design. This chip is a high-precision and high-speed four quadrant analog multiplier divider. This chip is stable and reliable, has high precision when used as multiplier, and is basically insensitive to power supply. The theoretical model includes the inclined linear in-phase axis in the upper left corner and the sine cosine in-phase axis in the lower right part. It can be seen that there is an unconformity between them. The data volume size of the theoretical model is 160×160, its time sampling interval is 1.5 Ms. According to the design requirements of weak signal enhancement method, the multiplier in the circuit should have the characteristics of high signal-to-noise ratio, low drift and small calculation error. The performance of stochastic resonance circuit is tested in Proteus. The input is a sinusoidal signal with amplitude of 0.8 V and frequency of 0.5 Hz, and the noise intensity is 1.5. In the above experimental environment, carry out the experimental test.

3.2 Experimental Result

In order to verify the effectiveness of the nonlinear time-varying weak signal enhancement method, it is tested experimentally. The nonlinear time-varying weak signal enhancement method based on wavelet transform and the nonlinear time-varying weak signal enhancement method based on oma-srm are selected for experimental comparison with the nonlinear time-varying weak signal enhancement method in this paper. Test the signal enhancement effects of the three methods under different carrier frequencies. The larger the value, the better the signal enhancement effect is proved. The experimental results are shown in Tables 1, 2, 3, 4 and 5:

It can be seen from Table 1 that when the carrier frequency is 2500 Hz and the number of experiments is 9, the signal-to-noise ratio of the wavelet transform method is 4.319 dB, the signal-to-noise ratio of the vector control enhancement method is 3.687 dB, and the signal-to-noise ratio of the method in this paper is 6.848 db; The average SNR of the nonlinear time-varying weak signal enhancement method in this paper and the other two nonlinear time-varying weak signal enhancement methods are 7.071 db, 3.769 db and 4.044 db respectively.

It can be seen from Table 2 that when the carrier frequency is 2000 Hz and the number of experiments is 6, the signal-to-noise ratio of the wavelet transform method

Table 1. Carrier frequency 2500 Hz signal-to-noise ratio (DB)

Number of experiments	Nonlinear time-varying weak signal enhancement method based on Wavelet Transform	Vector control enhancement method	Nonlinear time-varying weak signal enhancement method in this paper
1	3.154	4.215	6.649
2	3.448	4.331	7.121
3	3.025	4.587	6.874
4	4.112	3.615	7.205
5	3.697	3.697	7.116
6	4.055	4.005	6.874
7	3.871	3.474	7.252
8	4.216	4.259	7.316
9	4.319	3.687	6.848
10	3.788	4.571	7.451

Table 2. Carrier frequency 2000 Hz signal-to-noise ratio (DB)

Number of experiments	Nonlinear time-varying weak signal enhancement method based on wavelet transform	Vector control enhancement method	Nonlinear time-varying weak signal enhancement method in this paper
1	8.346	8.647	10.202
2	7.994	7.992	11.313
3	8.312	8.313	12.474
4	8.259	7.698	10.948
5	7.314	8.215	12.037
6	8.269	7.994	12.457
7	7.102	8.352	11.362
8	8.315	7.649	11.584
9	7.821	8.152	12.065
10	7.404	7.664	11.488

is 8.269 db, the signal-to-noise ratio of the vector control enhancement method is 7.994 db, and the signal-to-noise ratio of the method in this paper is 12.457 db; The average SNR of the nonlinear time-varying weak signal enhancement method in this paper and

the other two nonlinear time-varying weak signal enhancement methods are 11.593 db, 7.914 db and 8.068 db respectively.

Table 3. Carrier frequency 1500 Hz signal-to-noise ratio (DB)

Number of experiments	Nonlinear time-varying weak signal enhancement method based on wavelet transform	Vector control enhancement method	Nonlinear time-varying weak signal enhancement method in this paper
1	15.649	12.647	19.648
2	13.487	13.008	17.316
3	14.602	15.316	18.487
4	15.717	14.718	18.261
5	13.622	14.949	19.364
6	14.945	13.602	20.157
7	13.687	14.718	18.479
8	15.649	15.602	19.592
9	13.050	16.947	20.087
10	13.487	15.499	19.874

It can be seen from Table 3 that when the carrier frequency is 1500 Hz and the number of experiments is 10, the signal-to-noise ratio of the wavelet transform method is 13.487 db, the signal-to-noise ratio of the vector control enhancement method is 15.499 db, and the signal-to-noise ratio of the method in this paper is 15.499 db; The average SNR of the nonlinear time-varying weak signal enhancement method in this paper and the other two nonlinear time-varying weak signal enhancement methods are 19.127 db, 14.390 db and 14.701 db respectively.

It can be seen from Table 4 that when the carrier frequency is 1000 Hz and the number of experiments is 5, the signal-to-noise ratio of the wavelet transform method is 17.699 db, the signal-to-noise ratio of the vector control enhancement method is 19.648 db, and the signal-to-noise ratio of the method in this paper is 27.055 db; The average SNR of the nonlinear time-varying weak signal enhancement method in this paper and the other two nonlinear time-varying weak signal enhancement methods are 26.715 db, 17.992 db and 18.313 db respectively.

It can be seen from Table 5 that when the carrier frequency is 500 Hz and the number of experiments is 8, the signal-to-noise ratio of the wavelet transform method is 28.316 db, the signal-to-noise ratio of the vector control enhancement method is 23.515 db, and the signal-to-noise ratio of the method in this paper is 33.610 db; The average SNR of the nonlinear time-varying weak signal enhancement method in this paper and the other two nonlinear time-varying weak signal enhancement methods are 32.712 db, 25.514 db and 25.249 db respectively.

Table 4. Carrier frequency 1000Hz signal-to-noise ratio (db)

Number of experiments	Nonlinear time-varying weak signal enhancement method based on wavelet transform	Vector control enhancement method	Nonlinear time-varying weak signal enhancement method in this paper
1	17.366	18.516	25.613
2	18.205	20.061	26.104
3	17.649	18.364	25.818
4	18.215	19.154	26.377
5	17.699	19.648	27.055
6	18.612	18.466	26.145
7	18.347	17.055	28.319
8	18.105	18.031	27.144
9	17.697	19.611	28.069
10	18.021	14.219	26.071

Table 5. Carrier frequency 500Hz signal-to-noise ratio (DB)

Number of experiments	Nonlinear time-varying weak signal enhancement method based on wavelet transform	Vector control enhancement method	Nonlinear time-varying weak signal enhancement method in this paper
1	22.316	23.619	31.205
2	25.169	25.177	29.648
3	24.718	26.384	32.007
4	25.337	27.224	31.482
5	26.914	29.6161	32.907
6	26.822	23.487	33.642
7	25.677	23.441	32.544
8	28.316	23.515	33.610
9	24.018	24.919	34.464
10	25.848	25.106	35.614

It can be seen from Table 6 that when the carrier frequency is 250 Hz and the number of experiments is 10, the signal-to-noise ratio of the wavelet transform method is 28.316 db, the signal-to-noise ratio of the vector control enhancement method is 29.115 db, and

Table 6. Carrier frequency 250 Hz signal-to-noise ratio (DB)

Number of experiments	Nonlinear time-varying weak signal enhancement method based on wavelet transform	Vector control enhancement method	Nonlinear time-varying weak signal enhancement method in this paper
1	26.348	31.201	37.458
2	27.158	29.847	36.225
3	31.205	30.474	38.201
4	30.177	29.645	37.966
5	29.848	30.277	36.288
6	30.255	31.201	37.916
7	31.102	30.224	36.074
8	29.848	29.845	37.508
9	31.025	30.468	38.19
10	29.648	29.115	37.206

the signal-to-noise ratio of the method in this paper is 37.206 db; The average SNR of the nonlinear time-varying weak signal enhancement method in this paper and the other two nonlinear time-varying weak signal enhancement methods are 37.303 db, 29.661 db and 30.230 db respectively.

In order to further verify the signal enhancement effect of this method, wavelet transform method, vector control enhancement method and this method are used to verify the time-consuming signal enhancement. The results are shown in Fig. 2.

By analyzing Fig. 2, it can be seen that the signal enhancement efficiency is different under different methods. When the number of experiments is 20, the signal enhancement time of wavelet transform method is 36 s, the signal enhancement time of vector control method is 57 s, and the signal enhancement time of this method is 18 s; When the number of experiments is 60, the signal enhancement time of wavelet transform method is 60 s, the signal enhancement time of vector control method is 43 s, and the signal enhancement time of this method is 15 s; The method in this paper always has high efficiency of signal enhancement.

4   Concluding Remarks

In this paper, a nonlinear time-varying weak signal enhancement method based on particle filter is proposed. The particle filter is used to enhance the weak signal, extract the nonlinear time-varying diffusion coefficient, simulate the real distributed sampling process, build an adaptive neural network model, use the subtractive clustering algorithm to determine the selection and number of hidden layer neuron centers, and improve the weak signal enhancement mode. The experimental results show that the average signal-to-noise ratio of this method can reach 37.303 db at different carrier frequencies; In

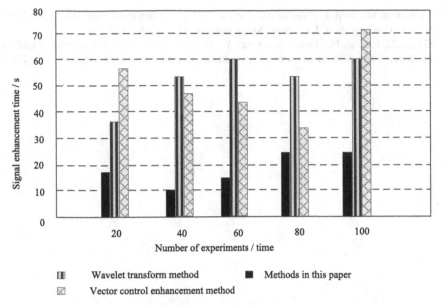

Fig. 2. Signal enhancement time

multiple iterations, the signal enhancement time of this method is no more than 25 s; This method always has high signal enhancement efficiency and improves the signal-to-noise ratio of weak signals. At the same time, it enriches the academic research on weak signal enhancement. In order to further improve the research of nonlinear time-varying weak periodic signals, we need to continue to improve the details in the future.

References

1. Huang, Y., Chen, J., Yan, N.: Adaptive enhancement method for communication weak signal based on single chip microcomputer. Bull. Sci. Technol. **36**(12), 23–26, 31 (2020)
2. Zhu, H., Liu, W.: Method and application of similarity detection for weak signals under impulse noise. Aerosp. Shanghai **37**(4), 141–147 (2020)
3. Zheng, Z.: Characteristic enhancement system for weak current signal of distribution device based on vector control. Electron. Des. Eng. **28**(7), 108–112 (2020)
4. Jin, Z., Cai, J., Yang, X., et al.: Research on weak signal amplification and processing technology applied to laser flash method for measuring thermal diffusivity. Metrol. Meas. Technol. **40**(1), 37–41 (2020)
5. Zhang, C., Sun, Q.: Multi-dimensional Taylor network identification and control of nonlinear time-varying systems with noise disturbances. Control Theory Appl. **37**(1), 107–117 (2020)
6. Chen, C., Wei, W.: Extraction of weak fault signal based on improved LMD and wavelet packet de-nosing. Mach. Des. Manuf. (1), 165–168, 172 (2020)
7. Zhong, W., Liu, C.: Fast multi band weak signal acquisition algorithm in LTE wireless network. Comput. Simul. **37**(12), 144–147, 203 (2020)
8. Wang, H., Li, Y., Li, C., Zhou, Z., Ma, Z., Tian, D.: Research on atomic fluorescence signal enhancement technology based on digital micromirror device. Anal. Chem. **49**(09), 1470–1479 (2021)

9. Li, S., Han, M., Wen, J.: Pipeline micro leakage signal enhancement method based on VMD-SVD self optimization. J. Electron. Meas. Instrum. **35**(12), 68–78 (2021)

10. Guan, Z., Huang, N., Dang, X., Xing, Y.: Blind separation of multi-source signals of mechanical vibration based on weighted distance. Comput. Simul. **38**(10), 397–400 (2021)

Design of Adaptive Multi Stream Transmission Control Method in Social Communication Network

Zhiming Li[✉]

Nanjing Vocational Institute of Railway Technology, Nanjing 210031, China
lizhiming_zhiming@163.com

Abstract. The traditional multi stream transmission control method does not allo-cate the transmission channel, which leads to the problem that the initial buffer loading time is too long. Therefore, this paper designs a new adaptive multi stream transmission control method for social communication networks. Build a direct communication link to obtain the received signals of relay users, build a channel bilateral matching model to allocate the transmission channels, design a multi stream distribution framework, balance the computing load of each edge server, optimize the transmission control mode, and achieve adaptive multi stream transmission control in social communication networks. Experimental results: the initial buffer loading time of the method in this paper is only 6.282 s, which proves that the application effect of the adaptive multi bit stream transmission control method in social communication networks designed this time is better.

Keywords: Social communication network · Adaptive · Multi stream transmission · Control method · Base station · Cellular user

1 Introduction

In the past decade, with the continuous development of mobile communication network and the rapid popularization of smart phones, Internet applications have penetrated into all kinds of life needs of users. Watching videos on mobile terminals through the Inter-net has long been an essential part of people's daily life. With the explosive growth of mobile multimedia services, text, picture, voice, video and other services have become the main body of mobile data service traffic. At the same time, with the rapid devel-opment of the Internet and the continuous popularization of high-performance mobile terminal devices, people's social scope has been expanded to the field of online social networking by using software such as making friends and instant messaging. Network and social software connect people and build a network service form with human society as the core - social network. The development of HD video technology also makes users have higher requirements for video fluency and clarity. In order to meet the increas-ing needs of mobile network video users, adaptive video streaming technology based on mobile terminals is also developing continuously, from the initial Microsoft smooth

W. Fu and L. Yun (Eds.): ADHIP 2022, LNICST 469, pp. 217–231, 2023.
https://doi.org/10.1007/978-3-031-28867-8_16

streaming technology to the current dash technology. Social networks provide a platform for mobile users to share interests, behavior dynamics, social relations and activity status. In social networks, by analyzing users' behavior, we can find some relationship between users, which can be used to describe users' choice of similar content, common interest or other similar behavior. The problem of adaptive multi code streaming transmission control in social communication networks is that HD streaming media content requires large storage capacity, large transmission bandwidth and high requirements for hardware computing power. If only upgrading the hardware equipment of the handheld terminal, it is not only difficult to support the strong business requirements, but also difficult to improve the performance of the hardware equipment to a large extent in a short time. The number of mobile network video users continues to grow rapidly, and the challenges related to device heterogeneity and network heterogeneity that adaptive video streaming technology needs to deal with are also more severe. In some hotspot areas, users are densely distributed, and multiple users may request the base station to download the same content. In the traditional communication mode, the base station needs to repeatedly send this content to these cellular users, but this will greatly increase the burden of the base station. In this case, using social traffic information network to push or share content among users with close social relationship has become a feasible method to reduce the burden of base stations. On the other hand, for video service providers, it is necessary to build a large-scale online video service platform to meet the demand of massive high-definition video. It is necessary to ensure the high reliability of the service and the scalability of the scale, which undoubtedly brings huge capital pressure to video service providers.

Relevant scholars have made some progress in the research of multi stream transmission control. For example, Luo Zhiwei et al. Proposed an embedded multi-channel wireless video transmission rate adaptive algorithm [1]. The dm368 chip acquires multi-channel wireless video data, calculates the video data transmission rate according to the Gaussian function, and realizes the bit rate equalization control of multi-channel wireless video transmission rate through the extremum suppression method. This method can improve the video transmission rate, and this method can improve the buffer capacity of wireless video transmission, But the delay jitter is high. Shao Ruirui et al. Proposed a measurement method of multi stream transmission rate of communication network based on toughness [2], constructed a toughness function to obtain multi stream transmission rate data, and realized multi stream transmission channel selection through transmission rate matching method, which effectively improved the accuracy of multi stream transmission rate measurement of communication network, but there was a problem of too long buffer loading time.

In view of the above problems, this paper designs an adaptive multi stream transmission control method for social communication networks.

2 Design of Adaptive Multi Stream Transmission Control Method in Social Communication Network

2.1 Prediction of Social Communication Network Bandwidth

The cellular communication and communication hybrid network architecture of social communication network is divided into two layers: social layer and physical layer [3]. In the social layer, the dynamic behavior of users on the social platform can be used to reflect the social connections between users. Therefore, users' behavior on social platforms such as microblog and twitter can be used to find the social relationship between users. Communication technology is a new communication technology that allows terminal equipment to directly communicate by reusing the spectrum resources of traditional cellular users in the cell under the control of the base station in the cellular system. It can not only improve the spectrum efficiency of cellular communication system and reduce the transmission power of terminal, but also solve the problem of spectrum resource shortage in wireless communication system. The direct communication link is built in the physical layer, and whether it can be built successfully mainly depends on the transmission distance between two mobile user devices. Each user in the social layer corresponds to a terminal mobile user in the physical layer. In order to realize information push or content sharing through communication links, it is necessary to comprehensively consider the social layer and physical layer information of the system [4, 5]. As we all know, WiFi technology or Bluetooth technology has strong interference with each other when users use it due to the use of unauthorized frequency bands, which will reduce the user experience. Different from the unauthorized frequency band used by WiFi technology or Bluetooth technology, the communication technology can use the authorized frequency band. When the communication technology uses the authorized frequency band to enable the two devices to communicate directly, the base station can assist in the coordination of interference. In channel modeling, Rayleigh fading is used to model small-scale fading, and free space propagation path loss is used to model large-scale fading. The received signals received by the receiving user in the transmission link and the base station in the cellular transmission link are:

$$E_0 = \frac{w \times \alpha^{-1}}{\beta} \times Z^2 \tag{1}$$

$$E_1 = \frac{w \times \frac{1}{\alpha}}{\beta} \times Z \tag{2}$$

In formulas (1) and (2), w represents the transmission power of the transmitting user, α represents the transmission power of the cellular user, β represents the channel response of the communication link, and Z represents the channel response of the cellular communication link. In general, if there is a strong social relationship between two users, the probability of establishing a direct communication link between them for content sharing will be higher, because their content preferences are more similar than two users with weak social relationships. In addition, having better channel conditions between two mobile user devices will promote the construction of direct communication links. In the physical layer, if the distance between two mobile users is within the transmission range required by the communication link construction, and the service quality of cellular users and communication users can be guaranteed, the communication link can be built smoothly. In this system, communication users can share the uplink spectrum resource blocks occupied by cellular users, and each cellular user can only occupy one spectrum resource block, and each resource block can only be multiplexed by one communication user at most, and vice versa. How to establish, continue and end the communication should be controlled by the base station. Communication from establishment to completion can be understood as a process in which users request resources from the cellular system, the base station allocates resources to communication users, the base station maintains the transmission service of communication, and the cellular system recovers spectrum resources. When the bandwidth formed by the transmitting user and the receiving user affects the multiplexing resource block of the communication link, the signal to interference plus noise ratio received by the receiving user and the received by the base station are respectively expressed as:

$$G = \frac{1}{l} \sum H_\delta^2 - H^\phi \tag{3}$$

$$G' = \sum \left| \frac{l - H^\phi}{H_\delta^2} \right| \tag{4}$$

In formulas (3) and (4), l represents the channel response of the interference link between the cellular user and the receiving user, H represents the channel response of the interference link between the transmitting user and the base station, δ represents the free space path loss factor, and ϕ represents the Rayleigh channel factor subject to complex Gaussian distribution. The multiplexing of uplink spectrum resources leads to the common channel interference between the base station and the receiving user. Therefore, the communication quality of cellular communication link and communication link is affected. In the social layer, the higher the similarity of two users' preferences for content, the closer the social relationship between users. However, due to the uncontrollability and uncertainty of users' social behavior, it is difficult to find an appropriate model to describe the characteristics of users' behavior. The user can also obtain the media content from the adjacent user terminal that has obtained the media service with the help of the communication network, so as to alleviate the downlink transmission pressure of the operator's cellular network. In addition, the traditional cellular communication between short-range users can also be switched to the communication mode to realize the unloading of cellular network traffic. Therefore, this section uses the probability of users selecting similar content to represent the similarity of users' social behavior. The higher the normalized correlation of the probability of users selecting similar content, the closer the social relationship between users. Communication technology can not only provide adaptive data service services, especially for data sharing and transmission services in local user communication services, but also flexibly meet business needs in terms of file sharing, information sharing and other services. In order to obtain the probability distribution of users choosing similar content, the system integrates the historical records of users' behavior on different social platforms, and obtains the probability density function of users choosing similar content by using Bayesian nonparametric model, so as to obtain the social relationship strength between users.

2.2 Build Channel Bilateral Matching Model

2.2.1 Meaning of Bilateral Matching

Bilateral matching means that the matching parties are two sets without intersection, and the elements of both parties are matched according to the preferences of each element, so that the elements in the two sets are related to each other on the basis of meeting a certain stability principle. The essence of matching is actually bilateral exchange. Therefore, both sides of matching will have their own sensitive preference list. Relay communication technology can be divided into single hop communication and multi hop communication. Single hop relay cooperation means that the destination can be reached through one relay cooperative communication from the source, while multi hop cooperative communication is the communication of nodes that can reach the destination through multiple relay cooperation. It mainly includes the following four models: single relay double hop model, single hop parallel model, multi hop serial model and multi hop cooperative model. According to the stable existence of bilateral matching, the matching problems can be divided into three categories: one-to-one matching, many to one matching and many to many matching. One to one matching means that each element in the two matching sets can match at most one element in the other set. From the perspective of signaling exchange and data exchange, the user still maintains the signaling link with the base station, which is no different from the traditional cellular user. The base station still allocates the wireless spectrum resources and manages the traffic of communication users, that is, traffic billing, mobility and security management. Many to one matching means that in one of the matching sets, at least one element can match multiple elements in the other set, while in another set, each element can match at most one element in the other set. For example, for school enrollment, a school can recruit multiple students, while a student can only select one school. Many to many matching means that at least one element in the two matching sets can match multiple elements in the other set.

2.2.2 Construction of Channel Bilateral Matching Model

In the first hop communication of cooperative communication, the signal size received by the relay user is:

$$R = \varepsilon\frac{1}{\eta} + \sqrt{F_2^{\eta-1} - r} \tag{5}$$

In formula (5), ε represents the fading coefficient, η represents additive Gaussian white noise, F represents the user distance, and r represents the distance between the user and the base station. However, different from traditional cellular users, the data transmission link between communication users does not need to be forwarded by the base station, but directly establishes a data transmission channel between two users.

This way of communication can not only get faster transmission rate and better user experience, but also reduce the burden of the base station to a certain extent. For example, some popular contents do not have to be downloaded from the base station repeatedly. Some previous work combining social network and communication technology focused on quickly sharing the same content to multiple users in the region, such as shortening the transmission time of the content in the whole region as much as possible. However, how to share and push the content when users have different preferences for the content has not been studied in detail. Social network is a new communication medium closely related to distance. It is a new communication service based on direct data transmission of short-range communication [6, 7]. Communication should meet a variety of different forms of service requirements. For example, communication technology can often be applied to a variety of local communication services, including communication in a small outdoor communication environment or indoor communication. Assuming the application scenario of a concert, in this case, the video service, especially the welcome of the audience, the concert organizer can apply for the communication spectrum resources from the cellular system, and then the audience participating in the concert can download the video service provided by the concert to each other by using the communication technology. In this way, it can not only meet the requirements of listeners, but also effectively reduce the load of base stations in cellular cells. For transmitting users and receiving users, the closeness of their social relationship is determined by the normalized correlation of their probability of selecting similar content, which is expressed as:

$$W_{pq} = \frac{(corr|p, q| - \eta)}{2} \tag{6}$$

In formula (6), p represents the transmitting user and q represents the receiving user. In order to use communication technology to push or share content among users with close social relationships. Firstly, social relations reflect the consistency of users' love for similar content. Using social relations can determine the transmitting users and receiving users of communication. This process can be regarded as a user discovery process. While communicating, the system can provide traffic and data sharing services of the Internet. With the increasing popularity of high-definition video and other media services, its large traffic characteristics also bring great challenges to operators' core network and spectrum resources. Secondly, communication users need to reuse the spectrum resources occupied by cellular users for communication, and the resulting co channel interference can not be ignored. Therefore, it is necessary to design an effective resource management scheme, so as to improve the system performance while ensuring the user communication quality. Using the local characteristics of communication, local media services can save the spectrum resources of the operator's core network. In the hot area, operators or content providers can deploy media servers to store the current popular media services in the media server, while the media server provides corresponding media services to users with business needs in an integrated mode. Based on the above two aspects, an effective content distribution scheme is proposed for cellular communication and communication hybrid networks, comprehensively considering the information of social layer and physical layer of the system, and optimizing user discovery and resource allocation, that is, the matching among users, shared content and spectrum resources, While improving users' satisfaction with the received distribution content, it can improve the performance of the system through an appropriate spectrum resource allocation scheme.

2.3 Design of Multi Stream Distribution Framework

Dynamic rate adaptive transmission control requires the base station or user terminal equipment to dynamically decide the video rate version sent to the user according to the dynamic network channel conditions and user buffer status, so as to maximize the user's viewing experience. In the wireless operator network, there is a scenario where multiple users request the same video URL at the same time. In this scenario, the dash transmission scheme considers that the "multicast" transmission mode of MBMS can be used for service transmission. We model the joint edge caching, transcoding and distribution decision-making problem and form it into a multivariable nonlinear integer programming problem, which aims to minimize the operating cost of the edge network and balance the computing load of each edge server under the constraints of the storage and computing resource capacity of the edge server. MBMS carries out the "multicast" transmission of dash content through MBSFN transmission mode. MBSFN mode has the following advantages: because the mobile terminal can effectively use the signal energy from multiple cells, the received signal strength of users can be effectively improved under MBSFN mode, especially the signal strength of users in edge cells in MBSFN area. Transmitting video with high-precision and panoramic characteristics brings a great burden to the existing network, including bandwidth resource consumption, storage resource consumption of relay cache, transmission delay, etc. [8–10]. Reduce signal interference, especially for users at the boundary of different cells. Because in the MBSFN transmission mode, the signal received by the user from the adjacent cell will be useful and will no longer be regarded as the interference signal of wireless transmission as in unicast. However, on the one hand, when users watch video, they will only watch the FOV area, so they can only transmit the video content in the FOV area or only keep the video content in the FOV area in a high bit rate version. On the other hand, edge caching and edge computing technology can effectively alleviate the transmission pressure of the return link, and the two resources can cooperate to share the video relay pressure. The diversity of wireless channel fading, because information can be received from several geographically separated places, usually the channel looks like time division multiplexing or frequency division multiplexing as a whole. In order to realize the above transmission framework, this chapter further optimizes the functional modules of the edge server. The execution edge server consists of four modules, as shown in Fig. 1:

According to Fig. 1, the functional modules of the edge server include: request collection module, request prediction module, cache decision module and distribution decision module. The request collection module is responsible for collecting the video slice requests of each user at the beginning of each transcoding and distribution stage. At the beginning of each cache stage, the request prediction module retrieves the user's historical request information from the request collection module and analyzes it to predict the video slice request of each user in this cache cycle time. The transcoding server is set up at the edge of the network (base station). For each video stream, it only needs to transmit a code rate version of the video stream from the core network to the base station once. After that, the transcoding server at the base station transcodes the

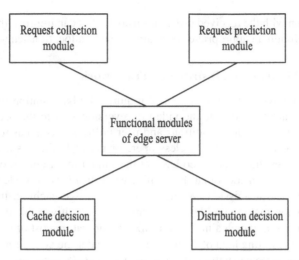

Fig. 1. Implementation of edge server function module

best code rate set by obtaining the user's demand and channel status in real time, which not only reduces the bandwidth consumption of the core network, but also improves the adaptability, system performance and user QoE of the system. At the beginning of each caching stage, the caching and distribution decision module makes a joint caching decision according to the predicted user request information, network topology information and the storage and computing resource capacity information of each execution edge server. At the same time, the edge server also has the function of wireless spectrum (bandwidth) adjustment, which can improve the bandwidth utilization by adjusting the user's bandwidth. The bandwidth adjustment module can also transcode a new code rate set again in combination with the transcoding server to further improve the system performance. At the same time, in each transcoding and distribution stage, the module makes transcoding and distribution decisions according to the real-time collected user request information and the current cache state of each execution edge server. Each execution edge server includes a cache module, a transcoding module and a distribution module. It is responsible for the caching, transcoding and distribution of video slices according to the decision command of the decision edge server. The bandwidth adjustment module is responsible for readjusting the bandwidth resources of users. While improving the bandwidth utilization, it can transcode a new code rate set in combination with the transcoding server to further improve the performance of the control method.

2.4 Optimize Transmission Control Mode

Combining adaptive streaming, edge caching and edge computing technology based on tile granularity, we can make full use of the local viewing characteristics of video and comprehensively utilize the multiple resources in the edge network to support the adaptive streaming of video with the minimum operation cost of the edge network. Adaptive transmission control technology includes two different driving modes, client-side driving and server-side driving. Client driven adaptive transmission control is also known as pull streaming media technology, and server driven adaptive transmission control is also known as push streaming media technology. An edge caching, transcoding and distribution framework for video adaptive streaming is designed, and the optimization method is used to model and solve the joint decision-making problem of caching, transcoding and distribution by using the storage and computing resources of the edge network in tile granularity coordination, so as to minimize the comprehensive operation cost of the edge network under the constraints of network resource capacity and ensuring user request response. In the adaptive transmission control driven by server, the most widely used is the real-time transmission protocol. RTP runs on the user datagram protocol, and UDP does not contain any rate control mechanism. This makes RTP more suitable for low latency and best effort streaming media transmission. In the usual server-side driven adaptive transmission control, the server uses the encoding rate as the transmission rate to match the consumption rate of the client. The expression formula of the effect function of maximizing users is obtained as follows:

$$V_g = \frac{|u - s|}{g} \times u^{-1} \tag{7}$$

In formula (7), u represents the number of code rate jitters, s represents the stability parameter, and g represents the penalty factor. In order to make full use of the storage and computing resources in the edge network to assist the video transmission, we need to consider not only the cooperation between multiple edge servers, but also the coupling and cooperation between multiple tasks (caching, transcoding and distribution). Considering that users directly obtain video slices from the remote server through the return link will cause large transmission delay, our proposed transmission framework follows the principle of allowing all users to obtain the requested video slices only from the edge server. This can ensure that the size of the client cache remains constant for a certain period of time, and make the best use of network resources. However, if the packet is lost or the network transmission is delayed, the recovery rate of the client packet will be lower than the consumption rate, resulting in buffer overflow and playback interruption. Here, adaptive transmission control technology can solve this problem well. In order to prevent buffer overflow, the server will automatically choose to send media streams with low bit rate. Moreover, in order to meet the feasibility of transcoding, we assume that the edge server already holds the highest bit rate version of each video slice. In the caching stage, the decision edge server makes the optimal caching decision by analyzing the storage and computing resource capacity and user request characteristics of each execution edge server, and sends the decision control information to each execution edge server. In this way, the media consumption rate of the client can be reduced, so as to offset the impact of the reduction of network bandwidth. When the network environment improves, the server will automatically choose to send media streams with high bit rate instead of before. By automatically monitoring the available bandwidth and buffer area and adjusting the transmission rate by selecting media streams with different bit rates, push streaming media transmission realizes the smooth playback of video with the highest quality level as possible. When the client plays video clips one by one, it can seamlessly reconstruct the original media stream. Each execution edge server downloads the video slices to be cached from the remote server according to the received cache decision control information. In the transcoding and distribution stage, the decision edge server decides the optimal edge server user connection pair (distribution decision) according to the cache content, calculation resource capacity and user request information of each current execution edge server. In the download process, the client automatically selects the appropriate bit rate video fragment according to the current available bandwidth. In this way, the client realizes adaptive transmission based on available bandwidth. Assuming that for any user, any dash video program can only be transmitted with one code rate value in one transmission mode at the same time, the expression formula of code rate transmission is:

$$T = \sum_{x=1} \frac{|x - y|^2}{M_{xy}} \times \sigma_y \tag{8}$$

In formula (8), x represents the dash video set transmitted in PTP mode, y represents the dash video set transmitted in SFN mode, M represents the total resource consumption, and σ represents the number of resource blocks. Although the partition structure of video files in different formats is different, the basic principle of partition is the same. When the audio data and video data are not interleaved, usually the audio frame is composed of some audio samples with the same time length, and each audio frame can be decoded separately by the audio codec. The execution edge server establishes a transmission connection with the user according to the distribution decision information, and distributes the requested video slices to the user. Specifically, for a specific user, if the execution edge server connected to it has cached the requested video slice, it will directly obtain the cached video slice; otherwise, the execution edge server connected to it will transcode the requested video slice using computing resources and transmit it to the user. Therefore, only a certain number of audio frames need to be combined into audio data with the same length of segmentation and filled in the segmentation. The processing of video data is completely different, because video frames cannot be decoded separately. Therefore, video partition exists in the form of picture group in segmentation.

3 Experimental Test

3.1 Experimental Preparation

This paper constructs the system in a LAN. The system consists of source, server and client. The client controls the network bandwidth between the server and the client through NetLimiter software. The slices in this paper adopt three quality levels: 350 kps, 750 kps and 1500 kps. Each slice is about 3–7 s long. Based on evalvid's open source video streaming architecture, rewrite the terminal and server-side programs. The traditional dash rate control algorithm and eams cloud algorithm are added to the terminal and server respectively. Add the energy model of NS3 to the terminal node to simulate the real-time energy consumption of the terminal equipment. The minimum playback time allowed for the media stream in the client cache is 3800 ms. After the system is started, it runs stably. We analyze the 1200 pieces sent.

4 Experimental Result

At the same time, the radius of the cell is set to 1200 m, and the total bandwidth of the system is 60 MHz. The adaptive multi stream transmission control method of social communication network based on cloud computing and the adaptive multi stream transmission control method of social communication network based on spatial Poisson point are selected for experimental comparison with the adaptive multi stream transmission control method of social communication network in this paper. Under different bandwidth conditions, the initial buffer loading time of the three methods is tested. The less the time, the better the performance is proved, The experimental results are shown in Table 1, 2, 3, 4 and 5:

Table 1. Initial buffer loading time with bandwidth of 500 kbps (s)

Number of experiments	Adaptive multi code stream transmission control method of social communication network based on cloud computing	Adaptive multi code stream transmission control method of social communication network based on spatial Poisson point	Adaptive multi code stream transmission control method for social communication network in this paper
1	1.998	2.765	1.233
2	2.342	1.276	1.098
3	1.672	2.868	0.868
4	1.801	1.766	0.965
5	2.553	2.671	0.575
6	1.166	2.337	1.232
7	1.869	1.976	1.004
8	2.224	2.117	1.673
9	1.867	1.988	0.984
10	2.653	2.202	1.223

It can be seen from Table 1 that the average initial buffer loading time of the social communication network adaptive multi stream transmission control method in this paper and the other two social communication network adaptive multi stream transmission control methods are 1.086 s, 2.015 s and 2.197 s respectively.

It can be seen from Table 2 that the average initial buffer loading time of the social communication network adaptive multi stream transmission control method in this paper and the other two social communication network adaptive multi stream transmission control methods are 2.141 s, 4.089 s and 3.809 s respectively.

It can be seen from Table 3 that the average initial buffer loading time of the social communication network adaptive multi stream transmission control method in this paper and the other two social communication network adaptive multi stream transmission control methods are 4.237 s, 6.868 s and 7.053 s respectively.

Table 2. Initial buffer loading time with bandwidth of 1000 kbps (s)

Number of experiments	Adaptive multi code stream transmission control method of social communication network based on cloud computing	Adaptive multi code stream transmission control method of social communication network based on spatial Poisson point	Adaptive multi code stream transmission control method for social communication network in this paper
1	4.979	3.099	2.099
2	4.883	3.234	2.356
3	4.652	4.543	2.673
4	3.562	3.742	1.977
5	3.867	4.987	2.390
6	4.456	3.234	1.793
7	3.982	4.549	2.212
8	4.459	3.339	1.994
9	3.030	3.320	1.868
10	3.018	4.038	2.052

Table 3. Initial buffer loading time with bandwidth of 1500 kbps (s)

Number of experiments	Adaptive multi code stream transmission control method of social communication network based on cloud computing	Adaptive multi code stream transmission control method of social communication network based on spatial Poisson point	Adaptive multi code stream transmission control method for social communication network in this paper
1	6.877	6.424	4.383
2	6.678	7.362	3.666
3	7.334	6.988	3.453
4	6.567	7.211	5.122
5	7.122	6.874	3.868
6	6.776	7.477	4.019
7	7.289	6.738	3.674
8	6.577	7.223	4.433
9	7.334	7.236	5.020
10	6.123	6.994	4.728

It can be seen from Table 4 that the average initial buffer loading time of the social communication network adaptive multi stream transmission control method in this paper

Table 4. Initial buffer loading time with bandwidth of 2000 kbps (s)

Number of experiments	Adaptive multi code stream transmission control method of social communication network based on cloud computing	Adaptive multi code stream transmission control method of social communication network based on spatial Poisson point	Adaptive multi code stream transmission control method for social communication network in this paper
1	15.567	14.277	9.004
2	14.673	15.938	9.863
3	13.535	14.261	10.028
4	14.679	13.304	9.378
5	13.464	13.346	8.433
6	14.266	14.099	8.646
7	14.309	13.017	9.893
8	15.206	13.048	10.017
9	14.217	14.183	9.363
10	15.236	14.125	10.738

and the other two social communication network adaptive multi stream transmission control methods are 9.536 s, 14.515 s and 13.960 s respectively.

It can be seen from Table 5 that the average initial buffer loading time of the social communication network adaptive multi stream transmission control method and the other two social communication network adaptive multi stream transmission control methods are 14.409 s, 19.774 s and 19.931 s respectively.

In order to further verify the effect of adaptive multi code stream transmission control in social communication networks of this method, cloud computing method, spatial Poisson point method and this method are used to verify the accuracy of multi code stream transmission control, and the results are shown in Fig. 2.

According to the analysis of Fig. 2, when the data volume is 100 GB, the multi stream transmission control accuracy of the cloud computing method is 70%, the multi stream transmission control accuracy of the spatial Poisson point method is 56%, and the multi stream transmission control accuracy of the method in this paper is 96%; When the data volume is 500 GB, the multi stream transmission control accuracy of cloud computing method is 74%, the multi stream transmission control accuracy of spatial Poisson point method is 63%, and the multi stream transmission control accuracy of this method is 98%; The multi stream transmission control accuracy of this method is always high, which shows that this method can improve the multi stream transmission control effect.

Table 5. Initial buffer loading time of 2500 kbps bandwidth (s)

Number of experiments	Adaptive multi code stream transmission control method of social communication network based on cloud computing	Adaptive multi code stream transmission control method of social communication network based on spatial Poisson point	Adaptive multi code stream transmission control method for social communication network in this paper
1	21.099	19.376	14.231
2	19.271	21.297	13.565
3	18.468	18.013	15.786
4	21.091	20.248	13.908
5	18.286	21.317	14.122
6	18.370	18.721	12.453
7	21.265	20.404	15.904
8	20.313	21.892	14.436
9	18.208	18.274	14.560
10	21.371	19.763	15.121

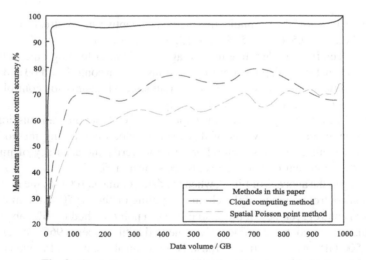

Fig. 2. Control accuracy of multi code stream transmission

5   Concluding Remarks

The adaptive multi stream transmission control method of social communication network in this paper can better adapt to channel changes and improve bandwidth utilization, and further improve system performance by adjusting user bandwidth. In this paper, the problem of adaptive video streaming is further divided into a problem of bandwidth and code rate adjustment. Through analysis, the rate allocation sub problem is transformed into user grouping problem, and it is proved that the user grouping problem is NP hard. According to the analysis of the energy consumption of smart phones in each link of video streaming processing, the mathematical model between the power consumption factors of smart terminal equipment and video codec parameters is established, and the improved adaptive algorithm of code rate and resolution based on energy consumption perception is used to solve it, which improves the performance of adaptive multi code streaming transmission control method in social communication network.

Although the method in this paper achieves good initial buffer loading efficiency, it still has high algorithm complexity. Therefore, the next research direction is how to reduce the complexity of the algorithm and improve the loading efficiency.

References

1. Luo, C., Qu, T., Deng, D.: Rate adaption algorithm for embedded multi-channel wireless video transmission. J. Comput. Appl. **40**(4), 1119–1126 (2020)
2. Shao, R., Fang, Z., Liu, S., et al.: Measurement and optimization of invulnerability of low-orbit satellite communication networks based on resilience. Oper. Res. Manag. Sci. **29**(7), 9–17 (2020)
3. Guo, T., Shen, P., Wang, L., et al.: Analysis of redundant structure of highly reliable train communication network based on ladder topology. Electr. Drive Locomot. (2), 81–87 (2020)
4. Li, S., Fan, X., Liu, Z., et al.: Status and analysis of wireless avionics intra-communications network protocol. J. Beijing Univ. Posts Telecommun. **44**(3), 1–8 (2021)
5. Zou, C., Lin, D., Yang, J.: Local route repair simulation of ad hoc communication network. Comput. Simul. **37**(1), 170–173, 225 (2020)
6. Gong, W., Shen, S., Pei, X., et al.: Clustering and associating method of dual heterogeneous communities in location based social networks. Chin. J. Comput. **43**(10), 1909–1923 (2020)
7. Cao, J., Gao, Q., Xia, R., et al.: Information propagation prediction and specific information suppression in social networks. J. Comput. Res. Dev. **58**(7), 1490–1503 (2021)
8. Chen, J.: Image transmission quality control technology considering visual masking characteristics. Sci. Technol. Innov. (18), 98–99 (2021)
9. Fu, W., Wu, L., Huang, L.: Construction of SoC FPGA based rate control transmission system. Radio Telev. Netw. **28**(06), 99–102 (2021)
10. Hu, W., Zhu, D., Mao, H.: ROI based rate control strategy for multi-channel video transmission. Radio Telev. Netw. **28**(07), 97–102 (2021)

Real Time Broadcasting Method of Sports Events Using Wireless Network Communication Technology

Xueqiu Tang[✉] and Yang Yang

Guangxi Sports College, Nanning 530012, China
tt57081637@163.com

Abstract. With the increase of time, the problem of image quality reduction caused by unbalanced network load will appear in the real-time broadcasting of sports events. A real-time broadcasting method of sports events is designed by using wireless network communication technology. Audio and video decoding and coding are divided into two independent threads working at the same time, which can make the frame rate reach the HD standard and enhance the stability of the encoding and decoding process. The GAN model is used to enhance the rate conversion. In the inter frame mode, integer transformation, quantization, reordering and entropy coding are performed on the residual block to complete the coding of the macroblock, which is stored or transmitted through the NAL layer. Wireless network communication technology is applied to distribute the number of channels in the space of mutual interference and balance the load of relay network. For the viewer, after receiving the streaming media data block, analyze the RTP packet, decode the video data, and then play the video. The test results show that the real-time broadcasting method of sports events using wireless network communication technology can improve PSNR, reduce the distortion of video sequence and ensure the stability of output picture.

Keywords: Wireless network communication technology · Sports events · Real time broadcasting · Audio and video decoding · Broadcasting service · Network transmission

1 Introduction

In recent years, China's sports broadcasting industry has shown a blowout development. Major online sports platforms such as Tencent sports, PP sports and broadcasting bar have purchased a large number of sports event broadcasting copyrights, such as Cup events represented by World Cup and track and Field Championships and regular events represented by NBA, Premier League, French open and Australian Open. China's sports event broadcasting is developing towards standardization and scale. In China, the right of real-time broadcasting of sports events has long been mainly controlled by traditional radio and television media, whose interests are monopolized and monopolized seriously. With the wide application of streaming media and other technologies, the rapid

© ICST Institute for Computer Sciences, Social Informatics and Telecommunications Engineering 2023
Published by Springer Nature Switzerland AG 2023. All Rights Reserved
W. Fu and L. Yun (Eds.): ADHIP 2022, LNICST 469, pp. 232–244, 2023.
https://doi.org/10.1007/978-3-031-28867-8_17

popularization of the new communication mode of Internet real-time broadcasting, the boost of policies, and the strong development momentum of network real-time broadcasting of sports event programs, great changes have taken place in the way and habit of spectators. Compared with traditional TV platforms, network broadcasting can not only provide more personalized and rich sports broadcasting services, but also expand the business model of sports event broadcasting industry, including paid viewing, copyright distribution, member ordering and video advertising, which further promotes the development of sports network broadcasting industry towards a healthy, large-scale and industrialized road [1].

The greatest value of sports event program communication lies in its timeliness. It is very important to master the technology and right of real-time broadcasting of sports event programs. Although the sports event broadcasting program is not the sports event itself, it is closely connected with the sports event. It comes from the real-time record of the sports event by the camera. It is the most intuitive expression of the ongoing sports event composed of a series of continuous pictures. It is not essentially different from the language, painting and photos of sports themes. In the past, the right and technology of real-time broadcasting of sports events programs were controlled by traditional media such as radio and television stations. However, with the wide application of streaming media technology, the popularization of the Internet and Internet access equipment, the combination of new technologies and the introduction of positive policies, this unreasonable phenomenon has undergone great changes. The programs of real-time broadcasting of sports events on the network have been pursued by more and more viewers. The core of sports broadcasting is the event content. There are great differences in the broadcast volume and user base between popular events and regular events. It is also directly related to whether the platform can provide users with high-definition, smooth and synchronous sports event broadcasting services. The key reason for the realization of network real-time broadcasting lies in the emergence and wide application of streaming media technology.

Traditional sports events are broadcast in real time through radio, cable or satellite broadcasting. Due to the growth of user scale and the demand for the improvement of service quality, it is urgent to optimize the real-time broadcasting method of sports events, so as to better improve the quality of service and user experience. In today's information explosion, users' time is occupied by a variety of services. Now users use a service from active inquiry to passive recommendation. The sports events are made into radio signals, cable TV signals, satellite signals and other program signals, and then these recorded signals are broadcast to the public. The public receives signals and watches the real-time broadcasting of event programs through television and other equipment. Streaming media technology refers to the technology of using streaming transmission to obtain continuous media data from the Internet. Through streaming transmission, users can watch programs in real time on the Internet. There are more and more data in the real-time broadcasting of sports events. The broadcasting service must consider a series of problems, such as the stable operation of the system, dynamic capacity expansion, the convenience of deployment, the efficiency of development, cost and so on. Therefore, using technical means to achieve real-time broadcast of live video has become the focus of current researchers. Reference [2] studied the bit rate adaptive algorithm for streaming media live broadcast scenarios. Based on the rate adaptive algorithm, the retransmission service can adapt to the time-varying characteristics of the channel by dynamically switching the video rate. Taking live video streaming as the background, aiming at improving the quality of user experience, and fully considering the delay requirements of services and the characteristics of transmission environment, a bit rate adaptive algorithm based on PID control is proposed. Reference [3] proposed a video transmission quality oriented opportunistic routing algorithm vor-mg based on multi person cooperative game to optimize video transmission quality and transmission overhead. The edge quality gain model of video packets is established; Multi platform video data transmission is modeled as a multi-user cooperative game; Each video packet is copied and forwarded based on its Nash optimal solution. However, the above method cannot guarantee the video stability of real time. To solve this problem, based on wireless network communication technology, this paper proposes a real-time broadcasting method of sports events to meet the requirements of customers for video quality and receive more video requests at the same time.

2 Real Time Broadcasting Method of Sports Events Using Wireless Network Communication Technology

2.1 Decoding of Broadcast Video of Sports Events

Streaming media coding mainly includes audio coding and video coding. Its purpose is to compress audio and video into a certain media format with appropriate coding and compression technology to reduce the occupation of traffic in the transmission process of streaming media. The decoded data is written into the audio and video cache block through the structure cache pointer. In the encoding thread, the audio and video data frames are read through cyclic query of cache block size for encoding. The video consists of one frame of images. After the image is divided into pixel blocks, it can be found that the brightness difference and chroma difference between adjacent images are small and the correlation degree is very high through pixel block scanning and pixel block search. The logical sequence of video decoding is to obtain the input video coding data, find and open the decoder according to the coding information, and finally input the input data into the decoder for decoding.

The decoding module first calls the stream information analysis function to input part of the data into the function, and the function will get the relevant video information through the input data stream. Buffer technology is an application level audio and video quality control technology widely used in audio and video broadcasting system at present. Through the setting of buffer, it can reduce the loss of audio and video data, smooth the jitter in the process of encoding and decoding, and improve the encoding and decoding performance and stability of broadcasting system [4]. The flow framework of the decoder forms a complementary structure with the encoder, and the decoding principle is shown in Fig. 1.

When the encoded code stream enters the decoder, the decoder performs entropy decoding and reordering on the received compressed data to obtain the residual coefficient, quantization coefficient and other relevant information of the macroblock, and

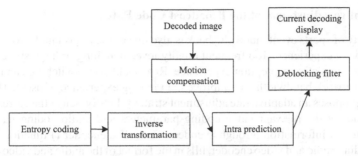

Fig. 1. Principle of decoding

performs inverse quantization and inverse transformation to obtain the residual macroblock. Through the obtained video information, assign a value to the structure object, and provide the required data information for operations such as finding and opening the decoder. In order to apply for cache space in the system for the received video data, this paper sets the size of the application space to the size of 25 video frames. Then, through the obtained video stream information, find the corresponding decoder for the H.264 video stream, and perform the H.264 decoding operation. The QP table stores the QP value of the video image macroblock in memory, and its label starts from left to right. Because the size of macroblocks is usually 16×16, the calculation formula for the number of macroblocks per line is:

$$p_1 = \frac{u_1}{16} + 1 \tag{1}$$

In formula (1), p_1 represents the number of macroblocks in each line; u represents the width of the video image. Further, the total number of macroblocks is, and the calculation formula is:

$$p_2 = (u_2 + 15)\frac{u_1}{16} + 1 \tag{2}$$

In formula (2), p_2 represents the total number of macroblocks; u_2 represents the height of the video image.

After obtaining the frame information corresponding to the video, apply to the system for decoding frame cache, which is set as the cache of 24 frame decoded video frame size, and set the pixel space format, frame size, resolution and other parameters of the cache block. Find and open the decoder according to the obtained stream information, and then send the audio AAC encoded data to the decoder as an input cycle for decoding. The PCM data decoded by the decoder is stored in the structure object, and finally the data is stacked into the decoding buffer to provide audio data for subsequent audio and video synchronization, audio playback and other operations. Finally, the recorded frame position is used to determine the corresponding decoder ID, and enter the initialization decoder process, assign and initialize the decoding structure cache, allocate memory space, and then open the decoder, read the video stream and start the cyclic decoding process. Then the prediction block is reconstructed according to the header information in the code stream, and the final decoded image is obtained after block filtering.

2.2 Adaptive Adjustment of the Broadcast Code Rate

The stability of sports event video playback is also one of the important factors affecting users' viewing experience. Too frequent quality level switching or large switching will affect users' subjective perception experience. Reasonable rate switching can make full use of the network bandwidth and improve the viewing experience of users. Therefore, this paper proposes an adaptive rate adjustment strategy. Firstly, select the corresponding encoder and set the relevant video coding parameters and audio coding parameters, and pass the set information through avcodec_ find_ The encoder () function finds the corresponding audio and video encoder, fills in the format of the audio and video data area after opening the encoder, compares the audio and video time information after setting the corresponding time benchmark, and obtains whether video coding or audio coding should be carried out at the moment, so as to enter the corresponding coding process. Because the past information also has certain reference value in the process of rate decision-making, in the research of some rate adaptive algorithms based on reinforcement learning, it is still considered to introduce a limited number of environmental information sequences collected in the past to represent the state for action decision-making. However, due to the limited sequence length of the introduced state sequence during each training, some key information in the past will be lost. This paper uses Gan model to strengthen bit rate conversion. In inter frame mode, firstly, the macroblock is obtained through motion estimation and motion compensation, and the residual is obtained by subtracting from the macroblock of the current frame. Then, the residual block is encoded by integer transformation, quantization, reordering and entropy coding, and stored or transmitted through NAL layer. The feature extraction network receives the input sequence with the length of 16 carrying the past key information, operates the sequence using the batch normalization algorithm, and then transmits it to the one-dimensional convolution network to extract the key information. The state characteristics of CNN network output are an important part of subsequent strategy network and value network input. The hidden features and the state features output by CNN network are spliced to generate a new feature expression, which is used as the input of strategy network and value network, as shown in formula (3).

$$\alpha_t = \chi(\beta_t, \gamma_t) \tag{3}$$

In formula (3), t represents the time; α_t represents feature expression; χ represents a discriminator function; β_t represents the hidden features; γ_t represents the state characteristics of the CNN network output.

The discriminator network is used to judge the probability that the newly generated hidden feature comes from the sample with positive reward value, so that the generator can generate the hidden feature vector with positive reward value and remember some environmental information lost due to limited sequence length, such as the control range of sports event broadcasting delay. The function uses 500K data by default, and the purpose of reading in size is to ensure that the video and audio information contained in the streaming media data can be completely analyzed, which is the later avcodec_ find_ Functions such as decoder () provide reliable guarantee. The FLV format has the advantages of small volume and fast data encapsulation. In order to keep the time information and frame sequence of audio and video from confusion, video coding and audio coding are combined into the same thread. In the audio and video coding thread, it is first determined to enter the video coding cycle or audio coding cycle according to the time reference and current timestamp information of H.264 and AAC. The reference time is set as {264.00, where h is set as {9001}. Copy the policy parameter amplitude of the current policy network to the periodic sampling data, and calculate the advantage function to update the policy network. The policy network update method is as follows:

$$w = \varphi\left[\min\left(bc, (1 - \delta)b'\right)\right] \tag{4}$$

In formula (4), w represents updating the network target; δ represents super parameter; b represents the strategic parameters of the value network; c represents the clip advantage function; b' means to limit the update range of the network. The update thread calculates according to the collected samples, and uniformly updates the network parameters of each thread agent.

2.3 Balance the Load of Relay Network Based on Wireless Network Communication Technology

In the wireless LAN environment, when the streaming end establishes a link with the server and receives streaming media data, the streaming end will wait for too long. The wireless network communication technology is used to balance the broadcasting network load and optimize the first frame delay of sports event broadcasting. The whole network is composed of multiple connection points (APS) responsible for data collection and transmission. Each sub network has its own AP, which is responsible for the collection and transmission of video data in its own process. The sub network APS are connected in a chain [5]. The data of the broadcasting system in this paper is mainly stored in the related system database and non related database credits. In order to ensure the high availability of the database and avoid the failure of single node MySQL and credits, this paper has carried out cluster deployment for both MySQL and credits. The sub network of wireless network is a star structure, and each ordinary node needs 1 hop to transmit data to AP [6]. The data transmission process can be regarded as a queuing process. It can be obtained that the average transmission delay of nodes is the queuing delay plus the service delay, which can be expressed as:

$$h_{xy} = \frac{1}{2\left(s_{xy} - g_{xy}\right)} \tag{5}$$

In formula (5), x, y represents two transmission nodes in wireless network communication; h_{xy} represents the time delay from the node to the control center; s_{xy} represents the data transmission rate; g_{xy} represents the average data generation rate.

To optimize load allocation and scheduling in wireless communication networks, it is modeled as an average delay minimization problem. Where, the objective function is averaged after summing the delay of each node in all subnetworks in the chain network, representing the average [7] of the delay of all nodes in the whole network. To achieve the goal of minimizing task response and completion time, the task life cycle is indicated. Suppose that the time point when the task request arrives is T_2, the time point when all the video files are successfully sent to the user side is T_1, the following relationships exist:

$$T = T_1 - T_2 \tag{6}$$

In formula (6), T represents the life cycle of the task.

This means that the duration from the time the system receives the request to the time the system satisfies the request is the life cycle of the task. Since the requested content is usually based on the number of videos, within the task cycle, the sports event broadcasting system also needs to realize the process of loading the video file from the auxiliary storage device and routing and transmitting the video file. In addition, the first constraint among the constraints is the minimum allocation of resources. In order to ensure that the data queue of the node will not overflow and ensure the stability of data transmission. The second constraint is to restrict each node to use at most one channel at the same time. The third constraint is to consider the limit of the number of AP interfaces in the sub network. The cluster built in this paper contains a total of 6 nodes, 1 management node, 2 data nodes and 3 SQL nodes. For the configuration file, it should be noted that "[ndbes mgmd]" represents the configuration block of the cluster management node, "[ndbd]" represents the configuration block of the data node, "and" [mysqld] "represents the configuration block of the SQL node. "Hostname" is the IP address of the node. The fourth constraint is to consider that the number of channel allocation of sub networks in the space of mutual interference will not exceed the total number of channels that can be used, which can be determined by the equivalent interference data. The data node is mainly configured with MySQL configuration file "my. CNF". You need to specify the root directory of the data node, the path of data storage, and the IP address of the server where the cluster management node is located. In order to facilitate troubleshooting, you can store the storage log file in the specified directory. The last constraint indicates that the resources allocated to each node in a subnet should not exceed the total resources available to the subnet. The implementation process of balancing the broadcast network load based on wireless network communication technology is shown in Fig. 2.

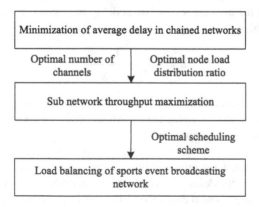

Fig. 2. Implementation process of load balancing in the broadcast network

As the number of channels in a wireless network increases, the throughput of the network increases, because the number of channels increases, the communication resources in the network increases. As can be seen when the number of nodes is small, this difference is not very big, because the communication resources in the network are enough to support the communication requirements of a small number of nodes, and when the number of nodes is large, the advantage of the number of channels will be reflected.

2.4 Optimization of Real-Time Broadcasting of Sports Events

On the basis of balancing the load of broadcasting network by using wireless network communication technology, the real-time broadcasting process of sports events is optimized. The data sending and receiving layer realizes the functions of transparent data receiving and sending and NAT circulation. The socket based on UDP is used to provide end-to-end datagram service. The processed data includes signaling messages and streaming media data. NAT traversal solves the problem that users in the subnet cannot provide P2P services. The data transceiver layer sends and receives data through sockets. In TCP/IP, sockets can be divided into streaming sockets and datagram sockets. Messages are sent in the form of HTTP requests. Typical message notifications include three types: cut-off, streaming and recording, which respectively mean that users stop pushing audio and video streams to the relay server, users start pushing audio and video streams to the relay server, and the relay server generates a new relay recording file. When the service server receives the streaming information from the relay server, it starts to record the real-time online number information of the relay. After receiving the streaming information from the relay server, the service server stops recording. The sports event broadcasting node and server adopt the datagram set over UDP protocol. The data transmission of the relay server is divided into two ways: (1) when receiving the data block sent by the relay node, select a fixed a node to send the chunk message directly; (2) Select <node, data block> pair through offer/select mechanism to send data block. Whenever the relay server receives new streaming media data from the relay node, it determines whether the data block with ID number is within the range of (106121). If so, it is required; If not, the data block is not required. The relay micro web page is connected with the back-end service through websocket. The back-end service records the contents of the table below according to the establishment and disconnection of the websocket connection of the relay micro web page. When the websocket is connected, record the basic information of the user's access, such as the user's ID, IP, UA identification, access time, websocket disconnection, including the user's active disconnection and passive disconnection, and record the user's departure time. After the relay server selects the peer nodes that need the data block, if the number of peer nodes is greater than the set value, it will randomly select a node from them to send a chunk message. The specific flow of the offselect mechanism of the relay server is shown in Fig. 3.

Parse the chunk message and read the ID number of the data block and the nodeid of the peer node. In the offer/select phase, the node selects the required data block according to the data block strategy and locks the ID of the data block. In the offer/select mechanism mentioned in the data block scheduling module, buffermaps are exchanged between nodes and peer nodes. After editing the template XML file, jasper uses jasper compile manager to compile it into binary "*. Jasper" file, and then jasper fill manager fills

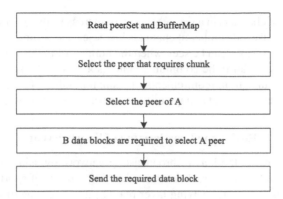

Fig. 3. The OFFER/SELECT mechanism of the broadcast server

the data source into the "*. It should be noted that the database can be either database data or Java Bean. At the same time, each node also maintains a buffermap for its peer node. Whenever an offer message is received, the node will update it. When a node periodically updates its cache, it first detects the time stamp of its neighbor. If the difference between the current time and the time stamp is greater than a fixed value, it indicates that the node has not updated data locally for a long time, it will determine the node as a failed node and delete it from the cache. Finally, you can choose to preview the report file with jasper, print the report with Jasper print manager, or directly export various forms of reports (PDF, HTML, etc.) with Jasper export manager. Such a network management method is more robust. The broadcaster collects video data through the capture device, compresses and encodes the video data in real time through the negotiated coding method, and then encapsulates the data packet in RTP format and sends it to the network. RTMP protocol is an application layer protocol. Therefore, RTMP block stream does not provide priority or similar reliability control to ensure information transmission. Therefore, it usually establishes a connection with a reliable transport layer protocol such as TCP. RTMP block stream ensures that all messages across streams can be transmitted according to the timestamp sequence. So far, the design of real-time broadcasting method of sports events based on wireless network communication technology has been completed.

3 Experimental Study

3.1 Experimental Preparation

The experimental test environment includes hardware environment and software environment. The hardware environment is mainly composed of multiple cloud servers on Tencent cloud. During the experiment, one windows device is used as streaming device, one windows device is used as standby streaming device, and two windows devices are used as streaming devices. The streaming device will be switched every once in a while. The software environment includes the main software environments in the whole system development, deployment and testing process, such as NGINX and keep alive for load balancing, Tomcat container for back-end code operation, Jenkins for continuous integration, etc. NGINX modular architecture allows developers to freely expand the functions of web server without modifying the core. It can be mainly divided into core module (core), event module (event), protocol module (HTTP), load balancer, etc. The code of each module is encoded together with the core code of NGINX. The server with public IP 148.70.221.38 is used as the master node, and the server with public IP 148.70.115.31 is used as the backup node. One active and one standby mode is set in kept. The virtual IP is 172.27.16.11, and the virtual IP is bound to the elastic public IP 148.70.136.51. Access the public IP, and the load balancer directs the video traffic of sports events to the master node. On the built server, the configuration file NGINX to configure video on demand service. First, set the port number and data block size of the configuration file. Add the video storage location in the application VOD field, and select the video storage location in the application VOD field_ HTTP add video on demand source address. After the VOD service is configured successfully, an IDO is placed in the directory of video storage location Mp4 playback file.

3.2 Results and Analysis

PSNR is an objective evaluation of the images in the video sequence. Through the frame by frame comparison between the reference frame of the original video sequence and the distorted video sequence, the video sequence is judged on the similarity of the distorted video. Generally, the larger the value of PSNR, the better the quality of the video sequence. Taking PSNR as the evaluation index, this paper tests the PSNR of the collected and encoded video sequence to test the effect of the real-time broadcasting method of sports events using wireless network communication technology. The real-time broadcasting methods of sports events based on data mining and random strategy are selected as the comparison methods. Test the PSNR mean value of each method in different broadcasting time, and the comparison results are shown in Tables 1, 2, 3 and 4.

In the test of continuous broadcast of sports events for 1 h, the mean PSNR of the real-time broadcast method based on wireless network communication technology was 34.61, which was 6.86 and 6.34 higher than the data mining and random strategy-based broadcast methods.

In the test of continuous broadcast of sports events for 2 h, the mean PSNR of the real-time broadcast method based on wireless network communication technology was 29.44, which was 4.20 and 5.37 higher than the data mining and random strategy-based broadcast methods.

In the test of continuous broadcast of sports events for 5 h, the mean PSNR of real-time broadcast methods based on wireless network communication technology was 26.29, which was 2.61 and 4.76 higher than data mining and random strategy-based broadcast methods.

In the test of continuous broadcasting of sports events for 5 h, the average PSNR of the real-time broadcasting method of sports events based on wireless network communication technology is 24.85, which is 4.51 and 4.43 higher than that based on data mining and random strategy. According to the above experimental test results, the video sequence output by the real-time broadcasting method of sports events using wireless network communication technology is less distorted, the output picture is stable, and the

Table 1. Comparison of PSNR for 1 h

Test times	The real-time broadcast method of sports events based on wireless network communication technology	The real-time broadcast method of sports events based on data mining	The real-time broadcasting method of sports events based on stochastic strategy
1	33.47	29.63	29.08
2	34.54	28.83	28.75
3	35.65	27.44	27.86
4	33.28	26.07	29.22
5	34.86	26.88	28.53
6	35.93	28.55	26.35
7	33.62	27.28	27.67
8	35.35	29.37	28.94
9	34.09	25.94	28.81
10	35.26	27.51	27.52

Table 2. Comparison of PSNR for 2 h

Test times	The real-time broadcast method of sports events based on wireless network communication technology	The real-time broadcast method of sports events based on data mining	The real-time broadcasting method of sports events based on stochastic strategy
1	29.47	24.66	23.74
2	30.87	25.88	24.88
3	28.56	26.77	25.66
4	29.34	25.14	23.33
5	28.55	24.01	24.40
6	29.22	24.29	23.32
7	28.71	25.79	24.34
8	29.84	25.49	23.78
9	29.63	25.85	23.45
10	30.25	24.52	23.84

broadcasting picture quality is improved to a certain extent, which can enable users to easily receive the high-definition real-time broadcasting picture of sports events. This is because this paper uses wireless network communication technology to divide audio and

Table 3. Comparison of PSNR for 3 h

Test times	The real-time broadcast method of sports events based on wireless network communication technology	The real-time broadcast method of sports events based on data mining	The real-time broadcasting method of sports events based on stochastic strategy
1	26.19	23.78	20.42
2	25.07	24.89	21.58
3	26.51	23.56	22.74
4	25.85	23.50	20.57
5	27.62	22.23	22.66
6	26.33	23.47	21.23
7	26.92	24.14	22.85
8	26.25	23.28	20.69
9	25.76	24.58	22.35
10	26.44	23.35	20.24

Table 4. Comparison of PSNR for 4 h

Test times	The real-time broadcast method of sports events based on wireless network communication technology	The real-time broadcast method of sports events based on data mining	The real-time broadcasting method of sports events based on stochastic strategy
1	25.73	18.46	20.40
2	24.87	19.88	20.72
3	24.92	20.65	19.89
4	25.65	19.22	19.65
5	25.20	20.59	18.37
6	24.01	20.38	21.44
7	23.50	21.06	22.43
8	24.57	20.46	21.16
9	25.74	21.73	21.85
10	24.28	20.97	18.29

video decoding and coding into two independent threads to work in parallel, so that the frame rate reaches the high-definition standard and improves the stability of the video image encoding and decoding process. Gann's model is used to improve the conversion rate. The wireless network communication technology is used to allocate the number of channels in the space of mutual interference and balance the load of the relay network, so as to improve the PSNR, reduce the distortion of video sequences and enhance the overall transmission image quality.

4 Conclusion

The development of real-time broadcasting technology of sports events not only enriches people's cultural life, but also plays an important role in some special occasions. However, with the large-scale application of broadcasting system, some deficiencies in its implementation are gradually revealed. This paper presents a real-time broadcasting method of sports events by using wireless network communication technology. After the test, the picture quality of sports event broadcasting has been improved to a certain extent. Due to the large number of users watching sports events and the complex broadcasting business scenario, it is difficult to formulate a unified user Qoe model that meets various requirements. However, the current research on broadcasting methods mostly aims at maximizing user QoE. Therefore, when the rate adaptive algorithm is trained, whether it can adapt to new business scenarios or meet the needs of new user groups needs further research and verification. This paper only completes the real-time broadcasting under the windows platform. However, in the field of mobile devices, it is still necessary to transplant and improve the video broadcasting system, such as dynamically reducing the real-time streaming media frame rate and improving the image quality in the 5g network environment.

References

1. Huang, H., Liu, L., Li, Z.: Technological transformation and content innovation of mobile communication in sports events of the 5G era: also on the enlightenment to Beijing 2022 Olympic winter games. J. Shanghai Univ. Sport **44**(5), 16–23 (2020)
2. Jin, Y., Wu, X., Zhang, Z., et al.: A rate adaptation algorithm for real-time video streaming. J. Commun. Univ. China Sci. Technol. **29**(1), 8–14 (2022)
3. Wu, H., Ma, H., Xing, L., Zheng, G.: Routing algorithm for video opportunistic transmission based on multi-player cooperative game. J. Softw. **31**(12), 3937–3949 (2020)
4. Huang, H.: Cross-network communication route automatic search method for heterogeneous wireless network. Comput. Simul. **37**(5), 259–262 (2020)
5. Cao, S., Jia, X., Lü, Y., et al.: Secure communication in cognitive radio network assisted by cooperative relay and UAV. Comput. Eng. **47**(6), 203–209 (2021)
6. Lin, M.: Study on application of streaming media static cluster based on open source architecture. J. Changchun Univ. **31**(2), 13–17, 22 (2021)
7. Ge, X.: Load balancing mechanism of server cluster for live sports event. Comput. Appl. Softw. **37**(6), 166–170 (2020)

Sports Event Data Acquisition Method Based on BP Neural Network and Wireless Sensor Technology

Yang Yang$^{(\boxtimes)}$ and Xueqiu Tang

Guangxi Sports College, Nanning 530012, China
gxtzyy2022@163.com

Abstract. The current data acquisition methods use sensors and fixed equipment to collect data. When applied to sports events, a large number of sensor deployment will lead to large data transmission loss and low data acquisition efficiency. In order to reduce the data transmission loss and improve the efficiency of data acquisition, the sports event data acquisition method based on BP neural network and wireless sensor technology will be studied. After building the wireless sensor network for sports event data acquisition, BP neural network is used to fuse the sensor node data. Compressed sensing is used to initially locate the sensor nodes in the network, and the sensor node data is gathered according to the honeycomb principle to realize the event data collection. The test results show that after the application of the proposed data acquisition method, the average energy loss ratio of the node is only 6.61%, the data acquisition efficiency is greatly improved, and the test effect is good.

Keywords: BP neural network · Wireless sensing technology · Sports events · Data acquisition · Data fusion · Routing protocol

1 Introduction

In the process of continuous practice and development of competitive sports, relevant practitioners gradually realize the guiding role of corresponding technical and tactical analysis and physiological and biochemical analysis for training and on-the-spot competition. Therefore, in the long-term evolution process, a set of objective competition data analysis system has been established. With the development and application of big data technology, data collection is more and more widely used in the field of sports events, such as improving athletes' competitive level, preventing sports injuries, measuring athletes' value, developing fan services, doping investigation and punishment, assisting referees in sentencing, etc. Using wearable equipment to collect athlete data is the most commonly used means at present. The use of sensors makes venue equipment such as camera and lighting become an important way of athlete data collection. A large number of athlete biometric data, technical action data and life data are obtained. In the field of world sports, data acquisition system is widely used, especially in ball games, such as

W. Fu and L. Yun (Eds.): ADHIP 2022, LNICST 469, pp. 245–256, 2023.
https://doi.org/10.1007/978-3-031-28867-8_18

football, basketball, volleyball, tennis, table tennis, badminton and so on. It can be said that the data statistics in ball games can not only collect the relevant information of the game, but also play a vital role in improving the sports level of athletes, the law enforcement level of referees and even the position of the sport in the world sports field. With the continuous improvement of digital technology, sports event data acquisition methods are also constantly updated [1]. In terms of collection means, the collection means that affect the normal life, training and competition of athletes should not be adopted. Usually, the collected sports event data can be mainly divided into technical data and non-technical data. Technical data mainly refers to the athletes' technical actions, personal performance, whether they break the rules, detailed single data, etc. Non technical data mainly refers to the data in the process of abnormal competition and some relevant data after the competition. It is mainly used to record the status information of athletes, referees, etc. during the competition, the violations after the competition and the corresponding punishment [2]. In sports event data acquisition, high-speed cameras are mainly used to capture the basic data of the movement track of the data acquisition object in the event from different angles at the same time: These data are generated into three-dimensional images through computer calculation; Finally, using the real-time imaging technology, the collected data information is clearly presented from the large screen.

Wireless sensor network is an intelligent self-test and control network system which is composed of a large number of ubiquitous micro sensor nodes with communication and computing capabilities densely arranged in the unattended monitoring area, and can independently complete the specified tasks according to the environment. The early sports event data acquisition mainly used the camera to assist the infrared sensor and pressure sensor to collect the movement start or trajectory change data of athletes in sports events. The accuracy and accuracy of the collected data are affected by the acquisition equipment. With the continuous maturity of wireless sensor technology, the accuracy of sensors is rapidly improved and the types of sensors are constantly enriched. The number of sensors available for sports event data acquisition is rapidly increasing rapidly, raising higher requirements for the accuracy of event data acquisition. Reference [3] studies the data acquisition method of underwater wireless sensor nodes. The self-organizing map is used to optimize the path of sensor nodes; Combining the optimized path graph and K-means algorithm to find the internal aggregation point of the path; The data collection points within the communication radius of the sensor are obtained by using the aggregation points and the nodes of the sensor. Finally, the optimal path for each data collection point to collect data is obtained by self-organizing mapping. Reference [4] designed a multi-channel synchronous data acquisition system suitable for magnetic anomaly detection sensors. The design system is based on the modular construction structure, and uses the parallel operation characteristics of field programmable gate array to ensure instruction synchronization; An analog-to-digital converter with a resolution of up to 32 bits is used in combination with digital filter and chopping zero stabilization technology to realize sensor data acquisition.

However, due to the design defects of the sensor and the weak ability of the sensor node processor, the accuracy of the sensor acquisition results is not high. In addition, the energy and communication capabilities of wireless sensor network nodes are very limited, and it is impossible to use existing wired network protocols for data transmission. In order to avoid wasting communication bandwidth and energy and improve the efficiency of information collection, this paper will study the sports event data collection method based on BP neural network and wireless sensor technology.

Establishment of Wireless Sensor Network for Sports Event Data Acquisition
Sports event data acquisition wireless sensor network is mainly composed of a large number of high-precision sensors, which are randomly deployed in the sensing area to collect environmental information. The high-speed and high-precision camera also uses sensors to track the data acquisition object and locate the object in time. In the wireless sensor network structure, sensor nodes are divided into cluster head nodes and cluster member nodes. The member nodes are responsible for collecting the environmental parameters around them, and the cluster head node is responsible for collecting the data of all member nodes in the whole cluster and forwarding these data to the base station node or mobile aggregation node.

Wireless sensor network node is the basic part of wireless sensor network. In different applications, the composition of nodes is different, but the basic composition includes the following units: sensor module, processor module, wireless communication module and energy supply module. In the whole network structure, the member nodes in the cluster are responsible for monitoring the surrounding environment data they are interested in. The member nodes can also cooperate with each other to remove the redundant data between adjacent nodes, and finally forward the data to the cluster head node in the current cluster. On the one hand, the cluster head node needs to indicate its identity to other nodes in the cluster by sending broadcast packets. On the other hand, it needs to process the received data and then send it to the mobile aggregation node or relay node. Cluster head nodes often need to consume more energy in the transmission process. In order to avoid premature death of some nodes, nodes with more energy are usually selected to play this role [5]. After the mobile sink node completes the information collection in the whole sensing area, it sends the data to the manager service node and data analysis terminal according to the network protocol. In terms of network topology structure, this paper mainly combines hierarchical and mobile network topology mechanism to realize an energy-saving and efficient wireless sensor network and maximize the life cycle of sports event data acquisition network.

The energy calculation formula of the sensor node sending and receiving data is as follows. E_d^{Tx} is the power consumed by the wireless sensor network transmitting a data packet, and E_d^{Rx} is the energy consumed by the data transmission mechanism. The energy E^{Rx} consumed by receiving information and the energy E^{Tx} consumed by transmitting information can be calculated by the following formula [6]:

$$E^{Tx}(l, r) = lE_d^{Tx} + l\varepsilon r^n \tag{1}$$

$$E^{Rx}(l, r) = lE_d^{Rx} \tag{2}$$

where l is the length of the transmitted or received information, r is the distance, and n represents the path loss index of the distance specified in the wireless signal propagation model; ε is the energy consumed by transmitting unit size data per unit distance. It can be seen that sending messages consumes energy to run wireless communication and signal amplifiers, while receiving messages only consumes energy to run wireless transmission. A large number of sensor nodes are randomly distributed in the network, and the network coverage is equal to 1. In the sensing range, sensor nodes can detect reliably. The sensing range is a circular area with radius R. Therefore, the effective area of each sensor node can be πR^2, but the range of simultaneous interpreting of different sensors may overlap. When calculating the coverage, we can't simply sum up the sensing range of each sensor node.

When deploying each sensor node in the wireless sensor network, it needs to be set according to the data collection range of sports events. When collecting sports event data, we should avoid loopholes in the sensor data collection range. Ieee802.0 is used in the wireless sensor network for sports event data acquisition 15.4 standard communication, and then set the routing protocol in the data acquisition network as LEACH protocol. In LEACH protocol, the sensing area is divided into multiple uneven clusters according to the Euclidean distance between nodes. In each cluster, a cluster head node is selected to collect and process the network information resources of its cluster members, and then each cluster head node transmits the data to the base station through a single hop. The protocol stipulates that new cluster head nodes will be reselected in each round, which avoids the problem of premature energy depletion in the long-term high load operation of cluster head nodes, thus prolonging the stability period of network transmission.

Nodes under LEACH protocol are divided into ordinary nodes and cluster head nodes. During the operation of the network, all nodes will generate a random number in the range of $(0, 1)$. if the random number of the node that has never served as the cluster head in the network is less than the set threshold $T(n)$, the node is determined as the cluster head, and the value of $T(n)$ is shown in the following formula [7].

$$
T(n) = \begin{cases} \dfrac{p}{1 - p * \left[t \bmod p^{-1}\right]}, & n \in G \\ 0, n \notin G \end{cases}
\tag{3}
$$

where, G is the node set that has not been the cluster head in the current round, t is the number of current round robin, p is the proportion of the number of cluster heads in all nodes, and $T(n)$ represents the probability threshold that the node will act as the cluster head. According to the above formula, all nodes will act as cluster heads once.

In this protocol, all nodes can compete for the cluster head in turn, which ensures that the energy consumed by each node is relatively balanced and can prolong the life cycle of the network. The cluster head sends an invitation message to other nodes in the network in the communication mode of NP CSMA. Other nodes choose to join the cluster with high signal strength according to the strength of the transmitted signal from the received cluster head, and send a join ACK message to the cluster head with NP CSMA to join the cluster. After the preliminary construction of sports event data acquisition wireless sensor network, data fusion is carried out.

1.1 BP Neural Network Fusion of Wireless Sensor Network Data

In practical application, the sink node can be placed on athletes. The sensing node cannot know the moving position of the sink node in advance. Once a person or animal enters the sensing area, the sink node can obtain the data of all sensor nodes within the communication range. In the aggregation node data collection algorithm based on random movement, the mobile aggregation node reaches the sensing range of the sensor node in a probabilistic way, which can be regarded as a Poisson arrival process.

In traditional static sensor networks, nodes mainly forward data to base stations by single hop or multi hop. This static network structure often has a typical energy hole problem, which can not guarantee the principle of balanced energy consumption of nodes in the network. In the data collection algorithm with mobile sink node, the mobile sink node collects the sensing data of sensor nodes in the cluster according to a certain moving track, which avoids the problem of high energy consumption and load of some nodes in the static network. However, for mobile sink nodes, designing an appropriate mobile path has also become an important problem. Generally, the mobile mode of sink node can be divided into fixed mobile mode and controllable mobile mode. A data collection algorithm of sink node with fixed moving track is adopted. On the one hand, the fixed moving mode is suitable for the network model of uniform clustering, which is easy to construct the moving track of cluster head node; On the other hand, the fixed mobile mode can save the computing cost of the network, including the planning of mobile trajectory and the later route reconstruction [8].

In this paper, BP neural network is used for data fusion of sports event data acquisition wireless sensor network. The model of wireless sensor network is similar to BP neural network. The node used to collect the surrounding environment information in WSN is equivalent to the neuron in BP neural network. WSN needs to transmit information through certain connection rules, just as BP neural network needs to transmit information through synapse. The whole process of wireless sensor network is to process a large amount of information collected and obtain the characteristics of data, which is the same as the function of data fusion based on BP neural network. Therefore, BP neural network can be applied to the data fusion of wireless sensor networks.

BP network model is usually composed of input layer with one or more nodes, output layer with one or more outputs, and one or more hidden layers. When the last layer of the network adopts curve function, the output is limited to a very small range, while the output can be any value using linear function. General neural networks are adjustable, or trainable, so that a specific input can get the required output data. The basic idea of BP algorithm: for an input sample, after the calculation of weight and excitation function, get an output, and then compare it with the expected sample. If there is a deviation, back propagate the deviation from the output, and adjust the weight and threshold to make the network output gradually consistent with the expected output.

The basic idea of bpnda is that firstly, after WSN forms a stable cluster structure according to the routing rules of LEACH protocol, the sink node collects the information tables of the nodes in the cluster and the cluster head, and constructs BP neural network. Then the sink node collects the samples matching the cluster member information in the sample database for training, and obtains the neural network parameters of the cluster member and the cluster head node. Finally, the bpnda data fusion algorithm is applied to each cluster. The cluster node transmits the original data to be fused to the cluster head. The cluster head node uses the BP neural network data fusion algorithm to fuse the collected information and transmits a small amount of eigenvalues of the reaction information to the sink node. Therefore, the application of bpnda in wireless sensor networks reduces the amount of data transmission between cluster head nodes and sink nodes, reduces the energy consumption of nodes, and prolongs the life cycle of WSN.

In the process of applying BP neural network algorithm to the data fusion of WSN, the wireless sensor network area has the following specified assumptions [9].

1. The ID of each sensor node is unique, its energy is limited, and the energy cannot be supplemented during the whole experiment. Its position will not move in theory after the deployment is completed.
2. The sink node is unique and fixedly distributed outside the region. Its energy can be continuously supplemented, and it has enough power to send data information to ordinary nodes, which cannot.
3. The position coordinates of all nodes can be obtained.

In BP neural network, the establishment of network model needs to set network parameters, such as the weight and threshold between neurons. These parameters can be trained by the information to be measured of BP neural network. After the wireless sensor network forms a stable clustering structure through LEACH protocol, before BP neural network is applied to wireless sensor network, BP neural network needs to be trained to obtain the weight and threshold value. Because the energy may be consumed greatly in the process of network training, resulting in the shortening of network life cycle, the data fusion algorithm bpnda will complete the training of neural network in the sink node of wireless sensor network and obtain relevant parameters.

In the process of data fusion of BP neural network, this paper selects a three-layer neural network: input layer, output layer and hidden layer. The input layer and output layer have only one node, and the hidden layer has k nodes.

The essence of data fusion using BP neural network is to train the weight by gradient descent according to the output error function, so that the error value tends to the minimum. In this process, the weight adjustment method is shown in the following formula.

$$W^{n+1} = W^n + \Delta W \tag{4}$$

$$\Delta W = -\eta \frac{\partial E}{\partial W} = \eta \delta O \tag{5}$$

where, W is the weight of neural network; E is the output deviation of neural network; η is the gradient descent learning rate; δ is the partial derivative ratio; O is the output of neural network. Formula (4) is a typical weight adjustment formula in BP neural network, which indicates that the next training weight W^{n+1} is the sum of the current weight W^n and the change rate ΔW of the current weight relative to the output error. Because the BP neural network algorithm adopts the gradient descent method, the weight adjustment in formula (5) is negative, indicating that the value of the error function gradually tends to decrease with the weight adjustment.

The error function of BP neural network is a multi-dimensional function about the weight W. The stereo image constructed by BP neural network has multiple extreme points, and the surface is in a steep state when it is adjacent to the extreme points. The change rate ΔW of the weight here with the error function is too large, resulting in too large adjustment range of the weight, and it is easy to miss the error minimum points in the adjustment process, resulting in oscillation. In the flat area of the image, if the change rate of the weight of the error function is too small, it may be mistaken that the error function has converged to the extreme point. This misjudgment will stop the training and not get the best value, and if the change rate of the weight is too small, the training time of the network will be prolonged.

On the basis of the adjusted weight term α, therefore:

$$W^{n+1} = W^n + \alpha \Delta W \tag{6}$$

Momentum term α is generally taken as a random constant between $(0, 1)$. Because the momentum term α is added on the basis of the weight change rate, the range of the change rate of the error function is relatively reduced. When the weight value converges on the steep slope of the surface, it can make up for the defect of missing the best minimum value or causing oscillation due to excessive adjustment range to a certain extent.

However, the defect improvement caused by flat area is not large, and the decrease of the adjustment range of weight will lead to slowing the adjustment speed and the longer time of weight training. This paper also considers the weight change rate of the first three times in the weight adjustment degree, and the improved weight adjustment degree is shown in formula (7).

$$\Delta W = \Delta W^n \pm \frac{(\Delta W^{n-1} + \Delta W^{n-2} + \Delta W^{n-3})}{\Delta W^{n-1}} \tag{7}$$

In formula (7), $\pm$ means that the positive sign is applied when the weight change rate ΔW^n is greater than 0, otherwise the negative sign is used to ensure that the value of the error function gradually tends to decrease with the weight adjustment. ΔW^{n-1}, ΔW^{n-2} and ΔW^{n-3} represent the weight change rates of the first three times, and its value can be obtained from formula (5).

After determining the parameters in the BP neural network, the data fusion of the sports event data acquisition wireless sensor network is conducted according to the flow chart shown in Fig. 1.

Firstly, the wireless sensor network is initialized to determine the initial state of each node in the network, including the initial energy, ID and location information of each node. Then, leach clustering routing protocol is used to select the optimal cluster head of the network, and a stable cluster model of wireless sensor networks is established. At this time, each node in the cluster sends its own information table to the cluster head of the cluster, and each cluster head transmits the information to the sink node. The sink node constructs a typical three-layer BP neural network model according to the specific requirements of the information to be fused. At the sink node, the relevant samples are trained by BP neural network to obtain the required parameter information. Finally, the sink node assigns the trained network parameters (weight and threshold) to each corresponding node, including nodes in the cluster and cluster head nodes. Finally, the wireless sensor network can use the trained BP neural network model to process

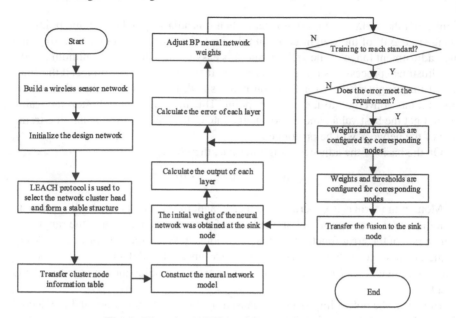

Fig. 1. Flow chart of BP neural network data fusion

the information collected by the wireless sensor network. The common nodes in the cluster are located at the bottom of BP neural network. The information collected by the wireless sensor network will be initialized by the input neuron function, and the processing results will be transmitted to the cluster head node. The cluster head node further fuses the information data according to the hidden layer function and the output layer function, and transmits the fusion results to the aggregation node. Based on the momentum term, the dynamic optimization of the weight adjustment degree Δw can make up for the defects of oscillation caused by excessive adjustment amplitude or missing the optimal minimum to a certain extent, make the final output result more accurate, and shorten the training cycle of BP neural network. After using BP neural network to fuse the data of sports events, the data collection of sports events is realized.

1.2   Realize Sports Event Data Collection

Sports event data collection needs accurate positioning information, and it is meaningless to study the lack of positioning information. This paper uses compressed sensing to locate the wireless sensor network nodes of sports event data collection. Compressed sensing can losslessly recover the signal from low rate sampling, and the probability of recovery is very high, mainly because it has the following two characteristics: the original signal has the characteristics of sparsity; Perception matrix and transformation basis are not related. As an important part of compressed sensing, signal reconstruction refers to the process of solving sparse signal X through observation matrix Y. Where Y is M dimensional, X is n-dimensional and satisfies $M \ll N$. The MP matching tracking algorithm is used to reconstruct the signals of wireless sensor network nodes for sports event data acquisition. The main idea of MP algorithm is to select an atom in the perception matrix after each iteration, which is the best matching item of the signal. The specific process is to calculate the residual after sparse approximation, select the obtained residual, and take the residual as the atom of the best matching item. After repeated iterations, the signal can be represented by the linear combination of atoms. The algorithm needs more iterations and is more complex, because the signal has the problem of non orthogonalization in the projection process. When the sparse matrix and perception matrix are not related to each other, the same solution will appear when solving the l_0 norm optimization. At this point, it can be transformed into linear programming.

The sensing area is set as a regular polygon structure, which is mainly to divide the sensing area into multiple honeycomb structures of equal size. Among them, k sensor nodes are deployed in the sensing area, and the specific IDs of these nodes are represented by set $\{S_1, S_2, \cdots, S_k\}$. These nodes are mainly responsible for monitoring and collecting the sensing signals in the target area.

In the process of network clustering, this paper mainly uses a uniform clustering method to divide the sensing area into multiple honeycomb clusters of equal size. Different from the previous clustering methods, this clustering method can be applied to the network environment with irregular shape. The area of the cluster is calculated by its fixed side length, which not only ensures that each sub cluster covers as many nodes as possible, but also avoids the communication blind area between clusters. In addition, in order to ensure full coverage of all nodes, the number of clusters is determined by the size of the sensing area.

After selecting the cluster head node of each cluster according to the reason protocol, first, the cluster head node will broadcast its ID number, specific location information and the remaining energy capacity of the node to all member nodes in the cluster by sending broadcast packets. After that, the ordinary member nodes in the cluster are responsible for receiving the broadcast packet and recording the information of the cluster head node. Finally, the ordinary node starts to forward the sensing data it obtains to the cluster head node. The greedy selection strategy can be adopted in the cluster. When the node is far from the cluster head node, the node will select the nearest node in the direction of the cluster head node as its next hop transmission relay node. This transmission method not only reduces the energy consumption of data transmission, but also avoids the packet loss caused by long-chain transmission. When the node is close to the cluster head node, the single hop transmission mode can be adopted. The energy calculation formula is as shown in formula (1) (2). In order to avoid long-chain transmission, the transmission distance in the cluster will not be greater than the set threshold value y_0, and the energy consumption calculation of free space model is adopted for data transmission. When the node is far away from the cluster head node, the multi hop transmission mode is adopted. The node will first send the data to the relay node st_j, which is responsible for receiving and forwarding the data to the cluster head node. After the data transmission in the cluster is completed, the sink node starts to collect the information of each cluster head node. Multi hop transmission mode is adopted between cluster head nodes. After minimizing the data transmission between cluster head nodes, it is necessary to forward the data to the sink node to realize the data collection in the sensor network. According to the above contents, the research on the sports event data acquisition method based on

BP neural network and wireless sensor technology is completed. The data acquisition method can improve the data transmission efficiency and reduce the energy consumption of wireless sensor network.

2 Test Research and Analysis

2.1 Test Preparation

This section will test and analyze the sports event data acquisition method based on BP neural network and wireless sensor technology. The simulation environment parameters of this test are: 200 identical network nodes (except geographical location) are randomly deployed in the 200 m * 200 m area, the sink node is located at the origin of the area, and 1200 cyclic experiments are carried out in wireless sensor networks. Distributed environment is configured with Hadoop framework and has Cloudera Hadoop version; serial environment is ordinary PC and Intel i5-9400F processor. The experimental platform was a Cluster consisting of multiple nodes consisting of 18 GB RAM and 2.98G 8 nuclear Intel Xeom X9870 CPU. Hadoop node cluster, and cluster communication is based on MPI library. The operating system of Win10 is used.

In order to simplify the test environment, a single type of sensor with the same parameter is used to obtain the test environment data. The specific sensor parameters are shown in Table 1.

Table 1. Test the sensor parameters

Number	Parameter	Numerical value
1	Sensing area side length	25 m
2	Number of sensor nodes	200
3	Communication radius of sensor node	80 m
4	Database size	10 MB
5	Cluster message length	50 bit
6	Baotou length	30 bit
7	Node initial energy	1.5 J
8	RF energy consumption coefficient	75 NJ/bit
9	Power consumption factor of the power amplification circuit	15 NJ/bit/m^2
10	Distance threshold	95.2 m

Reference [3] and Reference [4] are compared with this method. By comparing the energy loss of sensor nodes and the same amount of data acquisition time after using different methods to collect data, we can measure whether the performance of the data acquisition method can meet the requirements of sports event data acquisition.

2.2 Test Results

In the same experimental environment, the same group of sensors obtain the environmental data, and use three data acquisition methods to collect the data to the sink node. Take the average value of 100 tests as the final data of the current group of tests. Select Microsoft SQL Server 200 as the tool for building the data warehouse; and the data mining module developed by Microsoft Visual Basic6.0 is used to process the relevant data, and the processing results are stored in the SQL Server database. Table 2 shows the energy loss ratio of nodes and the comparison of data acquisition time in each group of data acquisition.

Table 2. Comparison of the energy loss ratio and the data acquisition time for the node

Group	Data acquisition method based on BP neural network		Reference [3] method		Reference [4] method	
	Energy loss ratio/%	Data acquisition time/ms	Energy loss ratio/%	Data acquisition time/ms	Energy loss ratio/%	Data acquisition time/ms
1	6.44	4.15	9.54	10.23	13.78	11.74
2	6.52	3.81	8.62	11.36	14.84	11.61
3	7.46	3.76	9.18	10.95	14.42	11.50
4	6.83	3.72	8.31	10.83	11.83	12.27
5	7.37	4.05	8.24	12.21	14.60	10.41
6	5.76	4.23	8.99	11.12	12.34	11.46
7	6.28	4.37	8.53	10.28	11.46	11.63
8	6.19	3.91	8.75	12.05	12.21	10.62
9	7.56	4.02	9.27	12.26	11.68	11.85
10	6.13	4.28	9.08	12.51	14.67	11.31
11	5.71	3.74	8.43	11.08	12.55	11.64
12	7.02	3.82	8.86	12.37	12.05	11.67

From the data analysis in Table 2, it can be seen that after the method collects data, the energy loss proportion of the sensor node is the smallest, with an average of 6.61%, which is higher than the average energy loss proportion of the reference [3] method of 8.82% and the average energy loss proportion of the reference [4] method of 13.04%. The data acquisition time of this method is significantly less than that of the other two comparison methods, which shows that the acquisition rate of this method is higher and the effect is better when applied to the data acquisition of sports events.

Summarizing the above test data, the sports event data acquisition method based on BP neural network and wireless sensor technology proposed in this paper can quickly collect sports event data, prolong the working time of sensor nodes and improve the efficiency of data acquisition.

3 Conclusion

In the process of sports events, the statistics, summary and analysis of the technical data of each game will help to improve the management and technical level of sports events. Athletes can find their problems in the competition by summarizing their various performances in each competition, so as to continuously improve their technical level; Referees can find their own mistakes in enforcing the competition by summarizing the data; Managers can find the lack of athlete training from the sports event data. The collection and statistics of technical data for sports events have been widely used in sports events in various countries. In this paper, a sports event data acquisition method based on BP neural network and wireless sensor technology is proposed. Through the comparison test with the current data acquisition methods, the effectiveness of the data acquisition method is verified.

Fund Project
2020 Research Foundation Ability Improvement project of Young and middle-aged Teachers in Guangxi Colleges and Universities: Research on the Integration development of Guangxi County tourism Industry and mass sports Events, Project number: 2020KY26007.

References

1. He, J., Kong, J.: Data acquisition system and its history, present situation and trends in the CFA super-league. J. Guangzhou Sport Univ. **41**(04), 67–70 (2021)
2. Cao, A., Zhang, H., Wu, Y.: A survey on visual analysis of ball games. Sports Sci. Res. **42**(03), 26–36 (2021)
3. Hong, Y., Guo, C.: Data collection method of underwater sensor based on K-means and SOM. J. Data Acquis. Process. **36**(2), 280–288 (2021)
4. Li, Q., Jia, Y., He, C.: Synchronous acquisition system for the high resolution fluxgate sensor. J. Nanjing Univ. Sci. Technol. **44**(1), 7–14 (2020)
5. Xu, W.: Protection of data rights of sports event organizers in the era of digital sports. China Sport Sci. **41**(07), 79–87 (2021)
6. Xu, Y., Duan, L.: Node redeployment of wireless sensor network based on leapfrog algorithm. Comput. Simul. **38**(10), 328–332 (2021)
7. Hong, Y., Guo, C.: Data collection method of underwater sensor based on K-means and SOM. J. Data Acquis. Process. **36**(02), 280–288 (2021)
8. Zhang, M., Cai, W.: Dubins curves based mobile data collecting algorithm for clustered wireless sensor networks. Chin. J. Sens. Actuators **32**(04), 603–609 (2019)
9. Zhang, L., Zhang, M., Ji, W., et al.: Virtual acquisition method for operation data of distributed PV applying the mixture of grey relational theory and BP neural work. Electr. Power Constr. **42**(01), 125–131 (2021)

Recommendation Method of Ideological and Political Mobile Teaching Resources Based on Deep Reinforcement Learning

Yonghua Wang[✉]

Sanya Aviation and Tourism College, Sanya 572000, China
wangyonghua00060@163.com

Abstract. In order to improve the quality of ideological and political education and achieve the goal of effective management of mass mobile teaching resources, this paper puts forward a recommendation method of ideological and political mobile teaching resources based on deep reinforcement learning. Firstly, based on the theory of deep reinforcement learning, the recommendation model of ideological and political mobile teaching resources is constructed, and the recommendation method of ideological and political mobile teaching resources is extracted effectively.

Keywords: Intensive learning · Ideological and political mobile teaching · Mixed resources recommendation

1 Introduction

The mixed resources of ideological and political education are the digital resources used to support the teaching activities of ideological and political education based on information technology. With the development of Internet and Big Data technology, the access to learning resources is becoming more diversified. Different learners can choose appropriate learning resources according to their preferences to carry out learning and achieve personalized learning goals. Many technologies are widely used in mobile ideological and political learning [1]. The common recommendation algorithms of ideological and political mobile teaching resources include content-based recommendation, in-depth reinforcement recommendation and mixed recommendation, etc. Content-based recommendation algorithm constructs learner feature model and resource feature model by recognizing and extracting resource content feature, and recommends learning resource to learners. Deep reinforcement learning algorithm divides learners into groups based on different preferences by mining their preferences, and recommends similar learning resources to each group [2]. Deep learning method has strong learning ability, wide coverage and good adaptability. The neural network of deep learning has many layers and wide width, which can be mapped to any function in theory and can solve very complicated problems. Deep learning is highly dependent on data, and the larger the

W. Fu and L. Yun (Eds.): ADHIP 2022, LNICST 469, pp. 257–272, 2023.
https://doi.org/10.1007/978-3-031-28867-8_19

data, the better the performance. Deep learning can use TensorFlow, Pytorch and other frameworks. Therefore, the application of deep learning can effectively improve the recommendation effect of mixed resources in ideological and political mobile teaching.

Deep reinforcement learning algorithm can effectively reduce the complexity of model building, but there are some problems such as sparse matrix and cold start. Based on the introduction of the theory of heat conduction and material diffusion, such as Liu Zhongbao and so on, a bipartite graph method is proposed to recommend learning resources to learners. Generally speaking, the rationality and scientificity of resource recommendation model is always a difficult problem for traditional recommendation algorithms to be applied to the field of hybrid learning resource recommendation [3]. Therefore, this paper proposes a recommendation method of ideological and political mobile teaching resources based on deep reinforcement learning.

2 Recommendation of Ideological and Political Mobile Teaching Mixed Resources

2.1 Mobile Ideological and Political Learning Content Extraction

The realization of blended learning resources recommendation is essentially an analysis of the relationship between learners and learning resources. In order to provide personalized resource recommendation service, the key is to collect the original data of the learning platform, analyze and mine the data effectively, and finally recommend appropriate learning resources to the learners [4]. The hybrid learning resource recommendation model can utilize the historical aggregation information of learners' learning resources. This information can be represented by the mxn matrix on the left side of the graph, where R represents the learning resource and L represents the learner. The shaded portion indicates the learner's learning resources. The white blank space indicates the unlearned resources. The problem to be solved is how to realize the recommendation of mixed learning resources through this historical information matrix, that is, to obtain the recommended resources that meet the learners' needs from the new learning resources. The principles of recommendation of mixed resources for mobile teaching are shown as follows (Fig. 1):

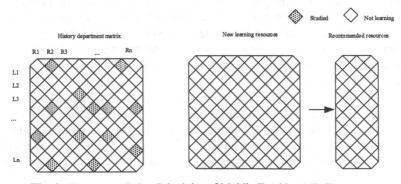

Fig. 1. Recommendation Principles of Mobile Teaching Mix Resources

The personalized recommendation method based on deep reinforcement learning can be summed up in two processes: model training process and resource recommendation proces. The model training process includes data processing of learning platform, algorithm design and so on, and the recommendation process is produced by the running of the recommendation model. The flow chart of teaching resource recommendation is shown in Fig. 2.

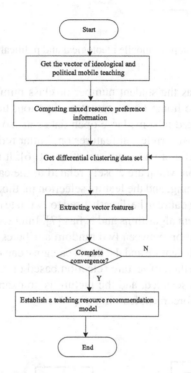

Fig. 2. Teaching resource recommendation process

The basic principle of the DSLL algorithm is as follows: a learner -learning resources scoring matrix is formed according to the learner's historical learning records, and the similarity measure method is used to mine the learner set similar to the target learner or the learning resources similar to the learner's historical preference by calculating the similarity between the learner and the learning resources, then a "neighbor" is formed based on the "neighbor" learner's scoring information to predict the target learner's predictive scoring value for the learning resources, and personalized recommendation is made for the target learner. If we strengthen the recommendation of deep learning, it will recommend the first N learning resources with the largest predictive score to the target learners. The specific steps of the content extraction method for ideological and political learning based on deep reinforcement learning are as follows (Fig. 3):

Mobile thought politics learning data contains learners' learning behavior records, there are also many implicit data, learning resource characteristics can also be obtained in the data [5]. There are irrelevant features or redundant features in the actual data

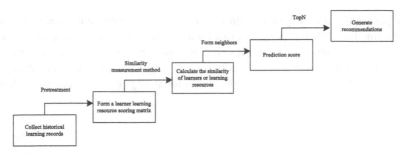

Fig. 3. Extraction steps of mobile ideological and political learning content

processing process, such as the student number or class number of a learner may be irrelevant features, and the home town, parent occupation, home address, etc. of the learner may also be recorded in some large open data sets, so it is necessary to screen or select features, and remove irrelevant features or similar redundant features to make the recommendation model more accurate and efficient [6]. It is necessary to select the features or data information which are closely related to the category in the practice of Deep Reinforcement Learning, and the feature selection method can be used. Generally, the performance of training data can be directly used to evaluate the characteristics, which is independent of subsequent algorithms and is fast [7]. Interactive information refers to the strength of the association between two random attributes or features. Judging the correlation between a single feature and the target category can reduce the redundancy of feature dimension. The method of feature selection based on mutual information (deep reinforcement learning) is selected, and the picture is an example of feature selection model based on deep reinforcement learning (Fig. 4).

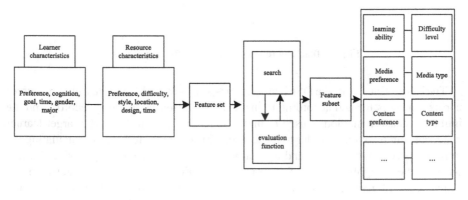

Fig. 4. Example of feature selection based on deep reinforcement learning

The Deep Learning Model consists of network nodes with multiple hidden layers. Networks with multiple hidden layers have powerful feature representation ability, and use multi-level nonlinear structure to abstract low-level features into high-level representation, so as to discover the internal relationship between data. The biggest difference between deep learning and multi-layer perceptron is that multi-layer perceptron needs to select feature input network manually. Deep learning can learn features autonomously, mine implicit feature representation from data, and depict intrinsic information of data. Deep learning generally constitutes a greedy hierarchical approach, with continuous learning and integration from the underlying inputs, and selection of effective features to improve the end result performance. For supervised learning tasks, deep learning analysis is similar to principal component analysis in that it converts data into compact intermediate classes, constructs hierarchical results, and eliminates redundant information. At present, deep learning has been applied in image recognition, intelligent speech, unmanned speech, natural language processing, medical health, etc.

2.2 Evaluation Algorithm of Ideological and Political Mobile Teaching Mixed Resources

Mixed-resource recommendation is one of the indispensable components of personalized learning, and it is also the key technology to realize personalized learning. Common personalized learning recommendation mainly realizes personalized recommendation process by establishing learner model, recommending algorithm processing and recommending result outputting. So, the structure of personalized learning recommendations, like this (Fig. 5).

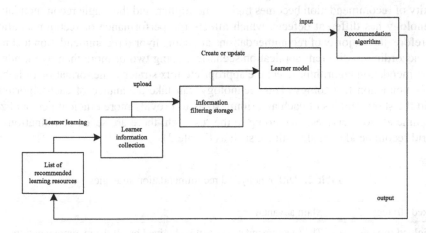

Fig. 5. Schematic of recommended structure for mixed resources

The recommendation algorithm of Deep Reinforcement Learning provides the core function support for recommendation, and the quality of recommendation results will directly affect the performance evaluation of recommendation. Based on the introduction of the main recommendation algorithms above, the advantages and disadvantages of these common recommendation techniques are compared as follows (Table 1).

The trust between learners in social network is difficult to be obtained by explicit way, because it takes the learners' time to input the trust of friends initiatively, and it can

Table 1. Common Recommended Techniques Comparison Table

Recommended technology	Advantage	Shortcoming
Recommendation algorithm based on Popularity	It is simple and easy, and there is no problem of cold start	Unable to make personalized recommendation
Collaborative filtering recommendation algorithm	It is simple and easy to operate with high recommendation accuracy	Rely too much on users' scores; cold boot; Low project coverage
Content based recommendation algorithm	No cold start problem	The recommended content is relatively simple
Model based recommendation algorithm	Fast and accurate, especially for businesses with high real-time performance	The model needs to be maintained frequently

not be replaced by fixed value because the trust between learners is in dynamic change. To some extent, the interaction between learners and friends in social networks reflects the trust relationship between learners and friends, which mainly includes comments and forwarding. The more times the learners comment on each other's messages, the more familiar they are with each other, and the more times the learners forward each other's messages, the more they agree with each other and trust each other. As the complexity of recommendation becomes higher and higher, and the single recommendation technology has different defects, which affects the performance of recommendation. Therefore, the majority of recommendations are using hybrid recommendation technology, according to different application scenarios, using two or more than two kinds of recommendation algorithm, select the appropriate mix strategy, the formation of hybrid recommendation technology. This technology can take advantage of each algorithm, avoid the shortcomings of each algorithm, so as to provide more efficient personalized recommendation services. According to the research, there are seven combinations of hybrid recommendation algorithms, such as (Table 2).

Table 2. Different hybrid recommendation strategies

Mixed strategy	Main advantages
Weighted mixing	The recommendation results obtained by all the recommendation algorithms used in the system are weighted and divided to obtain the final and optimal recommendation results, and then recommended to users
Switch mixing	When the recommendation system adopts the recommendation algorithm, it selects different recommendation algorithms according to different application scenarios and different users

(continued)

Table 2. (*continued*)

Mixed strategy	Main advantages
Cross mixing	In the recommendation system, the recommendation results obtained by various recommendation algorithms are mixed for recommendation
Feature combination	In the process of recommendation, the features of different data sources are combined to form vector features, and then the similarity is calculated and recommended
Feature Augmentation	The feature model of a recommendation algorithm is input into other recommendation algorithms as features to improve the recommendation performance
Graded mixing	The model constructed by one or more recommendation algorithms is used as the input of other recommendation algorithms, and then the recommendation results are generated
Series mixing	Multiple recommendation algorithms are used in series, that is, the later ones further screen and optimize the previous recommendation results to obtain the final recommendation results

In fact, learners usually comment on messages that they are interested in or unclear, but do not necessarily forward them, and the messages they forward are often self-selected and highly identified, so the weight of forwarding should be slightly greater than that of the comments. Inspired by the scholar Hu Xun and others on the method of computing the trust between mobile users, the trust between learners in the network environment can adopt the following formula

$$t_{u_1,v} = \frac{c_{u_1,v}}{\max c_{u_1,u}} + (1 - \alpha) \times \max f_{u_1,u} \tag{1}$$

In the formula, $f_{u_1,u}$ represents the trust between the learner u1 and the learner u; α represents the set of learners who interact with the learner, and $c_{u_1,v}$ the number of comments the learner has made on the message published by learner V. Since the forwarding behavior should be slightly more weighted than the commenting behavior, the $t_{u_1,v}$ value is 0.4. In the feature selection method based on deep reinforcement learning, information metric evaluation function is very important. Although the functions are various in form, the aim is to select the subset of features that are most relevant to the category, which is recorded as $g(C, f, S)$. The generalized information metrics evaluation function can be expressed as

$$J(f) = AB - g(C, f, S) - \delta t_{u_1,v} \tag{2}$$

Among them, A is the selected feature, δ is the candidate feature, B is the category, and the function is the information between C, F and S, that is, the correlation between the candidate feature and the category, β is the adjustment coefficient, which is used to adjust the information brought by the addition of W, and x is the punishment factor.

The simplest and most intuitive information metrics evaluation function can therefore be expressed as follows:

$$T = g(C, f) - \beta \sum_{s \in S} Wx - J(f) \tag{3}$$

The selected features indicate other features that will affect the learner's choice of resources, such as the knowledge content of the resources, the length of time of study, etc. Candidate features represent other features that are not known for the time being, such as the learner's age, professional background, gender, etc. The category is the extraction of the selected features, which is used to measure the relevance between the selected features and the candidate features. Finally, the evaluation function is constructed to judge the influence of candidate features on the evaluation results, and to filter out some redundant features so as to reduce the workload of deep learning training.

2.3 Realization of Mixed Resource Recommendation of Ideological and Political Education

Ideological and political education resources recommend to provide learners with a learning environment in line with their individual characteristics by analyzing the differences among learners. The meaning of personalized learning mainly includes the following aspects: First of all, personalized learning refers to taking flexible, appropriate and pertinent teaching resources, teaching strategies, teaching evaluation and other learning services according to the students' learning ability, learning style, learning attitude and other learning characteristics, so that students can get all-round development and progress. Secondly, the learning process advocated by personalized learning is to help students find their own strengths, highlight their own personality, the pursuit of self-realization. Finally, personalized learning embodies the essential difference between learning and education: learning is through the acquisition of knowledge and skills to achieve self-improvement. Education is to teach students some knowledge and skills, in accordance with the training objectives step-by-step training of students, to a certain extent, buried in the initiative and creativity of learners. In the study of recommendation of learning resources, the classical deep reinforcement learning algorithm can effectively train the history learning data, but the traditional deep reinforcement learning algorithm or simple network model can not meet the actual needs, and can not guarantee the convergence to an optimal solution. This paper designs a deep reinforcement learning model (see figure) to judge whether or not the learner learns a certain learning resource or how much attention is paid to the learning resource (Fig. 6).

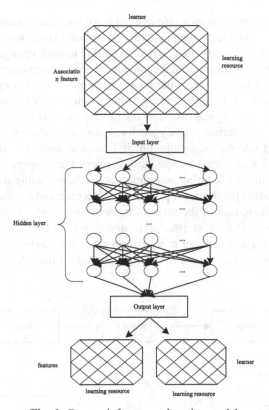

Fig. 6. Deep reinforcement learning model

The key of applying DSLL to the recommending scene of learning resources lies in modeling the history learning records of learners, mining the implicit features of the original data, and then standardizing the input layer and output layer in constructing the training model. Above, feature selection model based on DLL and learner-resource bipartite graph association model effectively solve the input and output of DLL. In the model training of deep reinforcement learning, a large number of transformations (s, a, r) will be obtained by interaction with the environment, but updating the parameter value of Q network according to each transformation will make the network fall into local optimization, and then forget the previous learning experience. At the same time, it takes a long time for the network to back-propagate and update the parameters, and the training efficiency of the whole model becomes low. So, in the deep reinforcement learning algorithm, experiential playback technology is used to solve this problem to break the dependence of time sequence and avoid the local optimization of the network. At the same time, it can accelerate the training by random microbatch data, break the similarities between samples, and accord with the learning and improving behavior of human using past experience. The use of experiential playback simplifies the debugging and testing of the algorithm, and makes the training task of the network model more similar to the common supervised learning mode.

In the model training of deep reinforcement learning, a large number of transformations (s, a, r) will be obtained through interaction with the environment. However, if the parameter values of Q network are updated according to each change, the network will be trapped in local optimization, and the previous learning experience will be forgotten. At the same time, it takes a long time for the network to back-propagate and update the parameters, and the training efficiency of the whole model becomes low. So, in the deep reinforcement learning algorithm, we use the experience replay technology to solve this problem, use the experience replay mechanism to break the time dependence, avoid the network falling into the local optimum, and can accelerate the training through the random microbatch data, break the similarity between the samples, at the same time, it accords with the human's learning and improving behavior using the past experience. The use of experiential playback simplifies the debugging and testing of the algorithm, and makes the training task of the network model more similar to the common supervised learning mode. The transformation is stored in the experience playback pool as experience, and a random number of transformations are randomly sampled from the experience playback pool for each run to form a random microbatch data for training the network (Fig. 7).

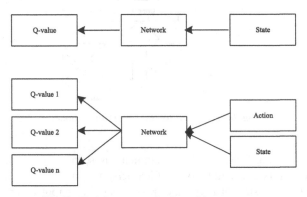

Fig. 7. Training network of teaching resources

Resource management subsystem is the management and maintenance part of adaptive learning. It is mainly used by teachers, educational administration and administrators. Specifically, teacher users shall realize the management of addition, deletion and check of curriculum related resources (courses, courseware, examinations, etc.); educational users shall realize the management of addition, deletion and check of curriculum resources and announcement information; and administrative users shall realize the management of addition, deletion and check of resources (users, logs, data, etc.). The resource management subprocess, shown in Fig. 8.

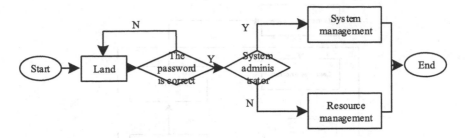

Fig. 8. Resource Management Subflowchart

For the users who use the resource management subsystem, the login verification of user account and password is carried out first, and then the identity is determined after the success. For the administrators, the management of all resources can be realized, while for other users, the corresponding part of the resource management is carried out according to the identity. Ideally the algorithm should be able to give moderately difficult resources as reflected in the answer rate that eventually converges to a fixed number. Let the learner do the resource search item N each time, the number of correct answers N each time, the construction deviation measurement index is.

For the users who use the resource management subsystem, the login verification of user account and password is carried out first, and then the identity is determined after the success. For the administrators, the management of all resources can be realized, while for other users, the corresponding part of the resource management is carried out according to the identity. Ideally the algorithm should be able to give moderately difficult resources as reflected in the answer rate that eventually converges to a fixed number. Let the learner do the resource search item N each time, the number of correct answers N each time, the construction deviation measurement index is

$$err = \left(\frac{n}{N} - \delta\right)^2 \tag{4}$$

The actual experiment looked at the mean of recommended deviations per turn (epoch or designated segment). Deep reinforcement learning is a learning method that combines deep learning and reinforcement learning to solve the problems in control and decision field and to realize one-to-one correspondence from perception to action. Deep learning is used to analyze environmental information and extract features from it, and reinforcement learning is used to further analyze environmental features and select corresponding actions to achieve target return. For the complex decision making problem of learner's cut-set selection strategy, the deep reinforcement learning method is used to extract effective information from the learner's historical query data environment, and the decision-making control is realized by combining the environmental information and the learner's response. The model is constructed as shown in Fig. 9.

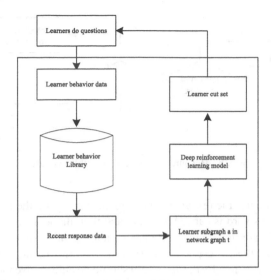

Fig. 9. Guided resource recommendation model based on deep reinforcement learning

With the development of "Internet Plus Education", the research on personalized learning recommendation algorithm in the field of education has become a hot spot in recent years. However, the current recommendation algorithms in the field of education rely more on the traditional linear recommendation methods, not taking into account the relevance between knowledge points and the important characteristics of the subject. At the same time, the current personalized learning recommendation algorithm can not recommend moderate difficulty for users, facing the recommendation efficiency is not high, the user answer rate is not stable. Due to the complexity of learners' learning ability level and the variability caused by the learning process, the processing ability and learning ability of deep reinforcement learning in large state space provide a new direction for learners to recommend learning resources.

3 Analysis of Experimental Results

In order to verify the conformity of personalized learning resources recommended by hybrid learning resources recommendation method based on deep reinforcement learning with learners' needs, a series of experiments were conducted. The experimental data includes not only the data of learning resources, but also the historical data of learners' learning. In the existing public datasets, such as edX, World Uc, and others, dozens of attributes are provided, including course data, learner information, and learner behavior data. Form the experimental data set. Guided resource recommendation model based on deep reinforcement learning algorithm doesn't know how to select learning area for users at the beginning of setting questions for users, and the parameters of deep reinforcement learning network are randomly initialized according to Gaussian distribution. Therefore, the selection of topics of deep reinforcement learning model at the beginning is also random. The model strengthens the ability of selecting test strategy in the process of giving test questions to users, updates the weight value in the network according to the return value returned from the user's historical data, and learns to choose the strategy of learning area questions for users after repeated training. This experiment sets 20000 rounds of training for each user, each model gives 10 questions to the user. The model parameter settings in the experiment are shown in the Table 3.

Table 3. Experimental parameter settings

Parameter	Size/content	Describe
Q	20000	Number of experiment rounds per user
E	110	Take the current status and answer the previous question
T	Comprehensive difficulty value	Comprehensive difficulty value of the first e question
Y	12	Number of topics selected each time
K	5	Number of topics with the highest ranking
B	6	Number of topics in the middle
L	5	Number of topics with the lowest ranking
M	0.9	Optimal Recommendation Index
P	5	Return value of err $= 0$
U	0	Return value at the end of turn

The learner -resource features, i.e., the subset of features to be input in the whole process, can be obtained by using the model processed by feature selection method based on deep reinforcement learning. Many deep reinforcement learning tasks have the characteristics of table discretization, as shown in the table. For example, a study record shows that the study resources belong to the computer category, the difficulty is easy, the media type is video format, and the learners study at 9: 00 AM. Description of resource association characteristics and their values are (Table 4):

Table 4. Part Description and Numerical Value of Resource Association Characteristics

Features	Numerical representation	Significance
Subject attribution/preference	1, 2, 3, …, 9	Including computer, economic management, literature and history, life science, art and design, etc.
Media type/preference	1, 2, 3, 4, 5, 6	Including video, audio, text, pictures and slides

(*continued*)

Table 4. (*continued*)

Features	Numerical representation	Significance
Difficulty level	1, 2, 3, 4, 6	Including easy, easy, moderate, difficult and difficult
Content type/preference	1, 2, 3, 4, 5	It includes concept explanation, test questions, cases, introduction and course review
Study time	1, 2, 3, 4, 5	Starting from the morning, every 5 h is a time period
Equipment terminal system	1, 2, 3, 4	ISO,Android,Windows

Accuracy refers to the proportion of the resources recommended to the learner's interest in the total resources recommended to the learner, while recall refers to the proportion of the resources recommended to the learner's interest in the total resources. The recommended list of recommendations is N = 5, 10, 15, 20, 25, 30, 35, 40, respectively. The recommended accuracy and recall rates of the traditional methods and the proposed methods are measured, as shown in the Fig. 10.

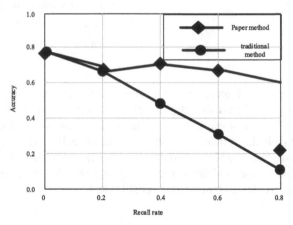

Fig. 10. Accuracy and recall of the two methods in different cases

In this paper, a scoring matrix of learners' learning resources is established according to learners' historical learning records, and the similarity measurement method is used to mine the target learners of similar learners. Therefore, the recall rate of this method is high (Fig. 11).

Because this method adopts hybrid recommendation technology, it can improve the mobile resource integration method according to different application scenarios, so it can effectively avoid the shortcomings of the algorithm and provide efficient personalized recommendation service. Therefore, the method of this paper carries out the coverage

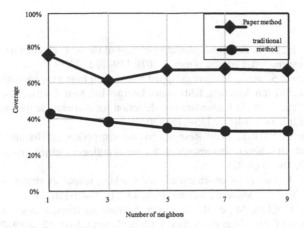

Fig. 11. Coverage of two methods with different number of neighbors

experiment under the condition of 9 neighbors. Experimental results show that this method is superior to traditional methods.

It can be seen that in the same environment, the accuracy and recall rate of the proposed method are higher than those of the traditional method, which shows that the proposed method has better performance and can recommend more accurate and more interesting learning resources for learners.

4   Conclusion

With the rapid development of "Internet Plus" education, the scale of mobile ideological and political learning resources expands rapidly, which makes it more difficult for learners to choose suitable resources. How to help learners acquire appropriate learning resources to carry out personalized learning has become a major research topic of intelligent learning. The key to realize the recommendation of mixed learning resources is to explore and mine the value of data application of mobile ideological and political learning platform. This method is superior to the traditional deep reinforcement learning algorithm in classification and regression ability evaluation index, which shows that it can provide better recommendation service of mixed learning resources in big data environment.

Fund Project. 1. Project of teaching reform of Sanya Aviation and Tourism College in 2020: On the teaching mode of "micro ideological politics" in Colleges and Universities(Project No.:SATC2020JG-12)

2. General topics of Educational Science Planning in Hainan Province: Investigation and Research on labor values of Higher Vocational College Students in the new era (Project No.: QJY20201018). Funded project for the construction of "double leaders" teachers' Party branch secretary studio in Colleges and universities in Hainan Province.

References

1. Jing, K.L., Si-, F., Ya-fu, Z.: Model predictive control of the fuel cell cathode system based on state quantity estimation. Comput. Simul. **37**(07), 119–122 (2020)
2. Su, J., Xu, R., Yu, S., et al.: Idle slots skipped mechanism based tag identification algorithm with enhanced collision detection. KSII Trans. Internet Inf. Syst. **14**(5), 2294–2309 (2020)
3. Su, J., Xu, R., Yu, S., et al.: Redundant rule detection for software-defined networking. KSII Trans. Internet Inf. Syst. **14**(6), 2735–2751 (2020)
4. Zhang, H., Chen, L., Tan, L., et al.: Research on co-construction and sharing of virtual simulation experimental teaching resources among provincial colleges and universities. Exp. Technol. Manag. **38**(05), 26–28 (2021)
5. Shi, Y., Zhang, J.: Design of information based teaching resources sharing system based on multimedia technology. Modern Electron. Tech. **44**(20), 32–36 (2021)
6. Luo, P., Yuan, J., Chen, M., et al.: Mean penalty random forest nonstationary time series prediction method for teaching re-sources. J. Chin. Comput. Syst. **42**, 2089–2094 (2021)
7. Sun, Y.: Design of practice teaching network platform for ideological and political teaching resources in colleges and universities. Microcomput. Appl. **36**(02), 152–155 (2020)

Personalized Recommendation Method of Nursing Multimedia Teaching Resources Based on Mobile Learning

Haitao Zhang[1]([✉]) and Yufeng Sang[2]

[1] Xinyu University, Xinyu 338031, China
zhanghaitao10054@163.com
[2] School of Beijing University of Technology, Beijing 100124, China

Abstract. Due to the variety and quantity of nursing multimedia teaching resources, the resource recommendation method has the problem of low recall rate. To this end, a mobile learning-based nursing multimedia teaching resource recommendation method was designed. First of all, this paper identifies the law of learning needs, annotates the keywords of teaching resources, and collects the data of students' learning records, so as to improve the recall rate of the recommendation results. Build a user interest preference model, improve the nursing multimedia teaching resource recommendation process, and optimize the mobile learning personalized recommendation model. The experimental results show that the recall rates of the proposed method and the other two methods are 78.627%, 70.615% and 70.200%, respectively, indicating that the proposed method has a high recall rate.

Keywords: Mobile learning · Nursing · Multimedia teaching resources · Personalized recommendation · Learning needs · Information resource system

1 Introduction

In the new century, the rapid updating of nursing knowledge and skills not only requires clinical nurses to have critical thinking ability and strong self-learning ability, but also good communication ability and cooperation ability to cope with the pressure from all aspects, which is also an inevitable trend in line with the requirements of higher nursing education. As the backbone course of nursing specialty, "Nursing Fundamentals" is also the core professional basic course of nursing specialty. It plays a role as a bridge between the preceding and the following in the subject system of nursing, and its practical training course has a strong operability and practicality. As a qualified nurse, you must master and be able to flexibly use the knowledge and skills of Basic Nursing. Due to the age, educational level and other factors, nursing college students may be inferior to nursing college students and postgraduates in theoretical knowledge, comprehensive ability and operating skills. In order to understand the principle, mechanism, nature and reason of nursing in theory, we should combine it with basic medicine and clinical medicine.

W. Fu and L. Yun (Eds.): ADHIP 2022, LNICST 469, pp. 273–285, 2023.
https://doi.org/10.1007/978-3-031-28867-8_20

At the same time, require students to nursing skills, to achieve accurate, standardized proficiency. This course has higher requirements for the quality of practical training teaching, teacher-student interaction and students' ability to learn independently. Basic nursing is an introductory course for students majoring in nursing. It is a subject that studies the basic theory, knowledge and skills of nursing. It plays a very important role in nursing education. Because of the limitation of teaching conditions, laboratory management and students' insufficient attention, there are many problems in the basic training teaching of nursing. Both from hospital feedback and students' point of view, there are lack of professional skills mastery and proficiency of clinical interns, leading to heavier work burden of clinical teachers, poor clinical adaptability of students, increasing the incidence of occupational injuries and operational errors. Under the guidance of the concept of holistic nursing, the students should have strong practical skills and basic knowledge of nursing, integrate theory with practice, cultivate the ability of observation, comprehensive analysis and problem-solving, critical thinking and innovation, and be able to make use of the knowledge and skills learned to serve the nursing objects. In order to improve the efficiency of nursing learning and improve the recommendation accuracy of teaching resources. This paper proposes a personalized recommendation method for nursing multimedia teaching resources based on mobile learning. This paper identifies the law of learning needs, marks the keywords of teaching resources, and collects the data of students' learning records. Based on this, the user interest preference model is constructed, and the mobile learning personalized recommendation model is optimized.

2   Personalized Recommendation Method of Nursing Multimedia Teaching Resources Based on Mobile Learning

2.1   Identify Learning Needs

In the law of learning, the progressive requirement of learning needs, information search is the first step of learners through the use of knowledge and information search. Firstly, learners can find useful information and resources through information search, and in the process of searching knowledge information, they can understand the scope of knowledge information resources more directly, deepen their understanding of knowledge information resources and select the knowledge information they need. Then learners will have to search the knowledge of the necessary additions and reductions and integration of information, many additions and reductions and integration of the final formation of their own concepts. Finally, learners fully digest and assimilate the integrated knowledge and information through search. Therefore, in the process of recommendation, educational resources and subject types need to be corresponding. The diversification of learning needs is mainly embodied in the diversification of learners and the diversification of information content.

Now learners are no longer a single existence of a certain industry and field, can be said to involve different fields of different industries, so the urgent need for information diversification, diversification. Whether rational research or empirical research, it is easy to find that learners' needs are the precondition of learning needs, which requires systematic service institutions to provide personalized knowledge and information environment and pay attention to communication and interaction with learners. Keywords of resource characteristics, resource retrieval and resource information sharing need to be tagged with Keywords. According to the learner's learning path and learning interaction to create a corresponding information environment, targeted, distinctive, comprehensive information resource system, to provide learners with a quick, easy learning effective access to learning information.

Different learners are not the same regardless of whether they are educated or have already acquired knowledge. For example, learners who graduated from the same school have different degrees of use of existing knowledge and different degrees of demand for new knowledge. How to choose various multimedia teaching resources and make multimedia teaching resources personalized recommend this modern scientific and technological means to serve school teaching, to improve teaching quality [1–3]. Keywords can reflect the important expression of the content of resources, in addition, the weight information of the keywords of resources, also can reflect the importance of the keywords in the resources. The learning needs pattern is shown in Fig. 1:

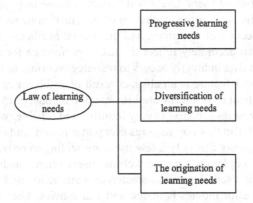

Fig. 1. Structure diagram of learning requirements

As can be seen from Fig. 1, the rules of learning needs mainly include: the progressive nature of learning needs, the diversity of learning needs and the initiation of learning needs. It is of great importance that how to introduce multimedia network technology into teaching so that teachers can deliver more, more intuitive and more accurate information in a specific time and space. The study difficulty descends, thus causes the classroom instruction effect to have a qualitative leap, realizes the multimedia network resources in the teaching optimized design [4].

Another example is that learners with low knowledge base mostly want to learn some universal knowledge, while learners with high knowledge level may pursue some high-level knowledge. Therefore, for learners, the existing knowledge base and level will also have an impact on the hierarchy of knowledge needs. The most objective difference between learners is their age, gender and other natural attributes. If two learners are not the same age, their needs for knowledge are necessarily different. Compared with an old person and a young person, their needs for learning are also completely different. The old person needs less knowledge and the young person needs more knowledge. Of course, the reason for this result has a lot to do with their age.

2.2 Build User Interest Preference Model

User preference information comes from User data, and User data basically includes: User basic information, User learning data, User behavior data. The earliest collection of user preference information is mainly based on user display data, such as: scoring, voting, forwarding, favorites, comments and so on, these data called display score. User's interest, user's characteristic, user's historical behavior and natural attribute are all part of the data feature in the system. After collecting the data, we need to extract the characteristics of the model data from the data. Later, with the development of technology and the need of recommendation precision, the recommendation based on data mining technology is widely studied and used. Through click-through rate, buying behavior, page stay time and so on, the implicit data is obtained, and it is weighted into the user's rating of resources, called implicit rating.

At present, the most popular classification is: explicit information extraction and implicit information extraction. Explicit information collection is a process that requires the active participation of users. Users will actively choose or provide the information they are interested in, such as the evaluation of information items, the sharing, recommendation and collection of resources, and the natural attributes of users. The former is more intuitive, can accurately reflect the user's preference for resources, the latter through the relevant data indirectly access to resources scoring, to make up for the lack of real data sparsity. A feature is an abstract result of an object or of the properties of a set of objects. A feature is a term used to describe a concept [5, 6]. The process of collecting user interest information is very intuitive and reliable, which is very helpful to the interest model. But this way requires user participation, and there is a certain user cost. And research shows that only a few users are willing to provide reliable, explicit information. For example, on a movie website, users often watch action movies and war movies, then the user's interest characteristics are action and war, indicating that the user's preference for movies is action and war movies. User interest characteristics are affected by many factors, such as gender, age, occupation, and behavior. User characteristics can be described in Fig. 2:

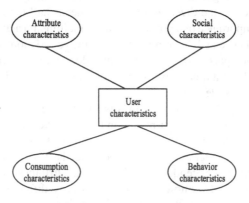

Fig. 2. User characteristics

As can be seen from Fig. 2, user characteristics mainly include: attribute characteristics, social characteristics, consumption characteristics and behavior characteristics. In order to effectively meet the personalized learning of students, personalized recommendation. User attribute features, which are collected from the students, include grade and interesting tags. These data are the basic information that users fill in when registering a website. This will have a certain impact on the establishment and optimization of the whole model, and directly affect the quality of the recommendation. Implicit information extraction process does not require the user's participation in the user's use of the system is automatically collected, there is no user cost. Popular technologies now generally use Web-side technologies such as Ajax or JS, and in the process of working with business logic, the user's behavior is also gathered.

User behavior characteristics is from the user login system, record every time the browsing records, purchase records, click records, collection records and other behavior data. In this system, the number of clicks on the course, the length of stay on the course page and the collection record are collected as behavior data. For example, user browsing, user collection, sharing, mouse trajectory and so on. Compared with explicit information extraction, implicit information extraction has obvious advantages, users do not have to participate in, reducing user costs.

At the same time, the information collected is generally not fabricated by the user. The quality of recommendation is closely related to the quality of user's interest model. User's learning characteristics, which are similar to the consumption characteristics mentioned above, is to collect the data of students' learning records, including the names of courses, grades of course resources, subjects to which the courses belong, and teachers' data.

User Interest Preference Model is the representation of a student's User Interest Preference, which mainly includes two types of features: attribute feature and learning feature. The contents of the attribute features are: students' current grade and interest tag. The contents of learning characteristics are user's curriculum preference, user's learning scope preference and teacher's preference. The quality of the user interest model directly determines the satisfaction of the recommendation results. In practice, user interest model is a kind of data structure. Using the traditional information retrieval technology, user interest model is expressed as a data structure that can be calculated and can also reflect the real interest of users.

2.3 Improved Recommendation Process for Multimedia Teaching Resources in Nursing

The essence of teaching resources recommendation method is to effectively organize the personalized information of users [7]. From the perspective of learner groups, learners' personality characteristics, knowledge background, emotional psychology and learning needs are different. For individual learners, their knowledge background, emotional psychology and learning needs develop dynamically in the process of cognition. Therefore, the student model should include students' static learning characteristics and dynamic changing knowledge structure, emotional psychology and learning needs. Learning process is a complex and continuous process, involving many factors such as psychology, emotion, cognition and so on. The core competencies of nursing students should include six aspects, as shown in Fig. 3:

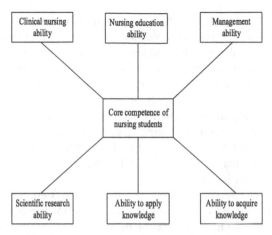

Fig. 3. Core competencies of nursing students

As can be seen from Fig. 3, the core competencies of nursing students mainly include: clinical nursing competency, nursing education competency, management competency, scientific research competency, ability to use knowledge and ability to acquire knowledge. But judging from these six core competencies, it is difficult to categorize them as soft or hard skills. In addition to defining the nurse's knowledge, competencies, and skills, it also covers individual characteristics, but core competencies also cover skill elements. Therefore, each kind of student model has its suitable scene and category.

On the one hand, the student model must be able to accurately express the characteristics and needs of students, and at the same time facilitate the technical implementation. In view of this, from the perspective of learners' style (mainly including: media preference, information processing method, cognitive personality style and learning tendency) and students' knowledge background (mainly including: existing cognitive level and ability level), as well as students' basic demographic information, students' model is constructed, and social annotation is adopted to enable learners to participate in the process of learners' model construction and increase the accuracy of recommended methods. The self-regulated learning ability of nursing students with high professional identity is higher than that of nursing students with low professional identity. Based on the practice of developing teaching resources, the teaching resources in various forms are characterized from the following aspects: serial number, name, brief introduction, keywords, subject attribution, applicable object, editor, creation time, media format, file size, etc. Among them, the educational resources of graphic image format add some technical attributes, such as preview file, resolution, scan resolution, color number and so on. Audio education resources add emotional type, sampling frequency, quantization digit, number of vocal tracks, playing time and other technical attributes.

The educational resources of video format add preview file, emotion type, frame number, frame specification, playback time, sampling frequency, sampling format and other technical attributes. Therefore, in order to improve the students' autonomous learning ability, we should strengthen their professional identity, encourage them to fully use modern information technology to acquire knowledge, change their teaching concepts, and make them change from passive receivers of traditional knowledge into active participants. The educational resources of animation format have added some technical parameters such as preview file, emotion type, frame number, frame specification and so on. Added for the network course script word number, image number, audio number, video number, animation number, page number and other technical attributes. For the question bank, the description of the type of question, knowledge points, difficulty, discrimination, cognitive classification, test requirements, reference answers, scoring criteria, suggested test time and other educational attributes are added.

2.4 Mobile Learning Optimize Personalized Recommendation Mode

In the process of mobile learning, the server can be used to actively push learning resources to provide students with a coherent and systematic mobile learning experience, improve the effectiveness of mobile learning, meet the needs of learners for timely access to resource services, and alleviate the problems such as easily interrupted learning and learners' distraction in the process of mobile learning. With the rapid development of mobile technology and wireless communication technology, ubiquitous learning and lifelong learning have penetrated into our daily life and changed our understanding and understanding of learning. Learner's cognitive level is the basis of the choice of learning starting point and learning resource difficulty coefficient. Learner's cognitive level reflects his mastery of current learning content. Individualized adaptive learning resources are based on the original knowledge of the learners, and the analysis of the original knowledge of the learners is conducive to the smooth follow-up learning.

In terms of network technology, resource pull mode adopts point-to-point transmission mode. When multiple users need a piece of information, it may lead to network congestion, prolong the time for users to obtain information and reduce the efficiency of users to obtain information. Moreover, because the server does not serve the users actively but passively, the timeliness of the information obtained by the users is poor. If the learner has no new knowledge, the recommended method should provide the knowledge in the previous section. If the learner has grasped the basic knowledge, then may enter the next node the study, simultaneously may provide certain expansion knowledge for the learner. Users in the resource pull mode must actively send out information requests and search for information on the server, which is a typical "people looking for information" way of information acquisition.

Therefore, in order to obtain information, the user must maintain a real-time connection with the server. But the network information is updated frequently. In order to obtain the latest information in time, the user must pay attention to the latest information in real time. Therefore, the adaptive learning resources provided by the method should be in line with the best development area of the learners, and the learning content is not available for the learners. Otherwise it will be too difficult to let the learners lose interest, too easy to let the learners without a sense of achievement, not conducive to the development and growth of learners. According to the behavior characteristics of the network learning behavior, some controllable behaviors can be selected to statistic behavior attributes.

User click-through rate and length of stay can reflect interest preference information from a certain angle, but the accuracy is not accurate enough, if the user clicks to view a course for 4 times, but each stay time is less than a few seconds, it can not reflect the strong interest motivation, the collection can reflect the interest preference of users more than the first two. The formula for calculating the user's curriculum preference is as follows:

$$W = \frac{P}{Q} + \frac{\beta}{\eta} \tag{1}$$

In formula (1), P represents the current number of clicks, Q represents the total number of clicks, β represents the length of stay, and η represents the total length of stay. User knowledge range refers to the stage and ability range of a user's current knowledge level. The expression formula is as follows:

$$\varphi = \frac{G \times \mu}{\gamma} \tag{2}$$

In formula (2), G represents the current course level, μ represents the number of levels of knowledge, and γ represents the total number of courses taken. The teacher is also one of the factors that affect students' interest in learning. Good teachers are knowledgeable and experienced in teaching, and students will love a course because of the teacher's teaching style. Teaching teachers also affect students' interest in learning one of the factors. Good teachers are knowledgeable and experienced in teaching, and students will love a course because of the teacher's teaching style. The calculation formula of users' interest preference for teachers is as follows:

$$\sigma = \frac{H \times \mu}{\gamma} \tag{3}$$

In formula (3), the H represents the teacher's scoring coefficient. Compared with the resource-pull technology, the resource-push technology is favored by users for its high efficiency, low cost and high timeliness. Information active push technology follows the preset technical specification or protocol standard to push the information that the user may need in real time and actively. Users can read the information directly or offline after receiving it. Through the statistics of behavior attributes and data, we can not only grasp the information characteristics of learners better, but also select more suitable learning resources for learners. Finally, the behavioral attributes and data are stored in the learner behavior characteristics database to prepare for the recommendation of later learning resources. The recommendation of digital resources is closely related to learners' learning behavior, learning style, cognitive level, knowledge level and learning goal. The server actively works for the user, which greatly reduces the time cost, energy cost and economic cost for the user to obtain information. Therefore, in many fields and environments, the demand for active service is becoming higher and higher.

Active push of information resources is a user-centered information service mode, which comes into being when intelligent technology develops to a certain stage. Active push of information resources involves the selection and dissemination of information resources. Without the dissemination of information resources, there will be no use of information resources, and it is difficult to form the value increment of information resources. Through judging the characteristics of learners' learning behaviors and learning styles, we can find the corresponding resource forms in the attributes of the resource bank, and then present the learning interface and navigation that learners are interested in according to their needs. The choice of information resources is the basis of information resources communication, it is to achieve the promotion of information resources communication efficiency.

Active push of information resources We shall, based on users' personal characteristics, historical behaviors and actual situations, analyze and mine users' needs, select content services and products consistent with users' needs, and make use of recommendation systems and automatic push technologies to provide users with resources and services in a timely and appropriate manner. At the same time, by testing the cognitive level, knowledge level and learning goal of the learners, we can find the content description of the resources in the resource database, and then recommend the learning content that the learners are interested in or the learning resources that are suitable for the current learning level of the learners.

Recommendation system is generally composed of user, project and scene. It mainly depends on data, algorithm and system. It pays attention to the user's character and scene, and recommends the most suitable project. The basic idea of resource recommendation is to use machine learning techniques to analyze new resources according to the student's evaluation of the resources he or she accesses. Currently, most resource recommendation services use traditional machine learning techniques. Individualized information proactive service can make different individuals have equal access to high-quality digital resources, integrate push technology into personal library expansion plan, push information to users via email by librarians to provide targeted information to meet user needs, strengthen the interaction between librarians and customers, and explain the importance of pushing effective information to users.

3 Experimental Test

3.1 Experimental Preparation

This experiment uses Eclipse integrated development environment, uses Java language coding to realize database operation, uses JavaScript to display the front-end interface and uses MVC and SSH framework technology. The crawling of educational resources is a combination of Python Beautiful packets.The environment of personalized recommendation method consists of a database server, an application server and a client. Database server to store permanent data, including the user's account information, all teaching resources data, etc. The database is MySQL version, in which data extraction, transformation, calculation and program deployment are written in Java language and data mining analysis module in Python. The final recommendation of personalized educational resources for the experiment was implemented on Tomact's server. Installs server-side application software on the application server. Users can access the system data after logging on through the client's computer.

3.2 Experimental Results

Selecting the method of personalized recommendation of multimedia teaching resources of nursing based on cluster analysis and the method of personalized recommendation of multimedia teaching resources of nursing based on genetic algorithm, and comparing with the method of personalized recommendation of multimedia teaching resources of nursing in this paper, the recall rates of the three methods are tested under the condition of different marked resources. The experimental results are shown in Figs. 4, 5, 6 and 7:

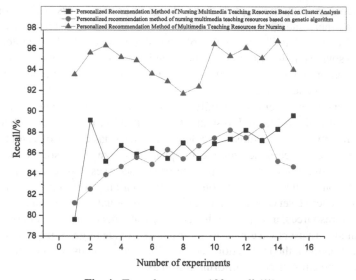

Fig. 4. Tagged resources 100 recall (%)

As can be seen from Fig. 1, the recall rates of the personalized recommendation methods of multimedia teaching resources for nursing, and the other two methods are 94.658%, 86.577% and 85.543% respectively; as can be seen from Fig. 2, the recall rates of the personalized recommendation methods of multimedia teaching resources for nursing, and the recall rates of the personalized recommendation methods of the other two methods are 83.44%, 75.430% and 76.103%; as can be seen from Fig. 3, the recall rates of the personalized recommendation methods of multimedia teaching resources for nursing, and the personalized recommendation methods of the other two methods of multimedia teaching resources for nursing, are 74.053%, 64.916% and 64.427% respectively; as can be seen from Fig. 4, the recall rates of the personalized recommendation methods of multimedia teaching resources for nursing, and the personalized recommendation methods of the other two methods of multimedia teaching resources for nursing, respectively, are 62.354%, 55.34% and 54.27%.

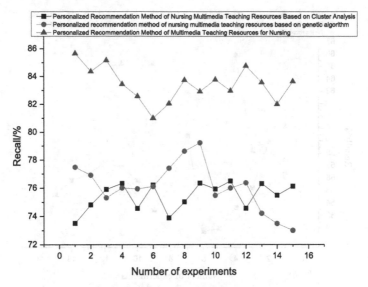

Fig. 5. Tagged resources 200 recall (%)

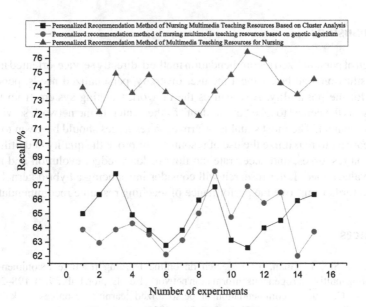

Fig. 6. Tagged resources 300 recall (%)

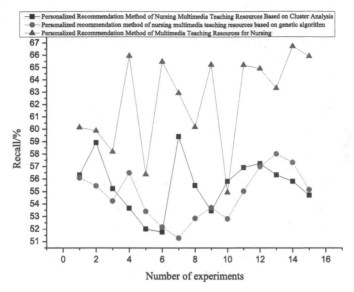

Fig. 7. Tagged resources 400 recall (%)

4   Conclusions

The design of personalized recommendation method, directly service-oriented individual learning situation, can better meet the user diversity, personalized needs, personalized learning for the possibility. Also causes the network teaching system from take "the resources" as the center to take "the student" as the center, to the network service higher level development. The most suitable information resources should be given to the most needed learners to maximize the use of resources, improve the quality and utilization of information resources, and accelerate the flow of knowledge, evolution and regeneration and value added. Later research will consider introducing a hybrid kernel function method to further improve the performance of teaching resource recommendation.

References

1. Zhong, Z., Shi, J., Guan, Y.: Research on online learning resource recommendation by integrating multi-heterogeneous information network. Res. Explor. Lab. **39**(9), 198–203 (2020)
2. Xu, Y.-J., Guo, J.: Recommendation of personalized learning resources on k12 learning platform. Comput. Syst. Appl. **29**(7), 217–221 (2020)
3. Tian, F., Li, X., Liu, F., et al.: The learner model construction of personalized learning resource recommendation system. Jiaoyu Jiaoxue Luntan (10), 304–305 (2020)
4. Chen, X., Du, Q.: Analysis of data mining technology of micro course mobile learning oriented teaching resource platform. Computer Era **1**, 62–65 (2020)
5. Wu, D., Chen, C.: Learning resource recommendation method based on social relationship. Mod. Comput. (1), 53–55,71 (2020)

6. Liang, C., Fan, R., Lu, W., et al.: Personalized recommendation based on CNN-LFM model. Comput. Simul. **37**(3), 399–404 (2020)
7. Yang, J., Hu, G., Wang, M.: Research on configuration effect of mobile learning adoption motivation. Libr. Tribune **40**(2), 64–73 (2020)

A Method of Abnormal Psychological Recognition for Students in Mobile Physical Education Based on Data Mining

Changyuan Chen[✉] and Kun You

Department of Physical Education, Xi'an Shiyou University, Xi'an 710065, China
chenchangyuan895@163.com

Abstract. At present, the abnormal psychological recognition of middle school students is mainly through psychological questionnaire, after data statistical processing, to assess whether there is abnormal psychological students. The recognition accuracy is strongly dependent on the reliability of the questionnaire, which leads to the poor recognition accuracy and stability. In order to solve these problems, the method of abnormal psychological recognition of students in mobile PE teaching based on data mining will be studied. After analyzing the influence of PE teaching on students' psychology, the behavioral characteristics that represent students' psychology are extracted. After constructing the students' psychological view, the students are classified preliminarily. Through constructing mental state mining decision tree, using iForest algorithm to realize abnormal mental recognition for middle school students. The test results of recognition method show that the accuracy of the mental recognition method is stable between 87.28% and 87.95%, and the recognition reliability is higher.

Keywords: Data mining · Physical education curriculum · Mobile teaching · Middle school students' psychology · Anomaly recognition · Decision tree

1 Introduction

All kinds of research and survey data show that the mental health problems of middle school students are becoming more and more serious at the present stage. Among them, sensitivity of interpersonal relationship, abnormal behavior and mental endurance are more common, and some other phenomena are depression, anxiety, weariness of study, suicide and so on. The rapid development of modern society, the intensification of competition, the acceleration of life rhythm, the increasing complexity of interpersonal relationships, the pluralism of values, make people face great pressure, people's mental diseases are also generally increasing, all kinds of mental diseases seriously trouble people, especially adolescents. Adolescence is the key period of students' physical and psychological development, as well as the high incidence of psychological conflicts and emotional and behavioral problems. In the field of psychology, the psychological characteristics of an individual are mainly described by the "state" psychological variables represented by mental health state and the "trait" psychological variables represented by personality. In the past, most of the methods to judge mental health status were self-assessment or other-assessment, but there were obvious deviations in this way, and the reliability of the evaluation results was poor [2].

Mobile teaching of physical education courses is another new attempt of physical education, which can enrich the physical education teaching methods and alleviate the dilemma of traditional physical education mode limited by venues. However, when the PE course adopts the mobile teaching method, it needs a lot of information technology, which can not only help to complete the basic teaching content, but also collect the students' behavior and feedback data. These data generated in the process of mobile teaching can not only be used to evaluate students' learning effect, but also be used to analyze students' psychological state. In related research, some scholars proposed a mental workload estimation method using deep BLSTM-LSTM network and evolution-ary algorithm [3]. This approach proposes a deep hybrid model based on bidirectional long short-term memory (BLSTM) and long short-term memory (LSTM) for workload-level classification, which can effectively monitor the mental state of mental activity. Other scholars have proposed a mental health sentiment analysis method based on deep convolutional networks [4]. This method is able to process facial images and account for the temporal evolution of emotions through a novel solution, and utilize standard linear discriminant analysis classifiers to obtain final classification results. The method can aid in the detection, monitoring and diagnosis of human mental health. However, when the above two methods are applied to the monitoring of students' mental health in mobile learning, there are still time and space limitations or untimely detection.

Therefore, how to effectively use a large number of data, through data mining technol-ogy to analyze the amount of information mining data, so that simple data can express the deep meaning of the data, to enhance the deep use of mobile teaching data, to strengthen the psychological state of students concern and guidance, is one of the important issues facing. Most of the traditional statistical analysis put forward deterministic analysis, which may not fully explain the extracted data, while data mining focuses on exploring the value information in the collected data. Using data mining technology can analyze the physiological data of students in the mobile teaching of physical education courses, thus improving the accuracy of abnormal psychological recognition. Based on the above analysis, this paper focuses on the mental health of students, and studies the method of abnormal mental recognition in PE mobile teaching based on data mining.

2 Abnormal Psychological Recognition Method for Students in Mobile Physical Education Based on Data Mining

2.1 Analysis of the Influence of Mobile Physical Education on the Psychological Change of Middle School Students

Antonouski presents the mental health model shown in Fig. 1. Through this model, the physical and mental health can be explained, and the necessary prerequisites for maintaining health and the necessary procedures for maintaining health are analyzed [5].

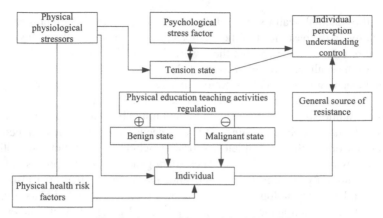

Fig. 1. Antonouski mental health model

From the health model, it can be determined: First of all, targeted physical activity directly contributes to the establishment of protective factors and resistance sources, thus resisting various pathogenic factors. Secondly, in addition to the related factors, physical activity has a direct impact on happiness and tension. The teaching process of physical education has the characteristics of intuition, which makes students have to use all kinds of senses synthetically: perceiving the action image through sight and hearing; perceiving the action essentials through touch and muscle itself, the degree of muscle exertion and the space-time relationship of the action, so as to establish the complete and correct action representation. In this process, the students' ability of perception, observation, image memory and action memory have been developed and improved.

Compared with the traditional physical education, the content of mobile education in physical education is more abundant, which can make students experience satisfaction, joy, tension, excitement, anxiety and so on. The team spirit of PE teaching and the mutual help and study among students can inspire students' social consciousness and enhance their self-esteem, confidence and sense of responsibility. The competitive nature of PE teaching can inspire students' enterprising spirit, inspire their will, and make various emotional experiences more profound. In this relaxing and autonomous activity, students are willing to participate in physical exercise actively and persist for a long time, so that the development of physical and mental health is more coordinated and lasting, and the control and regulation of anxiety, depression and other negative emotions are more significant [6].

Middle school is the key period to develop students' will quality. Mobile P. E. teaching can develop students' will quality such as consciousness, decision and self-control. In the process of PE teaching, students need to participate in various group activities, which provides conditions for cultivating students' social adaptability and promoting the diversified development of personality. Emotional infection, happy teaching, self-confidence training, willpower training, social contact training and psychological inducement are added to PE teaching to improve students' social adaptability, willpower, self-confidence, cooperation, physical and mental adjustment and self-evaluation.

Mobile PE teaching has a positive effect on students' mental health, but once the students' mental health is abnormal, the learning feedback collected by mobile PE teaching platform will appear abnormal. From the above analysis, the data that can reflect students' psychological state in the process of mobile physical education course teaching are students' learning enthusiasm, students' performance in team activities, students' emotion in physical training, students' will quality and students' specific learning situation. The premise of the application of data mining is to create a data warehouse with mass data and all the information. This paper uses the data stored in mobile teaching platform for data mining analysis. After determining the influence of mobile physical education on students' psychology and the data object of data mining analysis, the corresponding behavior data are collected and the psychological view of students is established.

2.2 Student Behavior Data Processing and Psychological View Construction

There are a lot of student data stored in PE mobile teaching platform, but the format of these data is not uniform, and there are dirty data and redundant data in this paper.

Because the data in this article comes from different terminals, data fields are different, and the format of data storage is different. So in this paper, heterogeneous data is stored in a database through data integration technology. A number of tables are associated with the desensitized student number as the main key, and gender, nationality and other characteristics are replaced by one-hot coding. In order to improve the efficiency of mining, it is necessary to simplify the dataset, that is, to delete the unused data according to the task of mining, and to decide which dataset to mine. The data selection work undertaken in this paper includes [7]:

1. School numbers, names and dates of birth are deleted directly, as these attributes are meaningless for mining.
2. Direct deletion of ethnicity, which has little impact on mining results.

The missing data in this paper is generally handled according to different situations. In the case of large enough data, some of the student's data is missing, this paper chooses to delete this data or delete the current student's information. If a student is missing a large number of course grades, delete the student from the student list in the grade module. In the part of noise anomaly, there are two kinds of anomaly data, one is the abnormal data beyond the fact, the other is the real anomaly data. Therefore, some of the data through visualization, according to the specific situation of noise data selection.

When students have psychological problems, their academic performance will be greatly changed, and generally with the straight line decline of academic performance, so the student's academic performance may be an indicator variable of psychological problems. In different mobile sports learning scene or student learning period, the student's psychological view may have mutual influence, and each individual's inner thoughts and habits will affect his performance in different views. Therefore, this article uses the MANE algorithm to fuse the different view performance, attempts to reconstruct each student this intrinsic thought, the custom. MANE algorithm can be formalized as: Given a network G, the multi-view network embedding goal is for each node $i \in U$, learning a low-dimensional vector representation of $d_i \in R^D$. To preserve the diversity of views as much as possible, for each node $i \in U$ and each view $p \in P$, multi-layer network embedding learns an intermediate representation of $d_i^{(p)} \in R^{D/P}$, which holds only the information within the view. After a series of intermediate processes (the intermediate representation of collaborative updates between views), this intermediate representation also preserves the potential information between views, and the final representation of a node is the splicing of the vectors of that node within each view, namely:

$$d_i = \oplus p \in d_i^{(p)} \tag{1}$$

Students' psychological view used in this paper is directed network, that is, the connection between nodes has directivity, node A connection B and node B connection A have different meaning, so in the network there are degrees and degrees. But each connection in the view p will have an Entry $I_d^{(p)}$ and an Exit $O_d^{(p)}$, so the average of the views is:

$$\bar{k} = \frac{N[E^{(p)}]}{N(U)} \tag{2}$$

The E represents the set of edges in a view and the U is the set of nodes. Different students have different psychological view, but the psychological view of abnormal students and other students psychological view of the similarity between the greater differences. Through calculating the similarity coefficients between student's psychological views, the students' psychological views with low similarity are separated. Jaccard coefficients are used to measure the similarity between different views. Because of the particularity of mobile PE teaching data, students have different behaviors. After setting up the psychological view of students in the process of learning PE, the decision tree algorithm is used to construct students' mental state decision tree.

2.3 Constructing Mental State Mining Decision Tree for Middle School Students

In the actual educational environment, the majority of students are self-care, and the abnormal students are only a small part. Therefore, there is interference in mining abnormal data. However, for schools and psychological institutions, psychological abnormality students should be more concerned about the object.

Decision tree is an inverted tree structure that can classify data automatically. Every node of decision tree is tested on its attribute by top-down recursion, and two or more branches are generated from different answers to questions on each node, which leads to different results. More common decision tree algorithms are the classic ID3 algorithm, C4.5 algorithm and CART algorithm and so on. This paper adopts C4.5 algorithm to construct decision tree of middle school students' psychological state.

1. The data with the lowest similarity coefficient shall be taken as the data to be divided according to the results of similarity calculation of students' psychological views. Calculate the information gain rate for each split attribute in the set as follows;

Set node N to store data dividing all samples of P. The expected information required for classification of samples in the P is given by the following formula:

$$I(P) = -\sum_{i=1}^{n} v_i \lg v_i \tag{3}$$

v_i is the probability that any sample in P belongs to C_i, and n is the sample serial number. $I(P)$ is also called entropy of P. Assume that the samples in the P are divided by attribute K, which has m values $\{k_1, k_2, \cdots, k_m\}$. If the value of the attribute K is discrete, then the attribute K can divide the P into a subset of m $\{P_1, P_2, \cdots, P_m\}$, where the value of the sample in the P_j is k_j on the attribute K. These subsets correspond to branches that grow from node N. The expected information required for classification of samples of P by attribute K can be derived from the following formula:

$$I(P)_K = \frac{|P_j|}{|P|} \times \sqrt[I]{|P_j|} \tag{4}$$

where $\frac{|P_j|}{|P|}$ is the weight of a subset of values of k_j on the attribute K. $I(P)_K$ is based on the desired information needed to classify the samples of P by attribute K.

The rationale behind C4.5 and ID3 is the same, except that C4.5 replaces the information gain as the attribute selection measure (splitting rule) in order to compensate for ID3's tendency to use information gain to select attributes with higher values. The information gain rate is defined as follows:

$$GR(K) = \frac{G(K)}{S(K)} \tag{5}$$

The entropy reduction is $G(K)$; split information is used in the above expression to normalize the information gain.

$$S(K) = -\frac{|P_j|}{|P|} \times \log_2\left(\frac{|P_j|}{|P|}\right) \tag{6}$$

2. By comparing the information gain rate of each split attribute, it is determined that the split attribute with the maximum information gain rate is the root node of the decision tree, and that the attribute has several values, and the data set is split into several subsets, and if there is only one value of the attribute, the split is ended;
3. Recursively perform steps 1 and 2 on each subset of the split data set.

Repeat the above steps to classify the sub datasets of each of the outgoing branches to lead to branching again. With the increase and extension of tree branches, the data set is divided into smaller subsets recursively, and finally, the decision tree of students' mental state is generated.

2.4   Realization of Abnormal Psychological Recognition for Middle School Students

According to the classification attributes, a tree-like structure is established, and a path from the root node to the leaf node forms a classification rule. Accordingly, the whole decision tree forms a set of disjunctive rules, which can be easily converted into IF-THEN classification rules, according to which it is easier to classify, identify and predict new data. After the decision tree of students' psychological state is formed, the trees are pruned. This paper chooses PEP algorithm to prune the decision tree of students' mental state.

PEP pruning algorithm is to overcome the shortcomings of REP algorithm that pruning dataset is not needed, but based on the false estimation of training dataset pruning algorithm. However, it also has some drawbacks, that is, it will lead to the large error of the estimation error rate of the algorithm. Therefore, the continuous correction in statistics is introduced to make up for this deficiency. That is, each leaf node is assumed to automatically misclassify the 1/2 instances it covers, and a constant is added to the subtree's training error. In calculating the standard error rate, the continuous correction follows the binomial distribution.

T represents the original tree, T_t represents the subtree with node t as the root, $r(t)$ represents the number of instances misclassified at node t, and $n(t)$ represents the number of all instances covered at node t. The classification error rate at node t is:

$$Y(t) = \frac{r(t)}{n(t)_v} \tag{7}$$

The PEP algorithm corrects it to:

$$Y'(t) = [r(t) + 1/2]n(t) \tag{8}$$

The PEP algorithm is faster and more effective than other algorithms because each sub-tree can be accessed at most once during pruning. After constructing the decision tree, we use the iForest algorithm to recognize the abnormal psychology of students.

The iForest forest is composed of decision tree units, which are constructed as follows:

(1)   Selecting abnormal characteristics of students' psychological state from the data set;
(2)   Randomly selecting a value of the feature;
(3)   Classify the data according to the characteristics, put the data less than this random characteristic value on the left, and put the data greater than or equal to the right;
(4)   Then the left and right branches are constructed recursively until they are satisfied. The height of only one data or tree in the input dataset has reached the limit.

The structure of iForest is similar to that of a random forest, all of which are randomly selected features that break each tree apart. There are differences between the trees, which are built by combining n iTree trees into iForest.

In this paper, due to the imbalance of student data, the use of undersampling will eliminate too much sample data, which may lead to the loss of key sample information and the lack of sample size, which will lead to the low accuracy and robustness of the training classification model. But only using over-sampling technology will expand a few samples and result in the over-fitting of classification. In order to solve the above problems, this section proposes a mixed sampling method for extremely unbalanced data. Through the mixed sampling of sample data, the problem of information loss caused by under-sampling and model over-fitting caused by over-sampling are avoided. There are two parts in the process of psychological state data collection. Firstly, the SMOTE algorithm is used to oversample the minority abnormal samples, that is, to interpolate each sample with its K samples to form a new minority sample. Secondly, the K-Means method is used to undersample most normal samples, that is, the data sets are clustered by K-Means method to eliminate redundant points in each class by calculating the cluster center.

Given a 2D array dataset, put the array into iForest for outlier identification. Build a binary tree with iTree n and iTree's sample count. The initial height of the tree defaults to 0. Through the mental view of students and the extraction of psychological characteristics of students, recursive, the distribution of psychological state of students under the abnormal point detection and recognition. Build n trees to form a forest. Because there are far fewer psychological anomalies than normal. Based on the principle of iForest priority, the algorithm will run quickly and the average depth of iTree will be shallow. Setting the number of samples properly can effectively control the maximum depth and reduce the running time under the condition of ensuring the recognition effect. After several iterations of classification output, we can get the recognition result of abnormal psychology of students in the process of mobile teaching of physical education courses.

3 Method Testing

Middle school students are in an important stage of physical and mental health development. While teaching students the theoretical knowledge and skills, they also need to pay attention to their mental health. In this section, we will design the test scheme of the recognition method, and through the analysis of the test data, we can verify whether the recognition method can be used in the management of students' mental health.

3.1 Test Readiness

Because of the large number of middle school students in China, their abnormal psychological tendency may develop into serious psychological problems. School education should guide students to master mental health knowledge and adjust psychological methods, and give understanding and guidance when necessary, so as to promote the healthy growth of middle school students. According to the professional psychological evaluation form, we choose to manually organize the data of students with abnormal psychology to avoid errors in data entry due to different data sources. Specific dataset parameters are shown in Table 1 below.

Table 1. Data Sets of Students' Learning Behavior in Physical Education

Field name	Field settings	Field type
User_ id	bigint	Student ID
Age	intefer	Age of student
class	varchar(10)	A student class
S_act	varchar(32)	Student behavior
S_t	varchar(32)	Physiological parameters of students
PE_S	char(8)	Achievement of students in physical education
PE_SP	varchar(32)	PE activity performance rating

Firstly, the original data is preprocessed by removing the heavy and empty data. Then, the abnormal psychological recognition methods based on data mining, SVM and feature extraction are used to analyze the artificial data and identify the students who have abnormal psychology. By comparing the recognition result of the method with the known result, the recognition accuracy of the method is obtained. By comparing the recognition accuracy of this method, we can verify whether this method is helpful to the construction of students' mental health in teaching activities.

3.2 Test Results

The accuracy of the two abnormal mental recognition methods for different grades of students is shown in Tables 2, 3 and 4.

Table 2. Comparison of Abnormal Psychological Recognition of Grade One Middle School Students

Serial number	Method in this paper	Recognition method based on SVM	Recognition method based on feature extraction
1	86.94	73.08	66.55
2	86.56	72.95	69.08
3	87.21	74.07	66.26
4	86.12	74.26	66.93
5	87.63	74.61	69.37
6	88.09	74.26	66.13
7	87.14	74.14	68.57
8	88.81	74.63	68.71
9	87.98	72.82	66.42
10	86.34	74.45	66.84

Table 3. Comparison of Abnormal Psychological Recognition of Second Grade Middle School Students

Serial number	Method in this paper	Recognition method based on SVM	Recognition method based on feature extraction
1	88.76	76.15	66.46
2	87.61	75.44	67.14
3	89.14	76.58	66.65
4	89.25	74.51	67.29
5	87.13	75.62	65.94
6	85.65	74.65	69.42
7	85.72	76.35	69.26
8	88.48	74.76	65.83
9	87.57	76.20	66.11
10	85.74	74.92	67.13

Table 4. Comparison of Abnormal Psychological Recognition among Junior Middle School Students

Serial number	Method in this paper	Recognition method based on SVM	Recognition method based on feature extraction
1	89.26	75.90	69.87
2	89.19	76.68	72.30
3	88.93	75.12	69.92
4	86.86	75.74	70.73
5	87.38	75.56	70.58
6	87.67	74.01	70.12
7	88.84	75.15	72.34
8	86.93	75.93	72.75
9	86.95	75.47	71.76
10	87.52	77.12	72.23

Analyzing the data in Tables 2, 3 and 4, we can see that the accuracy rate of abnormal mental recognition is higher than that of the other two methods in different grades. Among them, the recognition rate of the method based on feature extraction is obviously improved because of the psychological abnormality caused by the pressure of entering a higher school. The average recognition accuracy of the method is 87.28%, the average recognition accuracy of the method based on SVM is 73.93%, and the average recognition accuracy of the method based on feature extraction is 67.49%. The average recognition accuracy of the method is 87.51%, the average recognition accuracy of the method based on SVM is 75.52%, and the average recognition accuracy of the method based on feature extraction is 67.12%. The average recognition accuracy of the method is 87.95%, the average recognition accuracy of the method based on SVM is 75.67%, and the average recognition accuracy of the method based on feature extraction is 71.26%.

To sum up, in the process of PE mobile teaching, the method based on data mining proposed in this paper is more accurate and the evaluation result is more stable.

4 Conclusions

Mental health education in primary and middle schools is to meet the needs of the times and the healthy growth of students. It is a process in which the school exerts direct or indirect influence on students in a purposeful and organized way to improve their psychological quality and promote their all-round development. In this paper, we use data mining technology to study the method of abnormal mental recognition of students in mobile teaching of physical education. The test results show that the method has good recognition precision, and improves the reliability of the results of abnormal mental state recognition of students to a certain extent. And the recognition method proposed in this paper has strong reusability and extensibility.

Fund Project. Project of Shaanxi Provincial Department of Education: Research on the Intrinsic Mechanism of the Impact of College Students' Sports on Mental Health (Project No.: 20JK0288).

References

1. Xiang, S., Lin, W., Liu, J.: A randomized controlled study of effect of group counseling on facial expression recognition in college students with social delay [J]. Chin. Ment. Health J. **35**(03), 177–181 (2021)
2. Liu, D., Bao, L., Wan, C., et al.: Multi-layer partial information fusion model for psychological crisis identification of online forum users. J. Chin. Comput. Syst. **42**(04), 690–699 (2021)
3. Chakladar, D.D., Dey, S., Roy, P.P., Dogra, D.P.: EEG-based mental workload estimation using deep BLSTM-LSTM network and evolutionary algorithm. Biomed. Signal Process. Control **60**, 101989 (2020)
4. Fei, Z., Yang, E., Li, D., et al.: Deep convolution network based emotion analysis towards mental health care. Neurocomputing **388**, 212–227 (2020)
5. Su, Y., Liu, M., Zhao, N., et al.: Identifying psychological indexes based on social media data: a machine learning method. Adv. Psychol. Sci. **29**(04), 571–585 (2021)
6. Shang, Y., Yang, N.: An improved particle swarm optimization back propogation neural network algorithm for psychological stress identification. Sci. Technol. Eng. **20**(04), 1467–1472 (2020)
7. Zhao, T., Huang, Z., Wang, X., et al.: Recognition research on concealing behavior in psychological test. Comput. Eng. Appl. **56**(20), 158–164 (2020)

Design of Online Auxiliary Teaching System for Accounting Major Based on Mobile Terminal

Yanbin Tang[✉]

Jiangxi University of Applied Science, Nanchang 330100, China
yugsf574@163.com

Abstract. Online teaching is a common teaching form at present. In the process of using the online auxiliary teaching system for accounting majors, there is a defect that the memory occupies a large space. In order to solve the above problems, an online auxiliary teaching system for accounting majors based on mobile terminals is designed. Hardware part: The power supply design adopts the form of independent power supply in blocks, and configures the external memory interface of C6722B; the software part: builds a database of students' classroom behavior, migrates behavior attribute data, and uses the online teaching platform as a carrier to obtain the teaching objectives of accounting majors. The mobile terminal optimizes the data transmission function of the online auxiliary teaching system. Experimental results: The memory footprint of the online auxiliary teaching system for accounting majors designed this time and the other two online auxiliary teaching systems for accounting majors are: 357.42M, 484.96M, and 486.99M respectively. The online auxiliary teaching system for accounting majors is more suitable for use.

Keywords: Mobile terminal · Accounting major · Online auxiliary teaching · Teaching form · Data warehouse · Interactivity

1 Introduction

Modern mobile terminal technology has made rapid progress, with the terminal cost reduction and technology upgrading, function more perfect, a variety of products flooding the market. Take mobile phones as an example, China's current mobile phone users exceed 600 million, while the university campus penetration rate is basically 100%. The research on the application of mobile terminal in classroom teaching will help to open up new ways of classroom teaching, enrich classroom teaching mode and enrich educational teaching theory. As an assistant means of teaching, mobile terminal can provide classroom teaching resources more intuitively and comprehensively, and can detect students' learning status more conveniently and quickly through classroom learning tasks [1].

At the same time, the development of intelligent terminal and mobile network technology provides practical technical support for the new education mode. It is not only the

© ICST Institute for Computer Sciences, Social Informatics and Telecommunications Engineering 2023
Published by Springer Nature Switzerland AG 2023. All Rights Reserved
W. Fu and L. Yun (Eds.): ADHIP 2022, LNICST 469, pp. 297–311, 2023.
https://doi.org/10.1007/978-3-031-28867-8_22

need while social development, but also the trend of educational development to take auxiliary teaching as an extension of educational service. In addition, can realize the information technology and the classroom conformity, realizes the education informationization. At the same time, the application of mobile terminals in classroom teaching can change the classroom environment, change teaching and learning scenes, improve classroom teaching efficiency [2, 3]. Relevant scholars have made different degrees of research on the online assistant teaching system of accounting major. Genetic algorithm is used to speed up the construction effect of the hardware environment of accounting major, increase the coverage of online courses, simplify the teaching process through the design of functional modules, and improve the teaching effect of accounting courses. However, this method has redundancy when dealing with accounting information data. Cluster analysis can comprehensively use the information of multiple variables to classify the samples. Through cluster analysis, it can effectively reduce the complexity of the course content and results of network accounting major, make the teaching level clear, and enhance the teaching purpose. However, the computational complexity of this method is high, and singular values will have a great impact on the data clustering results, which may cluster into chains.

This paper will also study the current situation of mobile terminal used in education at home and abroad, sort out the relevant development process, systematically study the feasibility, related theory and effective operation mode of CAI, and provide experience for the further research and development of CAI based on mobile terminal from the direction of basic theory, which is of great significance to the development and popularization of CAI, and also to the teaching reform under the Internet age.

Presently, our main teaching organization form is the class teaching system. This kind of collective teaching form easily neglects each person's difference, neglects individual development. And the use of mobile terminals can give students more learning freedom and learning resources, so that students choose their own learning resources, so as to promote the development of students' personality. At the same time, teachers can further implement teaching students in accordance with their aptitude. Therefore, it is of practical significance to combine the traditional classroom with computer application technology and Internet technology, fully mobilize the atmosphere and enthusiasm of students in the classroom, improve the quality of classroom teaching, make use of the advantages of the Internet +, and make use of novel and vivid interaction to design and develop a well-interactive and functional online auxiliary teaching system for accounting majors that meets the needs of modern teaching.

2 Hardware Design of Online Auxiliary Teaching System for Accounting Major

For the designer, the key of the design is to set the resistance-capacitance parameters of the loop low-pass filter. If the parameters of the loop filter are not properly configured, the clock phase locking of the AD9517–3 will fail without clock output. Because the system adopts dual CPU core microprocessor, there are many peripheral circuits. In order to avoid the breakdown of the master-slave processor or the influence of the master-slave CPU on the peripheral circuit, the power supply design of the whole system adopts separate independent power supply. The AD9517–3 has an on-chip integrated phase-locked loop that integrates the phase detector and the VCO internally, requiring the user to design only the loop lowpass filter.

The PLL is a phase negative feedback control system. The system mainly consists of clock source crystal oscillator, phase discriminator, loop low-pass filter and voltage controlled oscillator. The specific scheme is as follows: LM2576: 7 ~ 40V input voltage, 5V output voltage, as the input voltage of the lower stage voltage chip. AP 1117: 5V input voltage, 3.3V fixed output voltage, power supply for STM32 and its peripheral devices. In the hardware of online assistant teaching system for accounting major, the configuration of register is shown in Fig. 1.

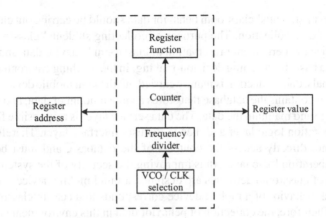

Fig. 1. Register Configuration Structure

As can be seen from Fig. 1, the register configuration structure includes: register function, counter, frequency divider, VCO/CLK selection, register value, and register address. After the error voltage output by the phase detector is filtered by the loop low-pass filter to filter out the high-frequency noise, the output signal frequency of the voltage-controlled oscillator is controlled to realize the feedback control of the phase. The frequency of the output signal is locked to the frequency of the input signal, and when the loop low-pass filter is stable, a stable sampling clock is output. TPS70345: 5 V voltage input, 3.3 V/1.2 V two-way output voltage, 1.2 V provides core voltage for C6722B, 3.3 V provides voltage for C6722B I/O port and its peripheral devices DCP0105: 5V voltage input, ±5 V Two voltage outputs, designed to power the TL084 operational amplifier.

Database server: TS300-ES is recommended for database server. It is a well-built vertical/rack-mount 5U server with built-in PSBP-E/4L high-performance motherboard and supports Xeon 7100 series CPU in Intel LGA775 package. The data storage capacity directly affects the data processing efficiency, and the core data processing tasks are all completed in the DSP data processing unit. Therefore, the configuration of memory resources is essentially to configure the external memory interface of C6722B.

Core Intel Xeon processor 7100, highly expandable up to 16GB memory type. The SATA controller adopts Intel ICH7R, supports 4xSATA2 300 MB/s, supports software

RAIDO, 1, 10 and 5 settings. Using the ADIsimCLK simulation tool, the user only needs to set parameters such as the input reference clock frequency, the charge pump current size, the VCO frequency division factor, and the order of the loop low-pass filter network. After running ADIsimCLK, the resistance and capacitance parameters of the loop filter network can be obtained.

3 Software Design of Online Auxiliary Teaching System for Accounting Major

3.1 Build a Database of Students' Classroom Behavior

The collection of students' classroom behavior data should be carried out closely around the purpose of data collection. The purpose of collecting students' classroom behavior data is to realize the permanent preservation of classroom behavior data and provide the necessary data basis for teaching decision-making. In the teaching environment based on mobile terminals, data collection is mainly carried out through mobile devices. In the data service layer, it contains the database that stores a large amount of data, and the software used to manage and maintain the data. The data service layer is responsible for providing data to the operation logic layer and then to the user interface layer. Therefore, the user terminal cannot directly access the contents of the database, and must be connected through the operation logic layer, thus improving the security of the system.

All kinds of classroom activities are designed around mobile devices, and a certain button or click behavior of a mobile device corresponds to a certain classroom activity or function. Therefore, the collection of behavior data in this environment is actually the collection of functional information of teaching software on mobile devices. The three-tier architecture adopts the client/server operation mode, and each layer can be developed by different teams with different programming languages and tools at the same time. Since changes in the functions of individual tiers do not affect other tiers, it is very easy to deploy in the enterprise. When it is necessary to modify or add new functions at any level, it will not affect the normal operation of other levels. The storage module in this process is mainly used for real-time transaction processing in the classroom, and can perform complex real-time insertion, deletion, update and other data operations.

The storage module provides the original data source for the data warehouse. During the data extraction process, the behavior attribute data corresponding to the storage module and the data warehouse can be directly migrated. The attribute data that is not in the storage module can be extracted and organized by the data conversion tool and stored in the data warehouse. Adopt 3-tie: framework and use database software to build dynamic website, in addition to making the website have the characteristics of interacting with users. In the system development stage, each level can independently use different tools for development in a modular way. The collection and analysis of data is one of the necessary skills in the Internet age. From individuals to large enterprises, they need to rely on data as the basis for decisions and choices. Classroom data occupies an important position in teaching research, and is an important reference for teachers to improve teaching strategies and improve teaching effects. The attribute map of students in the database is shown in Fig. 2:

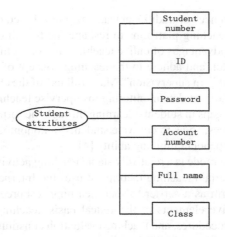

Fig. 2. Attribute map of students in the database

As can be seen from Fig. 2, the student attributes in the database include: student number, ID, password, account number, name, and class. In the future, if the functional requirements of any level are changed, it can be updated independently without affecting the operation habits of other levels. For website administrators, the information can be classified and stored in the database. And through the program to generate the web page, without the need to manually manage the content of the web page and generate new hyperlinks. This makes management more convenient and content updates faster. After determining the objects and methods of data collection, the preservation of data should also be considered. The data storage method is realized through the database.

There are many mature database management systems, and all of them can be connected with a variety of software development technologies and network technologies. The establishment of the data warehouse can be roughly divided into the following steps: System theme design, that is, the application goal of the data warehouse. Plan the physical structure and logical structure of the database table, so that the design of the database can reach a more optimized mode. It can better manage users' data, ensure that the system can manage data quickly and conveniently, and meet the needs of system functions. System logic model design, that is, the relationship between each entity table of the database and the event table. Relational database design, that is, the relationship between the database tables and the design of the attributes of each table. The design of the physical structure of the system, that is, the technical means and methods required for the realization of the data warehouse system. The last is the maintenance of the data warehouse system.

3.2 Acquire Accounting Teaching Objectives

In terms of teaching objectives, the accounting major focuses on the current economic society and the characteristics of various enterprise units and the requirements for accounting practitioners, and simultaneously pays attention to both the students' accounting professional ethics and professional knowledge and skills. Online teaching is not an "online recurrence" of traditional teaching, but is based on the design of online resources and online teaching platforms to reshape the teaching methods of teaching activities. Turn the disadvantages of online teaching into opportunities, form a student-centered teaching model, and adhere to the teaching concept of "student self-study + teacher guidance + platform supervision". Make full use of the characteristics of online teaching, use online platform data monitoring to supervise teaching, and do a good job of checking and filling gaps in students' learning through the formulation and release of online teaching resources, so as to cultivate students' autonomous learning ability and improve students' independent thinking ability [4].

The online teaching mode is a relatively stable teaching activity framework and program constructed by certain teaching theories. It uses the Internet as a medium and an online teaching platform as a carrier to share learning resources in online classroom teaching [5, 6]. The five elements of theoretical basis, teaching objectives, teaching conditions, teaching procedures, and teaching evaluation constitute a complete teaching model. These five elements with different functions are interconnected to form a complete teaching model. In terms of professional ability, accounting graduates should be able to adapt to the needs of career development, have a certain degree of autonomy in learning; have the basic information technology application ability to operate modern office software; Standardized writing and other basic skills; module 4 is to fill in accounting vouchers, the main content is to understand accounting vouchers, understand the business processing of various vouchers and prepare accounting vouchers for various economic businesses. Module 5 is the registration book.

The teaching content of accounting professional training involves the cultivation of students' basic professional knowledge, accounting practical skills and accounting professional ethics. The consolidation of knowledge and the cultivation of accounting professional ethics throughout. Master the steps to prepare a simple financial statement, and know how to view the main content of a simple financial statement. The seventh module is to sort out and keep accounting data. Mainly to understand the main accounting materials, classify and organize accounting materials, and module 8 is accounting assumptions and cycles. According to the training of skilled accounting personnel that meet the requirements of modern society. The school's accounting training courses mainly include post-intensive training, manual accounting simulation training, computerized accounting simulation training, and ERP sand table simulation training. The main teaching content of the accounting major is divided into three stages, and the teaching objectives of each stage are shown in Fig. 2:

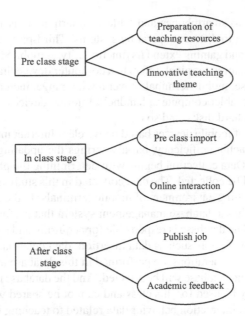

Fig. 3. Teaching goals for each stage

As shown in Fig. 3: The three teaching stages and contents of the accounting major are: pre-class stage: preparation of teaching resources, innovative teaching topics; mid-class stage: pre-class introduction, online interaction; post-class stage: homework release, learning situation feedback. At the same time, when cultivating accounting students, most colleges and universities combine vocational skill level certificates and graduation certificates to establish the training goal of "multi-certificate" to maximize the comprehensive vocational skills of students. In terms of professional quality, during the three-year education process, the school should focus on cultivating students' professional ethics and legal awareness, so that they can be honest and self-disciplined in their jobs, adhere to the guidelines, and have good communication skills and team awareness. At the same time have a healthy body, a healthy personality and good behavior habits.

Under the guidance of the concept of "online classroom", the author reconstructs and organizes the teaching content based on the students' ability goals. There are a total of 8 modules in the accounting online textbook, and the first module is to understand accounting. The second module is to fill in the business documents of the enterprise, mainly to understand various documents and the original documents. The third module is to understand the debit and credit bookkeeping method, the main subject is accounting elements, accounting subjects and accounts.

3.3 Mobile Terminal Optimizes Data Transmission Function of Online Auxiliary Teaching System

A mobile terminal refers to a computer device that can be used on the move. Broadly speaking, it includes mobile phones, laptop computers, and POS machines [7]. But most of the time, it refers to smartphones and tablets. Smartphone users can place desired applications according to different operating systems. Third-party applications such as education, shopping, and gaming extend its functionality. A mobile terminal, also known as a mobile communication terminal, refers to a computer device that can be used on the move. In a broad sense, mobile terminals cover a wide range, including mobile phones, notebook computers, tablet computers, handheld Internet devices, POS machines, and even U disks, mobile hard disks, and so on.

The introduction of mobile devices based on wireless Internet into the classroom not only improves the teaching efficiency, but also brings the updating of students' classroom behavior data. Data collection begins with determining the physical environment in which the data will be collected. The data collected in this study is the classroom data of students in the teaching environment of mobile terminals. In the course of classroom teaching, the system has a database management system that performs transaction processing in real time. The database is responsible for acquiring and saving real-time data, involving real-time operations such as data insertion, data deletion, and data update. If a large number of query operations are performed in this database management system, the running speed of the database will be affected. And the database management system in each class can only be used for this class and cannot be shared with other classes. It involves all classroom interaction behavior data related to teaching in this environment, including teacher-student interaction and student-student interaction behavior based on mobile devices.

Through the management and maintenance interface of this system, it is also possible to update and maintain website content from the remote end with browsers such as IE, so that the management of website content is not limited by time and space. And apply the MVC design style to design and implement. Second, determine the method of data collection. This needs to solve two problems, one is which data to obtain, and the other is how to obtain it quickly, that is, the data transmission method. Combined with the working principle of the clustering algorithm, the data matrix of the data object is calculated:

$$
L = \begin{bmatrix}
Y_{11} & \cdots & Y_{1q} & \cdots & Y_{1m} \\
\cdots & \cdots & \cdots & \cdots & \cdots \\
Y_{p1} & \cdots & Y_{mn} & \cdots & Y_{pm} \\
\cdots & \cdots & \cdots & \cdots & \cdots \\
Y_{q1} & \cdots & Y_{qm} & \cdots & Y_{pq}
\end{bmatrix}
\tag{1}
$$

In formula (1), Y represents the number of data objects, p, q represents an attribute value of each row and column data object, and m, n represents the data dimension respectively. Combined with formula (1), it is concluded that the most commonly used cluster evaluation method in cluster analysis is to use the square error criterion function, which is defined as follows:

$$
G = \frac{\left\| d_\beta - s_{\beta-1} \right\|^2}{2}
\tag{2}
$$

In formula (2), d represents the cluster center, s represents the cluster set, and β represents the sum of squares of errors between each data sample and the cluster center of the cluster where it belongs. The center point of the cluster in the square error criterion function is represented by the mean of the objects in each cluster, and the absolute error criterion is to select a representative sample in each class as the reference object. The latter can effectively reduce the impact of outliers and noise points on the clustering results. It is defined by the following formula:

$$K = \frac{\sqrt{\delta - 1}}{|d + s|} \times \frac{1}{\beta} \tag{3}$$

In formula (3), δ represents the absolute error sum of all data objects in the dataset. Combined with the calculation results, the data transmission mode in the system is optimized. At present, the mobile terminals used in classroom teaching mainly include tablet computers and smart phones, and the course knowledge is usually processed by certain system software. Enter some specific codes into the mobile terminal, then teachers and students set up their own accounts, and enter their own account information to enter their own systems and public platforms. This can be used in classroom teaching. But in most cases it refers to smartphones and tablets with multiple functions. The mobile terminal or mobile device in this study refers to the intelligent mobile terminal with wireless communication function and providing various computer application functions. Mainly include smartphones and tablets. The operating system of the mobile terminal is the basis for the operation of the mobile terminal.

A mobile terminal is not only a system that connects hardware and software and hosts applications. It is also developed from the traditional desktop operating system according to the different operating environment requirements of mobile terminals. And it plays an immeasurable role in the intelligent terminal, and becomes the control point for the development of the mobile Internet industry. The source data of the data warehouse comes from multiple such databases. The data in the database is converted into new data conforming to the logical model of the data warehouse through the data conversion tool, and stored in the data warehouse. Data warehouse systems have specific topics that facilitate quick on-demand queries. In addition, it can be distributed and stored and there is a large amount of redundant information, so there is no need to worry about the problem of equipment storage, ensuring the permanent preservation of classroom data. The Learning Module is the central point that connects the other modules, through which you can enter any other module.

In order to make the learner use the APP to learn without feeling boring and to improve the learner's interest in learning. This module also provides a variety of learning methods for learners to choose from. The highlight of this module is that it provides a variety of learning methods. The realization of database structure is a key step in the whole system, and the quality of data transmission can affect the bottleneck of the system to a certain extent. Moreover, in the design, the actual use situation should be fully considered, and the corresponding design and implementation should be carried out in accordance with the requirements of the paradigm. In addition, the main function of the learning module is to provide learners with various learning services and learning support. It is a multi-functional learning service hall, into which various learning services can be used. For example, viewing various learning resources, viewing previous learning records and notes, communicating with the learning community, sharing your own learning experience, answering questions raised by other learners, etc.

As each mobile device is connected via wireless Internet, data transmission is convenient and guaranteed by standardized network protocols. Therefore, the acquisition of data can adopt the network-based data acquisition method. It is necessary to analyze the functional modules of the system to determine the entities required by the database and the corresponding attributes to ensure that the information can be stored completely. Secondly, the storage size and storage type of the field need to be determined by considering the content of the word attribute.

4 Experimental Test

4.1 Experiment Preparation

The operating system of the test server is Windows, the development language Java uses the JDK development environment, and the Web server uses Tomcat. The system database environment is MySq. The server is distributed in the local PC. The web server of the server-side system architecture adopts the Tomcat host, which is currently the most widely used web server in the world. And the SQL SERVER2020 Database version built in the Myeclipse software suite is used as the database of this system, which is set up on the Microsoft Windows Professional computer host. Expand the subsystems such as Web Server, FTPServer, Mail Server, Database Server and other subsystems for the future needs of the host.

Considering the compatibility of different models, the client is divided into two versions: Android and iOS. The Android model is mainly used for testing. The main screen size is 5 inches, the main screen resolution is 1280*720, and the operating system is Android. The system requirements of the Sewer side are: Application software: Adobe Dreamweaver CS, PhotoImpact, Acorbat. Network programming language: The version that supports JSP web page format. Communication protocol: TCP/IP. The iOS model is iphone11, and the main screen resolution is 1334 * 750. The operating system is iOS. The network used for data transmission is the campus laboratory network. Website system/database: Apache Web Server Version, SQL SERVER2020, Database Version 5.3. Operating System: Microsoft Windows.

4.2 Experimental Results

In order to verify the effectiveness of the designed system, experimental tests are carried out. The online auxiliary teaching system for accounting major based on genetic algorithm and the online auxiliary teaching system for accounting major based on cluster analysis are selected to compare with the online auxiliary teaching system for accounting major designed this time. The memory footprints of the three systems are tested under the conditions of different online numbers, and the experimental results are shown in Tables 1, 2, 3, 4 and 5.

As can be seen from Table 1, the memory footprint of the online auxiliary teaching system for accounting majors designed this time and the other two online auxiliary teaching systems for accounting majors are 120.30M, 149.60M, and 155.46M respectively.

Table 1. Memory footprint of 200 users online at the same time (M)

Number of experiments	Online auxiliary teaching system for accounting major based on genetic algorithm	Online auxiliary teaching system for accounting major based on cluster analysis	The designed online auxiliary teaching system for accounting majors
1	153.61	149.36	112.36
2	142.06	152.09	124.19
3	135.88	156.34	119.25
4	149.21	162.04	121.66
5	151.26	158.49	118.69
6	153.20	161.03	121.44
7	145.11	148.79	118.06
8	156.93	146.82	125.18
9	143.15	156.31	124.71
10	159.34	155.49	126.08
11	153.77	158.07	118.45
12	148.26	162.37	119.74
13	153.19	152.84	121.03
14	147.05	157.09	116.11
15	152.01	154.74	117.58

Table 2. Memory footprint of 500 users online at the same time (M)

Number of experiments	Online auxiliary teaching system for accounting major based on genetic algorithm	Online auxiliary teaching system for accounting major based on cluster analysis	The designed online auxiliary teaching system for accounting majors
1	233.18	205.91	164.58
2	225.14	212.47	178.25
3	228.39	222.84	181.09
4	226.43	219.16	167.44
5	227.58	223.47	173.59
6	225.79	215.04	176.62
7	224.14	226.78	175.06

(*continued*)

Table 2. (*continued*)

Number of experiments	Online auxiliary teaching system for accounting major based on genetic algorithm	Online auxiliary teaching system for accounting major based on cluster analysis	The designed online auxiliary teaching system for accounting majors
8	226.20	225.01	168.33
9	218.55	229.56	172.15
10	225.05	216.84	165.46
11	231.74	205.77	159.25
12	226.94	231.99	164.88
13	226.48	215.18	163.24
14	228.15	220.49	164.29
15	215.02	224.11	157.61

As can be seen from Table 2, the memory footprint of the online auxiliary teaching system for accounting majors designed this time and the other two online auxiliary teaching systems for accounting majors are 168.79M, 225.92M, and 219.64M respectively.

Table 3. Memory footprint of 800 users online at the same time (M)

Number of experiments	Online auxiliary teaching system for accounting major based on genetic algorithm	Online auxiliary teaching system for accounting major based on cluster analysis	The designed online auxiliary teaching system for accounting majors
1	322.11	363.55	306.15
2	324.82	359.18	288.74
3	329.58	362.04	296.45
4	345.17	354.19	273.61
5	346.19	352.22	283.55
6	352.07	347.09	296.10
7	347.23	355.18	306.44
8	349.20	358.06	289.16
9	356.12	362.15	302.77
10	365.28	359.11	294.61

(*continued*)

Table 3. (*continued*)

Number of experiments	Online auxiliary teaching system for accounting major based on genetic algorithm	Online auxiliary teaching system for accounting major based on cluster analysis	The designed online auxiliary teaching system for accounting majors
11	359.03	348.71	296.13
12	367.15	342.66	313.22
13	373.48	359.18	311.47
14	382.07	361.02	266.71
15	325.04	355.13	259.48

It can be seen from Table 3 that the designed online auxiliary teaching system for accounting majors and the other two online auxiliary teaching systems for accounting majors have a memory footprint of 292.31M, 349.64M, and 355.96M respectively.

Table 4. Memory footprint of 1000 users online at the same time (M)

Number of experiments	Online auxiliary teaching system for accounting major based on genetic algorithm	Online auxiliary teaching system for accounting major based on cluster analysis	The designed online auxiliary teaching system for accounting majors
1	564.91	546.91	336.47
2	507.96	522.16	359.15
3	516.33	519.47	369.12
4	521.84	522.19	355.74
5	516.99	533.08	363.20
6	522.78	529.17	372.91
7	521.44	531.05	386.15
8	536.91	526.14	376.59
9	528.16	533.66	368.47
10	524.05	521.84	369.12
11	536.77	563.49	377.11
12	529.15	537.16	348.04
13	531.69	522.09	365.25
14	525.46	536.17	355.06
15	566.29	528.03	361.77

It can be seen from Table 4 that the designed online auxiliary teaching system for accounting majors and the other two online auxiliary teaching systems for accounting majors occupy 364.28M, 530.51M, and 364.28M of memory respectively.

Table 5. Memory footprint of 1500 users online at the same time (M)

Number of experiments	Online auxiliary teaching system for accounting major based on genetic algorithm	Online auxiliary teaching system for accounting major based on cluster analysis	The designed online auxiliary teaching system for accounting majors
1	725.31	716.22	556.91
2	736.26	723.31	545.63
3	722.13	733.15	539.21
4	746.19	741.14	518.88
5	725.23	719.08	520.16
6	733.14	726.05	523.17
7	731.05	731.29	516.25
8	729.18	725.22	548.38
9	731.02	736.13	520.76
10	729.04	740.17	522.92
11	733.01	739.14	531.87
12	729.04	735.20	529.25
13	733.07	733.13	526.14
14	724.18	746.17	532.19
15	742.29	744.16	529.70

It can be seen from Table 5 that the designed online auxiliary teaching system for accounting majors and the other two online auxiliary teaching systems for accounting majors occupy 530.76M, 731.34M, and 732.64M of memory respectively.

According to the experimental results, when the online assistant teaching system for accounting majors occupies 200–1000 users at the same time, the system memory will be improved. However, the memory occupied by method A is lower than that of method online auxiliary teaching system for accounting major based on genetic algorithm and method online auxiliary teaching system for accounting major based on cluster analysis. This is because method the designed online auxiliary teaching system for accounting majors adopts the form of segmented independent power supply, configures external memory interface, improves the calculation speed, migrates the behavior attribute data through the database, and optimizes the data transmission function.

5 Conclusion

The system designed here, on the basis of mobile terminals, improves the teaching mode of accounting major. We have highlighted the system's characteristics of modularization, miniaturization, prominent difficulties, contextualization, fun, clear content, and small capacity. At the same time, the performance of the system is improved. Due to limited capabilities, the article also needs to conduct in-depth testing of the accuracy of the system. However, due to the limited time and research conditions, the scope of the experiment is not wide enough, and the results are still limited. For example, this study only collected the accounting professional data of one school as the test object, and the data lacks universality. Therefore, the following experimental selection can be more in-depth and multi-dimensional, so as to consolidate the experimental research results and provide theoretical support for online assistant teaching of accounting major in the future.

Fund Project. 1. 2020 Jiangxi Provincial Department of Education Teaching Reform Project: The strategy research and practice of integrating socialist core values into the whole process of accounting professional teaching (Project number: JXJG-20–29-3)

2. 2019 Jiangxi Provincial Department of Education Science and Technology Project: Research on systemic risk identification and countermeasures of industry-university-research cooperation projects (Project number: GJJ191199).

3. 2020 Jiangxi Provincial Culture and Art Science Planning Project: Research on policy paths for Jiangxi cultural enterprises to solve difficulties under the new crown epidemic(Project number: YG2020154).

4. 2020 Jiangxi college of application science and technology-level Humanities and Social Sciences General Project: Application Research on Fuzzy Risk Calculation of Industry-University-Research Cooperation Project Based on FMEA (Project number: JXYKRW-20–1).

References

1. Zhang, J., Huang, S., Liu, L., et al.: Research on equipment identification based on machine vision in mobile terminals. Fire Control Command Control **45**(2), 155–159,165 (2020)
2. Zhang, H., Wang, H., Guo, J., et al.: Online teaching of embedded system based on Tencent classroom and virtual simulation technology. Exp. Technol. Manag. **37**(12), 170–174
3. Han, L., Yu, X., Wu, H.-Y., et al.: Chin. J. Immunol. **36**(20), 2516–2519, 2523 (2020)
4. Mu, S., Wang, Y.: Turning "Crisis into opportunities": how emergency online teaching moves towards systematic online teaching. Mod. Dist. Educ. Res. **32**(3), 22–29 (2020)
5. Wen, X., Xu, L., Chen, Y.: Research on the mixed online and offline teaching mode under the background of "golden course" ——taking the course of web application system development as an example. Computer engineering & Software **41**(7), 292–296 (2020)
6. Peng, W.: Research on online teaching system and its key technologies [J]. comparative study of cultural innovation, 4(24): 130–132 (2020)
7. Yao, K., Li, L.: Mobile terminal network survivable database security anti-tampering simulation. Comput. Simul. **37**(1), 456–459,483 (2020)

Mobile Teaching Quality Evaluation Model of Industry-University-Research Education Based on Data Mining

Yanbin Tang[✉]

Jiangxi University of Applied Science, Nanchang 330100, China
yugsf574@163.com

Abstract. Teaching quality is the main indicator for evaluating teaching level. But it is affected by a number of contributing variables. To address existing issues in teaching quality evaluation and boost the accuracy of teaching quality evaluation, a data mining-based teaching quality assessment model is developed. To begin, this model investigates and analyzes the relevant literature on the present evaluation of teaching quality, generate evaluation indicators of factors affecting teaching quality, and gathers data on teaching quality influencing factors. And creates research samples for evaluating teaching quality at schools of higher education as well as determines the grade of educational effectiveness through specialists. And applies data mining technology to train study samples, forming the model of university teaching quality assessment. Analyzes the superiority of the college and university teaching quality model using real instances. The results reveal that data mining can represent the disparities in quality of instruction grades in universities and produce high accuracy quality of instruction assessment results.

Keywords: Data mining · Industry-University-Research education · Mobile teaching · Teaching quality evaluation

1 Introduction

Education in a broad sense refers to education with three functions: teaching, scientific research and service. In senior high school education, the primary goal is to make use of the different educational resources and environment of schools, industries, enterprises and research institutions to cultivate applied talents suitable for the needs of industries and enterprises, that is, to make use of the respective advantages of schools, enterprises and scientific research institutions in personnel training, and to organically combine the educational environment mainly focusing on classroom teaching of indirect knowledge with the production site environment mainly focusing on direct acquisition of practical experience and capability. With the continuous increase in the number of students, teaching quality has become more and more important in the quality of higher education. It has become a measure of teaching effectiveness and talent evaluation. In a university, it is a challenging problem to evaluate the teaching quality because of many subjects,

© ICST Institute for Computer Sciences, Social Informatics and Telecommunications Engineering 2023
Published by Springer Nature Switzerland AG 2023. All Rights Reserved
W. Fu and L. Yun (Eds.): ADHIP 2022, LNICST 469, pp. 312–324, 2023.
https://doi.org/10.1007/978-3-031-28867-8_23

some overlapping subjects, flexible teaching methods and so on. At present, teaching quality evaluation in colleges and universities can be divided into two branches: one is based on qualitative analysis, including expert system, association rules and so on.

The evaluation system of teaching quality includes some quantitative factors and non-quantitative factors, so it is difficult to describe the teaching quality of colleges and universities by qualitative analysis. Teaching quality evaluation indicators include useful information, interactive content, content page length, etc. The evaluation index is the same for all types of organizations. The other is the teaching quality evaluation method based on quantitative analysis, which is subdivided into traditional statistical evaluation method and machine learning algorithm evaluation method. Traditional statistics mainly includes linear regression and gray theory, which can only describe the simple and linear statistical relationship between influencing factors and teaching quality, so that the accuracy of teaching quality evaluation can not meet the practical requirements. Moreover, it can make full use of expert knowledge and experience to get a better evaluation result of college teaching quality. Relevant scholars build teaching quality evaluation models by different methods to improve mobile teaching quality. The quality evaluation model is constructed by fuzzy theory. Combining with the actual and comprehensive consideration of teaching software, the fuzzy theory is used to evaluate the quality of teaching software, which can improve the unreasonable and complex disadvantages of traditional evaluation methods, and get scientific, accurate and objective evaluation results conveniently and quickly. However, this method takes a long time to calculate. The evaluation model of auxiliary teaching quality is produced using an active learning support vector machine, and the current appraisal system of classroom teaching quality is built. However, the computation of this method takes a long time. It is extensively applicable to utilize BP neural network to increase the effect of teaching quality analysis and evaluation, as well as to design the model structure of BP neural network teaching quality analysis and assessment. However, this method's generalization performance is low.

In order to improve the accuracy of teaching quality evaluation, this paper establishes a teaching quality evaluation model based on data mining technology. The research results show that the accuracy of teaching quality evaluation based on data mining technology is high.

2 Mobile Teaching Quality Evaluation Model of Industry-University-Research Education

2.1 Evaluation Index of Mobile Teaching Quality in UIC

The training data produced by random sampling comprises a lot of redundant or even irrelevant data in the actual training and testing data. When confronted with this type of training data, the normal data mining approach based on passive learning will be badly impacted, resulting in a decline in the accuracy of final test data. As a result, in order to successfully cope with unpredictable and irregular training datasets, the active learning method is presented. The most significant advantage is that the training process is interactive. Figure 1 depicts a schematic of the active learning concept.

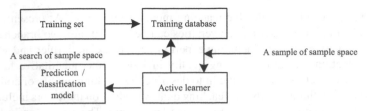

Fig. 1. Mobile teaching management system of industry-university-research education

The main body of teaching quality evaluation is composed of teachers themselves, students, teaching departments and peer teachers. The improvement of teaching quality can not be decided by a single subject. Some scholars believe that the main body of teaching quality evaluation should include administrators, students, colleagues and teachers themselves. Some scholars also propose that students, teachers, schools, society and employers should be the main body of evaluation and the corresponding evaluation model. At present, management evaluation and student evaluation are still the most common, but there is no conclusion on who is the most important subject of evaluation, but it is certain that these directly determine the results of evaluation. In general, the subject analysis and assessment themselves are quite important. According to the actual situation, we classify the evaluation users into 4 categories in the evaluation system, namely, supervisor, peer teacher, teacher and student, as shown in Fig. 2.

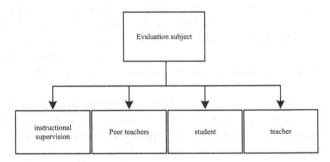

Fig. 2. Teaching quality evaluation system

College and university teaching is a methodical effort with complicated relationships. Its system is clearly graded. Educational status is a reflection of the quality of education, and the quality of an education is the overall level of a school's education. Figure 3 shows such a progressive integration structure.

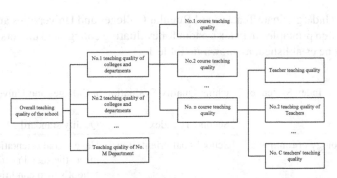

Fig. 3. Teaching Evaluation System Structure Diagram of Colleges and Universities

The teaching quality of teachers is evaluated by establishing teaching quality evaluation standards, curriculum evaluation standards, department evaluation index system, department teaching quality evaluation system, department teaching quality, overall teaching level, and teaching quality evaluation index system.

2.2 Evaluation Algorithm for Mobile Teaching Quality in UIC

In order to evaluate teaching activities fairly and effectively, a scientific evaluation System must be established to increase the accuracy and effectiveness teaching Evaluation. The pyramidal index system is commonly used in the evaluation and study of teaching quality. That is, the teaching evaluation objectives are first decomposed into several first-grade indexes, and then the first-grade indexes are decomposed into several second-grade indexes in accordance with the actual situation…. By analogy, according to the complexity of the specific problems, they are decomposed into several levels, as shown in Fig. 4. The more levels of decomposition, the more specific the indicators.

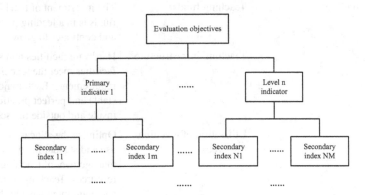

Fig. 4. Tower structure of evaluation index system

Based on the above analysis of the main factors influencing college and university teaching quality, and in accordance with the Ministry of Education's Excellent Evaluation

Scheme for Undergraduate Teaching of Regular Colleges and Universities and the index system's design principle, an index system for evaluating college and university teaching quality may be established, as shown in Table 1.

Table 1. Index System of Teaching Quality Evaluation in Colleges and Universities

Primary index	Secondary index	Quality standard
Guiding ideology of running a school	School goal orientation	Have a goal orientation that meets the social needs and the school's own conditions; Practical development and construction planning and implementation; The overall layout and structure of majors and disciplines are reasonable
	Teaching center status	Teachers at all levels attach importance to teaching and devote themselves to it; There are policies to encourage teaching work, and the effect is good; The development of graduate and undergraduate education at all levels is in line with the positioning of the University
	Educational ideas	Pay attention to the study of educational ideas and often organize the discussion of educational ideas
Teaching conditions	Teaching funds	The investment of teaching funds is in a leading position and continues to grow
	Teaching infrastructure	Hardware facilities and sports facilities meet the teaching requirements; Each major has a stable and perfect practice base inside and outside the school
	Utilization of teaching resources	Optimize the allocation, efficient use and standardized management of teaching resources; Teaching instruments and equipment have high efficiency and can give full play to their role

Because the students collected the secondary index input using the centesimal approach, the size of each component is quite varied. If the original data is quoted verbatim, without any alteration, the absolute value of the original value may be too great and fall outside of the neuron's effective processing interval, resulting in the so-called "saturation phenomena." Even though the absolute value of the original data is not overly big, the effect on the network may be significantly larger than that on other components, causing the other components to lose control of the network. As a result, the input samples must be normalized. Students evaluate instructors' teaching skill, innovative thinking, teaching impact, and so on, whereas teachers evaluate their own teaching attitude, initiative, and curriculum teaching effect. While instructors are accountable for assessing their students' teaching consciousness, uniformity, and originality.

The maximum and minimum methods are utilized to normalize the data in this research. Because the approach is a linear transformation, its original meaning is preserved and no information is lost. The input normalization formula is as follows:

$$R = Fw - \frac{I}{p \prod I_{\max} - I_{\min}} \tag{1}$$

In the formula, w is the input value I after normalization, which is the unprocessed input value $I_{\max}$, and the input value $I_{\min}$ is the minimum. The original p indices are made a linear combination as a new comprehensive index. If the first selected linear combination, the first composite indicator, is recorded as F, it is natural to want F to reflect as much information about the original indicator as possible. The information here is expressed in terms of the variance of F, that is, the greater the F, the more information it contains. Therefore, the first principal component selected in all linear combinations should be the one with the greatest variance. If the first principal component is not sufficient to represent the information of the p indices, the second linear combination, i.e. the second principal component, can be used to construct the 3rd, 4th, and pth principal components, which are independent of each other and the variance decreases. There are n samples, each of which is represented by p indices x1, x2, ..., xn, Description, Raw Data Matrix:

$$X = R \begin{bmatrix} x_{11} & x_{12} & \cdots & x_{1p} \\ x_{21} & x_{22} & \cdots & x_{2p} \\ \vdots & \vdots & \vdots & \vdots \\ x_{n1} & x_{n2} & \cdots & x_{np} \end{bmatrix} = R(X_1, X_2, \cdots, X_p) \tag{2}$$

In the formula, R is the cognitive level of students. Use the p vectors $X_1, X_2, \cdots, X_p$ of the data matrix a_{pi} as a linear group. Combine the following:

$$F_i = a_{1i}X_1 + a_{2i}X_2 + \cdots + a_{pi}X_p \tag{3}$$

In terms of data forecasting, it is excellent and unique to assume that the sample set $\varphi(x) \in R\varpi$ is the input data and the $B \in R$ is the output data. Suppose:

$$y(x) = X\varphi(x) + FB \tag{4}$$

In the formula, B is the difficulty index. To find vectors and scalars, define the optimization problem:

$$\min J = \frac{1}{2}F_i R\varpi + \frac{1}{3}X\sum_{k=1}^{M}\xi_k^2 \tag{5}$$

In this equation, ϖ is the regularization parameter, M is the total amount of teaching data and ξ_k is the relaxation variable. The data processing process shall be simple as follows: (1): mining and obtaining data sources from various data sources (paper materials, text documents, databases, etc.), carrying out simple analysis and formatting of the data obtained, so that it basically has a certain form and then classified storage; and (2): cleaning and processing the dirty data obtained in (1), analyzing the data, establishing data structures, classifying and sorting out different data, and finally storing the data in the text documents or databases. (3) Establish a quality evaluation model and select quality evaluation tools. (4) Read the data to be evaluated, analyze the correlation, select the appropriate form of quality evaluation, and present the data in the form of a view, as shown in Fig. 5:

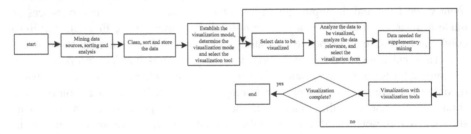

Fig. 5. Flow chart of educational data quality assessment

Different quality evaluation forms show different data meanings, different dataset structures, different data relationships, and different analysis results. The objective of data quality assessment is not to present the simple, single-dimensional data, but to collect relevant data through statistics, mine the data information from different dimensions, and analyze and compare the intricate relationships among the data. The results are then presented to the user to reveal the potential value between the data. During the evaluation of dataset quality, it is necessary to analyze and process dataset, use interactive charts, data maps and other forms of expression to carry out targeted and feasible quality evaluation. Through data quality evaluation, we can explore and reveal the multi-level meaning of each data in different data relations, find out the potential value hidden among the data, and show us the hierarchical correlation among educational data.

2.3   Construction of Educational Data Quality Assessment Model

Due to the differences in the objectives and types of courses, teaching environment, class size and other external conditions, and the lack of clear boundaries between the organization and implementation of mixed teaching activities, this study incorporates evaluation types, nodes, indicators, content and evaluation subjects into the evaluation system based on the above contents (as shown in Table 2). The standardized weights of specific dimensions need to be discussed by teachers combined with the actual situation of mixed teaching. But what can be sure is that the function of mixed teaching quality evaluation is no longer the only way to judge the value of students' knowledge and skill proficiency.

Table 2. Mixed Teaching Quality Evaluation System

Evaluation type	Evaluation node	Evaluating indicator	Evaluation content	Evaluation subject
Process evaluation	Pre class learning evaluation	Student activity; Video viewing times; Video viewing duration and number of replies	Landing times of the platform; Completeness and frequency; Duration and rumination ratio; Number of Posts and replies	teacher
	Classroom activity evaluation	Classroom performance; Collaborative learning outcomes	Quality of questions and answers; Content, technology and creativity	Teachers and students
	After class learning evaluation	Achievement display effect; Team contribution	Content, technology and creativity	Teachers and students
Summative evaluation	Final evaluation	Group summary report; Course assessment test	Team and individual reflection; Understanding and mastery	Teachers and students

Confirmatory factor analysis is to determine the ability of a secondary indicator system to simulate real data. According to the two-level evaluation index system in Table 3, we constructed a model with five factors.

Table 3. Index System of Classroom Teaching Quality Evaluation

Primary index	Secondary index
Teaching attitude	Fully prepared for teaching, proficient in content and positive emotion; Strict management and good classroom order
Content of courses	The teaching content is systematic, substantial, rhythmic and reasonably arranged; Integrating theory with practice and reflecting the frontier of the discipline

(continued)

Table 3. (*continued*)

Primary index	Secondary index
Teaching method	Use heuristic teaching such as facts to help students accept and understand relevant knowledge and cultivate students' ability of independent thinking; Timely and appropriate use of blackboard writing and modern teaching technology means, the effect is good
Teaching effectiveness	Encourage pupils to actively participate in the teaching process, and create a welcoming environment in the classroom. Students have a high level of mastery of course material. To encourage healthy study habits and methodologies in pupils
Teaching quality	Clear organization, accurate teaching knowledge, highlight key points and clarify difficulties; Fluent expression, concise language and strong attraction

When using the data mining principle to construct the quality evaluation model of distance education, we must first select the most crucial quality evaluation index. The particle swarm optimization technique is then utilized to optimize the data mining settings. Finally, the model of distance teaching quality evaluation is established by using the principle of optimal parameter data mining and important distance teaching quality evaluation indexes.

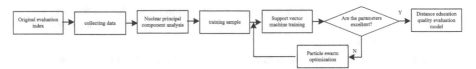

Fig. 6. Running process of distance learning quality assessment model

In addition to the perfect structure, the complete index system also needs to give the corresponding weight to the index, and make clear the importance of each index, so as to make a quantitative analysis of the classroom teaching quality and a scientific and correct evaluation. Data mining principle is a common weighting method, is a mathematical means, by experts at all levels of indicators for pairwise comparison operation of the weight method. This method can effectively solve the multi-index and multi-level weight problem. The multivariate hierarchical model built is shown in Fig. 7.

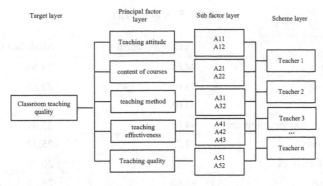

Fig. 7. Data Mining Based Hierarchical Model of Teaching Evaluation

The consistency of other decision makers' decision matrices is checked by the same principle, and the corresponding weight vectors are calculated. The weight vectors of the decision-maker are averaged and normalized to get the final index system, which realizes the realization of teaching evaluation.

3   Analysis of Experimental Results

A case study is carried out to assess the efficacy of the data mining-based teaching quality rating methodology. Table 4 shows the experimental platform as one of them.

Table 4. Experimental Analysis Platform

Platform parameters	Specific configuration
Programming tools	Python
CPU	Pentium IV, 2.8G
Memory	32GB
Hard disk	SAMSUNG 256G SSD
Display card	HDMI

Table 5. Generality analysis of teaching quality evaluation model

Course name	Evaluation accuracy /%	Modeling time / ms
College English	95.85	22.62
C language programming	95.79	23.78
Communication principle	95.29	25.26

(continued)

Table 5. (continued)

Course name	Evaluation accuracy /%	Modeling time / ms
Linear algebra	94.65	23.65
Machine learning	93.65	22.85
College Physics	93.05	26.52
Fundamentals of college computer	94.65	23.84
Physical chemistry	94.35	25.32
Engineering Mechanics	95.71	26.05
College Chinese	95.65	26.32

To assess the universality of a data mining-based teaching quality evaluation model, 10 university courses are chosen as the test object, and their accuracy and modeling time are measured.

Using the SPss software, this paper analyzes and compares the score distribution of teachers' classroom teaching quality evaluation results before and after the revision of the index system. The results are shown in Table 6.

Table 6. Comparison of segment distribution before and after revision of index system

Fraction	Original proportion	Current proportion
Below 80	1.6	3.1
80 ~ 85	4.6	35.6
85 ~ 90	25.2	58.2
90 ~ 100	68.6	3.1
Total	100%	100%

This set of indicators in the trial period has also been the evaluation of the quality of classroom teaching organizers, evaluators and those who are highly recognized. From the perspective of organization managers, the system matches the orientation of research universities and keeps pace with the teaching reform. And the index system is concise, clear and easy to operate. Teaching quality evaluators believe that the system pays attention to the principles of completeness, guidance, measurability and independence of the indicators, which is conducive to scientific evaluation. The most important thing is that the evaluation results are highly recognized by the evaluated teachers, who believe that the evaluation results are fair, just and credible. Because of the wide variety of the experimental data set, the domestication approach is employed to quantitatively assess the model's correctness. Furthermore, the conventional SM model and the E-data mining model are evaluated on the same experimental dataset for performance comparison and analysis. The graphic depicts the experimental findings of the three models.

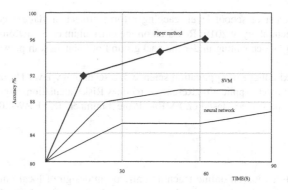

Fig. 8. Three Methods for Evaluating Accuracy Compare Detection Results

Compared with the typical SVM model and the neural network model, the model based on data mining theory has some advantages in the accuracy of teaching quality evaluation. In addition, in the run-time, the typical data mining model needs the least time, but the accuracy is the worst, the practicability is not high. Compared with E-data mining model, this paper proposes an evaluation model, which needs less time and takes into account both accuracy and time efficiency, without complicated evidence theory.

4 Conclusion

In order to get a better result of teaching quality evaluation, data mining technology and adaptive a genetic algorithm is used to create a model for evaluating teaching quality. Experts determine the teaching quality grade based on the evaluation index of influencing factors and the data of influencing factors of teaching quality, and the teaching samples are trained based on the research samples of teaching quality evaluation in colleges and universities. The results show that the model is a high-accuracy and efficient model. Although some work has been done in the research, due to the constraints of level, time and resources, there are inevitably many shortcomings in the research, which need to be further improved. There is not much discussion on the construction of teaching evaluation subjects, such as peer evaluation and flexible transformation of evaluation objects in the system. The research on students' evaluation of teaching evaluation has not been carried out in depth, and the follow-up research can list it as the research direction to build a more complete mobile teaching quality evaluation model of Industry-University-Research education.

Fund Project. 1. 2019 Jiangxi Provincial Department of Education Science and Technology Project: Research on systemic risk identification and countermeasures of industry-university-research cooperation projects (Project number: GJJ191199)

2. 2020 Jiangxi Provincial Culture and Art Science Planning Project: Research on policy paths for Jiangxi cultural enterprises to solve difficulties under the new crown epidemic (Project number: YG2020154)

3. 2020 Jiangxi Provincial Department of Education Teaching Reform Project: The strategy research and practice of integrating socialist core values into the whole process of accounting professional teaching (Project number: JXJG-20–29-3).

4. General project of school level teaching reform project of Jiangxi college of application science and technology in 2019: Research on the curriculum construction of comprehensive simulation training of accounting under the background of "Shuangwan plan" (Project number: JXYKJG-19–21).

5. 2020 Jiangxi college of application science and technology-level Humanities and Social Sciences General Project: Application Research on Fuzzy Risk Calculation of Industry-University-Research Cooperation Project Based on FMEA (Project number: JXYKRW-20–1).

References

1. Zhang, Y.: Discussion on online teaching curriculum design of local university of finance and economics based on industry-university-research cooperation mode. Curriculum Teach. **4**, 38–44 (2021)
2. Yang, X., et al.: Industry-university-research interaction model for teaching quality evaluation of economic management courses. Int. J. Emerg. Technol. Learn. (iJET), **16**(23), 202–215 (2021)
3. Yongzhou, L., Yinghuan, Z., Teng, F.: Research on the blended experiential learning mode of business administration talents in universities. In: Proceedings of the 2020 the 2nd World Symposium on Software Engineering, pp. 231–235 (2020)
4. Shuo, W., Ming, M.: Exploring online intelligent teaching method with machine learning and SVM algorithm. Neural Comput. Appl. **34**(4), 2583–2596 (2022)
5. Liu, S., He, T., Dai, J.: A survey of CRF algorithm based knowledge extraction of elementary mathematics in Chinese. Mob. Netw. Appl. **26**(5), 1891–1903 (2021). https://doi.org/10.1007/s11036-020-01725-x
6. Cheriguene, A., Kabache, T., Kerrache, C.A., et al.: NOTA: a novel online teaching and assessment scheme using Blockchain for emergency cases. Educ. Inf. Technol. **27**(1), 115–132 (2022)
7. Gao, P., Li, J., Liu, S.: An introduction to key technology in artificial intelligence and big data driven e-learning and e-education. Mob. Netw. Appl. **26**(5), 2123–2126 (2021). https://doi.org/10.1007/s11036-021-01777-7
8. Zhao, H., Zhang, C.: An online-learning-based evolutionary many-objective algorithm. Inf. Sci. **509**, 1–21 (2020)
9. Wang, S., Liu, X., Liu, S., et al.: Human short-long term cognitive memory mechanism for visual monitoring in IoT-assisted smart cities. IEEE Internet Things J. (2021). https://doi.org/10.1109/JIOT.2021.3077600
10. Jing, K.L., Yu, F.S., Fu, Z.Y.: Model predictive control of the fuel cell cathode system based on state quantity estimation. Comput. Simul. **37**(07), 119–122 (2020)

Classified Evaluation Model of Online Teaching Quality in Colleges and Universities Based on Mobile Terminal

Lei Han[1](✉) and Qiusheng Lin[2]

[1] Department of Primary Education, Yuzhang Normal University, Nanchang 330103, China
hanlei10023@163.com
[2] Guangzhou Huali College, Guangzhou 511325, China

Abstract. At the moment, only students' academic achievements or questionnaire statistics are used to assess teaching quality. Its accuracy and efficiency are limited when applied to online evaluation of teaching quality. To address the aforementioned issues, this work investigates and develops an online teaching quality categorization evaluation model based on a mobile terminal. The crawler crawls the necessary data after collecting the evaluation of teaching excellence data on the mobile terminal. The online teaching quality categorization and assessment dimension is built based on data climbing, allowing for multi-dimensional teaching quality evaluation. On this basis, the teaching quality classification and evaluation index system is constructed. An adaptive variant genetic algorithm was used to improve the BP neural network and establish a classification and model for assessing teaching quality. The model test results show that the average evaluation accuracy of the model is 88.16%, and the model has good evaluation efficiency and stability.

Keywords: Mobile terminal · Teaching quality · Classification evaluation · Model construction · BP neural network

1 Introduction

In the process of development, the demand for talents is increasing day by day. High-quality talents reserve is related to a country's comprehensive competitiveness in the world to a certain extent. The impact of science and technology on higher education is increasingly far-reaching. For example, with the popularization of the Internet, teaching tools have become diversified. Educators and learners no longer simply refer to teachers and students, but have become more flexible in their objects and content. In this learning environment, both educators and students must adapt to and learn new technologies. In order to promote the development of modern higher education and integrate scientific and technological means into it, we must carry out modern reforms in education contents, means and methods so as to effectively improve the quality of education. Online teaching is the main way to ensure the smooth teaching plan in university. Therefore, online teaching quality evaluation is very important for college teaching research [1].

W. Fu and L. Yun (Eds.): ADHIP 2022, LNICST 469, pp. 325–335, 2023.
https://doi.org/10.1007/978-3-031-28867-8_24

The evaluation of teaching quality is an important way to reflect whether the teaching goal is realized or not, and whether the curriculum is valuable or not. The existing problems of traditional teaching quality evaluation methods are mainly embodied in the following aspects: emphasizing management, ignoring the inadequacy of teaching process and evaluation index, pursuing quantity excessively, and imperfect evaluation subject. Compared with the statistical evaluation method of questionnaire, the reliability of the evaluation method of AHP and fuzzy hierarchy is relatively improved. Reference [2] method using the concept of student-centered, this paper analyzes the three indexes of students' learning input, learning support and learning effect, and puts forward a scientific and reasonable evaluation system of experimental teaching quality by combining learning motivation, refining resources support and incorporating multi-dimensional perspective into the evaluation path. Reference [3]. After controlling for the characteristics of students, instructors, and schools, the difference between anticipated values of students' accomplishments and average values of classes may be viewed as the net influence of teachers' instruction on students' academic progress. The influence of teachers' teaching quality is represented by adding the average values of these residuals at the class level, and the assessment results of teachers' teaching quality are derived.

When employing mobile terminals for effective online teaching, controlling instructor teaching quality becomes more challenging, and there are several constraints in the actual application of existing teaching quality evaluation methodologies [4]. Therefore, this paper will study and build a model of online teaching quality classification and evaluation based on mobile terminal to solve the operational problems of online teaching quality evaluation and evaluation system, and continuously improve online teaching quality through quality evaluation, stimulate teachers' enthusiasm and responsibility.

2 Construction of Classified Evaluation Model of Online Teaching Quality in Colleges and Universities Based on Mobile Terminal

2.1 Access to Online Teaching Data by Mobile Terminals

Teachers' teaching quality is finally embodied by students' performance and feedback, but teachers can't get students' feedback on teaching process accurately and timely when teaching online. Therefore, this paper uses the corresponding function modules of mobile data terminal to collect the data of students in real time.

Students use mobile terminals to log on to the online teaching platform of colleges and universities, and then enter the corresponding teaching port. When the teaching activity begins, the backstage of the online teaching platform stores all the data from the authorized mobile terminal into the corresponding database. The data collected by mobile terminals usually include the students' facial expressions, attendance, enthusiasm for class and feedback after class. These data are often cluttered and unordered, and this paper uses distributed crawlers to obtain data that will ultimately be used to classify and evaluate teaching quality [5].

The crawler searches for data through the mobile terminal identification code stored in the background and the corresponding student ID, reads the contents of a data table from a certain data table in the database, finds the link addresses of other data tables that exist in the data table, then jumps to the next linked data table and continues the search until all the data tables in the online teaching platform database are grabbed. According to the breadth priority strategy, the priority of layer 1 is the highest, and then according to the depth of the layers, the priority is: layer 2, layer 3, layer 4, layer 5. Because the breadth-first fetching strategy is decided according to the level of the target link, it does not need to record the branch node of the last crawl, which reduces the control difficulty. This paper introduces a Bayesian classifier to define the type and direction of crawler data acquisition, which is not suitable for teaching quality classification and evaluation.

For category dc in each data type, a Bayesian classifier can be constructed to compute the probability dc of data table e belonging to category $P(dc|e)$. Crawling allows you to pick a list of interesting analogies for $d\hat{c}$. Each data table that is crawled down is given a correlation score [6]:

$$R(e) = \sum_{dc \in d\hat{c}} P(dc|e) \tag{1}$$

There are two different strategies for qualifying a reptile. In the soft-qualified strategy, the crawler uses the score $R(e)$ of e for each crawled table as a priori value for all unaccessed table links extracted from the e, which are then enqueued with priority. In the hard-qualified strategy, when the data table e is crawled down, the classifier first finds the leaf node $d\widehat{c}(e)$, which is most likely to be the e of the category:

$$d\widehat{c}(e) = \arg\max P(dc|e) \tag{2}$$

When the above requirement is met, the crawled data table is added to the queue, otherwise it will be dropped. The maximum and minimum normalized function, which is used after the data of online teaching quality is crawled, can transform the data linearly without losing the original meaning and information. The normalized formula for the maximum and minimum method is as follows [7]:

$$P' = \frac{P - P_{min}}{P_{max} - P_{min}} \tag{3}$$

P' is the normalized data of teaching quality evaluation, P is the collected raw data, P_{min} is the minimum of the original data, and P_{max} is the minimum of the original data.

2.2 Establishment of Dimensions for Classified Evaluation of Online Teaching Quality in Colleges and Universities

Scientific classification is of great significance. It can summarize, consolidate and improve the results of comparison, systematize complex things, reveal the internal structure and proportional relationship of things, reveal the relations and differences between all kinds of things and things at all levels, improve the consciousness of scientific research, and provide clues for finding the law of development of things. In college and university online teaching quality categorization and evaluation, we should stimulate teachers' potential, innovate consciousness, promote teachers' comprehensive ability, and make the evaluation methods and classification system more objective, real and reasonable.

Scientific classification must be carried out according to some attributes and relations of objects themselves. Because the objective things have many attributes, therefore, the classification standard is also many aspects. It is generally believed that there are two basic types of classification: phenomenal classification and essential classification. Online teaching has four main characteristics: dialogue, direct participation, support and control of the learning process, abbreviated as DISC characteristics. The characteristics of DISC vary with two variables, namely whether intentional learning is teacher-controlled or student-directed, and whether the learning activity is strictly defined or open. According to the two dimensions of online teaching (the openness of learning tasks and the degree of students' self-control), in the setting of teaching method indicators, the pertinence of design, the target of thinking, the vividness of language, the interaction of communication and other factors shall be integrated; in the setting of teaching content indicators, the transmission and integration of new knowledge, new ideas and new professional trends in teaching and education shall be integrated; in the setting of teaching effect indicators, the teaching effect shall not be measured simply by examination or by the department, but by the understanding and mastery of knowledge by students in multiple aspects [8].

Classified evaluation system must be able to promote teachers' teaching ability. Some of them belong to phenomenal classification and some to essential classification. The phenomenon is classified as follows: according to the level of online education, classification or purposes; according to the use of network courses for the classification of media; according to teaching information transmission and interaction patterns for classification. Among the essential classifications are: the classification from the perspective of teaching, the classification according to the openness of learning tasks and the degree of students' self-control, and the classification according to the relevance between learners in learning activities. Based on the teaching attitude, teaching method, teaching content and teaching effect, this paper constructs the online teaching quality classification and evaluation index system.

2.3 System of Online Teaching Quality Categorization and Evaluation Index Construction

When developing an online teaching quality assessment index system, teachers should consider not only teaching materials and techniques, but also students' learning effect and excitement. Effective integration of online and offline materials, not only in accordance with national standards, but also in accordance with local features of teaching resources, resulting in a more content-rich course.

Among them, teacher literacy can have a direct and key impact on the quality of teaching, teachers need to assume a diverse identity. At the same time, teachers need to use a variety of management tools to check students' practice results, more flexible and more diverse teaching methods, methods and means, so as to achieve better teaching results. The influence of teachers' accomplishment on teaching quality does not interfere with the teaching mode. In the practice of online teaching, we should not only accomplish the teaching task smoothly and achieve the given teaching goal, but also guarantee the teaching effect. In online teaching, teachers should increase or decrease the teaching content of the platform, so that students can keep up with the development of the subject, master cutting-edge knowledge. In fact, teaching process is the key factor that affects teaching quality.

This paper establishes the index system of online teaching quality classification and evaluation, as shown in Table 1 below [9], based on the dimensions of online teaching quality classification and evaluation and the factors that affect online quality of instruction.

Table 1. Classified Evaluation Index System of Online Teaching Quality

Categorical evaluation dimension	Grade I evaluation indicators	Secondary evaluation index
Teaching attitude D1	Teacher Teaching Literacy C1	Professional teaching ability C11
		Online Teaching Ability C12
		Professional background C13
	Instructional Reflection C2	Professional Teaching Thinking C21
		Basic teaching ideas C22
Teaching methods D2	Teaching Skills C3	Mobile terminal operating capacity C31
		Teaching Aid Data Collation Ability C32
	Lesson plan C4	Course Schedule Reasonable C41
	Diversity C5	Is it possible to use multiple teaching methods for teaching C51
Content D3	Rationality C6	Student Learning Speed C61
		Fit with teaching method C62
	Scientific C7	Most Difficult Points C71
		Grasp the front of science, introduce the latest trends and academic thought C72
		Theory with practice C73

(continued)

Table 1. (*continued*)

Categorical evaluation dimension	Grade I evaluation indicators	Secondary evaluation index
Teaching effect D4	Teaching Management C8	Ability to accurately monitor students' online learning C81
		Be able to combine daily teaching management with student performance assessment C82
	Teaching Feedback C9	Ensure that students are communicated with C91 after each class
		C92 Bc able to improve the teaching process in real time based on teaching feedback

After establishing the classified evaluation model of online teaching quality shown in Table 1 above, according to the actual teaching experience, there are obvious differences in the weights of different evaluation indexes. Therefore, after constructing the online teaching quality classification evaluation system in Table 1, define the importance of the classification dimension of the classification evaluation is equal.

On the basis of the constructed indicators, the expert opinion form shall be formulated, and experts shall be invited again to assign the importance of the indicators to the Starr Relative Importance Table shown in Table 2 for each indicator item at the same level, and create the corresponding judgment matrix to obtain the weight of each indicator item [10].

Table 2. Stahl relative importance scale

Relative importance	Grade	Relative importance	Grade
A is as important as B	1	A and B are of equal importance	1
A is slightly more important than B	3	A is slightly less important than B	1/3
A is obviously more important than B	5	A is obviously less important than B	1/5
A is more important than B	7	A is less important than B	1/7
A is more important than B	9	A is less important than B	1/9
Intermediate value of adjacent degree between A and B	2, 4, 6, 8	Intermediate value of adjacent degree between A and B	1/2, 1/4, 1/6, 1/8

From the comparison table of relative importance, a judgment matrix can be obtained. For the evaluated judgment matrix, the next step is to sum the rank of each row of the judgment matrix, and then get the initial weight column vector. However, it is still not clear that this weight is a standard and meaningful weight. The consistency test of the judgment matrix must also be performed. The following is the test formula:

From the relative importance comparison table, a judgment matrix J can be obtained. For the evaluated judgment matrix J, the next step is to sum the rank of each row of the judgment matrix, and then get the initial weight column vector. However, it is still not clear that this weight is a standard and meaningful weight, and the consistency test of the judgment matrix needs to be conducted. The formula for the test is as follows:

$$CR = CI/RI \tag{4}$$

Among them, CI represents the consistency index, and λ_{max} is used to get the maximum eigenvalue of the judgment matrix, and n represents the order of the square matrix. RI represents the average random consistency index, and its values can be found according to the corresponding table. In the case of $CR \leq 0.1$, it indicates that the judgment matrix passes the consistency test and has consistency, whereas in the case of $CR > 0.1$, it indicates that there is a big deviation in the judgment matrix and the score needs to be modified until it passes the consistency test.

2.4 Construction of Classified Evaluation Model for Online Teaching Quality

When there are enough samples to train, the network may adjust the proper weights and then forecast the teaching quality assessment outcomes based on the sample data. Because the standard BP neural network is prone to falling into the local minimum and has a poor convergence speed, this research uses an adaptive mutation genetic method to enhance the BP neural network. The following are the processes for developing the BP neural network model based on the adaptive mutation genetic algorithm:

(1) The design of the input layer: according to the secondary indicators in the classification and evaluation system of online teaching quality of colleges and universities established above
(2) Design of output layer

The number of output neurons is one because the evaluation results are used as the BP network outputs in this paper.

(3) Design of hidden layers

The more hidden layers inside a neural network, based on its structure and training process, the more difficult the BP neural network is. Kosmogorov's theory selects the BP network with just one hidden layer structure.

(4) Determination of the number of neurons in the hidden layer

The most appropriate number of neurons in the hidden layer is determined by empirical formula and several experiments (the quality of network convergence performance). The empirical formula selected in this paper is as follows:

$$l = \sqrt{m+n} + \alpha \tag{5}$$

l is the number of neurons in hidden layer, m is the number of neurons in input layer, n is the number of neurons in output layer, and α is the constant between 1 and 10. The calculation suggests that the number of neurons in the hidden layer should be between 6 and 14. The learning error and fitness functions are shown below.

$$\begin{cases} E = 0.5 \sum_{k=1}^{p} \sum_{j=1}^{l} \left(i_j^k - o_j^k \right) \\ fit = E^{-1} \end{cases} \tag{6}$$

Among them, i_j^k and o_j^k are input and output vectors of neural network, E is learning error of neural network, fit is fitness function. Evaluate the quality of online teaching in colleges and universities according to the flow chart shown in Fig. 1.

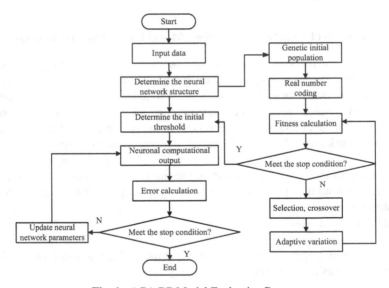

Fig. 1. AGA BP Model Evaluation Process

Among them, the iterative selection of genetic algorithm is roulette. Firstly, the fitness of neural network is calculated, and then the proportion of the fitness is calculated, which is the probability of the individual being selected in the selection process. After calculating the likelihood of each individual being chosen, the accumulated probability must be used to determine if the chosen individual may be passed on to the next generation. The genetic algorithm proposed in this paper crosses genetic operators by heuristic crossover. According to the input data of neural network, the final evaluation results of online teaching quality are output according to the process in Fig. 1. So far, we've finished building the online teaching quality categorization assessment model based on mobile terminals.

3 Evaluation Model Test

3.1 Test Content and Preparation

This section will test the accuracy of the online teaching quality classification and evaluation model based on mobile terminals. AHP -based teaching quality evaluation model and fuzzy theory based teaching quality evaluation model are selected as the comparison, and the evaluation accuracy of the model is calculated.

There are 337 data sets in this experiment, of which 287 are training data sets and the remaining 50 are testing data sets. The dataset is evaluated using three teaching quality evaluation models. The evaluation results are compared with the simulation results. The accuracy of the model is obtained by calculating evaluation errors. Simultaneously, a statistical evaluation model is used to evaluate the time cost of each subset of data in distinct data sets in order to determine the model's processing efficiency.

3.2 Test Results

Table 3 below shows the comparison of the accuracy of online teaching quality evaluation using different teaching quality evaluation models.

Table 3. Comparison of Model Evaluation Accuracy/%

Groups	Classified Evaluation Model of Online Teaching Quality in Colleges and Universities Based on Mobile Terminal	Teaching Quality Evaluation Model Based on AHP	Teaching Quality Evaluation Model Based on Fuzzy Theory
1	87.18	81.38	79.51
2	89.03	84.63	74.30
3	88.36	79.45	70.55
4	87.92	81.51	78.57
5	89.34	79.52	78.04
6	86.89	79.96	80.22
7	89.55	80.23	81.35
8	88.21	82.64	77.77
9	87.12	80.27	78.63
10	87.97	76.91	79.69

From the analysis of the data in Table 3, we can see that under the same test standard, the evaluation accuracy of the online teaching quality evaluation model based on mobile terminal is higher than that based on AHP and fuzzy theory. Moreover, from the perspective of data change, the evaluation accuracy of online teaching quality classification model based on mobile terminal is more stable and not affected by data grouping and data differences. The accuracy scores of different evaluation models are 88.16%, 80.65% and 77.86%, respectively, according to the data in Table 3. The evaluation result is more accurate, the evaluation performance is more stable and the reliability is better.

Figure 2 below shows the comparison of time cost of online teaching quality evaluation using different teaching quality evaluation models.

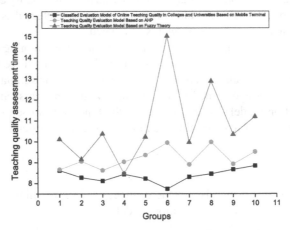

Fig. 2. Time cost comparison of teaching quality evaluation model

From the analysis in Fig. 2, the evaluation time cost of the model is significantly lower than that of the other two models when the data sets with different degrees of intra-group differences are evaluated. It shows that the evaluation model constructed in this paper can be applied to different evaluation links, and the evaluation efficiency is higher. The necessary data are crawled after collecting the teaching quality assessment data on the mobile terminal. We created an online teaching quality classification and evaluation dimension and a teaching quality classification and evaluation index system. We can improve our assessment of teaching quality. To summarize, the mobile terminal-based online teaching quality classification assessment model offers the benefits of accurate evaluation outcomes, high evaluation efficiency, and broad application.

4   Conclusion

In recent years, with the high speed popularization of computer and network, a new educational mode, namely online education, has emerged. Using online teaching can facilitate teachers to answer students' questions in time, teaching is not limited by time and space, can improve students' learning efficiency, promote the rapid formation of students' knowledge system, and integrate as soon as possible. Under the current special background, there are obvious differences between the teaching activities carried out by mobile terminals and traditional teaching activities in universities. This paper constructs a classification and evaluation model of online teaching quality in colleges and universities based on mobile terminal, and verifies the feasibility and accuracy of the model.

References

1. Yang, H., Yanfang, F.: Construction of blended teaching quality evaluation index system in higher vocational colleges in the New Era. Vocational Tech. Educ. **42**(35), 67–72 (2021)
2. Ou, J., Fugen, W., Wenbin, Y.: An empirical study on experimental teaching quality evaluation based on student-centered concept. Res. Explor. Lab. **40**(07), 209–212+224 (2021)
3. Lei, W., Hongmei, M., Huaming, H.: Evaluate teaching performance by residuals of student achievement. J. East China Normal Univ. (Educ. Sci.) **39**(07), 84–91 (2021)
4. Shuai, W., Liu, X., Liu, S., et al.: Human short-long term cognitive memory mechanism for visual monitoring in IoT-assisted smart cities. IEEE Internet Things J. (2021). https://doi.org/10.1109/JIOT.2021.3077600
5. Zhao, M., Wei, Z.: Teaching quality evaluation system based on deep learning algorithm. Mod. Electron. Tech. **43**(13), 143–146+149 (2020)
6. Shuai, L., et al.: Human memory update strategy: a multi-layer template update mechanism for remote visual monitoring. IEEE Trans. Multimedia, **23**, 2188–2198 (2021)
7. Yun, X., Ji, W.: The evaluation system of online teaching quality in universities: value orientation and construction strategies. Heilongjiang Res. Higher Educ. **38**(10), 141–144 (2020)
8. Gao, P., Li, J., Liu, S.: An introduction to key technology in artificial intelligence and big data driven e-Learning and e-Education. Mob. Netw. Appl. **26**(5), 2123–2126 (2021). https://doi.org/10.1007/s11036-021-01777-7
9. Wang, B., Tonghai, W., Yugang, D., et al.: Experimental teaching quality evaluation system based on engineering education accreditation. Res. Explor. Lab. **39**(05), 149–152+181 (2020)
10. Weijie, W., Yasheng, Z., Hong, Y., et al.: Research on spacecraft attitude dynamics simulation experiment teaching based on simulink. Comput. Simul. **38**(12), 176–181 (2021)

Dynamic Recognition Method of Track and Field Posture Based on Mobile Monitoring Technology

Qiusheng Lin[1]([✉]) and Lei Han[2]

[1] Guangzhou Huali College, Guangzhou 511325, China
linqiusheng66@163.com
[2] Department of Primary Education, Yuzhang Normal University, Nanchang 330103, China

Abstract. The recognition technology using conventional sensors or image processing is effective for static gesture recognition, but for dynamic gesture recognition, the moving object can not be tracked in time, resulting in low recognition accuracy and efficiency. In order to optimize the above problems, the dynamic recognition method of track and field posture based on mobile monitoring technology is studied. Set up mobile monitoring equipment in the movement area and the movement track to acquire the data of track and field movement posture. After de-noising the track and field posture data, a Gaussian model is established to segment the image background. Based on the human skeleton model, the motion posture features are extracted. Using BP neural network improved by artificial fish swarm to classify the input movement posture data, the recognition of track and field movement posture is realized. The test results show that the recognition accuracy of the proposed methods is higher than 95%, the recognition efficiency is greatly improved, and it has good practical value.

Keywords: Monitor · Track and field sports · Dynamic attitude recognition · Gaussian model · BP neural network · Artificial fish swarm algorithm

1 Introduction

There are many kinds of track and field sports, and the training methods of each sport are different in different degrees, but the most important thing in track and field sports training is whether the athletes' movement posture is standard or not. The standard and scientific posture can improve athletes' competitive ability and physical quality, and prevent sports injuries, which is the fundamental guarantee for athletes to achieve good competitive results. The continuous development of science and technology has promoted the diversification of training methods and the scientific process of training means in track and field events. Using related technologies to guide and adjust athletes' body posture can avoid physical injuries caused by improper sports posture, and lay a foundation for improving athletes' track and field performance.

The definition of standard movement posture during exercise is mostly based on pictures or oral guidance, which leads to the lack of quantitative evaluation criteria for

W. Fu and L. Yun (Eds.): ADHIP 2022, LNICST 469, pp. 336–348, 2023.
https://doi.org/10.1007/978-3-031-28867-8_25

standard movement posture. At present, there are many methods to study the human motion posture recognition, and the two main methods are: human posture recognition based on image and video analysis and human posture recognition based on motion sensor [1]. Early human motion recognition needs the assistance of external equipment to perceive the change of human posture and then recognize human motion. With the development of machine learning and deep learning, there are many different research directions in academic circles, such as image processing, SVM classifier and deep neural network. With the development of these technologies, computers can sense human movements only through cameras and other devices, thus greatly reducing the number of external sensors. In document [2], the laser sensor is used to collect the motion signal data of the moving target, and the characteristic values of the initial motion signals after segmentation are extracted by the time domain signal analysis method, which are input into the BP neural network to obtain the motion recognition results. Literature [3]. Machine learning is used to identify the local feature points of human movements. By using the differences of human bodies in space-time state and the changes of motion frequency, multi-scale local space-time domain features are constructed, mathematical models of human behaviors are constructed, and neural network parameters are trained to reduce the recognition error of local feature points of human movements. The above methods are effective for static movement recognition, but there are some limitations for dynamic track and field movement recognition with space-time continuity.

When athletes do track and field sports, the posture change from the beginning to the end can be regarded as a dynamic sequence composed of several static postures [4]. Although using sensors to monitor the movement can accurately obtain the movement posture data, wearable sensors will affect athletes' movements and are not suitable for daily training. The types of data collected by non-contact sensors are relatively simple, and it is difficult to meet the identification requirements. Using mobile monitoring technology to collect images of athletes' postures in track and field training can simultaneously achieve the acquisition of athletes' postures from the whole to the local, and reduce the influence of data collection stage on posture recognition. Based on the above analysis, this paper will aim at improving the efficiency and accuracy of athletes' attitude recognition in track and field training, and study the dynamic recognition method of track and field sports attitude based on mobile monitoring technology, which can be used to assist athletes' training and improve the efficiency of track and field training. Aiming at the problem of insufficient efficiency and accuracy of dynamic motion recognition, this paper applies mobile monitoring equipment to collect motion trajectory information. When processing the information, it first removes noise to avoid noise interference, and then improves the recognition accuracy. It uses the improved BP neural network algorithm of artificial fish to solve the pose, so as to avoid falling into local optimization and improve the efficiency of motion pose recognition.

2 Research on Dynamic Recognition Method of Athletic Attitude Based on Mobile Monitoring Technology

2.1 Acquisition of Dynamic Monitoring Data of Athletic Attitude

Athletes' body posture changes rapidly in training or competitive state. Using mobile monitoring equipment to monitor athletes' posture in sports can quickly and accurately acquire the whole process of athletes' movements in the whole track and field events. In order to meet the requirement of high-speed capture of athletes' movement posture in track and field, this paper uses mobile monitoring technology to obtain the dynamic posture of athletes when they make corresponding track and field movements.

According to the difference of track and field events, mobile monitoring terminals are set up within the effective training range of track and field events. The mobile terminal consists of signal transceiver, high-speed camera, communication module, etc. When athletes enter the effective monitoring range of track and field sports, the tracking camera is started, and a series of movement posture data of athletes within the monitoring range are collected, and a dynamic movement sequence composed of several static image frames is generated [5].

The process of human movement is the process of human posture change, which is continuous. The camera with higher frame rate can capture more images of athletes' postures per second, which will not miss the postural changes of athletes' postures during strenuous exercise and reduce dynamic blur. Therefore, the posture change of human body is not obvious in the images collected by the camera with high frame rate in continuous time intervals. According to the sampling theorem, in the process of discretization of continuous signals, when the sampling frequency is more than twice of the highest frequency in the signal, the sampled discrete signal can completely retain the information in the original signal. Therefore, in the design process of the human posture evaluation system, by selecting the appropriate frame rate and setting the appropriate sampling interval. Try to make the sampling frequency more than twice the frequency of human motion.

The acquired motion posture data is transmitted to computer equipment for subsequent processing by WIFI and ZigBee wireless transmission. In order to realize the dynamic real-time monitoring of track and field posture, H.264 video compression format is selected as the basic format of mobile monitoring video compression, and RTP/RTCP advanced protocol is used to assist wireless transmission UDP protocol when monitoring video transmission, so as to provide some data traffic congestion adjustment services and network traffic deployment control services in the process of information and image transmission [6]. The transmission flow of H264 mobile surveillance video format in IP/UDP/RTP protocol is shown in Fig. 1 below.

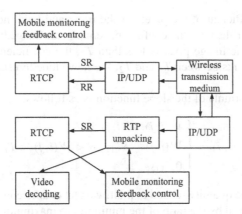

Fig. 1. Mobile surveillance video transmission process

As can be seen from Fig. 1, H.264 video data is first encapsulated by RTP protocol. After encapsulation, the merged data packet is transmitted through the network through the appropriate network protocol, and the data information is unpacked and decoded by RTP protocol to obtain the decoded form of the collected data. RCTP protocol, as an auxiliary protocol of UDP protocol, can adjust and control the traffic in real time during the process of data encapsulation and unpacking, ensure the quality of data encapsulation and unpacking, and improve the fault tolerance of data. After monitoring and collecting the athletes' track and field postures in the training area, the monitoring data is transmitted to the data processing computer according to the above protocol flow, and then used to identify the movement postures after processing.

There are many kinds of human body postures, each of which seems relatively simple from the visual point of view, but actually quite complicated. Human body movements involve many parts (such as legs, waist, arms, etc.), and the intensity and characteristics of each posture are different, and each person's physiological data and exertion methods of track and field events are also different, which increases the difficulty of track and field posture recognition. Therefore, it is necessary to preprocess the image data of track and field movement monitoring before the movement gesture recognition, so as to improve the efficiency of the movement gesture dynamic recognition.

2.2 Image Processing of Mobile Monitoring Posture

Track and field training is usually conducted outdoors. Outdoor environment interferes with the motion video collected by mobile monitoring technology to varying degrees, which affects the accuracy of gesture recognition. Therefore, firstly, the track and field movement monitoring video is denoised. In this study, the threshold method of multi-wavelet transform is used to smooth the motion posture and movement surveillance video. Because multiwavelet does not have translation invariance, it will produce obvious Gibbs phenomenon in the neighborhood of signal singularity (image discontinuity). Two different thresholds T_1 and T_2 are adopted, of which $T_1 = \sigma(2\ln N + 2\ln\ln N)^{1/2}$, $T_2 = \mu\sigma(2\ln N + 2\ln\ln N)^{1/2}$, of which μ is an adjustable parameter, $\mu \in [0, 1]$. If the multiwavelet coefficient of the pixels of the track and field movement monitoring video image is greater than T_1, the coefficient remains unchanged; If the multiwavelet coefficient value of the image pixel is less than T_2, the coefficient is set to zero. For multiwavelet coefficients between T_1 and T_2, the new slope threshold function is used to shrink [7].

The calculation formula of the slope function is as follows:

$$\dot{W}(i,j) = \begin{cases} W(i,j), W(i,j) > T_1 \\ W(i,j) - \delta T_1, T_2 < W(i,j) < T_1 \\ 0, \ else \end{cases} \tag{1}$$

where, δ represents the quantity related to multiwavelet coefficients of pixel points. The coefficient is determined by the ratio of the minimum to maximum value of the median value of the pixel in the 3×3 neighborhood. After the original image is converted into a gray image, the decomposition number of multiwavelet is $K = 2$, and the odd/even pre-filtering is used to convert two adjacent pixels in a row or column of the monitoring image $J(i,j)$ with size $M * N$ into a vector pixel of $C(i,j)$, which is 2×1 or 1×2. The row-column conversion formats are as follows:

$$C(i,j) = \begin{vmatrix} J(i, 2j) \\ J(i, 2j + 1) \end{vmatrix} \tag{2}$$

$$C(i,j) = \begin{vmatrix} J(2i, j) \\ J(2i + 1, j) \end{vmatrix} \tag{3}$$

After preprocessing, the noise image is subjected to K-times multiwavelet transform, and the multiwavelet transform coefficients of the image are obtained. The threshold of wavelet denoising is obtained by thresholding the coefficients of multi-wavelet transform with the existing wavelet denoising threshold method. The threshold process is reconstructed by wavelet, and the intermediate image is obtained. According to the inverse process of pre-filtering, the intermediate image is post-filtered, and the final denoised motion monitoring image is obtained. After the image is denoised, Gaussian background segmentation is performed on the image where the gesture recognition object, that is, the individual athlete, is located to reduce the influence on the recognition accuracy.

2.3   Attitude Recognition Background Segmentation

In track and field training, athletes' instantaneous explosive force is strong, and their postures change rapidly, so the recognition of athletes' postures will be influenced by shadows. Therefore, Gaussian background model is used to segment the detected object and background in surveillance video.

When building Gaussian background model, we should also consider the influence of background shadows on the detection effect. When building background model in RGB color space, if the pixels in RGB space are covered by shadows, in which the values of R, G and B are linearly attenuated, the shadows can be detected by calculating the posterior probabilities of pixel background, moving objects and moving shadows. However, this method is not suitable for rapid and accurate detection of targets because of its large computation. However, it is easier to remove the influence of shadows when building a background model in HSV space. In HSV space, when the pixels are covered by shadows, the brightness wallpaper of the pixels is approximately linear, and shadows will not greatly change the chroma of the background pixels and can reduce the saturation of the background pixels. This method can simply and quickly suppress shadows. Therefore, this paper establishes a background model in HSV color space to suppress shadows, and the specific algorithm is as follows [8]:

$$SP(i,j) = \begin{cases} 1, r \leq \dfrac{I_V(i,j)}{B_V(i,j)} \leq R \\ |I_S(i,j) - B_S(i,j)| \leq -1 \\ |I_H(i,j) - B_H(i,j)| \leq 4 \\ 0, \ else \end{cases} \qquad (4)$$

where, $I_H(i,j)$, $I_S(i,j)$, $I_V(i,j)$ respectively represents the component of the pixel; $B_H(i,j)$, $B_S(i,j)$, $B_V(i,j)$ represents the component of the background pixel respectively. In which parameters $0 < r < R < 1$, The value of R is related to the light intensity in the environment. The stronger the light, the smaller the value. The value is taken according to the light intensity of the environment. $SP(i,j)$ is the shadow pixel mask at pixel coordinate point (i,j) of the image, $SP(i,j) = 1$ if pixel $J(i,j)$ is judged as shadow, otherwise $SP(i,j) = 0$.

Gaussian background model is established. For each pixel in the image, Q states are used to represent the color of the pixel. Each of the Q states is represented by a corresponding Gaussian function. If each pixel is represented by J_t, its probability density function is represented by Q three-dimensional Gaussian functions as:

$$g(J_t = j) = \sum_{i=1}^{Q} w_{i,t} \eta\left(j, \bar{j}_{i,t}, \overline{G}^2\right) \qquad (5)$$

Among them, $\eta\left(j, \bar{j}_{i,t}, \overline{G}^2\right)$ is the first Gaussian distribution, $\bar{j}_{i,t}$ is the mean value of the i Gaussian model at t times, $\overline{G}^2$ is the covariance of the Gaussian model, and $w_{i,t}$ is the weight of the i Gaussian distribution at t times. A Gaussian mixture model is established in each color channel to improve the real-time performance of the algorithm. The mixed Gaussian model is initialized, and the gray mean and variance of each pixel in the video sequence image in a period of time are calculated as the next Gaussian distribution parameters. When establishing the initial background model, if the pixel satisfies the following formula, it is judged as the background pixel, otherwise it is the foreground pixel.

$$|J_t - \bar{j}_0| \leq 2/5 * \overline{G}_0^2 \qquad (6)$$

When the background light or background objects have not changed, the background model is consistent, and the corresponding background model can be obtained without realizing the background update process. During the detection process, the background model should be updated due to the changes of the moving target and the background light in the environment, and the updating process is updated by updating the model parameters. The updated production data of the model is updated by the downward gradient method. The larger the value of the update learning rate, the faster the update speed, and the smaller the value, the slower the update speed. After segmenting the background in the surveillance image, the gesture features of track and field movements are extracted for gesture recognition.

2.4 Feature Extraction of Athletic Attitude

People are non-rigid objects, and different people have different postures at different times and places, so it is necessary to extract human posture features. Feature extraction is a primary operation on the original data, the purpose of which is to transform the original information representing the categories of sports behaviors into features with more obvious physical or statistical significance. Feature extraction is a very important step in the whole process of motion gesture recognition. It converts the collected original behavior data into the data form after preprocessing, and obtains the feature vector that can better represent the behavior category, which is used by the classifier to learn, identify and classify. Therefore, the selection of features directly affects the classification performance and accuracy of the classification model. Features that are widely used now mainly include three categories: time domain features, frequency domain features and time-frequency features.

Time domain features show the characteristics of the movement behavior information in the time dimension, which can be obtained by directly calculating the features of the movement data collected by sensors. Time domain features are widely used in human motion behavior recognition because of their simple calculation and small amount of computation. Frequency domain feature is also a common feature, which reflects the frequency domain features of motion signals. Generally, it is necessary to transform the original signal from time domain to frequency domain by fast Fourier transform, so as to extract the corresponding frequency domain features. In this paper, the feature extraction method based on principal component analysis is used to extract the posture features of track and field sports.

Because the information between features overlaps to some extent, this paper uses principal component analysis (PCA) to analyze the extracted features. When PCA is used, it is necessary to calculate the covariance matrix and its eigenvalues of data, and then select the eigenvector with the largest eigenvalues to form a new matrix and transform it into a new space, so as to realize the dimension reduction of data features. The main process is as follows:

Before extracting human posture features, firstly, the coordinate system of human posture skeleton points under different track and field events is established. Firstly, the human contour model is established according to the feature vectors of the human contour, and the skeleton model under the human motion posture is obtained according to the skeleton information. As the support of the human body, bones are an important part of the human movement system, providing support in movement is the foundation of movement. Human skeleton sequence can be modeled by joint points and bones. There are 18 skeleton points in the human skeleton point model, which refers to the traditional human posture model. Including nose, neck, right shoulder, right elbow, right wrist, left shoulder, left elbow, left wrist, right hip, right knee, right ankle, left hip, left knee, left ankle, left eye, right eye, left ear and right ear. According to the sample image data of track and field movement posture, the skeleton feature sequence of track and field movement posture is obtained according to the human skeleton state under different movement postures.

In order to extract the motion characteristics of skeleton, it is necessary to extract skeleton sequence from video, and construct Shi Kongtu to input the nodes related to skeleton sequence according to the connection between joints. There are two kinds of joint connection edges, one is the skeleton edge formed by the internal connection nodes according to the natural connection order of human bones, the other is the inter-frame edge formed by the connection of the same joint point in consecutive frames according to the time order, thus obtaining a Shi Kongtu containing the natural connection of human bodies and the same joints between frames.

The gesture data of human body is expressed as matrix form X, and the initial matrix is linearly transformed to obtain a new matrix Y. Calculate the orthogonal matrix of the new attitude feature matrix, $Y = UX$. Therefore, matrix U is a matrix composed of the eigenvectors of the correlation coefficient matrix of l random variables.

Because the transformed points have the maximum variance on the y_1 axis and the minimum variance on the y_1 axis. At the same time, the covariance of all points for different y_i axis and y_j axis is zero. Let all l eigenvalues be non-negative 0, the eigenvector corresponding to λ_i is ζ_i, let $U = [\zeta_1, \zeta_2, \cdots, \zeta_i]$, then the variance of y_1 is:

$$Var(U_1 X) = U_1 X X^T U_1^T = \zeta_1 \tag{7}$$

y_1 has the largest variance, y_2 has the second largest variance, and there is covariance:

$$\text{cov}\left(U_i^T X^T, U_j X X\right) = U_i^T R U_j \tag{8}$$

If the eigenvalues are arranged from large to small, the expression of the contribution rate of principal components is:

$$P_i = \frac{\zeta_i}{\sum\limits_{k=1}^{l} \zeta_k} \tag{9}$$

Generally, the principal components with cumulative contribution rate greater than 80% and characteristic value greater than 1 are selected. For the original feature set, after dimensionality reduction by PCA, the new feature vectors obtained are pairwise uncorrelated. By orthogonal transformation of high-dimensional features, a new feature set is obtained. Then use formula (8) to calculate the contribution rate of different new features, and select new features with high contribution rate to form new feature vectors. The extracted motion posture features are used as the input of recognition classifier, and the dynamic recognition of track and field motion posture is realized through classifier processing.

2.5   Realize Dynamic Recognition of Motion Posture

When BP neural network algorithm adjusts the weights and thresholds of the network based on the negative gradient direction of error, it is easy to fall into local extremum. Artificial fish swarm algorithm is used to optimize the neural network. The core idea of artificial fish swarm algorithm is to simulate the behavior of fish swarm. It is a new strategy that can be used for global optimization. The realization of the algorithm is mainly to simulate the behavior of fish looking for food, which is an application based on animal behaviorism. When fish are looking for food, they mainly swim to places where there is more food. In the process of searching, they usually take the initiative to search or follow other schools of fish. In AFS algorithm, its main purpose is to search for places where there is more food, and the highest density through searching is the global optimal value. The algorithm mainly achieves the result of local optimization through several behaviors of fish, which mainly include foraging, clustering and rear-end collision. AFS algorithm has good convergence, can save the whole process time, has strong robustness and can achieve the effect of overcoming local extremum.

When using AFS to optimize BP neural network, it is necessary to combine the advantages of both. The process of fish searching for food in AFS algorithm is to find the global optimum. Combining it with the local searching ability in BP algorithm can achieve the effect of overcoming the local extremum. At the same time, it can accelerate the convergence speed of BP neural network and have better generalization ability. The implementation steps are as follows:

(1) Parameters of BP neural network are randomly initialized. After determining the main parameters, network training can be carried out.
(2) The parameters of AFS algorithm are randomly initialized. The main parameters include state s_i, quantity N_s, dimension D, crowding factor ϑ and so on. Each artificial fish represents a neural network, and its initial solution is a D-dimensional vector. If the number of neurons in the input layer is N_i, the number of neurons in the hidden layer is h and the number of neurons in the output layer is m, then the dimension of D is as follows:

$$D = (N_i + 1) \times h + (h + 1) \times m \tag{10}$$

Calculate the fitness of food concentration of artificial fish.

The fitness of artificial fish's food concentration ρ_F is set as the reciprocal of total error E in BP algorithm, and the point with the largest food concentration is the point with the smallest error in BP algorithm. Fitness value of food concentration:

$$\rho_F = E^{-1} \tag{11}$$

(4) Each artificial fish simulates the second step and the third step of AFS, calculates the fitness of food concentration in the current position, and performs the behavior with better fitness, and the default way is foraging behavior.
(5) After each action, each artificial fish will be calculated and compared with the adaptive value of the previous step, and if it is better than that, it will be replaced by itself.

(6) When the number of cycles exceeds the maximum number of cycles, the training is finished. Otherwise, return to step 4.

Optimize the BP neural network, form a network identification model, input the processed track and field movement monitoring movement data, and identify the human movement posture data. According to the above process, the purpose of dynamic identification of athletes' track and field posture by using mobile monitoring technology is realized.

To sum up, the design of dynamic recognition method of track and field posture based on mobile monitoring technology is completed. The process of this method is shown in Fig. 2 below:

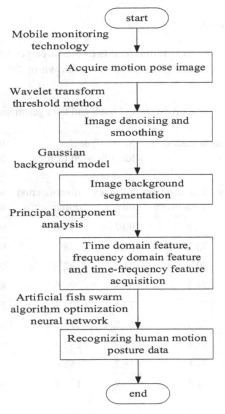

Fig. 2. Flow chart

3   Test Experiment

A dynamic identification method of track and field posture based on mobile monitoring technology was proposed above. Before applying this method to actual track and field training, the usability of this method was tested as follows.

3.1 Experimental Content

In this paper, the dynamic attitude recognition method of track and field based on mobile monitoring technology is compared with the attitude recognition method based on machine learning and sensor. Three kinds of recognition methods are used to dynamically recognize the posture of the motion posture data set with known results, and the recognition accuracy and recognition efficiency are selected as the comparison indexes. The experimental data set is the posture images of track and field sports shot by professional athletes. In the shooting process, the computer is used to establish the theoretical optimal model according to human physiological data to guide athletes' movements, so as to avoid interference with the experimental results. By analyzing the index data, the experimental verification is completed.

3.2 Experimental Result

The comparison data of the recognition results of the three gesture recognition methods on the gesture images of track and field sports are shown in Table 1.

Table 1. Comparative results of track and field gesture recognition

Number of identification objects	Identification method in this paper		Sensor-based identification method		Recognition method based on machine learning	
	Accuracy rate/%	Identification duration/ms	accuracy rate/%	Identification duration/ms	accuracy rate/%	Identification duration/ms
10	99.5	65.8	90.2	73.6	93.5	78.1
20	98.7	72.3	90.3	78.1	94.7	79.6
30	96.8	70.9	84.5	82.4	90.1	84.3
50	97.4	74.2	83.1	96.5	86.7	106.7
60	95.6	75.5	73.4	107.3	88.4	119.9
70	96.3	78.6	79.8	119.9	85.9	125.3
80	97.2	83.4	87.9	137.4	89.6	139.8
90	98.1	84.1	78.2	152.6	87.3	141.3
100	97.5	84.8	79.9	174.8	85.8	146.9

From the data analysis in Table 1, it can be seen that the identification accuracy of the identification methods proposed in this paper is higher than 95%, while the identification accuracy of the sensor-based identification method is obviously reduced in track and field events with strong light interference, which lowers the identification accuracy of the overall method. The recognition accuracy of the recognition method based on machine learning is affected by the training samples, and it is impossible to recognize the dynamic and continuous motion posture. From the recognition time of the method, the recognition time of this method is shortened by about 31.56% compared with the other two comparison methods, and the recognition efficiency is higher.

Taking the average recognition rate as the index, on the basis of the above experiments, we continue to use different methods to test, and the results are shown in Fig. 3 below:

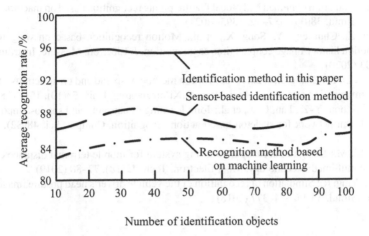

Number of identification objects

Fig. 3. Average recognition rate of different methods

It can be seen from Fig. 3 that when using this method for action gesture recognition, it can still maintain a high level, with an average of 95.6%, under different numbers of recognized objects, while the average recognition rate of other methods is below 89%, and the recognition effect of dynamic actions is far less than that of this method.

According to the above analysis, it can be seen that the dynamic recognition method of track and field movement posture based on mobile monitoring technology has higher recognition effect for track and field movement posture.

4   Concluding Remarks

If track and field athletes want to achieve excellent results, they must go through systematic training for many years. Among many high-level sports teams, many excellent athletes who can win gold medals are plagued by injuries and injuries, which is the most difficult problem that coaches need to face. The standard and scientific sports posture is the foundation of athletes' sports career, which can maintain and expand the competitive state of excellent athletes and achieve more excellent sports results in future competitions. In this paper, a dynamic identification method of track and field movement posture based on mobile monitoring technology is proposed. By using the mobile monitoring technology, the moving posture of athletes during track and field training in the training ground can be effectively dynamically identified after image analysis and processing. Compared with other gesture recognition methods, the experimental results show that the proposed gesture dynamic recognition method has higher recognition accuracy and efficiency, and can assist the daily development of track and field training.

References

1. Yang, Y., Sun, G., Wang, Z.: Human activity recognition based on android sensors. J. Nanchang Univ. (Natl. Sci.) **43**(06), 616–620 (2019)
2. Lin, H.: Research on intelligent recognition of motion based on laser sensor. Laser J. **42**(07), 84–89 (2021)
3. Liu, W.: Simulation of human body local feature points recognition based on machine learning. Comput. Simul. **38**(06), 387–390+395 (2021)
4. Hongyu, Z., Chunfeng, Y., Song, X., et al.: Motion recognition based on weighted three-view motion history image coupled time segmentation. J. Electron. Measure. Instrum. **34**(11), 194–203 (2020)
5. Wang, B., Li, D., Zhang, J., et al.: A recognition method of spread and grasp actions combining motion imagination and action observation. J. Xi'an Jiaotong Univ. **53**(10), 151–158 (2019)
6. Ye, S.-T., Zhou, Y.-Z., Fan, H.-J., et al.: Joint learning of causality and spatio-temporal graph convolutional network for skeleton-based action recognition. Comput. Sci. **48**(S2), 130–135 (2021)
7. Jia, X., Li, M.: Design of security monitoring system for mobile terminal distance education based on multimedia technology. Mod. Electron. Tech. **42**(18), 77–80 (2019)
8. Juan, L., Ying, R.: Simulation of recognition of the vehicle driver's head posture image feature. Comput. Simul. **35**(1), 374–377 (2018)

Personalized Recommendation Method of Online Music Teaching Resources Based on Mobile Terminal

Hui Lin[1](✉), Ying Lin[2,3], and Hongping Huang[4]

[1] College of Art, Xinyu University, Xinyu 338004, China
linhui66621@163.com
[2] School of Psychology and Education, University Malaysia Sabah, 88000 Sabah, Malaysia
[3] College of Foreign Languages, Xinyu University, Xinyu 338000, China
[4] School of Literature and Communication, Xinyu University, Xinyu 338004, China

Abstract. Due to the large number of users of the mobile teaching terminal and the many types of music teaching resources, the recommendation accuracy is low. To this end, this paper proposes a personalized recommendation method for online music teaching resources based on mobile terminals. This paper identifies the characteristics of online music teaching resources, connects the resources through knowledge points, and optimizes the streaming media storage format using mobile terminals. The time continuous signal is converted into discrete time signal, and the user interest model is constructed by collaborative filtering, and the favorite resources of neighbor users are recommended to the current user. The experimental results show that the accuracy of this method is 75.694%, 66.669% and 66.350%, respectively, which shows that the performance of this method is better than the other two methods.

Keywords: Mobile terminal · Online music teaching · Teaching resources · Personalized recommendation · Audio materials · Information age

1 Introduction

Personalized recommendation method of online music teaching resources based on mobile terminalThe huge demand of online education promotes the development of education informationization. Compared with the traditional teaching mode, the network teaching has gradually become a new and widely used mode. The existing music teaching resources present the characteristics of huge quantity, numerous types, wide distribution, repeated development and strong dynamic change. Teachers can obtain rich teaching resources through network teaching, students also effectively reduce the burden of books, is a useful complement to the traditional teaching mode. In order to solve the contradiction between the massive teaching resources and the user's individualized needs, and further improve the utilization of educational information resources, so that learners become passive in the learning process. However, with the rapid increase in

W. Fu and L. Yun (Eds.): ADHIP 2022, LNICST 469, pp. 349–361, 2023.
https://doi.org/10.1007/978-3-031-28867-8_26

the number of teaching resources, users find it more and more difficult to find the real need of teaching resources, and spend more and more time, that is, the so-called "teaching resources overload", "information lost" phenomenon [1, 2]. Based on the idea of personalized service, this paper adds the function of personalized recommendation to the traditional network teaching resources, so as to solve the problem of users' finding teaching resources and the low utilization of teaching resources. Using traditional information search technology to obtain resources has some problems, such as low accuracy, more redundant information, and there must be a general direction of information search. However, in many cases, it is difficult for users to express their needs clearly or do not know their specific needs, so the traditional search technology has become increasingly difficult to help users find their true useful information. In order to solve this problem, personalized recommendation is applied to the teaching resources platform to provide personalized services for users.

In order to realize personalized recommendation service, user interest model is established by collecting and analyzing user registration information and behavior history data. Based on the user's interest, the teaching resources are pushed to the user actively, so that the teaching resources are more pertinent, personalized learning is realized, and the learning interest, quality and efficiency are improved. This undoubtedly has a certain promoting significance to the development of distance education. Personalized recommending service of teaching resources is to provide users with teaching resources that they may be interested in automatically by analyzing their personalized information. Different from the "one to many" mode of resource search service, the resource personalized recommendation service can meet the user's personalized needs better, and the system users don't need to participate in it too much, which greatly reduces the cost of resource search and makes the user more convenient to use. In addition, it also has rich academic value and practical significance for the research and practice of developing network teaching resource system. Personalized recommendation service changes the mode of "people looking for resources" into the intelligent mode of "people looking for resources, resources looking for people". In order to improve the accuracy of online music resource recommendation, this paper proposes a mobile terminal-based personalized recommendation method for online music teaching resources. This paper clarifies the characteristics of online music teaching resources, and optimizes the streaming media storage format by using mobile terminals. Convert a time continuous signal to a discrete time signal. The user interest model is constructed through collaborative filtering, and the resources of neighbor users are recommended for users.

2 Personalized Recommendation Method of Online Music Teaching Resources Based on Mobile Terminal

2.1 Identify the Characteristics of Online Music Teaching Resources

The network music education resources are opposite to the traditional music education resources. Generally speaking, the traditional music education resources include books, newspapers, magazines, CD, tapes and other physical material carriers, or radio and television carriers, which carry and spread music education resources. With the development of information technology in education, teaching resources platform is also increasing, promoting the wide dissemination and sharing of high-quality resources [3, 4]. Different from the merchandise in E-commerce, the digital teaching resources in the platform have their own characteristics. Network music education resource is a new type of music education resource based on virtual digital technology and Internet. It is the combination of music education, modern digital technology and network technology. Although there are a variety of resources classification: according to the level of teaching, can be divided into basic teaching resources and higher teaching resources. According to specialty, it can be divided into modern teaching resources and classical teaching resources. Generally speaking, the network music education resources have two kinds of broad and narrow sense. In a narrow sense, network music education resources refer to the resources that specially serve music teaching in schools at all levels. But the broad sense network music education resources are then refers to all and music education related but can serve in each kind of type music education resources. The characteristics of online music teaching resources are shown in Fig. 1:

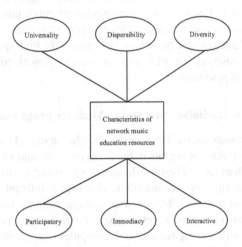

Fig. 1. Characteristics of network music teaching resources

From the Fig. 1, we can see that online music education resources include universality, decentralization, diversity, sharing, immediacy and interaction. Network music education resources generally exist in the following forms: network music education curriculum. However, no matter what the classification of teaching resources is used to teach knowledge, each teaching resources have their own knowledge and skills points, these knowledge and skills points constitute the core of this resource, representing the essential nature of teaching resources [5]. Network Music Education Resource Base and various resource sites based on Network Music Education Resource Base. Scattered personalized network music resources, such as personal music blog, blog, etc. Moreover each teaching resources knowledge point information is carries on the annotation by the expert scholar, has certain authority. Among them, the resource sites based on the online music education resource database can often act as the resource center within a region (such as a prefecture-level city) and effectively consolidate and connect the online education resource database of the subordinate regions (such as the districts under the city, counties and cities), thus playing the role of an information hub for the transmission of music education resources. The knowledge points of resources may overlap, so we can link the resources by the knowledge points, and the user's preference for the resources also represents the user's preference for the knowledge points.

Because the Internet is distributed all over the world and there is no entity boundary, all kinds of music education resources are widely distributed. Music learners and music teaching researchers can search for their own resources across cities, provinces and even across countries. Users share digital resources through the teaching platform, learning related knowledge, is the user of teaching resources. Users generally have some pertinence in learning, and the classification is clearer, that is, the user's own characteristic information and the selected resources have a certain relationship. On the other hand, it is difficult to establish an overall index catalogue of online music education resources because of the loose distribution of independent education resources. Therefore, the similarity of users who choose the same resource is much higher than that of users who choose different resource. Because a static web page of music education resources can be browsed by thousands of learners, the network music education information resources break the limitation of entity music resources. Because of the high speed of network transmission, learners can easily and quickly browse, access, download, real-time access to online music education resources.

2.2 Mobile Terminal Optimized Streaming Media Storage Format

Traditional teaching resources, such as video and audio tapes and videotapes, are simply preserved in their form. Analog signals in the form of audio and video tapes are used to record information. Therefore, the establishment of video and audio analog signals in the first version library records this information into the digital information of the computer, namely, video and audio data collection and storage. Mobile terminals, also known as mobile communication terminals, refer to computer equipment that can be used in mobility, including mobile terminals, notebook computers, multifunctional terminals and even on-board computers [6, 7]. One classification can decompose the implementation of a functional method into a series of scattered files. Program developers should put a series of related methods into a category to make the code more readable. But most refer to smart mobile terminals or have a variety of applications.

With the development of network technology for broadband capacitance enhancement, the mobile communication industry will move towards the real information age. The collection of video and audio materials is a complex work process. The whole process can be divided into three main steps. One is sampling, the other is quantification, and the third is coding. It is wrong to think that the data collection is divided into three steps. The principle of the whole process is simple, and it is very complicated to realize on the network. For example, you could add a category called "spell checker" to the string class and then put the code associated with the spell checker into the category. In addition, with the development of electronic technology, mobile terminals are becoming more and more powerful, and mobile terminals are changing from a simple call tool to an integrated information processing platform. Sampling, also known as sampling, refers to the periodic scanning of a certain time interval of analog signals, time continuous signals into discrete time signals of continuous and amplitude pulse modulation information, which also gives mobile terminals to add a broader space for development.

Mobile terminal communication mode is very rich, can be through GSM, CDMA, edge, 3G wireless network communication, can also be through WLAN, Bluetooth and infrared. The size of the sampling frequency is a key parameter in sampling, which means that the sampling frequency of the analog signal, noise, and will not produce a second overlapping fold. The sampling frequency generally requires the highest frequency of the analog signal, although the value of the sampling frequency may not be too high, because the total data rate after sampling increases exponentially with the increase of the sampling frequency, thereby increasing the requirements for data processing, transmission bandwidth, and memory capacity. For example, if the system's original string class implementation method does not have spell checking settings, the programmer can add such methods without changing the original string class code. Digital video and audio need to be further compressed and edited in order to adapt to network operations.

According to the different broadcast quality, there are different editing schemes for digital video and audio. Modern mobile terminals, like our personal computers, have a stable miniaturized operating system with powerful information processing capabilities, memory and memory cards, can complete complex processing tasks. Audio for the sound of the general participation of Effect Audio, Cool Edit and other tools, especially video editing is more complex, the need for better hardware conditions. The simplest is to use audio and video editing, more professional there are professional video editing software such as Primier. Based on the principles of mobile terminals, four main modes of mobile learning are presented, as shown in Fig. 2:

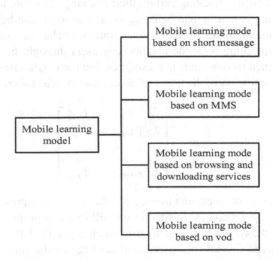

Fig. 2. Schematic diagram of mobile learning mode

According to Fig. 2, the main modes of mobile learning are short message-based, multimedia message-based, browse-based, download-based, and video-on-demand. When the program runs, the method in the class is the same as the original method in the class. The code in the class can access all the member variables, including the private member variables. For image processing in general sampling photoshop for processing, if it is batch compression can be used, such as optical magician processing tools. Through the use and disposal of these tools, the collected materials can be stored in the original ecological teaching resource management system to ensure quality, improve resource storage efficiency and save system space. If a class declares a function with the same name as the original method in the class, the system chooses to call the method in the class. Therefore, classification can not only add class methods, can replace the existing methods. Streaming media (streaming media) is a multimedia format used for transmission on the network, mainly including streaming media video, audio, or animation, etc. Streaming media is the main multimedia technology used on the Internet. If the methods in both categories have the same name, then it is unpredictable which method is invoked at runtime.

2.3 Collaborative Filtering Build User Interest Model

The most common method of log mining tracking is to evaluate the user's interest in the page according to the number of clicks. Implicit tracking is a practical and effective way of tracking, which is automatically completed by the system, reducing the burden of users. Because the teaching and learning of online music courses are often not in the same place, online music courses must build effective communication, exchange and interaction mechanisms between teaching and learning in various ways with the help of rich and diverse man-machine interfaces and webpage (website) functions. The process of a user's interest tracking is the process of obtaining the user's interest, which generally includes two ways: display tracking and implicit tracking. The core idea of user-based collaborative filtering is that human behavior is similar to human behavior, and users with similar behavior will make similar choices, and user behavior contains user interest information. The algorithm finds the neighboring users through the user's preference for resources, and then recommends the resources that the neighboring users like to the current user. The expression of the user evaluation matrix is as follows:

$$L(p, q) = \begin{bmatrix} L_{1,1} & L_{1,2} & \cdots & L_{1,q} \\ L_{2,1} & L_{2,2} & \cdots & L_{2,1} \\ \vdots & \vdots & \vdots & \vdots \\ L_{p,1} & L_{p,2} & \cdots & L_{p,q} \end{bmatrix} \tag{1}$$

In the formula (1), rows represent users, p in total, columns represent items, and q in total. Display tracking is a process in which users fill in personal information or answer questions raised by the system by filling in forms, such as personal information and user's evaluation of resources. Consider the user rating as a vector in the multidimensional item space, and set the rating to 0 if the user does not rate an item. The similarity between user a and user b is obtained by calculating the cosine angle between vectors. The higher the cosine value is, the higher the similarity is. Given that the scoring vectors for user a

and user b are a and b respectively, the similarity between users is:

$$sim(a, b) = \begin{cases} \cos\left(\overline{a}, \overline{b}\right) \\ \dfrac{\overline{a} \times \overline{b}}{\|a\| \times \|b\|} \end{cases} \tag{2}$$

Since the VCC does not consider the different user's rating scale in computing the user's neighborhood, the modified VCC algorithm improves the problem by subtracting the average user's rating. Based on the calculation results of formulas (1) and (2), the nearest neighbor set of the target user can be obtained, and then the user's scoring on the item can be predicted. The specific formula is as follows:

$$G = \overline{\delta} + \frac{sim(a, b)}{\phi} \tag{3}$$

In formula (3), ϕ represents the number of users, and $\overline{\delta}$ represents the average user rating of the project. Through the support of rich media technology, the network music education not only achieves, but also surpasses the teaching interaction effect of the entity music course in some aspects. Display tracking is completely dependent on the user, and is likely to reduce the user's enthusiasm to use the system. Even if the user is willing to enter the user model by hand, it is difficult for the user to list all the keywords he is interested in, which leads to the inaccuracy of the user model. For online music learners, online courses provide them with a technical platform that can maximize their learning initiative, rather than a simple music teaching video. The user model built using the display trace is static and will not change once the user model is completed. Therefore, in the design of the webpage (website) of the online music course, the function of prompting learning objectives, disaggregating and displaying learning content and navigating learning paths must be added. The longer the time elapsed, the greater the difference between the displayed tracked user model and the real user interest. Implicit tracing does not require information from the user, and all tracing is done automatically by the recommended method. User behavior records such as browsing the Web, clicking a mouse, marking a bookmark, and dragging a scroll bar can all indicate a user's potential interest. The aim is to enable music learners to get a clear understanding of the objectives, contents and learning strategies of online music courses, and to arrange the preparation of music learning materials and the process and steps of music learning. Implicit tracing can be divided into two ways: behavior record tracing and log mining. Among them, the former one is that the system reveals the user's interest effectively by recording the user's behavior. In order to optimize the allocation of curriculum resources to the greatest extent, we can effectively grasp the key points of knowledge taught by teachers in the limited and centralized network music teaching time, and interact with music teachers in a targeted manner, thus achieving twice the result with half the effort. The latter method of tracing is to extrapolate user interest from server log information statistics.

Log mining tracking method can be used to create and update the user's interest model by obtaining the number of page clicks, page stay time and page visit order. In order to determine the overall teaching style of online music course, it is necessary to select representative specific "teaching units", determine the elements such as knowledge system construction, interface style and navigation strategy, and solicit the opinions of music learners through activities such as "trial teaching" and "trial listening" within a certain scope. But the data collected by implicit tracking may contain too much redundant and irrelevant information, which will increase the computational cost and complexity in the model learning process. For example, "syncopation" can be selected as a specific "teaching unit", first clear this "teaching unit" goal is to master the basic form of syncopation, singing and its role. In order to grasp user's interest better, display tracking and implicit tracking can be combined to obtain static user's interest and dynamic user's interest by implicit tracking. Then cut into the specific teaching process, we can first explain the concept of syncopation, pointing out that its role is to change the rhythm of the intensity of the law. However, the concept is far from enough. Three basic forms of syncopation, such as "within one bar", "between two bars" and "with body stop", must be enumerated, and the relevant notation examples must be attached.

2.4 Design Personalization Recommendation Pattern

Choose specific songs and instrumental music fragments containing syncopation as teaching cases, and choose different vocal and instrumental music types, such as art songs, melodic music, popular music and other vocal music types, as well as piano music, guzheng music and other different instrumental music types, so as to help students understand the different forms of syncopation in various music concretely and vividly, so as to understand the role of syncopation in a more comprehensive and profound way. There are not only registered users but also unregistered users in the personalized recommending method of teaching resources, which can be searched, browsed and non-personalized. This paper introduces the singing method of syncopation, that is, beginners can break the syncopation of the quarter note into two octaves, or break the syncopation of the quarter note into two sixteenth notes, and then unify the syncopation after being skilled, so as to enable students to freely master the singing method of syncopation. For the registered users, in addition to the rights of unlogged users, we can also evaluate the resources and personal information management. At the same time, the recommendation policy for new registered users is different from that for graded users. After introducing the three basic forms of syncopation and showing the relevant notation examples, you can also add a personalized teaching link, for example, can guide students to imagine the kind of "Ouch Hey" chant shouted when the two sides compete in a tug-of-war game, which is a similar form of syncopation rhythm in daily activities. There are many kinds of teaching resources in the network, and it is difficult to recommend the unstructured resources such as audio and video based on the content. The recommendation based on association rules has low personalization degree, and it is difficult to get good personalized recommendation results. By citing examples from daily life, it is easy to bridge the gap between students and the boring knowledge of music theory. The recommendation method inputs the basic information of users and resources and scores, and outputs the recommendation results. Personalized recommendation can be divided into three steps: data processing, recommendation calculation and forecasting recommendation. The architecture of the personalized recommendation model for teaching resources is shown in Fig. 3.

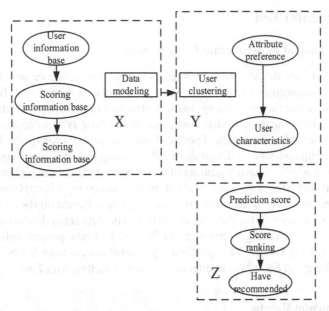

Fig. 3. Personalized recommendation method structure diagram

According to Fig. 3, X part is to preprocess the data in user information database, resource information database and rating information database, and construct data model, namely, to get user feature vector, resource attribute vector, resource attribute preference vector and so on for the subsequent calculation of user similarity. After making this kind of experimental teaching unit, we can check and accept the teaching effect through the activities of "trial teaching" and "trial listening" in a certain scope, and solicit students' opinions so as to determine the overall style of network music course. The Y part calculates the similarity of the recommendation model based on the X part to find the nearest neighbor, in which the new user is based on the user characteristics and information entropy model, and the other user is based on the rating and resource attribute preference model. Z is responsible for the user's forecast score and the final Top-N recommendation. At the same time can also join the tug-of-war when the demonstration audio, and tug-of-war animation video, in order to enhance the visual effect of music teaching. Collaborative filtering can handle complex unstructured objects without considering the form of recommendation resources, and it has high degree of automation, can mine the potential interest of users, and the recommendation quality is high. If the tug-of-war cases and animated video can achieve good teaching results, this teaching method should be retained and carried forward, and the overall style of online music course will tend to be lively, picturesque and interesting.

3 Experimental Test

3.1 Construction of Experimental Environment

This experiment uses B/S structure, B/S refers to the browser/server structure, that is, only one server installed and maintained, while the client uses the browser. B/S structure is with the rise of Internet technology, the C/S structure of a change and improvement. B/S structure adopts three-tier architecture, that is, database system, application server and client browser. After the design of personalized recommendation method of teaching resources is completed, 694 grading data of 300 teaching resources were collected from 80 users, among which the user's grading of resources is 1–5, and the higher the grading is, the more favorable the resource is. Most of the transaction logic is implemented on the server side, and very few of the transaction logic is implemented on the client browser. Meanwhile, the sparsity of the data set is 94.69%. The data set is divided into training set and testing set according to the scale of 3: 1. C #, a new programming language designed specifically for the. NET platform by a development team led by Microsoft's Anders Hejlsberg and Scott Willamette, has many similarities with Java.

3.2 Experimental Results

Different users have different understandings of the same word, which results in different evaluation of the filtering results. At present, the representative evaluation standards adopted take the precision rate as the test standard for testing the personalized recommendation method of online music teaching resources. The calculation formula is as follows:

$$T = \frac{H}{\eta} \times 100\% \tag{4}$$

In formula (4), H represents the number of information items in the filter results that meet the user's interests, and η represents the number of information items in the filter results. Choose the personalized recommendation method of online music teaching resources based on neural network and machine learning, and compare with the personalized recommendation method of online music teaching resources in this paper, and test the accuracy of the three recommendation methods under different number of users. The experimental results are shown in Table 1, 2 and 3:

Table 1. Number of users 100 Recommended method accuracy (%)

Experimental rounds	Personalized recommendation of online music teaching resources based on neural network	Personalized recommendation of online music teaching resources based on machine learning	Personalized recommendation method of online music teaching resources
1	76.615	76.331	82.313
2	77.845	74.205	83.155
3	76.312	76.894	84.619
4	77.948	78.211	85.207
5	75.062	77.302	86.131
6	78.009	77.166	87.449
7	77.315	78.299	86.512
8	76.452	75.416	87.619
9	78.299	75.487	86.233
10	76.154	76.945	85.199

Table 2. Number of users 200 Recommended method accuracy (%)

Experimental rounds	Personalized recommendation of online music teaching resources based on neural network	Personalized recommendation of online music teaching resources based on machine learning	Personalized recommendation method of online music teaching resources
1	69.487	65.288	72.616
2	65.219	66.918	73.198
3	66.177	65.317	75.232
4	69.347	66.203	74.951
5	67.515	67.858	75.129
6	66.974	66.123	76.313
7	65.398	68.544	77.209
8	66.152	69.202	75.114
9	64.399	65.337	76.980
10	63.251	62.130	75.318

Table 3. Number of users 300 Recommended method accuracy (%)

Experimental rounds	Personalized recommendation of online music teaching resources based on neural network	Personalized recommendation of online music teaching resources based on machine learning	Personalized recommendation method of online music teaching resources
1	53.615	56.487	64.317
2	57.818	59.154	66.659
3	56.319	54.198	65.419
4	55.825	55.316	66.286
5	56.322	54.811	67.344
6	55.848	52.319	68.259
7	59.120	56.474	67.188
8	56.487	57.822	65.286
9	59.416	59.316	66.310
10	55.377	55.422	67.251

As can be seen from Table 1, the average accuracy rates of the personalized recommendation method and the other two methods are 85.44%, 77.001% and 76.626% respectively, and from Table 2, the average accuracy rates of the personalized recommendation method and the other two methods are 75.206%, 66.392% and 66.292% respectively. Because this method identifies the characteristics of online music teaching resources. Connect the resources through the knowledge points, and optimize the mobile terminal streaming media storage format. In addition, we converted the temporal continuous signal into the discrete time signal, constructed the user interest model by collaborative filtering, and recommended the favorite resources of the neighbor users to the current users.

4 Conclusion

Designing and developing personalized recommendation method of teaching resources to meet the personalized learning needs of online music users will help the effective use of teaching resources. At the same time, users need to choose more teaching resources to meet the different personalized needs of users, so that each learner can be fully developed in the field. Users are clustered offline based on resource attribute preference, and users with similar preference are partitioned into the same cluster and recommended in several clusters similar to the target user. This not only promotes the perfection of each learner's personality, but also meets the needs of social development for music talents. In the future, we need to do more research on cold start. Although this article has made a suitable summary and construction of classification information for Chinese songs. But the included classes and instances are far from the real ontology of Chinese songs, and there is still a lot of work. We need continuous improvement and timely updates.

References

1. Bing, C.: Distance teaching system of public music course based on SOA service framework. Mod. Sci. Instruments **6**, 27–30 (2020)
2. Wang, Z., Jianhua, L.: Research on the rapid recommendation model of online teaching resources in colleges and universities. Inf. Stud. Theory Appl. **44**(5), 180–186 (2021)
3. Chen, X.: A methodology study of enhanced college education reform powered by online education resourcing. Guide Sci. Educ. **27**(1), 8–9 (2020)
4. Geping, L., Xing, W.: Reshaping online education by virtual reality: learning resources, teaching organization and system platform. China Educ. Technol. **11**, 87–96 (2020)
5. Zhang, J., Hao, W., Ban, W., Rong, J.: An optimal design of vocal music teaching platform based on virtual reality system. Comput. Simul. **38**(06), 160–164 (2021)
6. Zhang, J.-X., Huang, S.-L., Liu, L.-J., et al.: Research on equipment identification based on machine vision in mobile terminals. Fire Control Command Control, **45**(2), 155–159 (2020)
7. Nie, L., Juan, F., Chengqi, Y., et al.: Measuring enterprise's offline resumption with mobile device positioning data. Data Anal. Knowl. Discov. **4**(7), 38–49 (2020)

Dynamic Evaluation Method of College English Cross-Cultural Teaching Based on Mobile Terminal Technology

Ying Lin[1,2(✉)], Hui Lin[3], Megawati Soekarno[1], and Yangbo Wu[4,5]

[1] Faculty of Psychology and Education, University Malaysia Sabah, 88000 Sabah, Malaysia
Linying0813@126.com
[2] School of Foreign Languages, Xinyu University, Xinyu 338000, China
[3] College of Arts, Xinyu University, Xinyu 338004, China
[4] Faculty of Computing and Informatics, University Malaysia Sabah, 88000 Sabah, Malaysia
[5] College of Mathematics and Computer, Xinyu University, Xinyu 338000, China

Abstract. The existing studies on intercultural communicative competence have explored the constituent elements, theoretical models, training models and evaluation methods of intercultural communicative competence, but the basic intercultural communicative competence is the main research object, which is not entirely applicable to the practice and research of the cultivation of intercultural communicative competence of English majors.In order to improve the practical applicability of dynamic evaluation of College English cross-cultural teaching, a dynamic evaluation method of College English cross-cultural teaching based on mobile terminal technology is proposed. Combined with mobile terminal technology, this paper constructs the evaluation system of College English cross-cultural teaching, standardizes the evaluation indicators of College English cross-cultural teaching, and optimizes the evaluation algorithm of College English cross-cultural teaching. Finally, the experiment proves that the dynamic evaluation method of College English cross-cultural teaching based on mobile terminal technology has high practicability in the process of practical application and fully meets the research requirements.

Keywords: Lack of culture · Mobile terminal technology · College English · Cross cultural teaching

1 Introduction

In view of the lack of bilingual education culture in recent years, the Ministry of Education recently promulgated the national standard for the teaching quality of English Majors in Colleges and universities, which is a programmatic document for the admission, construction and evaluation of English Majors in Colleges and universities in China, The training objectives are pointed out "English majors aim to cultivate compound and Applied English with solid basic language skills, international vision and humanistic quality, master the basic theories and knowledge of foreign language and literature, applied economics, business administration, law and other related disciplines, have English application ability, cross-cultural communication ability, practical ability, speculation and innovation ability, autonomous learning ability, and be able to engage in international work Talent ". For the first time, the English national standard clearly puts forward the quality, knowledge and ability that English majors should have, and establishes a complete construction system for English majors [1]. Among the necessary abilities, language application ability and cross-cultural communication ability are the core abilities, which should become the main observation point to test whether the teaching of English majors meets the national standard of teaching quality, and also an important means to solve the lack of culture. The cross-cultural communicative competence of English majors includes two parts: basic cross-cultural communicative competence and cross-cultural communicative competence. The existing researches on intercultural communicative competence have explored the constituent elements, theoretical models, training schemes and evaluation methods of intercultural communicative competence, but taking basic intercultural communicative competence as the main research object, it is not fully applicable to the practice and research of intercultural communicative competence training of English majors [2]. On the basis of summarizing the relevant studies on intercultural communicative competence and combining with the requirements of English national standards for intercultural communicative competence, this paper constructs a framework for English Majors' intercultural communicative competence, and puts forward corresponding training approaches from five aspects: curriculum setting, teaching concepts, teaching methods, teacher allocation and evaluation mechanism. So as to promote the practice and research development of English Majors' cross-cultural communicative competence.

2   Dynamic Evaluation of College English Cross-Cultural Teaching

2.1   Evaluation System of College English Cross-Cultural Teaching

Although the current academic circles have different views on the elements of intercultural communicative competence, its core part can be summarized into three levels: cognition, emotion and behavior. Therefore, based on the target quality, knowledge and ability of English Majors in the English national standard, and combined with the relevant research results of cross-cultural communicative competence at home and abroad, we put forward the framework of cross-cultural communicative competence of English majors [3]. In terms of theoretical framework, cognition, emotion and behavior constitute three basic elements of cross-cultural communicative competence in the international context. Each element can be subdivided into different sub elements to build a college English cross-cultural teaching evaluation system, as shown in Fig. 1.

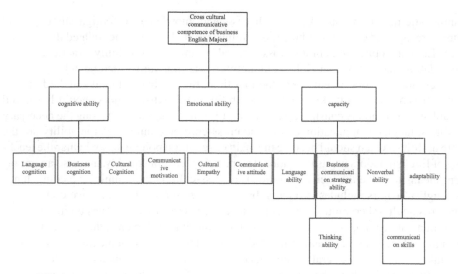

Fig. 1. Framework of College English cross-cultural teaching evaluation system

In the evaluation system of College English cross-cultural teaching, language teaching is the noumenon and culture teaching is the auxiliary. Scholars who hold this view believe that language teaching is the starting point and purpose of foreign language teaching and has an unshakable dominant position. Culture teaching is an auxiliary, which can help learners better understand the values and customs of the target language countries, so as to better carry out cross-cultural communication. If there is only language teaching but no culture teaching, then language teaching will be "malnutrition". Language teaching is the first and culture teaching is the second [4]. Scholars who hold this view believe that students must first understand a language before they can really understand the culture behind the language, rather than understanding a culture before they can understand the language expressed by the culture. Language teaching and culture teaching are you with me and I with you. There is neither language teaching without language teaching nor culture teaching without language teaching. To some extent, language teaching is culture teaching [5]. Culture teaching is a part of language teaching. Most scholars who hold this view, based on the narrow definition of culture, believe that grammar and vocabulary are not the whole content of language teaching. The teaching of cultural knowledge is also an important part of English language teaching. Culture teaching plays a complementary and promoting role in language teaching. Language teaching is a part of culture teaching. Most scholars who hold this view support the broad definition of culture. They believe that because the content of culture is almost all inclusive and culture includes everything a person wants to understand, learners can fully communicate with other language users in a way they can accept [6]. In this sense, the language of a society is an aspect of its culture, so language teaching is also a part of culture teaching. In addition, from the perspective of teaching content, language teaching and culture teaching in English teaching are distinguished, which can be shown in Fig. 2:

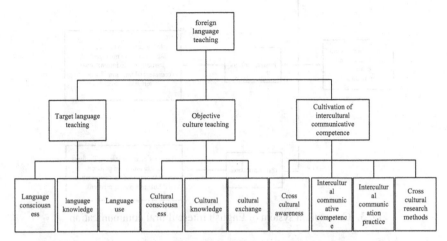

Fig. 2. College English cross-cultural teaching structure and Curriculum

The curriculum should first deal with the four pairs of relations between cultural education and professional education, language skill training and professional knowledge teaching, elective courses and compulsory courses, practical (experimental) teaching and theoretical teaching. Secondly, the curriculum system must cover teaching contents at all levels of cross-cultural communication, and communication, and teachers should actively guide students to understand the dynamic relationship of various influencing factors of cross-cultural communication through theoretical teaching and practical teaching, so as to promote them to understand and deal with international communication problems from a more dialectical and sensitive cross-cultural perspective [7]. There are three levels of courses, namely, cross-cultural communication, communication economics and cross-cultural communication system, as shown in the following Fig. 3:

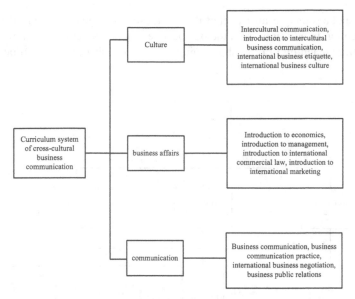

Fig. 3. Curriculum system of English intercultural communication

Intercultural communicative competence is one of the most important competence elements of English majors, and it is also one of the core objectives of English Majors' training. Constructing a complete, systematic and operable evaluation system of cross-cultural communicative competence has important practical guiding significance for evaluating the effect of English teaching and perfecting the evaluation system of English teaching. Many domestic scholars have built an evaluation system for college students' cross-cultural communicative competence. These evaluation studies have explored the constituent models and evaluation methods of cross-cultural communicative competence from different perspectives, and consolidated the basis of theoretical and practical research on cross-cultural communicative competence, but did not involve relevant indicators and quantitative standards of cross-cultural communicative competence [8]. Moreover, using the evaluation scales of other professional talents training for reference to measure the cultivation of cross-cultural communicative competence of English majors will affect the accuracy of the comprehensive evaluation index and is not conducive to the realization of the training objectives of English majors. At present, there is no evaluation system and scale of cross-cultural communicative competence at home and abroad, However, the research results related to the evaluation of cross-cultural communicative competence can provide some enlightenment and reference for the evaluation of cross-cultural communicative competence: cross-cultural communicative competence can be evaluated; The assessment of intercultural communicative competence must adopt a combination of quantitative and qualitative methods from different angles; The evaluation of cross-cultural communicative competence must consider many factors, such as evaluators, evaluation objects, evaluation objectives, evaluation use, specific situations, even social conditions, historical background and so on. We call on the academic circles to speed up the research on the evaluation system of cross-cultural communicative competence and improve the evaluation system of English teaching as soon as possible.

2.2 Evaluation Index of English Cross-Cultural Teaching

The evaluation index system is an organic whole formed by different levels of evaluation indicators according to the logical structure of the evaluation object itself. It is a set of indicators or a set of specific indicators. The evaluation index system of College English classroom culture teaching is a collection of a series of indicators to judge the value of middle school English teachers' classroom culture teaching according to the characteristics of middle school English culture teaching. It is the embodiment of the concept of middle school English culture teaching [9]. According to the characteristics of English classroom culture teaching, the evaluation index system of College English classroom culture teaching is a detailed, clear and operable organic whole. It is an evaluation system that can effectively measure the development of middle school English culture teaching. The evaluation index system of College English classroom culture teaching takes culture as the main axis, Through a number of interrelated, complementary and organic whole with a certain structural level connected by teachers' cultural teaching activities, these indicators can not only reflect the characteristics of cultural teaching, but also take into account the relationship between various elements in the teaching system. This study adopts the methods of literature analysis, theoretical analysis and expert consultation, Select indicators to meet the scientificity and integrity of the evaluation index system of culture teaching in English classroom teaching. The process of constructing the evaluation index system of culture teaching in English classroom is shown in Fig. 4:

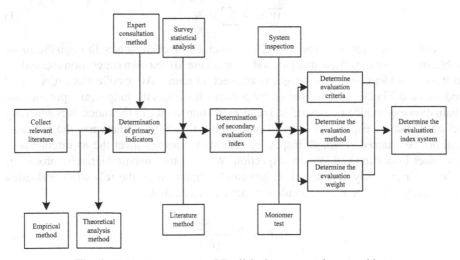

Fig. 4. Evaluation process of English classroom culture teaching

The Enlightenment of the sociological analysis of cultural differences is to promote social unity and enable students to obtain cultural capital through cross-cultural education, and cultivate students' cultural skills. The teaching and evaluation of cross-cultural education must also be closely related to daily life experience. The significance of psychology's interpretation of cultural differences to cross-cultural education is to let students master the framework of understanding cultural differences and take cultivating students' psychological adjustment ability as an important goal. The author also believes that the curriculum of cultivating cross-cultural ability in school education can be divided into two ways: integrating discipline curriculum and developing comprehensive curriculum [10–12]. The subject curriculum needs to reorganize the existing subject curriculum content or add the content of cross-cultural ability training, while the comprehensive curriculum is specially designed to cultivate cross-cultural ability. In terms of teaching, cross-cultural education takes group discussion and activity teaching as the main teaching strategy; in terms of evaluation, performance evaluation is adopted. At the same time, the practical implication of cross-cultural education is based on the dynamic and non essentialist view of cultural differences and identity, and takes cultivating learners' cross-cultural ability as the basic goal. It is a new interpretation of the thought of "learning to coexist". The enlightenment to China's cultural diversity education is to treat cultural differences and identity from a non essentialist and dynamic standpoint: build an educational process to promote cross-cultural communication; pay attention to the cultivation of learners' common values in the educational process; incorporate the cultivation goal of cross-cultural ability in general education; gradually implement the cultivation of Teachers' cross-cultural ability. Distribute the weight allocation W to the evaluation experts, and invite the experts to independently assign the weight N_{ij} of each index in the first round, and fully explain the reasons. To recover the consultation form, it is necessary to make statistics on the data n and calculate the average value and deviation of each index.

$$\overline{W} = \frac{1}{n} \sum W_j N_{ij} \tag{1}$$

Feed back the statistical results and the two eigenvalues calculated through the above publicity to experts, and ask them to modify or adhere to their own suggestions according to their own ideas, and give weight in the second round. After collection, it is sorted and counted. Finally, after weighing, the person who makes the judgment represents the evaluation expectation of experts and reflects the importance of the index. It is generally believed that the expectation should reach more than 60%. Otherwise, the indicator will not be retained. Through inspection, the above indexes meet the requirements of monomer inspection and system inspection. Whether the analytic hierarchy process is reliable for the analysis result of weight can be measured by the following evaluation consistency ratio C.R.). The calculation method is as follows:

$$C.R. = \overline{W} - \frac{\frac{\lambda_{max}}{n-1}}{(R.I.)_n} \tag{2}$$

where: $(R.I.)n$ is the evaluation consistency index of n-order matrix; The average consistency index of λ_{max} matrix.

2.3 Realization of Dynamic Evaluation of English Cross-Cultural Teaching

According to the measurement standard, the correlation below 0.5 is very weak, which is not suitable to be used as a secondary evaluation index. It can be seen from the statistical results that the correlation test of evaluation factors such as cultural teaching background, cultural teaching process, cultural teaching style, cultural teaching style, cultural teaching management, cultural teaching attitude and teaching reflection has not been passed. Therefore, the above evaluation factors will not be included in the evaluation system of English classroom cultural teaching. Therefore, this study finally determines that there are 9 secondary evaluation indicators for English classroom culture teaching evaluation. The evaluation indicators are as follows:

Table 1. Evaluation index system of College English cross-cultural teaching

Teaching stage	Primary index	Secondary index
Teaching preparation stage	Cultural teaching preparation	Cultural teaching concept, goal and design
Teaching process stage	Cultural teaching process	Cultural teaching contents, methods and atmosphere
Teaching effect stage	Cultural teaching effect	Teaching feedback, classroom response and compliance

The evaluation object in the preparation stage of teachers' cultural teaching is mainly the preparation of teachers for the upcoming cultural teaching. The second level evaluation indicators include teachers' cultural teaching ideas. Cultural teaching objectives and cultural teaching design. Cultural teaching ideas are important factors affecting the implementation of cultural teaching. Cultural teaching ideas mainly refer to two aspects: one is whether teachers fully understand the necessity of cultural teaching, and the other is whether teachers correctly understand the internal relationship between language teaching and cultural teaching. The goal of cultural teaching refers to the expected results of students' cultural learning through teachers' cultural teaching activities. Cultural teaching goal is not a unilateral goal of teachers or students. It includes both teachers' cultural teaching goal and students' cultural learning goal. From the perspective of cultural teaching, the evaluation of cultural teaching objectives should mainly consider the following two points: first, the description of cultural teaching objectives should be clear and specific, reflecting the integration of three-dimensional objectives; Second, whether the cultural teaching goal pays attention to the cultivation of students' cultural awareness and cultural understanding, and the cultural teaching design is the arrangement of cultural teaching activities under the guidance of the teaching goal. Specifically, it includes teachers' treatment of cultural teaching content, whether the introduction method of cultural teaching is appropriate, and the process arrangement of cultural teaching. The evaluation points of this index mainly include: first, we should teach students relevant cultural content based on teaching materials; Second, whether the design of cultural teaching scheme is reasonable, whether the content is appropriate and easy for students to understand; Third; Whether the cultural teaching scheme has strong practicability. The evaluation indexes of College English classroom culture teaching methods mainly include three aspects: first, the creation of culture teaching environment; The key points of this evaluation mainly depend on whether teachers choose specific cultural teaching situations in cultural teaching, whether it is in line with teachers' teaching style and students' psychology of cultural cognition. Second, the choice of cultural methods; If teachers use appropriate teaching methods according to different teaching contents and teaching situations, such as guidance method and cultural comparison method, they can promote students' cross-cultural understanding, enhance students' cultural experience and enhance the effect of cultural teaching. Third, the use of cultural teaching resources; It mainly depends on whether teachers make full and effective use of various cultural teaching resources in the process of cultural teaching, because all kinds of teaching resources are silent culture. Of course, there is no best method for cultural teaching, only the method suitable for specific cultural teaching activities. English classroom cultural atmosphere affects the cultural interaction between teachers and students and students' perception of culture. The research shows that a good classroom culture teaching atmosphere is conducive to the cultivation of students' cultural learning ability. For the evaluation of this index, we need to investigate the key points: first, whether teachers and students often have cultural interaction in culture teaching; The second is to investigate whether the culture teaching classroom has a democratic, open and classroom cultural atmosphere conducive to students' cultural perception.

3   Analysis of Experimental Results

According to the survey data of personal information in the first part of the survey of College English cross-cultural teaching, this paper analyzes the specific situation of the respondents' gender distribution, age distribution and educational background distribution. The purpose is to analyze whether the samples involved in the survey are representative and can represent the basic situation of this kind of personnel, so as to ensure the scientificity of the research results. See Table 2 for the specific distribution of the samples involved in the survey (Table 1).

Table 2. Statistical table of survey sample distribution

Project	Category	Number of samples	Percentage
Gender	Male	1	2.98
	Female	36	98.20
Age	30–35	23	62.35
	36–40	9	23.65
	41–45	3	5.68
	46–50	3	5.68

(continued)

Table 2. (*continued*)

Project	Category	Number of samples	Percentage
	50–55	3	5.68
Education	Undergraduate	12	28.65
	Graduate student	28	73.68
	Doctor	0	0

According to the cross-cultural teaching situation of English teachers, this paper mainly investigates three questions. Teachers' understanding and influence on cross-cultural teaching includes the following aspects: the reserve of teachers' own cultural quality: Teachers' understanding and understanding of cross-cultural communication knowledge learning; Teachers' understanding of cross-cultural teaching content; Teachers' choice of culture teaching mode; Ways of students' cultural learning; Students' interest and attitude towards cultural learning: the way teachers deal with the relationship between language skill training and cultural knowledge learning; The occupation of teachers' culture teaching time. The content of cultural content in teaching materials. It is the problems existing in cross-cultural teaching in the eyes of teachers and the suggestions for cross-cultural teaching. With the help of spss17 0 software package for data statistical processing of the first two problems. The survey scores of the above aspects are shown in Table 3 (Table 4).

Table 3. Scores of teachers and teaching materials on the impact of cross-cultural teaching

		Teachers' own knowledge level	Teachers' attitude towards cross-cultural learning	Teachers' understanding of cross-cultural teaching	Ways of students' cross-cultural learning	Students' interest in cultural learning	How teachers deal with language skills	Occupation of culture teaching time
N	Effective	38	38	38	38	38	38	38
	Defect	0	0	0	0	0	0	0
	Mean value	1.65	26.33	9.15	6.24	1.77	5.65	3.65
	Standard error of mean	.102	.415	.218	.095	.129	.128	.181
	Median	2.00	25.00	11.00	18.00	2.00	6.00	3.00
	Mode	3	28	11	26	7	3	5
	Standard deviation	.608	2.652	1.625	5.683	.771	.765	1.086

(*continued*)

Table 3. (*continued*)

	Teachers' own knowledge level	Teachers' attitude towards cross-cultural learning	Teachers' understanding of cross-cultural teaching	Ways of students' cross-cultural learning	Students' interest in cultural learning	How teachers deal with language skills	Occupation of culture teaching time
Variance	.398	6.268	1.826	.365	.596	1.165	1.152
Minimum	0	18	7	8	5	2	1
Maximum value	4	28	12	27	4	5	5

Table 4. Evaluation results of English cross-cultural teaching

Teaching stage	Primary index	weight	Secondary index	weight	Comprehensive evaluation results (Miss Sun)	Comprehensive evaluation results (Mr. Wang)
Teaching preparation stage	Cultural teaching preparation	0.3	Teaching concept	0.3	A	B
			Teaching objectives	0.4	B	A
			instructional design	0.4	C	B
Teaching process stage	Cultural teaching process	0.5	content of courses	0.3	B	A
			teaching method	0.4	C	B
			Teaching atmosphere	0.3	B	A
Teaching effectiveness	Cultural teaching effect	0.3	Classroom effect	0.4	A	A
			Teaching feedback	0.3	B	A
			Compliance status	0.4	B	A

According to the relationship between statistical Cronbach a coefficient and reliability, Cronbach a coefficient greater than or equal to 0.9 means that the evaluation result is very reliable. It can be seen from the above results that the evaluation results have high reliability. This is because the method in this paper combines mobile terminal technology, constructs a college English cross-cultural teaching evaluation system, standardizes the College English cross-cultural teaching evaluation indicators, and optimizes the College English cross-cultural teaching evaluation algorithm. Through the above evaluation analysis, combined with the opinions of five evaluation experts and the comprehensive calculation of the evaluation data, the final evaluation results of English cross-cultural teaching are as follows:

Using spss17 0 kmo and bartlette ball test for factor analysis to judge whether this study can do factor analysis. This paper uses the statistical method of factor analysis to extract multiple factors affecting cross-cultural teaching through principal component analysis, and then determine the main factors affecting cross-cultural teaching. First, test them. The test results are shown in Table 5.

Table 5. Results of kmo and Bartlett's test and spherical test in English

	Kaiser Meyer Olkin measure of sampling adequacy	.625
Bartlett's sphericity test	Approximate chi square	79.652
	df	38
	Sig	.000

By calculating the characteristic roots and arranging them according to their size, the main scattered point diagram (scattered point diagram) is output. In the diagram, the ordinate is the characteristic root and the abscissa is the number of factors. The output results are shown in Fig. 5.

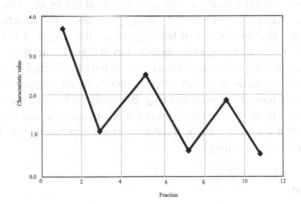

Fig. 5. Statistical data

It can be seen from the figure that except the first four principal components, the characteristic roots of other principal components are lower than 1, so the number of factors is determined to be 4. With the help of spss17 0 software package to process statistical data. This is because according to the target quality, knowledge level and ability level of English majors, combined with the relevant research results of cross-cultural communication ability at home and abroad, this paper puts forward the framework of cross-cultural communication ability of English majors.The histogram of student achievement distribution is shown in Fig. 6, and the test diagram of normal distribution is shown in Fig. 6.

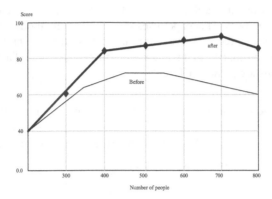

Fig. 6. Changes in school rules of English cross-cultural school

The chart shows that under the traditional method, students with scores below 60 account for the majority, that is, most students fail. The reason is that there are problems in cross-cultural teaching, resulting in poor students' scores. After using a certain terminal technology for teaching, students' teaching performance is significantly improved. Cross cultural education is an educational activity about personal world outlook, values, identity and cross-cultural awareness and ability carried out by the school through the determination of training objectives, the setting of courses, the selection of teaching contents and materials, the renewal of teaching ideas, the design of teaching methods and teaching activities, and the organic combination of school education and social practice, It is necessary to integrate cross-cultural education into foreign language teaching. Firstly, it can meet the needs of the interdependence between language and culture and reflect the process of natural language learning. Secondly, adding cultural taper to foreign language teaching will make language learning more meaningful and fun; Furthermore, it is recognized that the development of cross-cultural communication is the ultimate goal of foreign language teaching.

4 Conclusion

On the basis of summarizing the representative research results of cross-cultural communicative competence at home and abroad, combined with the ability requirements of English national standards, this paper constructs the framework of cross-cultural communicative competence of English majors, and puts forward the corresponding training ways from five aspects: curriculum, teaching ideas, teaching methods, teacher allocation and evaluation mechanism. The in-depth research on the cross-cultural communicative competence system of English majors is in line with the fundamental needs of China for the cultivation of cross-cultural English talents under the background of economic globalization. It can also provide theoretical and practical guidance for the teaching reform of cross-cultural courses and the construction of English Majors in Colleges and universities, and effectively promote the realization of the cultivation goal of English majors.In the future development, advanced science and technology and information technology will be introduced and combined with the evaluation system to effectively improve the accuracy and practicality of evaluation.

References

1. Lian, J., Fang, S.Y., Zhou, Y.F.: Model predictive control of the fuel cell cathode system based on state quantity estimation. Comput. Simul. **7**, 119–122 (2020)
2. Hu, D.F.: College English teaching and the cultivation of international talents. J. Foreign Lang. **5**, 8–20 (2020)
3. Zhang, H., Zhang, W.X.: English teachers' perceptions of NMET washback on senior high school English teaching and learning —— based on a large-scale nationwide survey. Foreign Lang. Teach. **3**, 36–45 (2020)
4. Zhang, L., Zhang, Y.: Professional English teaching mode for export-oriented acupuncture and massage students in TCM colleges and universities under the background of the belt and road initiative. West. J. Tradit. Chin. Med. **7**, 62–65 (2020)
5. Yang , K.C., Lu, Y.: A New approach of instruction evaluation: justifying an instruction by its own process. Mod. Dist. Educ. Res. **6**, 49–54 (2021)
6. Mao, G., Zhou, Y.L., He, W.T.: Development trend of teaching evaluation theory under background of educational big data. E-educ. Res. **10**, 22–28 (2020)
7. Guo, X.L.: Research on innovation of evaluation mechanism of practical teaching of ideological and political theory courses in colleges and universities. J. Anshun Univ. **4**, 58–61 (2020)
8. Zhao, Z., Yang, J., Xi, H., et al.: Research on mobile terminal technology supporting intelligent maintenance of substation. J. Phys: Conf. Ser. **3**, 032139 (2021)
9. Zhang, C., Tong, H.: Research the application of artificial intelligence technology in mobile phone terminal. J. Phys: Conf. Ser. **2**, 022024 (2020)
10. Hou, T., Zhang, T., Huang, H.: Design of wine grape mobile terminal scheduling platform architecture based on IOT. J. Phys. Conf. Ser. **1**, 012009 (2021)
11. Feng, Y.: Mobile terminal video image fuzzy feature extraction simulation based on SURF Virtual Reality technology. IEEE Access **99**, 1 (2020)
12. Xu, J.: Big NB-IoT data: enhancing portability of handheld narrow-band internet of things performance on big data technology. Mob. Inf. Syst. **5**, 1–6 (2021)

Blockchain-Based Social Media Software Privacy Data Cloud Storage Method

Jianjun Tang[✉] and Xiaopan Chen

Jiangxi University of Software Professional Technology, Nanchang 330041, China
tangjianjun857@163.com

Abstract. At present, the leakage of private data stored on social media software has attracted the attention of the academic community. Therefore, it is necessary to discuss the cloud storage method of private data of social media software in detail. In the process of using the social media software privacy data cloud storage method, there is a problem that the file storage takes too long. In order to alleviate the above problem, a blockchain-based social media software privacy data cloud storage method is designed. Obtain the identity management mechanism of social media software, formulate identity restriction plans according to the available scope of accounts, classify the risk factors of privacy information, improve the data reading mode, adjust the independence of data, and use the blockchain to set up the cloud storage structure. Experimental results: The file storage time of the social media software privacy data cloud storage method and the other two social media software privacy data cloud storage methods are: 2.672 s, 4.229 s, and 4.727, respectively. After the cloud storage method of software privacy data is combined, the application effect of the method is better.

Keywords: Blockchain · Social media software · Private data · Cloud storage · Information technology · Personal information

1 Introduction

With the development of information technology and human civilization, the amount of data produced is increasing explosively. The traditional data storage and analysis model can not meet the existing needs. Social media has quickly become the dominant media on the Internet, thanks to the craze for social software such as Facebook and the development of computer and Internet technologies. From home to abroad, all kinds of social applications are coming out. However, the leakage of privacy information often causes users into a dilemma, and the information security problem has become an important factor restricting the development of cloud storage. Therefore, it is particularly important to take some protection measures for the privacy information of individual users under the cloud storage environment to effectively avoid the risk of privacy disclosure [1].

Speaking of the domestic, along with the new media technology development and the intense market competition, from initial person net, happy net, skyline, cat flutter. Up

W. Fu and L. Yun (Eds.): ADHIP 2022, LNICST 469, pp. 376–389, 2023.
https://doi.org/10.1007/978-3-031-28867-8_28

to now, the development of Weibo, WeChat, Mo, Line, Youjia, social media from product positioning design, positioning and services, interesting, personalized features are increasingly evident. In order to get close to the users and realize the win-win of word-of-mouth and influence communication, many enterprises take advantage of the opportunity to infiltrate social elements into the communication of marketing and publicity. Secondly, the protection of citizen information security depends on the cooperation of all parties to build and maintain a harmonious social network environment, so it is necessary to build a user privacy information protection system. From the multi-dimensional point of view, we can comprehensively enhance the security of user privacy information. To the user, convenient, individual application experience, have powerful attraction. Whether from the perspective of work or life, most users of social applications, is not satisfied with the simple online communication and exchange, but look forward to online and offline integration. At the same time, the user's privacy security issues have become increasingly prominent.

From the multi-dimensional point of view, we can comprehensively enhance the security of user privacy information. To the user, convenient, individual application experience, have powerful attraction. Whether from the perspective of work or life, most users of social applications, is not satisfied with the simple online communication and exchange, but look forward to online and offline integration. At the same time, the user's privacy security issues have become increasingly prominent.

The essence of cloud storage is that users upload a large amount of information to the cloud, and can access information resources at any time without the limitation of devices and time. Therefore, the cloud storage platform has a large amount of user privacy information [2, 3].

In order to improve the efficiency of social media privacy data storage, this paper designs a blockchain-based social media software privacy data cloud storage method. First, the identity management mechanism of social media software is acquired. Secondly, an identity restriction scheme is formulated according to the available scope of the account, and the risk factors of private information are classified. Finally, improve data reading methods, adjust data independence, and use blockchain to build cloud storage structures.

1.1 Access to Social Media Software Identity Management Mechanisms

With the popularity of social media and the development of e-commerce, targeted marketing, precision marketing has been sought after. Over-collection of personal privacy means that the merchant collects personal information in many ways in order to store and develop more potential users for the need of self-profit. Mainly in two ways, one is to register or fill in the form of the relevant forms. Second, through the form of technical tracking collection. Digital identity is identity, too. Its difference is that it is virtual, invisible and untouchable. Therefore, against the digital identity may be false identity attacks, identity theft attacks and witch attacks. As far as the current usage of social media is concerned, whether it's a simple social networking application or a forum, a shopping site, etc., you have to provide personal information: name, phone number, email address, home address, like WeChat Wallet, etc., you have to bind personal bank card information. False identity, refers to the attacker using other people's information, or using incorrect information to obtain the account. Identity theft, refers to the attacker in the case of unauthorized use of other people's accounts, as long as you know that other people's accounts can embezzle their identity. Witch attacks are attacks where the attacker actually controls many identity accounts, and you might think that there are many accounts "flooding" the forums, but there may be only one person behind those accounts.

In order to prevent these three attacks, we need to restrict the digital identity. Traditionally, an identity manager distributes digital identities. Users need to submit personal information and set an account password to obtain digital identities. Only after the manager confirms, can the account be owned. The secondary development and utilization of personal privacy data refers to the use of data processing or information mining and other methods to process the collected user information by adopting the CRF algorithm and search for the content with commercial value [4, 5]. In the field of marketing, the secondary development and utilization of personal information has been highly praised as a new business model. Personal information submitted by users is typically identified or stand-alone, such as an ID number, a mobile phone number, or an email address registered on a different server, thus ensuring that different users have different accounts that can only be used by the owner through a password. Depending on the available scope of the account, these identity restrictions are shown in Fig. 1:

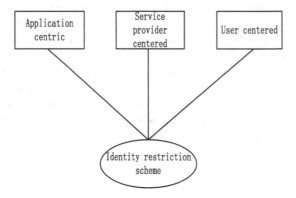

Fig. 1. Schematic diagram of identity restriction scheme

As you can see from Fig. 1, there are three identity restriction schemes: application-centric, service-provider centric, and user-centric. Application-centric identity management scheme: This scheme refers to the way people interact socially, and allocates accounts around different activities. In real life, people go to school, drive a car, work, get married respectively need a student card, driver's license, work permit, marriage certificate. For some businesses, through consumer information analysis and secondary development, we can better understand user needs, and then perfect and optimize their own services, or launch new services, for their own and users are win-win. Each application only needs to manage its own account, and users can only use the corresponding service after registering their account, and these accounts can only be used in the application services. Because of its simplicity, each application can embed the identity management module directly and provide services directly without maintaining the identity management module separately.

Currently, privacy deals in social media and even the entire Internet are conducted in two ways. One is information-sharing through business cooperation, each taking what he or she needs. Service-centric identity management solutions: As the business expands, the same service provider may develop multiple applications, such as Alibaba has Taobao, Alipay, Aliwanwang, Tencent has QQ, WeChat and many games. From the user's point of view, they are separate, different, and provide different services for the user. But we do know that there is more than one partner, and there are other partners, and so on, and we see the horror in the process, that if the scope of information sharing is not effectively regulated and controlled, the user's information will be known to more merchants. This is undoubtedly a disguised invasion of users' personal privacy. But from the service provider point of view, they are unified, the same, are controlled and managed by the same service provider. User-centric identity management: As long as a user has an account, it's like getting an ID card and can log in to all applications without having to sign up for a separate account.

1.2 Categorizing Privacy Information Risk Elements

The main function of cloud storage is to help users to store information, manage data and share information. So it is different from other Internet related services. As far as the privacy protection policy is concerned, most of our social media privacy protection policies have no obvious location, which is difficult to attract users' attention. In particular, with the development of mobile Internet technology, users mainly download App to the mobile terminal for direct use, with little attention to privacy protection policies. In order to better identify the risks of user privacy information in the cloud storage environment and conduct in-depth research on the protection of user privacy information, we classify the user privacy information involved in the process of cloud storage services into basic personal information and personal activity data [6]. This paper considers that the risks of user privacy information in cloud storage include.

As can be seen from Fig. 2, the types of user privacy information risk in a cloud storage environment include: management risk, technical risk, and legal risk. The management risks of users' privacy information in cloud storage environment refer to the risk factors that lead to the disclosure of privacy information in the process of cloud storage services due to poor management and poor security quality of the relevant organizations and individuals that have the management responsibility for users' privacy information. As far as the content is concerned, it is a pile of terms, and users need enough time to understand and digest it. But the author also found that nearly 40% of users never read the privacy policy, and directly accept it, which is related to the piling up of these terms.

Cloud service is a kind of network technology service, which has been springing up in recent years. Its government management has overlapping functions, unclear rights and responsibilities, and the relative lack of laws and regulations has brought difficulties to the supervision and management of law enforcement departments. Privacy policies rarely describe the approach to the use of personal data and the privacy management of

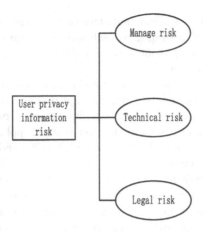

Fig. 2. Risk types of user privacy information in cloud storage environment

the platform. As far as privacy permissions are concerned, the default privacy permissions for social media platforms are mostly open to everyone. The management risks of cloud service providers refer to the operational errors caused by the server interruption caused by the lack of management and maintenance of servers, networks and other equipment, the malicious theft of user privacy data by internal employees caused by the lax internal management of employees, and the non-standardized internal operation and inadequate staff training of enterprises. And the location of the privacy settings feature is also different.

At present, WeChat, Weibo, Tencent and other large enterprises, in the settings function, there are obvious "privacy", "privacy and security" settings. Like Hungry, this location-based, commonly used ordering software doesn't have a "privacy" setting. In order to obtain convenient services, cloud storage users often ignore the privacy information security considerations, lack of sensitivity and protection skills necessary to face information risks. Therefore the user to the personal privacy protection omission is also causes the information disclosure the important factor. Similarly, people in cyberspace to use QQ, Weibo, Alipay and other applications to provide services, chat with QQ to register the QQ account, Weibo messages will be registered Weibo account, Alipay money will be registered Alipay account. The technical risks of users' privacy information in cloud storage environment refer to the technical risk factors that lead to the disclosure of privacy information in the process of cloud storage services, such as the limitations of cloud storage security technology and the lack of research and development capability for emerging security technologies. But do not think so to the user however, the secondary development goal that the businessman develops to information above all whether proper infer hard. Second, when a user's mailbox is full of ad emails. When users in the process of browsing the Web to see the scrolling ads have been browsed on the product information, it will have a personal privacy leak, personal domain is violated feeling. A computer expert, especially a programmer, who used to be keen on computer technology. After deduction, especially refers to the use of system security vulnerabilities on the network to attack and destroy or steal data.

Wireless Internet is the most common way for cloud storage users to connect to the Internet. Even if the transmission signal adopts information encryption technology, but the signal exposed in the air will be intercepted. The virtual storage unit technology used in cloud storage makes it impossible to detect the data exchanged between virtual systems, and also enhances the possibility of Trojan virus attack due to virtualization. Most of the information stored in the personal cloud is stored in the public cloud, if the data is not isolated properly, the information between users will interfere with each other. Hacker's existence, no matter to the enterprise, the individual even the country, is the enormous threat. Hackers can invade computer systems, steal and tamper with user information, through the design of the Trojan program into the target system, break the permissions directly log in. Survey found that nearly 40% of users have encountered the problem of social account password theft. But the address book or the document divulges, phenomenon and so on handset computer poisoning also has the occurrence. Therefore, the unstable Internet environment and virtualized storage technology increase the risk of leakage of privacy information of individual users in cloud storage.

1.3 Improve Data Read Mode

The data in the program exists in the form of objects, but the additional data of the blockchain transaction only accepts the form of hexadecimal strings. Therefore, the program cannot directly read and write data on the chain, but requires a conversion process from hexadecimal strings to objects and objects to hexadecimal strings. This section will describe the entire process of reading and writing blockchain data. And the data conversion mechanism is introduced. The SQL language covers many functions such as data query, data manipulation, data definition and data control. The language style is unified, and all activities of the database life cycle can be completed independently. Users can use SQL language to define, modify and delete relational schemas, define and delete views, insert data, and create databases. At the same time, the data in the database can be queried and updated.

Data writing is initiated by a program to store objects in memory at run time on a chunk chain. The first step in writing data to a chunk chain is to serialize the in-memory object into a recognizable JSON string, and add the state information of the data header to form a complete data structure. Because of its high degree of non-procedural, users in the process of using SQL can directly complete specific operations without concern about the underlying complex encapsulation operations, reduce user burden, adjust the independence of data. The second step transcodes the entire data structure into a hexadecimal string, and adds the identifier "0x" before the hexadecimal string to form a form that can be stored by a block chain. The final step is to construct a new transaction in which the converted data structure is attached to the transaction and submitted to the block chain to save the data. In the data reading mechanism, the expression formula of the encryption parameter is.

$$D = \frac{\varepsilon}{W} \times \exp\|W - 1\|^2 \qquad (1)$$

In formula (1), W represents the original plaintext document set, and ε represents the number of documents in the original plaintext document set W. In an encrypted database, data is stored in ciphertext. So it brings great technical challenge to the calculation and retrieval of data. Traditional ciphertext data retrieval based on ciphertext matching can not meet the requirements of database such as conditional query, ranking, fuzzy query, and so on. Specifically, the program first selects a node in the blockchain network in the final step. Using the key file corresponding to the built-in account of the node, the private key of the account is calculated. Then a new transaction is constructed using the address of the account and the converted data. Finally, the transaction API of Ethereum is called to submit the transaction to the block chain to write the data, and the hash of the transaction returned by the API is obtained. On the basis of the above description, the formula for calculating the keyword set generation process is derived.

$$\beta = \sum_{j=1}^{i} \frac{\sqrt{d_i + t_j}}{2} \times \mu_{ij} \tag{2}$$

In formula (2), d represents the topic set extracted from W. t represents the number of topics in the topic set d. μ represents the set of keywords extracted from W. i, j represents the number of keywords in the keyword set μ and the total number of keywords, respectively. Therefore, according to the different security levels of database columns, an adaptive and customizable data column security level encryption method is designed to balance data security and system efficiency. Through the self-defined encryption interface, under the premise of ensuring data security, the operability of ciphertext is improved, and the running efficiency of database system is ensured.

Block chain access to data is public, and each transaction in the chain has a unique hash value. As long as you hold the hash of a transaction, you can access the data stored on that transaction. Thus, the first step in reading the data from the blockchain is to use the transaction hash to get the transaction data through the Ethernet API and read the value of the input field in the transaction. Because users have different security requirements and operation requirements for different columns in the data table. Therefore, the data table should support the different strength encryption algorithm to the different column according to the user demand, or uses the encryption algorithm which supports the inquiry and the computation. The second step removes the flag "0x" from the hexadecimal string header and decodes it as a stored data structure. Finally, according to the description of the data structure, the JSON string stored in the data body is deserialized to the object in the program context, and the data is read.

1.4 Blockchain Setup Cloud Storage Structure

Block chain is a combination of cryptography, consensus mechanism and other mature technologies. It works in a decentralized mode, in which all the nodes in the network participate in the billing process. It also provides sophisticated scripts to support different business logic. Cloud server is only responsible for the cloud server data storage operations, and provide retrieval services, it will not take the initiative to update ciphertext document collection, will not actively update the security index. A blockchain is a chain-like structure composed of a series of blocks. Blocks are the places where the data are packaged and stored, and are the basic units of a block chain, which are added to the chain in the order in which they are generated [7].

Blockchain runs on a point-to-point distributed network, in which all nodes are peer to each other and there is no central node with special authority that can control other nodes. Each node completes the data processing through the consensus mechanism. Given the corresponding set of ciphertext, the server cannot learn anything about the original plaintext document from these ciphertext sets. This is typically done by using symmetric encryption schemes to encrypt documents or message blocks. In addition to the search results, the server can not learn more information about the open text accounts of the completion of verification written to the block chain data can be all nodes publicly accessible. But public access does not mean that the information itself is open to all nodes. Transaction data cannot be tampered with. No one can delete or modify the data once it has been saved to the chain through validation in the block chain. The cloud server cannot retrieve any keywords without the authorization of the data owner, which can be achieved by generating a keyword retrieval token using a pseudo-random function. The server cannot generate a valid retrieval token without the key provided by the data owner.

Collective maintenance, storage, transmission and verification of data in the chain are carried out by all the nodes in the blockchain under the mechanism of consensus. Even if some nodes in the blockchain network go offline or fail, the whole system will not be paralyzed. In addition, if two different miners perform the workload proof process for a new block almost simultaneously, the two blocks may be validated and accepted by two different subsets of the bitcoin network. Then a bifurcation is formed, and other miners need to choose the bifurcation with more blocks and longer chains for subsequent additions. The server cannot learn any information about the keyword to be retrieved from the retrieval token provided by the authorized user. If the server is a malicious server, in addition to the above requirements, the server can not forge encrypted data and related metadata. Then the addition field calculation formula for the metadata is.

$$F = \left| \frac{\varphi}{d} \right|^2 - \left| \frac{t}{\mu} \right| \times \varphi^{-1} \tag{3}$$

In formula (3), φ represents an identifier generation function. On the basis of formula (3), the calculation formula of the length of the retrieval vector is obtained as:

$$Q = \frac{1}{(H)} \times \sqrt{\frac{\lambda}{\mu}} - \varepsilon \tag{4}$$

In formula (4), H represents the set of topic distribution vectors, and λ represents the number of distributions in the vector. In order to adapt to the increasingly complex network environment, the traditional relational database has gradually developed into a distributed data management system through network interconnection. A single data storage node gradually becomes a cluster of distributed servers. In a multi-user scenario, a malicious user should not be able to query with any keyword without the authorization of the data owner, nor should he be able to learn additional information from the query submitted by the user. Key exchange, authorization control, access control and other technologies have been more mature, and widely used. Therefore, this chapter does not consider these security models. Compared with centralized database, distributed data management system has more advantages in expansibility, reliability, availability and cost-effective.

The most important feature of distributed data management is the multi-node backup of redundant data to deal with the data recovery in the case of failure. Given a ciphertext model, an attacker can retrieve ciphertext information stored by the data owner on a cloud server. These include encrypted document collections, secure indexes, and retrieval tokens. But the attacker can not get the encryption key, and can only attack through the ciphertext information.

However, there may be conflicts between the redundant data backed up, and data consistency mechanisms are needed to ensure that the data backed up by different nodes are consistent, so data consistency mechanisms can be considered as the core of a distributed data management system. Given the background information model, the cloud server can not only obtain all the information in the known cryptograph model, but also analyze the query records, the relationship between different search tokens, mathematical statistics, etc. For example, according to the length of the search vector to get the total number of keywords and other information, so as to construct keywords for further attacks. It is guaranteed by the properties of the hash function that a lot of calculations must be made to get the required blocks, and the faster the calculations are, the earlier the results are likely to be. Nodes can only add blocks if they have "book-keeping rights," and all other nodes are updated according to the data of the node that has the book-keeping rights.

2 Experimental Test

2.1 Build an Experimental Environment

In order to avoid the impact of external network fluctuations on the experiment, all experiments in this chapter are carried out in a distributed system composed of multiple virtual hosts, each virtual machine configuration is the same. The cloud encrypted database system consists of three parts: user application client, encrypted database agent and cloud database server. Client and cryptographic database agents for user applications are deployed on workstations, cloud database systems are deployed on Alibaba cloud servers, and network bandwidth is 100 MB/s. Three physical hosts were used to virtualize 8 virtual hosts, and 4 virtual hosts were distributed on each physical machine. Among them, 4 virtual hosts are used to construct distributed file storage module, 2 virtual hosts are used to construct blockchain network, 1 virtual host is used to run Web server, and the other one is used as test machine. Using Openstack to set up the cloud environment, the encrypted database agent Qin-Router is deployed on the private cloud and is responsible for encrypting and decrypting the data. The MySQL database stores the user's information. The virtualization software used in the experiment was the Oracle VM Virtual Box.

2.2 Experimental Results

This experiment uses the experimental comparison way to verify the performance of the privacy data cloud storage method of the social media software. Choose the privacy data cloud storage method of social media software based on encryption algorithm and the privacy data cloud storage method of social media software based on neural network. Taking the case of opening block chain storage as an example, the storage experiment selected files of 64 MB, 128 MB, 256 MB and 512 MB in size as subjects, and the storage time consumption of the three methods were tested. The experimental results are shown in Tables 1, 2, 3 and 4.

Table 1. 64 MB file storage time (s)

Number of experiments	Social media privacy data cloud storage method based on encryption algorithm	Storage method of social media privacy data cloud based on neural network	This paper describes the privacy data cloud storage method for social media software
1	0.851	0.845	0.313
2	0.748	0.769	0.225
3	0.916	0.855	0.364
4	0.815	0.771	0.205
5	0.744	0.692	0.308
6	0.569	0.701	0.231
7	0.899	0.825	0.305
8	0.736	0.644	0.199
9	0.812	0.793	0.347
10	0.903	0.851	0.258
11	0.688	0.962	0.367
12	0.716	0.774	0.284
13	0.814	0.855	0.369
14	0.951	0.946	0.422
15	0.879	0.831	0.361

As can be seen from Table 1, the storage time of the privacy data cloud of the social media software is 0.304 s, 0.803 s and 0.808 s, respectively, compared with the other two methods.

Table 2. 128 MB file storage time(s)

Number of experiments	Social media privacy data cloud storage method based on encryption algorithm	Storage method of social media privacy data cloud based on neural network	This paper describes the privacy data cloud storage method for social media software
1	2.645	2.255	0.988
2	2.066	2.361	1.032
3	2.482	3.055	1.525
4	1.306	3.444	1.204
5	1.987	2.616	1.516
6	1.008	2.152	1.377
7	1.254	3.008	1.259
8	1.366	2.145	1.306
9	1.487	2.263	1.425
10	1.288	1.546	1.228
11	1.316	2.162	1.137
12	1.219	3.554	0.998
13	1.015	2.845	0.857
14	2.114	3.116	0.966
15	1.352	2.582	0.963

As can be seen from Table 2, the storage time of the privacy data cloud of the social media software is 1.185 s, 1.594 s and 2.607 s, respectively, compared with other two storage methods.

As can be seen from Table 3, the storage time of the privacy data cloud of the social media software is 1.908 s, 3.925 s and 4.024 s, respectively, compared with other two storage methods.

Table 3. 256 MB file storage time(s)

Number of experiments	Social media privacy data cloud storage method based on encryption algorithm	Storage method of social media privacy data cloud based on neural network	This paper describes the privacy data cloud storage method for social media software
1	4.615	3.131	2.615
2	3.485	4.206	1.948
3	4.163	3.054	2.331
4	3.848	4.163	2.664
5	3.919	3.456	1.564
6	3.055	4.912	2.006
7	4.647	3.714	1.948
8	3.612	5.002	2.174
9	4.001	4.693	1.566
10	3.665	3.855	2.133
11	4.163	4.162	1.847
12	3.787	3.787	2.021
13	4.699	4.336	1.225
14	3.105	3.855	1.021
15	4.112	4.027	1.554

Table 4. 512 MB file storage time(s)

Number of experiments	Social media privacy data cloud storage method based on encryption algorithm	Storage method of social media privacy data cloud based on neural network	This paper describes the privacy data cloud storage method for social media software
1	10.65	11.312	7.615
2	9.316	10.879	6.102
3	8.237	12.645	6.887
4	9.487	11.011	7.411
5	10.122	10.561	8.025
6	11.066	11.217	7.163
7	12.447	12.699	6.548
8	11.555	11.502	6.906

(*continued*)

Table 4. (*continued*)

Number of experiments	Social media privacy data cloud storage method based on encryption algorithm	Storage method of social media privacy data cloud based on neural network	This paper describes the privacy data cloud storage method for social media software
9	12.061	10.377	5.877
10	11.499	12.515	6.199
11	12.554	11.026	6.311
12	10.948	10.369	8.463
13	9.648	11.549	9.745
14	8.697	12.566	8.642
15	10.612	11.825	7.480

As can be seen from Table 4, the storage time of the privacy data cloud of the social media software is 7.292 s, 10.593 s and 11.470, respectively, compared with the other two methods.

The above experimental data show that the present method has a good cloud storage function of private data. We formulated the user identity restriction scheme according to the scope of account availability, and classified the risk factors of privacy information. We have improved the data reading mode, adjusted the data independence, used the blockchain to build the cloud storage structure, and improved the cloud storage performance of the data.

3 Conclusion

The design of social media software privacy data cloud storage method, improve the data in the block chain storage structure. This paper clarifies the connotation and classification of privacy information of individual users in cloud storage environment, and summarizes its risk categories as management risk, technical risk and legal risk. In addition, the data and interface of block chain transaction are adapted and encapsulated, and the method of reading and writing data on the chain is defined. On this basis, we have improved the way data is read and adjusted for data independence. Using the blockchain to build a cloud storage structure, an identity management mechanism for data access on the chain is designed to strengthen the protection of related data. In the cloud storage environment, the main ways of users' privacy information disclosure are overingestion by cloud storage service providers and theft of privacy information caused by hacker attack or management vulnerability. To strengthen the internal staff's professional ethics education, in the business level to balance commercial profits and privacy protection of the contradictory two levels have a role in promoting. At the same time, the technical level to establish a self-inspection mechanism to eliminate hidden dangers. We define the sharding storage method and optimize the access mode of file data, improve the efficiency of data synchronization of distributed file storage module, and reduce the time consumption.

References

1. Hai-chun, Z., Xuan-xia, Y., Xue-feng, Z.: Cloud storage data integrity audit based on an index-stub table. Chin. J. Eng. **42**(4), 490–499 (2020)
2. Shi, C., Lai, M., Li, S., et al. Integrity verification of dynamic multiple-replica data in cloud storage. J. Chengdu Univ (Natl. Sci.) **39**(1), 64–68 (2020)
3. Li, S.-Q., Liu, L., Zhu, D.-Y., et al.: Protocol of dynamic provable data integrity for cloud storage. Comput. Sci. **47**(2), 256–261 (2020)
4. Wang, L., Xu, Y., Kang, Y.: Simulation of node-level data privacy protection mining method in cloud computing. Comput. Simul. **37**(10), 433–436, 460 (2020)
5. Liu, G.-J., Xiong, J.-B., Zhang, L.-N., et al.: An efficient privacy-preserving data auditing scheme for regenerating-code-based cloud storage. J. Southwest Univ. (Natl. Sci.) **42**(10), 37–45 (2020)
6. Zhu, Y.-J., Yao, J.-G., Guan, H.-B.: Blockchain as a service: next generation of cloud services. J. Softw. **31**(1), 1–19 (2020)
7. Wang, Q.-C., Chen, Q.-Y., Zhang, C., et al.: Study of medical privacy data security protection in 5G cloud-side collaboration scenarios. Telecom Eng. Tech. Stand. **33**(12), 64–67 (2020)

Research on Abnormal Behavior Extraction Method of Mobile Surveillance Video Based on Big Data

Liyong Wan[✉] and Ruirong Jiang

Jiangxi University of Software Professional Technology, Nanchang 330041, China
wanliyong258@163.com

Abstract. In the application of mobile surveillance video anomaly behavior extraction method, there is a problem of high unlocking rate. Therefore, a big data-based anomaly extraction method is designed. Segmenting moving surveillance video dynamic image, representing human body action in the form of mathematical symbols, extracting target feature key frames, matching two adjacent frames in video sequence, using big data technology to detect behavior trajectory, defining and distinguishing abnormal behavior, and improving abnormal behavior extraction process. Experimental results show that the average unlocking rate of the proposed method and the other two methods are 2.920%, 5.564% and 5.890% respectively, which shows that the proposed method is more effective.

Keywords: Big data · Mobile surveillance video · Abnormal behavior · Image quality · Dynamic image · Key frame

1 Introduction

In recent years, intelligent video surveillance technology is still the focus of research and discussion in the field of computer vision. From the view of video quality, high resolution and high frame rate of video has become the demand of users. Video monitoring technology is continuously developing towards network, high definition and intelligent direction [1, 2]. However, with the increase of emergencies, people's awareness and attention to social security have increased, and various surveillance cameras can be seen everywhere. This is because the higher the quality of the video data, the more useful the analysis will yield. Intelligent video analysis technology has been widely used in traditional fields such as traffic management, security and case investigation, finance and commerce, and in emerging industries such as unmanned driving technology, robotics and VR. It can be said that the camera has been all over our city in every corner. Intelligent monitoring system is mainly used in crowded or high security places, such as museums, warehouses, banks, residential areas, supermarkets, subway, stations and so on. For humans, more than 70% of the information comes from the visual senses, and for this large web of perception, the camera will be the most significant source of information.

W. Fu and L. Yun (Eds.): ADHIP 2022, LNICST 469, pp. 390–402, 2023.
https://doi.org/10.1007/978-3-031-28867-8_29

Audio-visual perception network is gradually becoming a living reality, as the physical basis for audio-visual perception of the camera is widely deployed and rapidly growing. In addition, the intelligent monitoring system in vending machines, ATMs, intersections, nursing homes, smart homes and other aspects of a wide range of use. Surveillance is ubiquitous, extending from the public to the interior of a home. The research of human abnormal behavior detection technology can promote the development of intelligent technology, which is of great significance to promote social harmony [3]. In the field of medical care, to be able to monitor the patient's collision or fall and can immediately alarm, can make the injured in the first time to be treated, in a critical moment can even save the lives of patients. Abnormality detection and analysis in video, this research can be used in many fields, and has immeasurable application prospects, and its commercial value has unlimited potential. At present, it is mainly used in the following areas: monitoring of human behavior, smart home health monitoring, motion analysis, video surveillance systems. In the field of traffic supervision, collision between cars, speeding, drunk driving or pedestrians violating traffic regulations, they can be closely monitored and managed, so as to avoid the emergence of uncertainties that can threaten human safety. At present, there are two main research directions of human abnormal behavior recognition: one is the study of human posture and movement based on human model, such as standing, running, jumping, squatting, sitting, falling, climbing, etc. The second is the analysis of some simple abnormal behavior, such as beyond the boundary, remnants, lingering and so on. In the aspect of social security, the technology of human abnormal behavior detection can also be applied to banks, electric power, residential property management and other areas.

2 Research on Abnormal Behavior Extraction Method of Mobile Surveillance Video Based on Big Data

2.1 Segmentation of Mobile Surveillance Video Dynamic Image

Dynamic image is a sequence of a series of interrelated still images combined in time, so it contains much more information than a single still image. In addition, the dynamic image is complex, so it is necessary to select a reasonable dynamic image segmentation algorithm, which considers the details in the image sequence. There are some pseudo-moving points in the original image captured from camera or mobile surveillance video recording device because of the inevitable noise in the process of camera and transmission. Therefore, it is necessary for us to preprocess the input motion monitoring video image before extracting abnormal motion behavior to remove noise, improve image quality and make the image clear. Moreover, the background of dynamic image sequence will change in real time, so it is necessary to construct dynamic background. However, due to the moving of the target body (such as human body), there may exist the situation that the target color is similar to the background color in each frame forming the dynamic image, or the target edge in individual frames is rather fuzzy, so the situation of target body local missing may easily occur. Image preprocessing also provides a certain guarantee for the subsequent processing, such as correctness of abnormal behavior extraction. Because the human body abnormal behavior recognition studied in this paper is based on the mobile surveillance video, but generally, the mobile surveillance video is affected by the shooting environment and equipment itself, there are more or less noise. In this paper, the concept of wavelet transform is introduced, and the expression formula of kernel function of image processing is obtained as follows:

$$G_\beta = \int_{e=1}^{\beta} \frac{\beta}{\sum \|e - 1\|^2} de \qquad (1)$$

In formula (1), β represents the basis wavelet function and e represents the scale expansion coefficient. At present, there are many methods for extracting dynamic image saliency objects, which can be classified according to different standards. If there is manual participation in the process of object extraction, it can be divided into automatic and semi-automatic methods. Noise, which is an unpredictable random error and can only be analyzed by the method of probability and statistics, refers to the factors that hinder the senses from understanding the information they receive. Among them, the automatic way is the system runs the segmentation procedure to complete the target extraction automatically, does not need the manual intervention. The semiautomatic approach requires more or less manual participation. According to the use of dynamic images, it can be divided into segmentation for moving surveillance video compression coding and segmentation for content-based interactive applications. The latter generally does not require real-time and automatic extraction of moving surveillance video objects, but it requires higher accuracy in identifying the contours of moving surveillance video objects [4] Gaussian filtering is a linear filter based on space. Its weights are selected according to the shape of Gaussian kernel function. Gaussian smoothing filter is very effective to the noise which obeys the normal distribution. In image processing, the filter based on 2D zero-means discrete Gaussian function is usually used. The expression of 2D zero-means discrete Gaussian function is:

$$h(p, q) = t \times \frac{|p^2 - q^2|}{2\phi^2} \qquad (2)$$

In formula (2), p, q represents the pixel value of two pixels respectively, ϕ represents noise subject to normal distribution, and t represents dark point noise. It is necessary to preprocess these uncontrollable noises in the image before extracting the abnormal behavior.

Image filtering is a common method to remove noise in images. Human behavior is a combination of some basic actions and corresponding transitional actions in space and time. These basic actions mainly include walking, jumping, sitting and running, etc. From the technical point of view of human behavior recognition, any kind of human behavior can be decomposed into the above basic actions and their spatiotemporal relationships. The aim of moving object detection is to extract the changed region from the immutable background in the image sequence so as to simplify the object of study. Moving object detection based on mobile surveillance video is one of the research emphases and difficulties in intelligent surveillance system, machine vision and other fields. So, the aim of dynamic image segmentation is to separate these basic actions from the original dynamic image data, so that we can get smaller granularity data with enough motion information, and finally express the human body movements in the form of mathematical symbols. The quality of moving object extraction is crucial to the classification, behavior understanding, tracking and recognition of moving object. Whether the moving object can be accurately extracted from the video is the key to the success of target tracking, behavior understanding and recognition. Because the dynamic image contains more information than the static image, the dynamic image is rich in feature to recognize the range of moving object in the background when segmenting and extracting dynamic video object. Due to the interference of light, noise and background in the real scene, the accuracy of moving object detection is affected. In the current dynamic image segmentation, most of the standard algorithms are used to extract abnormal motion behavior. According to the different algorithms, the existing systems have different detection effects and computational complexity.

2.2 Extracting Target Feature Key Frames

Key frame refers to the frame image representing the main information of the image, which can be one or more images and can express the shot content concisely. At present, the selection of key frames follows the principle of "better wrong than less", and removes the redundancy or repetition when the representative features are not specific. This key frame extraction method is realized by analyzing the spatiotemporal correlation of the target. In this method, optical flow method and block matching method are used to match the two neighboring frames in video sequence, and the corresponding relation based on motion field estimation is established. Based on this principle, different selection algorithms can establish their own criteria according to different principles. At the same time, key frame extraction can reduce the amount of video operations. At present, key-frame extraction technology plays an important role in content-based video retrieval. Key frame extraction methods can be divided into the following four categories:

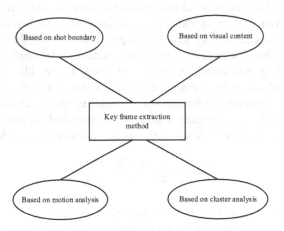

Fig. 1. Key frame extraction method

As can be seen from Fig. 1, there are four key frame extraction methods: key frame extraction based on shot boundary, key frame extraction based on visual content, key frame extraction based on motion analysis and key frame extraction based on clustering analysis. In this method, the video is segmented into several shots, and then the first frame, the middle frame and the last frame are extracted as the key frames in each shot. This method can be used in complex scenes with changing background elements, and has high application value, but it also has some problems such as insufficient tracking accuracy and inability to ensure the integrity of target features. When the optical flow method is used for matching, the motion field of the target can be estimated by moving the spatiotemporal gradient of the surveillance video image. This method is easy to design and computationally simple, and is suitable for mobile surveillance video with simple contents or fixed scenes. However, when the shot changes frequently and the transformation methods are various, the extracted key frames may not necessarily represent the main content information of the mobile surveillance video, which may lead to that the extracted key frames cannot accurately describe the contents of the mobile surveillance video. Therefore, this method is suitable for mobile surveillance video with simple scenes and not much content change [5]. But the optical flow method is very easy to be disturbed by various kinds of noises, and it is difficult to guarantee the precision when it is applied to the tracking of moving objects. At the same time, this method can not locate the contour of the target accurately, which will result in the loss of the important features of the target. Key frame extraction based on visual content. This method extracts key frames according to the changes of shape, texture and color of each frame. This method can flexibly determine the number of key frames to be selected according to the significance of the change of video content, but the disadvantage is that when the shot changes frequently, the selected key frames are too many and the information is redundant. Decomposing the moving object into several small blocks and replacing the pixel with these small blocks can simplify the algorithm and achieve better matching effect. This method is called block matching method. Based on motion analysis and key frame extraction technology, Wolf calculates the motion of the shot by optical flow analysis, and selects the key frame at the local minimum of the motion, which reflects the motion-surveillance video data at rest, in which the importance of the motion surveillance video is emphasized by the camera staying at a new position or by the short stay of a certain action of the character. The concrete method is to set up a priori motion model, and use the computational optimal matching method to match the neighboring two frames in the moving surveillance video, obtain the motion parameters of the small blocks, and finally realize the detection and tracking of the target in blocks. In this method, firstly, the optical flow is calculated by horn Schunck algorithm, and the modulus sum of each pixel optical flow component is taken as the motion amount $H(w)$ of frame w, that is:

$$H(w) = \sum_{m} \sum_{n} \frac{|f_w(m, n)|}{2} \tag{3}$$

In formula (3), m represents the m component of the pixel flow in frame w and n represents the n component of the pixel flow in frame w. Based on the key frame extraction technology of clustering analysis, this method fully considers the correlation between shots and intra-shot, and cluster the moving surveillance video frames according to the similarity between frames. This method is also called image segmentation based tracking algorithm. The principle is to classify all pixels by judging whether they belong to moving foreground or not, so as to separate moving objects and background. In most cases, the extraction of key frames based on clustering analysis can accurately describe the main content of mobile surveillance video. Similar to image segmentation, this method extracts pixels with special features according to the color, displacement and texture of pixels, and the classification method is the same as image segmentation.

2.3 Big Data Detection Behavior Trajectory

In the era of big data, there are more and more complex network data. Due to the limitation of traditional platform's processing capacity and storage space, the efficiency of machine learning algorithm is obviously reduced. Therefore, it is necessary to provide a method for anomaly detection which can deal with a large amount of secure data. Abnormal behavior is a broad definition, according to the different scene, the definition of Abnormal behavior and the definition of standards are not the same, in some occasions, the definition of Abnormal behavior is more severe, for example, in an important place of the gate or access, in addition to the normal walk through, other behavior can be considered abnormal behavior. Human behavior is a subject studied in many disciplines. According to physiologists, behavior is the response of human organs to external stimuli. There are many human actions, such as walking, running and jumping, squatting, standing, fainting and so on. The language of walking is that a person keeps his body upright at a slower pace than running. Walking is done with both feet, and only one foot leaves the ground during the walk. In public places such as shopping malls and squares, the definition of abnormal behavior is relatively loose. Except for theft, robbery and other crimes and violence such as sabotage and brawl, other human activities can be defined as normal behavior. In contrast, when running, are generally both feet off the ground. Then how to use computer readable abstraction, digital language expression behavior will be the focus of this chapter, difficult. Therefore, in the aspect of feature selection, this paper mainly focuses on object motion features. Because of the ambiguity of the definition, the definition of abnormal behavior is aimed at specific occasions, and it is very difficult to use a system to meet the needs of various applications. The basic task of feature selection and extraction is to select the most representative information from the multitudinous data and then describe the object. Therefore, the selection and extraction of features is a key step in anomaly detection. In the common scenes, the probability of abnormal behavior is relatively low, and most of the behavior in the moving surveillance video images belongs to normal behavior. In order to identify abnormal behavior effectively, it is necessary to define and classify normal behavior in addition to abnormal behavior, in order to define and distinguish abnormal behavior. The basic information for detecting targets is as follows:

As can be seen from Fig. 2, the basic information of detecting targets includes the number of targets, the area of target domain, and the aspect ratio of targets. In order to describe the behavior accurately and effectively, there are several criteria for judging the characteristics of abnormal behavior: distinguishability, and different types of images have different processing purposes, so the selected characteristics must have obvious differences. For example, the central feature that distinguishes motion from rest is velocity, which is zero and not so regarded as motion. Because of the complexity and variety of human behavior, how to effectively define and classify various behaviors has been a

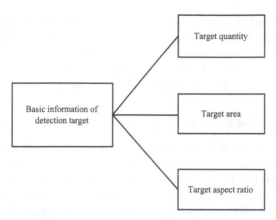

Fig. 2. Basic information of detection target

serious problem in related fields. Generally speaking, compared with the normal human activities, abnormal behavior has some remarkable characteristics. First, the probability of abnormal behavior is relatively low. Secondly, the occurrence of abnormal behavior is unpredictable. Finally, the abnormal behavior usually includes some special actions that have different normal activities. Therefore, speed is the most powerful feature to distinguish whether a target is moving or not. Independence, no matter what characteristics they choose, they must have a strong independence, so as to prevent unnecessary waste generated. Fewer features, because the number of existing technology is not unlimited, the fewer features to select, then the easier the classifier design, then easier to implement. The composition of human behavior is very complex, and has great unpredictability, even if the same type of behavior, often by the combination of different actions. So the method of training by collecting samples has great limitations in practical application. Firstly, the background subtraction method based on the improved hybrid Gaussian background model is used to get the foreground, and then the people in foreground image are tracked by Kalman filter to get the displacement information. If a large number of features are selected, the complexity and complexity of classification system will be greatly reduced. The number of targets is determined by the distance between the connected area, area and centroid coordinates. Finally, through the analysis of the trajectory, the behavior of the target is identified, so as to achieve the normal activities and suspicious activities in the mobile surveillance video. Based on the process of trajectory recognition, it is mainly used to judge whether the trajectory curve of the target center point is closed or spiral. If the foreground is simply connected and the area is larger than the corresponding Min value, then it can be identified as a single target. If there are multiple connected areas in the foreground, it is necessary to check the distance between the coordinates and the centroid coordinates of the corresponding connected areas to see whether they stay in the range of defined thresholds, and then determine the number of targets according to whether they are single targets or multiple targets. If the trajectory is not closed and not in a spiral shape, the target is moving normally. Otherwise, the target is in a wandering state, and it is necessary to analyze the characteristics of regional optical flow to further judge whether abnormal behavior occurs.

2.4 Improve Abnormal Behavior Extraction Process

Feature extraction is the use of computer to extract image information, the result of Feature extraction is to divide the image points into different sets, specifically to abstract and digitize the image. Real objects have many characteristics, such as shape, space position, mass, temperature, color, sound, surface texture, motion (track, speed, acceleration), etc. Anomaly detection methods based on mobile surveillance video data include a variety of definitions of anomaly behaviors of targets, and different scenes have different definitions of anomaly behaviors for different targets, such as falling, intruding into fixed areas, and losing objects, etc.. For different objects and different requirements, the choice and description of features is particularly important. The study of the daily behavior of the human body, we can find that the daily walking or running and other movements, generally with periodicity. Therefore, the periodicity of motion can be regarded as a behavioral characteristic to define normal human behavior. In this study, we investigate the trajectory based pedestrian abnormalities such as pacing, prowling, and sudden acceleration in an open parking lot. The definition of normal behavior, according to the actual, we study the population is mainly living alone elderly, patients, so this article will only walk this behavior defined as abnormal behavior, other behaviors are regarded as abnormal behavior. Based on KNN classifier, this paper classifies and identifies several indoor abnormal human behaviors, mainly for special groups of people living alone, including the elderly, patients and so on [6, 7]. In this paper, we do not classify the abnormal behavior of the specific target, but compare the normal trajectory with the target's trajectory, so that the trajectory which is much different from the normal trajectory can be judged as abnormal. By learning the normal trajectory, we can find the abnormal trajectory, that is, the abnormal behavior of pedestrians. The intelligent monitoring system can detect the abnormal behavior of the human body when they fall down or get sick, so as to alarm and remind the relevant personnel to deal with it in time. The main definition of normal behavior in the study was walking. Abnormal behavior included falls, squatting, chest pain, and headaches. Abnormal behavior must correspond to abnormal trajectory, abnormal trajectory accumulates to a certain extent, pedestrian's behavior is abnormal, abnormal trajectory and normal motion trajectory must be different, if the scene area is divided, the normal trajectory in the area is usually similar. The abnormity is judged by the rule of the center of mass changing with time, the ratio of height to width of the rectangle, and the minimum inclination of the rectangle. By judging these three kinds of human body characteristics, we can judge whether the human body behavior is normal or not. Through the analysis of a large number of behavioral motion surveillance video sequences, we conclude that the position of the human centroid changes in a specific range under normal circumstances, and the change curve presents periodicity. For example, the normal trajectory of the road in the picture is usually close to a straight line, so we only need to analyze the trajectory features detected in the scene where the target is located to find out the different trajectories to detect the abnormal behavior. In mobile surveillance video, pedestrian trajectory points are different from those in normal walking. We analyze the trajectory points, extract their characteristic vectors and substitute them into SOM. When walking, the person's tilt angle is perpendicular to the ground and does not exceed 10°. Aiming at the behaviors of walking, falling down, squatting, chest pain (bending down) and headache (leaning back), the video sequences are walking sequence, moving monitoring video sequence of sudden chest pain during walking, squatting sequence during walking, backwards and backward sequence of sudden headache during walking, and suddenly falling forward during walking. The process for extracting abnormal human behavior is shown in Fig. 3:

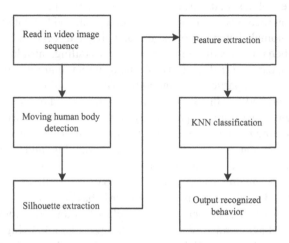

Fig. 3. Extraction process of human abnormal behavior

According to Fig. 3, the process of abnormal human behavior extraction mainly includes: reading video image sequence, moving human detection, silhouette extraction, feature extraction, KNN classification and output recognition. Then, the Euclidean distance between the winning neuron and the trajectory feature point can be calculated and the threshold can be predetermined. When the calculated distance is greater than the set threshold, the track point is regarded as an anomaly, that is, the behavior of pedestrians is abnormal. At the end of the training, the sample vector of the maximum distance from each winning neuron and the corresponding Euclidean distance are obtained and the distance is set as the threshold. For each track point extracted from the point, the above distance needs to be calculated, if the given threshold is exceeded, the track point is considered abnormal. Based on the improved Vibe algorithm, the method of multi-feature fusion is adopted in this chapter. The shape feature, geometry feature and motion feature of the moving human body are selected as input of recognition feature, and KNN is used as classifier to train and recognize the abnormal behavior. In a trajectory, if there are 5 or more trajectory points, the trajectory is regarded as an abnormal trajectory, and the corresponding behavior of the trajectory is an abnormal behavior.

3   Test of Application Effect of Research Method

Because the existing human behavior database does not include some of the behaviors studied in this paper, we need to collect our own database. This experiment uses a fixed camera, set the environment is indoor, camera shooting angle is perpendicular to the human movement route, that is, the main body side information extracted for experiments. In order to explain the problem better, in this experiment, the normal track and abnormal track of the test data were 50 and 30 respectively. The database, based on the existing Weizmann database, was completed by a total of 12 people, including 6 men and 6 women, of representative height and size. Each action was shot four times, in two different directions into the camera, a total of 180 video clips. Randomly extracting 80 video sequences from video as training samples and 100 as testing samples are used to verify the effectiveness of the proposed method. A total of 6,895 frames of 25 videos of running, fainting, meeting, fighting and other behaviors were selected for the test. Among them, there are 963 frames for scene without target, 1661 frames for single normal behavior, 801 frames for single abnormal behavior, 2573 frames for double normal behavior, and 897 frames for double abnormal behavior. In order to reduce the computational complexity, we use the above method to extract key frames from each video, and ensure that the selected frame image can represent a complete human motion cycle. The number of neurons in the output layer within the SOM is too small to describe the trajectory points. If the number of neurons is too large, the training time will be too long, and the anomaly detection will take too long. It is also necessary to analyze the aspect ratio of video images. The hardware environment used in this experiment is a Hongxian desktop computer, the operating system is Win7 Ultimate Edition 32-bit, memory 4 GB, processor is Core i7, the programming environment is Microsoft Visual Stdio 2020 Opencv and Mlabat2020.

3.1   Test Result

Select the mobile surveillance video abnormal behavior extraction method based on deep learning and the mobile surveillance video abnormal behavior extraction method based on wavelet transform to compare with the mobile surveillance video abnormal behavior extraction method in this paper, and test the lock loss rate of the three methods under different human action conditions. The experimental results are shown in Tables 1, 2, 3 and 4:

3.2   Experimental Analysis

(1)   Table 1 Analysis of experimental results

It can be seen from Table 1 that the average loss rate of down motion locking of the mobile surveillance video abnormal behavior extraction method proposed in this paper is 1.092%, while the average loss rate of down motion locking based on the deep learning method and the wavelet transform method are 2.966% and 3.125% respectively. It can be concluded that the method in this paper has a low loss rate.

Table 1. Locking loss rate of falling action /%

Number of experiments	Mobile surveillance video abnormal behavior extraction method based on deep learning	Method for extracting abnormal behavior of mobile surveillance video based on Wavelet Transform	The abnormal behavior extraction method of mobile surveillance video in this paper
100	3.215	3.664	1.055
200	2.487	2.878	1.166
300	3.619	3.066	1.012
400	2.847	2.877	1.334
500	3.010	3.619	1.128
600	2.554	2.548	1.025
700	3.008	3.647	0.997
800	2.946	3.004	0.946
900	3.157	2.846	1.254
1000	2.819	3.105	1.006

Table 2. Locking loss rate of bending action/%

Number of experiments	Mobile surveillance video abnormal behavior extraction method based on deep learning	Method for extracting abnormal behavior of mobile surveillance video based on Wavelet Transform	The abnormal behavior extraction method of mobile surveillance video in this paper
100	5.649	5.649	2.649
200	4.497	6.4863	3.022
300	6.648	5.216	2.914
400	6.747	6.301	2.649
500	5.418	5.869	3.665
600	6.251	5.612	2.782
700	6.317	6.008	3.108
800	5.208	5.949	2.878
900	6.997	6.258	3.146
1000	5.848	6.313	2.051

(2) Table 2 Analysis of experimental results

It can be seen from Table 2 that the average loss rate of the mobile surveillance video abnormal behavior extraction method proposed in this paper is 2.886%, while

Table 3. Locking loss rate of recline action/%

Number of experiments	Mobile surveillance video abnormal behavior extraction method based on deep learning	Method for extracting abnormal behavior of mobile surveillance video based on Wavelet Transform	The abnormal behavior extraction method of mobile surveillance video in this paper
100	4.152	6.132	1.485
200	3.216	5.848	3.151
300	5.155	6.119	2.548
400	4.124	5.203	2.007
500	3.997	5.715	2.136
600	4.687	4.331	2.548
700	5.231	5.022	2.310
800	4.978	6.949	1.978
900	5.162	5.288	2.644
1000	4.886	4.172	2.005

Table 4. Locking loss rate of squat movement/%

Number of experiments	Mobile surveillance video abnormal behavior extraction method based on deep learning	Method for extracting abnormal behavior of mobile surveillance video based on Wavelet Transform	The abnormal behavior extraction method of mobile surveillance video in this paper
100	9.468	9.488	5.489
200	8.647	8.647	5.062
300	8.776	8.449	4.987
400	7.794	9.672	5.036
500	9.023	8.553	5.112
600	8.466	9.878	5.033
700	9.577	9.021	6.416
800	8.346	8.789	5.223
900	9.525	9.125	6.720
1000	8.114	8.664	5.124

the average loss rate of the down motion locking based on the deep learning method and the wavelet transform method are 5.958% and 5.966% respectively. The loss rate of this method is obviously lower than the other two methods, so it can be concluded that this method has a lower loss rate.

(3) Table 3 Analysis of experimental results

It can be seen from Table 3 that the average loss rate of the mobile surveillance video abnormal behavior extraction method proposed in this paper is 2.281%, and the average loss rate of the down motion locking based on the deep learning method and the wavelet transform method is 4.559% and 5.478% respectively. It can be seen that the average loss rate of this method is lower than the other two methods, which proves that this method has certain advantages.

(4) Table 4 Analysis of experimental results

It can be seen from Table 4 that the average loss rate of the mobile surveillance video abnormal behavior extraction method proposed in this paper is 5.420%, the average loss rate based on the deep learning method is 8.774%, and the average loss rate based on the wavelet transform method is 9.029%, which is significantly lower than the other two methods. It can be concluded that the method in this paper has a low loss rate.

4   Conclusion

Based on the analysis and comparison of common foreground detection technologies, the shortcomings of traditional methods are studied. Aiming at the characteristics and requirements of mobile surveillance video anomaly recognition, a mobile surveillance video anomaly detection method based on big data is proposed. By optimizing target prediction, using big data technology to detect behavior tracks, defining and identifying abnormal behaviors, improving the extraction process, and reducing the loss rate of human action locking, it has certain practical application value.

References

1. Zheng, M., Qian, H., Zhou, X.: Human abnormal action detection based on the Farneback optical flow arithmetic of surveillance video. Foreign Electron. Measur. Tech. **40**(3), 16–22 (2021)
2. Du, Q., Huang, L., Tian, L., et al. Recognition of passengers' abnormal behavior on escalator based on video monitoring. J. South China Univ. Technol. (Natl. Sci. Ed.) **48**(8), 10–21 (2020)
3. Zeng, T., Huang, D.: A survey of detection algorithms for abnormal behaviors in intelligent video surveillance system. Comput. Measur. Control, **29**(7), 1–6, 20 (2021)
4. Zhang, H.: Research on anomaly recognition method of video surveillance in smart community based on deep learning. J. Xi'an Polytec. Univ. **34**(2), 103–109 (2020)
5. Liu, J., Chen, P.: Abnormal behavior recognition based on key frame location and spatial-temporal graph convolution. Mach. Electron. **40**(1), 48–53, 58 (2022)
6. Wang, L.: Community abnormal behavior intelligent identification anti-theft alarm system simulation. Comput. Simul. **37**(4), 393–396 (2020)
7. Hu, Z., Zhang, L., Li, S., et al.: End-to-end SSD real-time video surveillance anomaly detection and location. J. Yanshan Univ. **44**(5), 493–501 (2020)

Network Information Security Risk Assessment Method Based on Machine Learning Algorithm

Ruirong Jiang[✉] and Liyong Wan

Jiangxi University of Software Professional Technology, Nanchang 330041, China
jiangruirong632@163.com

Abstract. The current computer network information security risk assessment methods have the problems of low assessment accuracy, which seriously restricts the assessment effect. In order to solve this problem and improve the effect of network information security risk assessment and the level of network information security, this paper designs a network information security risk assessment method based on network learning algorithm. Describe the risk calculation form, extract the performance characteristics of network information, identify the network risk factors, draw conclusions according to logical reasoning, adopt computer network risk control and defense measures, use machine learning algorithm to build a security system model, and optimize the security risk assessment mode. The experimental results prove that the highest accuracy rate of the network information security risk assessment method is 95.612%, indicating that the network information security risk assessment method is more practical after combining the machine learning algorithm.

Keywords: Machine learning · Network information · Security risk · Risk assessment · Security defense · Security risks

1 Introduction

With the rapid development and popularization of computer technology, computer network has increasingly become a bridge for people to learn and communicate, and gradually become an indispensable tool in people's daily life. Information resources can be shared through the network, which not only expands the space and time, but also improves the efficiency of work or learning. But with the rapid expansion of the network, the number of cyber security accidents and the losses are also spreading. However, in the field of computer and network security, it is difficult to symmetry the security risk assessment research and results of network system with its importance. Network information security has gradually become the focus of people's attention, and the core of network security is the risk assessment of the network system [1–3]. Network risk assessment, also known as security risk assessment and network security risk assessment, refers to the whole process of detecting, identifying, controlling and eliminating the known or potential security risks and security risks in the network, and is one of the necessary measures for enterprise network security management.

W. Fu and L. Yun (Eds.): ADHIP 2022, LNICST 469, pp. 403–416, 2023.
https://doi.org/10.1007/978-3-031-28867-8_30

Risk assessment is from the perspective of risk management, using scientific methods and means to systematically analyze the threats to the network and information system and their vulnerabilities, and evaluate the degree of harm caused by security events. Through the network security assessment, we can comprehensively sort out the assets in the network, understand the existing security risks and hidden dangers, and carry out targeted security reinforcement to ensure the safe operation of the network. Network security refers to the information security on the network, which means that the hardware and software of the network system and its data in the system are protected, the system operates continuously, reliably and normally, there is no damage, change or leakage due to accident or malicious reasons, and the network service is not interrupted. Therefore, in the face of the rapid growth of computer network information security needs, only passive defense technology can not meet the network security defense, and can not fundamentally solve the network security defense problem. It is necessary to comprehensively prevent and control the security of information systems.

Network security is essentially the information security on the network. In a broad sense, all the related technologies and theories involving the confidentiality, integrity, availability, authenticity and controllability of the information on the network are the research fields of network security. Therefore, the computer network information security risk assessment can effectively analyze the current and future risk development trend and location of the network information system, evaluate the risks to the computer network information security threat and its impact degree, in order to better develop security defense strategy, provide safe operation guarantee for the computer network information. Network security involves content, technical and management issues, which are interrelated and indispensable. The technical focus is on how to prevent external illegal attacks, while the management side focuses on the management of internal human factors. How to protect the important information data more effectively and improve the security of the network system has become an important problem that must be considered and solved in the computer network applications. Put forward targeted protection countermeasures and rectification measures to control the risk at an acceptable level, so as to maximize the security of the network information system. Network security risk assessment is the basis and premise of ensuring network security, and it is of important research significance. However, the current computer network information security risk assessment methods have the problem of low accuracy, which seriously restricts the assessment effect. In order to solve this problem and improve the effect of network information security risk assessment and the level of network information security, this paper designs a network information security risk assessment method based on network learning algorithm.

2 Network Information Security Risk Assessment Method Based on Machine Learning Algorithm

2.1 Extract the Network Information Performance Features

Network security in essence, network security is the information security on the network, refers to the network system hardware, software and its data in the network system is protected, not because of accidental or malicious factors were destroyed, change, leakage, so as to make the system operate continuously, reliably and normally, the network service is not interrupted. To achieve this purpose, it is of vital significance to effectively evaluate the risks of information security. Only by effectively evaluating the security risks faced by the information system, can we fully grasp the security state of the information system, and take targeted risk control measures, so that the information security risks are within a controllable range.

Network security risk assessment refers to the process of scientific evaluation of the processing, transmission and storage of network information, as well as the confidentiality, integrity and availability of information, according to the relevant national network information security technical standards. In a security defense system, a perfect and reasonable security architecture is the core and foundation. The security architecture is usually based on the corresponding security conceptual model. As the primary link of the risk assessment project, the security model can be regarded as the first stage of the whole security framework, which is of great significance to the establishment of the whole security framework. To evaluate the vulnerability of the network information system, the threat faced, and the actual negative impact generated after the vulnerability is utilized by the threat source, and to identify the security risk of the network information system according to the possibility of security events and the degree of negative impact. Based on the assessment results, propose effective safety measures to eliminate or minimize the risk [4, 5]. Threat is the potential cause of accidents that may cause damage to assets or organizations, and risk assessment is concerned about the possibility of threat. Fulnerability is also known as vulnerability, which is the shortcomings of assets or asset groups that can be threatened. Frailty itself does not constitute harm, but once vulnerability is threatened, it may cause damage to the assets. It is not appropriate to temporarily put forward some information security risks in the information security risk assessment. We must determine the four elements of assets, threat, vulnerability and risk and their mutual relationship before the risk assessment under the premise of information risk management. The formal risk calculation principle is described as:

$$T = \frac{\delta(E, G, H)}{\sum_{i=1} |\delta_i - 1|^2} \tag{1}$$

In formula (1), δ represents the impact of an asset security event on the business of the institution, E represents the asset, G represents vulnerability, H represents threat, and i represents the vulnerability of an asset itself. Risk is the potential possibility of damage to the vulnerability of a specific threat. Risk is the result of the combination of the likelihood and impact of threat events. Assets are anything of value to an organization, including computer hardware, communications facilities, databases, document information, software, information services, and personnel, etc. It spreads throughout the project work as the task is executed, and is consumed with the effective control measures taken by each sub-task. Through this model, some risks that are easily overlooked by traditional methods can be found. Therefore, determining the information security risk assessment model is the basis of risk assessment and the basis for effective risk assessment. It can provide methods and benchmarks for determining the risk size of the system. Asset value refers to the importance and sensitivity of assets. Asset value is the attribute of assets and also the specific content of asset appraisal. Security requirements refer to the requirements put forward in the information security measures to ensure the normal implementation of the business strategy of the unit. Security measures refer to various events, regulations and mechanisms implemented to deal with threats, reduce vulnerabilities, protect assets, limit the impact of accidents, discover and respond to accidents, promote disaster recovery and combat information crimes.

2.2 Identify Network Risk Factors

After statistical analysis, the current computer network security may encounter threats mainly from the following aspects: virus, Trojan, attack, vulnerability, encryption, eavesdropping. The complexity of information system determines the diversity, dynamics and uncertainty of security risk factors. However, this does not mean that the factors causing the safety events are completely unknown, and the rules can be found through the observation, statistics and analysis of a large number of risk event data. Usually, eavesdropping on the network, do not need to interrupt the network to transmit information, is called negative criminals. Malicious attackers often use this as a basis to recycle the remaining tools for more destructive attacks. Because network security risks involve a wide range of areas, complex nature and risk characteristics throughout the whole life cycle of the information system determine the great difficulty and complexity of their identification, and many interrelated and penetrating factors cannot be identified by observation and touch [6, 7]. Tampering is a legitimate user on the offense modifying, dele, deleting, inserts, and then sends forged information to the recipient, which is purely damaging between information communication, such cyber criminals are known as active criminals. The information package on the positive criminal Web is intercepted and modified, so that the information that favors oneself or is added deliberately is displayed. Risk analysis and assessors in unclear, incomplete, inaccurate, cannot completely according to logical reasoning and concluded to take some qualitative, fuzzy, effective risk factor identification method to as far as possible to collect, sorting, speculation data, and analysis of the actual situation, simulation, simulation and judgment, so as to find the potential, existing in the incomplete information. The basic structure of the safety risk assessment elements is shown in Fig. 1:

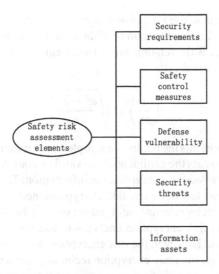

Fig. 1. Safety risk assessment elements

As shown in Fig. 1, security risk assessment elements include security requirements, security control measures, security threats, defense vulnerabilities, and information assets. A denial of service attack is an attack system that is slow or even paralyzed by some other means that can prevent access by legitimate users to the service. The conduct denial is that the communications entity denies that the act has occurred. Electronic deception is the purpose of killing the user's identity through counterfeiting legitimate cyber attacks, thus concealing the identity of the real attacker and blaming others. Then the obtained information is analyzed and sorted out, so as to determine the risk cause and the contribution of each factor to the system risk. According to the different starting points of risk factor identification, the identification methods are generally divided into two categories: one is to derive information and information system risks based on the security events that have occurred, and the other is to conduct direct analysis. Direct analysis can be divided into two categories: the method using scanning analysis tools and the method using factors. Unauthorized access is where no pre-agreed use of network or computer resources is seen as unauthorized access. It mainly has the following forms: fake identity attack, illegal users into the network system, illegal operations, legal users without authorization. Transmission of virus computer virus, through the network transmission is very destructive, and the user is difficult to prevent, seriously can paralyze the whole network.

2.3 Machine Learning Algorithm Build a Security System Model

The purpose of machine learning is to solve the dependencies between the input and output of the system based on a given training sample, enabling it to make an as accurate prediction as possible for the unknown output. Therefore, in the process of research, it is necessary to obtain the effective information contained in it from the actual data containing a large amount of useless information. Therefore, the measure of traditional

European space is difficult to use for real-world nonlinear data, which requires the use of a new treatment method for the distribution of the data. In machine learning algorithms, the generalization bound is the relationship between empirical risk and actual risk, and the specific expression formula is:

$$\eta\left(\frac{1}{\varpi}\right) = \sqrt{\frac{\varpi - \lambda^2}{\eta}} \tag{2}$$

In formula (2), η represents the number of samples, ϖ represents the dimension of the function set, and λ represents the confidence interval. The confidentiality of the network information can be guaranteed by encrypting the information. There are three main kinds of information encryption technologies: link encryption, node-to-node encryption, and end-to-end encryption. Protect the link information security between network nodes can be used link encryption, link encryption encryption data message, routing, check and other control information. However, the link encryption does not encrypt the data in the network nodes. Therefore, the node encryption technology arises. The node encryption uses the protection device installed for encryption and decryption in the middle node to realize the encryption between the nodes. In the traditional calculation method, the relationship between data and data is defined in the European space, but in practice, these data points may not be distributed in the European space, rapid development in recent years, with the development of the Internet technology, we live in the world produces a lot of data, we also filled with a lot of information, information leads to data explosion but effective knowledge is poor. Network security threats are very different. Based on the basic idea of identifying unknown factors without difference, we should not only pay attention to the traditional various types of network threats, but also pay attention to some constantly emerging and dynamically changing various types of network threats. Network security threats can come from internal and external aspects. To sum up, they can be classified into four types, as shown in Fig. 2:

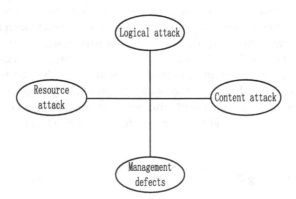

Fig. 2. Type of network security threats

As shown from Fig. 2, network security threats can be classified into logical attacks, resource attacks, content attacks, and management defects. Among them, many specific attack threats simultaneously have a variety of attack attributes, belonging to the compound security threats. Other disasters, such as natural disasters and human error, are a special type of security threat. Although most natural disasters are force majeure, they can be prevented, avoided and reduced in many cases, so they should also be included in the category of artificial management defects.

In the process of security risk assessment of computer network information system, the components of computer network information system can be evaluated, and these components are closely related, so it is very important security in the process of risk assessment. Therefore, in the process of risk assessment, various attributes such as asset value, security events, security needs, business strategy and participation risk should be fully considered. The attack object of logical attack can be various aspects, it is to find and use the logic defects and loopholes in the existing system or application, through technical means to obtain system control rights, obtain illegal access rights, affect system performance or system functions, cause system crash, etc. Among such attacks, the public is most familiar with the various attacks against operating system vulnerabilities, such as stack overflow in Windows PnP services, ping of death attacks, red code attacks, shock wave attacks, panda incense virus attacks, which are all developed through the vulnerabilities of the system. Risk assessment can include the following contents: the relevant risk assessment agencies to develop detailed business assessment development needs strategies, and to effectively evaluate their assets. Attacks against the defects of network protocols such as TCP/IP also belong to the type of logical attacks, such as ARP deception using half-open connection to consume server resources, DNS hijacking, etc. These attacks are technically "legal", but they are illegal. Computer network information system assets use unit has relevant value, institutions to carry out specific business assets have a high degree of dependence, if the assets have very large value, so these assets will face very big risk: in the process of computer network information system execution, risk mainly includes two kinds, respectively is human risk or natural risk, assets face security events are very big threat, lead to computer network information system assets face greater risk. Common injection attacks against the databases of WEB servers are also logical attacks by exploiting vulnerabilities in the WEB service design. Other such as Trojan, spyware, password scanning attacks, also mostly to use the system or application defects to attack. Computer network threats need to be caused or generated by using system vulnerabilities or defense technology vulnerabilities, while assets have high vulnerability, and the stronger the risks faced by computer network information systems.

For the computer network information system, risk is inherent and unavoidable. People's cognition of risk is the need for security and risk assessment. The main attack object of resource attack is mainly the target system resources or network resources, such as a lot of server CPU and memory, a lot of bandwidth or connections, a lot of storage space, etc. The most typical of such attacks is the denial of service attacks, such as various Dos. Ddos (Distributed denial of service). The security requirements of assets can be adopted to ensure that the computer network information system has high security, and can ensure the inherent value of computer network assets, using computer network risk control and defense measures, in order to avoid being detected by the computer network information system. The current zombie or zombie network attacks have been developed on this basis, which can have a variety of effects at the same time. The malicious resource occupation of viruses is also an important manifestation of resource attacks, such as spam and excessive network advertising not only waste network bandwidth and storage space, increase system memory footprint, but also waste a lot of users' working time, and may include virus attacks. Content attack is the deletion, modification, theft, deception, drowning, mining of the target, etc. Information mining for the target system is a very hidden type of attack. Computer network information system implement security protection measures, can effectively reduce the threat of assets, reduce the vulnerability of security events: implementation of computer network security measures, can reduce the threat of assets, reduce threat vulnerability, serious security risk events, so as to reduce the network information system damage as far as possible. The risk of computer network information cannot be reduced to zero. In fact, the risk cannot be reduced to zero. Therefore, it is necessary to take strict risk prevention measures to ensure that the residual risks of computer network information systems are identified and strictly control the residual risks.

2.4   Optimize the Safety Risk Assessment Mode

Information security risk assessment work is an extremely complex and extremely challenging work, heavy workload, need a lot of professional knowledge to support. Therefore, a very practical assessment tool is required for completing the risk assessment work. Traditional content attacks such as network listening, network message sniffing, etc., in recent years, some more popular content attacks, such as IP address deception, it does not change the original correct IP corresponding content, but to cheat, through technical means to get the wrong feedback results. It can not only free the technical personnel from the heavy asset statistics and risk assessment work, but also complete some work that manpower can not complete.

Risk assessment tools are an important factor to ensure the credibility of the risk assessment results. Phishing takes advantage of their psychological weaknesses by using similar domains, IP turns and other decoys to get the wrong page and the trust of visitors. According to the different focus of the risk assessment tools in the assessment activities, the risk assessment tools are divided into three categories: management risk assessment tools, technical risk assessment tools and risk assessment AIDS. Management risk assessment tool is a comprehensive assessment management tool, which focuses on security management, comprehensively considers the security risks faced by information, and finally gives corresponding control measures and solutions. Some attacks at the application layer also belong to the type of content attacks, such as malicious (rogue) software will steal user privacy, collect user habits, forcibly push non-requested content, and will consume a lot of users' system resources and working time. Such assessment tools are usually based on some kind of model or expert system. These tools are mainly divided into three categories: risk assessment tools based on information security management standards or guidelines issued by the state or government, risk assessment tools based on expert systems, and risk analysis tools based on qualitative or quantitative algorithms. The tool focuses on collecting the data and data required for evaluation and establishing the corresponding information base and knowledge base. It is a management information system that integrates various risk assessment knowledge bases and guidelines. Spam also plays an important role in content attacks, not only forcing non-requested information, wasting bandwidth, user time and storage space, but also including attacks such as cheating and viruses.

The purpose of machine learning is to minimize the expected risk function, but the available information is the sample, so it is difficult to calculate the defined function. In order to overcome this difficulty, in practical application, empirical risk is often used instead of the expected risk function for risk calculation, in which empirical risk is defined as:

$$Y_\mu = \frac{1}{\mu} \sum \left\| \mu^2 - \lambda \right\| \tag{3}$$

In formula (3), μ represents the weight vector and λ represents the training sample set. With other attack types of attack source from external, management defect is due to their intentional or unintentional management defects, errors and make information system under the security threat, such as mismanaged system of internal information theft, security measures deployment is not in place or deployment error lead to other attacks, etc., these management defects objectively lead to information system attack. Commonly used risk assessment AIDS include checklists, personnel interviews, asset information questionnaire, intrusion detection tools, security audit tools, and so on.

The above three kinds of tools have different priorities. In the complex process of risk assessment, these three kinds of tools should be used comprehensively to better improve the efficiency of information security risk assessment and the correctness of the assessment results. The classification of attack types is mainly to investigate the convenience of analysis. In the actual security research, various security threats are complex, so we must identify these problems with no different unknown factors, and solve the problems with elastic closed structure ideas. Network assets include computer hardware, communication facilities, databases, document information, software, information services and personnel. Asset transfer is the evaluation of the safety value of assets. The grade appraisal of the final asset value is based on the assignment level of the confidentiality, integrity and availability of assets. After comprehensive evaluation, it can be divided into five different grades from 1 to 5. It mainly includes five stages: risk assessment preparation, risk factor identification, risk determination, risk assessment and risk control. The preparation of risk assessment, judging whether the risk is acceptable, maintaining existing control measures, and implementing risk management need to be experienced by the assessors, and the rest can be done with auxiliary tools. The higher the level, the more important the asset is. Threats to network security can be roughly divided into two types: one is the threat to information in the network. The second is the threat to network equipment. The final threat is assigned in a qualitative relative manner. The level of threat is divided into five levels, the higher the level, the more likely the threat occurs, so as to complete the assessment of network information security risk.

3 Experimental Test

3.1 Experimental Preparation

Because MIPS64 series can support up to 16 cores, efficient resource scheduling management system can dynamically allocate these cores, according to the network data flow, arrange a certain number of cores for underlying package processing, TCP inspection, and QoS function, so as to meet the requirements of basic network data processing/firewall. The AEEEM dataset, collected by D'Ambros et al., contains information about five Eclipse projects supporting the Eclipse network projects Eclipse Kura, Ecfipse Paho and EclipseOM2M, each including several samples with one software module in each sample. The information database adopts the SQL Server2020 database under the Windows platform, which not only covers the evaluation elements of various evaluation criteria, but also can provide customers with standard evaluation application, questionnaire survey and other forms, but also provides previous evaluation experience, historical data and expert experience for the risk assessment process. The system will also arrange a certain number of cores to complete the encryption and decryption function of network data to meet the needs of VPN function. At the same time, a certain number of cores should be arranged for packet depth detection to meet the requirements of intrusion prevention and intrusion management function (IMS/IPS).The knowledge base can also enrich the information database by accepting the risk information collected by the client machine and the evaluation results of the application server. Each software module consists of 61 attributes, containing information based on the software code, development process, entropy, biweekly system resolution, historical information extracted from the cvs logs, etc. Running environment: the computer is configured with a CPU of 2.80 GHz, 215 MB of memory, Windows XP environment, V C++ 6.0 version compiled toolbox. The operating environment of ANN is: the computer is configured with a CPU of 2.80 GHz, 215 MB of memory, and Windows XP environment, MATLAB.

3.2 Experimental Results

To test the effectiveness of this design method, the experimental test was conducted. Literature [6] and literature [7] methods were selected as comparison methods for experimental comparison. The accuracy rates of the three network information security risk assessment methods were tested with different numbers of training samples. The experimental results are shown in Table 1–4:

Table 1. Training sample 20 accuracy (%)

The number of experiments	Literature [6] methods	Literature [7] methods	This article, method
1	84.646	86.144	92.613
2	85.314	85.991	91.808
3	86.002	81.205	90.554
4	84.997	86.717	92.617
5	85.131	84.310	89.665
6	84.206	85.441	92.388
7	85.449	86.212	93.445

(*continued*)

Table 1. (*continued*)

The number of experiments	Literature [6] methods	Literature [7] methods	This article, method
8	86.778	85.411	94.618
9	85.001	82.337	95.612
10	86.314	85.912	93.005
11	86.188	83.776	94.119
12	85.474	83.210	93.552
13	86.313	84.914	94.663
14	85.909	80.208	93.574
15	84.555	82.494	92.151

Table 2. Training sample 80 accuracy (%)

The number of experiments	Literature [6] methods	Literature [7] methods	This article, method
1	62.515	63.004	74.551
2	63.848	65.214	75.612
3	62.191	66.997	76.994
4	64.774	67.855	75.313
5	65.021	68.549	76.205
6	66.988	66.303	77.946
7	64.192	65.714	78.112
8	63.550	66.122	79.645
9	64.548	65.977	76.484
10	63.215	64.515	77.699
11	64.829	68.312	78.505
12	63.774	69.553	77.646
13	64.152	65.225	78.985
14	65.901	67.306	79.633
15	66.324	64.121	75.001

3.3 Experimental Analysis

From Table 1, At the training sample of 20, The designed network information security risk assessment method, The average accuracy of the other two network information security risk assessment methods is 92.959%, 85.485% and 84.285% respectively; From Table 2, At the training sample of 80, The designed network information security risk

Table 3. Training sample 150 accuracy (%)

The number of experiments	Literature [6] methods	Literature [7] methods	This article, method
1	56.948	55.649	66.588
2	57.644	56.344	67.915
3	58.499	55.124	66.922
4	57.214	56.942	65.212
5	58.331	55.812	66.748
6	57.029	56.919	65.201
7	58.677	55.701	66.339
8	57.123	56.228	65.201
9	56.911	55.322	66.819
10	55.744	57.415	63.155
11	56.998	56.021	64.498
12	55.303	55.947	62.815
13	56.472	56.319	61.402
14	55.988	57.414	60.994
15	56.461	55.168	59.166

Table 4. Training sample 300 accuracy (%)

The number of experiments	Literature [6] methods	Literature [7] methods	This article, method
1	32.152	33.144	52.131
2	36.519	34.505	49.878
3	35.597	36.917	53.009
4	34.008	35.222	49.677
5	33.455	34.919	51.088
6	34.813	35.060	52.119
7	35.914	36.977	53.477
8	36.477	34.181	49.512
9	35.219	35.199	48.667
10	36.088	34.162	47.101
11	35.164	36.121	46.922

(*continued*)

Table 4. (*continued*)

The number of experiments	Literature [6] methods	Literature [7] methods	This article, method
12	35.007	32.849	45.833
13	34.151	33.744	50.004
14	33.088	38.542	52.825
15	33.121	32.179	51.147

assessment method, The average accuracy rate of the other two network information security risk assessment methods is 77.222%, 64.388% and 66.318% respectively; From Table 3, At the training sample of 150, The designed network information security risk assessment method, The average accuracy rate of the other two network information security risk assessment methods is 64.598%, 57.023% and 56.155% respectively; From Table 4, At a training sample of 300, The designed network information security risk assessment method, The average accuracy rate of the other two network information security risk assessment methods is 50.226%, 34.718% and 34.915% respectively.

4 Conclusion

The previous computer network information security risk assessment methods have the problems of low assessment accuracy, which seriously restricts the assessment effect. In order to solve this problem and improve the effect of network information security risk assessment and the level of network information security, this paper analyzes the basic principles and analysis steps of the computer network information system security risk assessment method. This method takes non differential identification of unknown factors as the object, does not forcefully divide risk factors and non risk factors, that is, non differential identification of unknown factors. Through time, environment, object and dynamic transformation, The unknown factors are transformed from a whole into a ring-shaped closed structure. This paper introduces in detail the specific characteristics and existing problems of the introduction of machine learning algorithm into the traditional network information security risk assessment method, so as to reduce the impact of assessment factors, improve the objective accuracy of risk assessment, greatly reduce the blind areas and errors of risk assessment, and determine the determination principles and methods of unknown factors, which can solve the problems existing in the current computer network information security risk assessment, To help ensure the information security of the computer network.

References

1. Mao, Z., Hong, M., Xiao, Y., et al. Risk assessment of smart city information security based on bayesian network. Mod. Inf. **40**(5), 19–26, 40 (2020)
2. Kong, S., Zhao, Y.: Research on web-based network information security risk assessment model. China Comput. Commun. **32**(9), 200–202 (2020)
3. Liao, Y., Wang, J., Tian, K., et al.: Dynamic information security risk assessment for railway signaling safety data network based on Bayesian inference. J. China Railway Soc. **42**(11), 84–93 (2020)
4. Ren, J.-W.: Research on AHP model of computer network information security risk. Adhesion **43**(9), 157–160 (2020)
5. Liu, D.-W.: Risk assessment of network security technology based on fuzzy MCDM. Inf. Technol. **10**, 82–86 (2020)
6. Liu, J., Meng, X.: Survey on privacy-preserving machine learning. J. Comput. Res. Dev. **57**(2), 346–362 (2020)
7. Wang, Y., Liu, L.F., Huang, D.Z.: Simulation of network information trustworthiness based on machine learning algorithm. Comput. Simul. **37**(8), 239–242, 445 (2020)

A Dynamic Monitoring Method of Social Network Worm Attack Based on Improved Decision Tree

Wei Ge[✉]

Jiangxi University of Software Professional Technology, Nanchang 330041, China
gewei01475@163.com

Abstract. In the network security, worms are a kind of more aggressive virus, it is necessary to carry out a detailed discussion of worm attacks. There is a problem that the number of immune nodes is small in the application of worm attack dynamic monitoring method of social network, so a new method based on improved decision tree is designed. According to the attack infiltration theory and propagation mechanism, the worm type is identified, the worm propagation path is extracted, the copy of the worm program is transmitted to the adjacent nodes, and the heterogeneous model of topology structure is constructed by using the improved decision tree. Then the control is transferred to the function called to execute, and the dynamic monitoring mode of the worm attack is optimized. Experimental results show that the average immune nodes of this method and other two methods are 505, 363 and 373 respectively, which proves that the performance of this method is more outstanding than that of other two methods.

Keywords: Improved decision tree · Social network · Worm attack · Dynamic monitoring · Network security · Network nodes

1 Introduction

Among the network security problems, malicious code to the network causes by the largest proportion of economic losses. It mainly includes: computer viruses, a program that can infect other programs by modifying other programs to copy itself or its variants. Worms have been one of the greatest threats to network security. Every outbreak of worms has done great harm to the whole network since it first appeared. The worm belongs to a kind of computer virus, but it and the ordinary computer virus also have a very big difference. A social network worm, a program that sends itself from one node to another and launches itself through the network's communication capabilities. This paper makes a comparative analysis between worm and common computer virus from the aspects of existence form and spreading mechanism. In order to exist, ordinary computer viruses need to parasitic in the host program or file, and only when the host program or file is running, the virus is activated, with the ability to infect. A Trojan horse, a program that executes beyond the definition of a program. A compiler is a Trojan horse

W. Fu and L. Yun (Eds.): ADHIP 2022, LNICST 469, pp. 417–430, 2023.
https://doi.org/10.1007/978-3-031-28867-8_31

if, in addition to performing compilation tasks, it secretly copies down the user's source code. A worm, on the other hand, is a self-executing code file that does not need to be stored or activated.

In the transmission mechanism, the common computer virus is mainly infected with the files on the host, and the worm attack target is not only a single host, but also spread the attack program to the whole network. A logic bomb is a program that performs other special functions when the environment under which it runs satisfies a particular condition. A denial of service attack is an attack in which a user occupies a large number of shared resources so that the system does not have any remaining resources available to other users. This attack reduces the availability of system resources, including network server processors, disk space, printers, etc. The result of the attack is to reduce or lose service. Generally speaking, worms do much more harm to computer systems than ordinary computer viruses. From a technical point of view, the worm is described: a worm can run independently, and can send their own copy of the file to other computers as a virus. In addition, many people put forward the malicious code as a means of attack in cyber warfare, when network security has risen to the height of national security. The characteristics of worms are mainly determined by the attack program, so in order to better understand the characteristics of worms, the researchers have carried out a detailed study of worm attack program.

2 A Dynamic Monitoring Method of Social Network Worm Attack Based on Improved Decision Tree

2.1 Identifying Social Network Worm Types

The biggest difference between social network worms and viruses is that viral infections require users to distribute infected files to susceptible hosts. Therefore, the spread of the virus will be much slower than the spread of social network worms, this is because the spread of the virus requires manual user participation. However, social networking worms can spread very quickly and can infect a large number of hosts within a day or even a few hours. Later, it is found that the worm attack program mainly consists of four modules, which are information collection module, scanning detection module, infiltration module and self-propelling module. Because of the variety of worm propagation modes, it is necessary to divide the worm into several groups, so as to facilitate researchers, study of the worm propagation models and detection methods. Scanning worm generates the IP address of the network through a target selection algorithm, and then detects the vulnerability of the host on the IP address by vulnerability scanning, if the vulnerability exists, it will infect the host directly. Therefore, the spread of scanning worms does not require human intervention. The self-replicating nature of worms is influenced by the self-propelling module that helps worm programs form multiple copy files in preparation for a new attack [2, 3]. Active transmission of worms is affected by information collecting module, scanning detection module and attacking penetration module, which the information collecting module and scanning detection module are the preparatory stages. Scanning worm attacks have the following characteristics: attackers do not need to collect information on the network host vulnerability, easy to write, relatively short infection time and so on. The scanning strategies include random scanning, selective random scanning, linear scanning, target list based scanning, divide and conquer scanning, route based scanning, DNS based scanning and so on.

Through the research on the function modules of worm program, the understanding of worm propagation mechanism in the network has become more in-depth. Topology worms propagate in a logical topology communication network by collecting the topology information of hosts or the hosts in the network that they carry. Morris worms collect topology information, including the Web Yellow Pages/etc./hosts and other resources to find social network worms. Mail worms are topology worms that propagate through mail. When an email user opens attachments in an email or when a vulnerable email client receives an email that has been infected by a worm, the user's host computer is infected by the worm. The process of worm propagation in the network can be divided into four stages, which are the information collection, scanning detection, attack infiltration and self-promotion. Among them, penetration is the main part of worm propagation. Therefore, it is important to study the state change of nodes in the stage of attack and infiltration. Based on the propagation mechanism of social network worms, we divide social network worms into four types, as shown in Fig. 1:

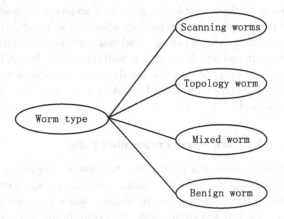

Fig. 1. Types of social network worms

As can be seen from Fig. 1, depending on the propagation mechanism of social network worms, the types of network graph worms include scanning worms, topology worms, benign worms, and hybrid worms. Based on this, the focus of the research on worms has shifted to the quantitative characterization of worm propagation. At present, the research focuses on establishing worm propagation model by mathematical modeling method, predicting worm propagation trend by simulation experiment, and finding the key factors to prevent and control worm propagation. As a kind of P2P worm, IM worm propagates through IM network. Firstly, it collects the contact information of the infected host user, then attacks the contact by IM protocol or flaw, and propagates the copy of IM worm to the attacked contact.

The behavior of attacking and infiltrating worms is called worm attack, which includes two processes: initiating attack and infiltrating. In the process of infection of worm attack, worm program will make use of the vulnerability of node to infiltrate it. According to attack infiltration theory, the result of attack infiltration is the increase of node vulnerability and the promotion of user authority. A benign worm is a controllable, stand-alone program that can be run without the intervention of a computer user. It obtains some or all control over a computer that has a loophole in the network, and then uses the transport function to obtain assistive tools for tasks such as immunization, repairing, clearing worms, and closing the back door. Finally can safely self-destruct. Like drugs in medicine, benign worms can spread like worms, patching or removing worms from infected hosts, or both. It is found that when a node is attacked by several neighboring infected nodes at the same time, the increased vulnerability of some neighboring nodes and the promotion of user privileges may become the precondition for the node to be attacked by other neighboring nodes, so that the intrusion infiltration of the latter neighboring nodes can be completed more easily.

Hybrid worms can be spread using a combination of social network worm attack strategies, such as mail worms, scanning worms, and social network worms using network sharing attack strategies. Hybrid worms require worm writers to have a comprehensive knowledge of the attacks, and are more difficult to write than social network worms that use only one attack method. At the same time, according to the theory of cooperative intrusion, multiple attacks can constitute an attack sequence, and the information transfer between them will continuously affect the vulnerability of nodes. Based on the above theoretical analysis, it can be found that multiple attacks have correlation in the process of intrusion, and attack correlation will increase vulnerability and infection probability of nodes. In order to describe the dynamics of infection probability more accurately, the attacking correlation should be taken into account in the construction of worm propagation model.

2.2 Extract Social Network Worm Propagation Paths

Worm attacks on social networks involve two processes: transmission and infection. Propagation is the process that a worm attacks a neighbor node with the help of the host node. Infection refers to the process in which a worm program makes use of the vulnerability of the attacked node to infiltrate the worm. In the process of worm attack, worm attacks all neighboring nodes by infective nodes. The main behavior of worm attacks is to establish information transmission channels between nodes and send copies of worm programs to neighboring nodes.

Social network worm is a kind of autonomous agent, which can spread rapidly in the network, especially in the high-speed network environment, the worm outbreak rate is greatly increased, the coverage is more extensive, and the loss is greater and greater. There are three factors that affect the speed at which social network worms can spread. The first is the speed at which loopholes are discovered. The second is how many potential "fragile" hosts can be exploited. Third, the social network worm infection rate of these targets. In addition to the division based on the propagation mechanism, social network worms can be classified differently from different perspectives. According to the transport mode of social network worms, social network worms can be divided into three kinds of social network worms: automatic transmission, second channel and embedded transmission. The main determinant of worm propagation speed is the speed at which a vulnerable host can be found, that is, how many efficient host systems can be found, which is done by the worm's scanning module. Therefore, in order to defend social network worm more quickly and effectively, we must study the propagation mechanism of social network worm from the scanning and attacking methods of worm. Assuming that the social network worm no longer accesses the visited host system, the propagation rate of the divide-and-conquer scanning worm can be expressed in the following formula:

$$L_{p+1} = (S - L_p) \times \left(1 - \frac{10^9 - \varepsilon}{\sum_{\varepsilon=1}^{n} p + 1} \right) \tag{1}$$

In formula (1), S represents the total number of host systems with exploitable vulnerabilities of social network worms, L represents the scan rate, ε represents the granularity of the address list, and p represents the number of hosts that have been infected by time p. According to the activation methods of social network worms, it can be divided into three types: manual activation, scheduled process activation and self-activation. The propagation process of worm attack can be regarded as the information transaction between two nodes. The infective status node is the initiator of the transaction and the neighbor node is the receiver. According to the analysis, it can be seen that the trust evaluation of the transaction receiver to the initiator of the transaction will affect the receiving behavior of the transaction receiver. Therefore, in the process of worm attack propagation, the neighbor node will decide its receiving behavior according to the trust degree of the infected state node.

Social network worms that use a random scan strategy scan for unassigned IP addresses, most of which are non-routable, greatly slowing down the spread of social network worms. Before the social network worm is released, knowing which addresses are routable can greatly reduce the worm's scanning space, speed up propagation and effectively evade detection. In social networks, the buffer size for most applications to hold data is fixed [4, 5]. If an attacker sends excessive data to one of these buffers, and the program does not check the size of the data, the buffer overflows. According to payload partition, it can be divided into empty payload, Internet remote control payload, spam forwarding payload, HTML proxy payload, Internet denial of service attack payload, data collection payload, sales access payload, data destruction payload, remote control payload, denial of service attack payload, search payload, damage payload, worm maintenance payload. From formula (1), the expression formula for worm propagation on mobile devices is derived:

$$\frac{S}{L} = \varphi^2 + \sqrt{(h - \varphi)^2} \tag{2}$$

In formula (2), φ represents the forward invariant set of the host, and h represents the removal rate of the virus on the mobile device. The target computer then executes the "overflow" data, just as it executes the program. If the attack buffer is in a legitimate process, the malicious program takes full control of the target computer and then does whatever it pleases, including executing commands on the target computer, stealing passwords or other confidential information, changing the system configuration, and/or installing a back door. Based on the classification of attackers, we can divide social network worms into: curiosity, business interests, mischief, political purposes, terrorism, and cyberwar. From the point of view of worm attack, the receiving behavior of neighbor nodes affects the propagation probability of worm replicas propagated by infected nodes. In order to describe the dynamics of the propagation probability more accurately, trust should be taken into account in the construction of the worm propagation model.

2.3 Construction of Topological Heterogeneous Model by Improving Decision Tree

In the improved decision tree algorithm, entropy is used to represent the degree of confusion and the degree of possible states [6, 7]. In the field of information theory, information entropy indicates the degree of confusion, indicating the extent to which states may occur. In the field of information theory, information entropy indicates the probability of discrete random events.

The reason why buffer overflow attacks have become a common means of attack is that buffer overflow vulnerabilities are too common and easy to implement. Moreover, buffer overflows are the primary means of remote attack because they give an attacker everything he or she wants: the ability to seed and execute attack code [8]. Random scans give infected computers a chance to scan the Internet for vulnerabilities when looking for new targets, so scans target all IPv4 address spaces. The implanted attack code runs the program with buffer overflow vulnerability with certain authority, thus obtaining the control of the attacked host.

Buffer overflow is mainly divided into two types according to its overflow form: stack-based Buffer overflow and heap-based Buffer overflow. Most Buffer overflow attacks are stack based. Based on the analysis of worm propagation process, it is found that the trust of peers will affect the information transaction between peers and thus the worm attack propagation probability. The higher the trust degree of the node is, the greater the probability of successful information transaction and the greater the probability of information transmission. Based on this, this paper establishes a positive correlation between transmission probability and trust, as shown in Formula (3):

$$G_{mn} = \eta \times \frac{1}{\sqrt{W_{mn}}} \tag{3}$$

In formula (3), η represents the propagation probability, W represents the correlation coefficient, and m, n represents the trust degree of node m to node n, respectively. Both codeRed and Stammer worms use random scans, which generally have large scan address space and slow propagation. But the Stammer worm spreads very fast, mainly because it uses the UDP protocol for disconnected scanning, and uses a large number of threads to scan, making its scan mainly limited by bandwidth. In the improved decision tree algorithm, the information entropy data is used as the parameter to reflect the minimum randomness of the partition, and the amount of information of the sample is minimized. Represents the probability of the occurrence of a discrete random event. In the decision tree algorithm, the information entropy is used as a parameter to reflect the minimum randomness of the partition, and the minimum information is needed to show the probability of discrete random events.

Stack-based overflows take advantage of buffers stored on the program's stack. The stack is an area of memory where local variables are stored within each program. It is used as a buffer storage area for program subroutines. When a function is called in a program, the system first puts the arguments required by the called function on the stack in reverse order, and then puts the address of the instruction following the call on the stack. In the decision tree algorithm, the information entropy is used as a parameter to reflect the minimum randomness of the partition, and the information needed by the sample is minimized. Once the first root node is calculated, the data set can be divided into subsets, and then the information gain can be calculated on the basis of the subset, and the subset can be recursed when all the attributes have been divided, or when the subset belongs to the same category, or when the recursion is over. Sequential scanning is when a worm on an infected host randomly selects a network address to propagate. According to the local priority principle, a worm typically selects an IP address in its network. If the target address IP of the worm scan is A, then the next address IP of the scan is A 1 or A-1.Once scanned to the host network with many vulnerabilities will achieve a very good dissemination effect.

The aim of the improved decision tree algorithm is to get a set of decision rules from the training data set, so that the training samples can make correct decisions and deal with the new input samples well. There are three processes to find the optimal decision tree: attribute selection, decision tree generation and decision tree pruning. Attribute selection is the key to decision tree learning, that is, how to select the optimal attribute for partitioning nodes. The control then goes to the called function to execute, and the program typically allocates the required storage space for local variables within the called function after stacks of values that require registers to be saved.

2.4 Optimize the Dynamic Monitoring Method of Worm Attack

In recent years, heap overflows have become very common for several reasons: the lack of stack execution mechanisms, and breakthroughs in new attack technologies. If a heap buffer overflow occurs, it provides an unusable stack protection mechanism that is currently not executable, which the system can ignore, but which makes the social network vulnerable. Based on the SEIR model of worm propagation, the Markov chain of state transition is constructed, and the transition probabilities from susceptible state, latent state and infective state to recovery state are calculated.

Based on the above calculation results, the dynamic monitoring mode of the whole social network is optimized. These average times suggest the best time to take defensive measures in time. Routing worms use addresses in the BGP routing table to reduce the address space to be scanned. Because only 28.6% of the IPv4 address space is included in the BGP routing table, the probing address space will be smaller than the random scan, so the propagation speed will be increased. For the state transition dynamic equation of the susceptible state node, see Formula (4):

$$D_{ij}^e(\eta + 1) = \frac{e + T_{ij}}{\sum\limits_{i=1}^{n} T_{ij} - \eta} \tag{4}$$

In formula (4), e represents the number of attacks, T represents the infection probability of the node, and i, j represents the initial value of the infection probability and the change rate of the infection probability, respectively. The essence of worm attack behavior is to promote worm propagation from different angles or at different costs to achieve different objectives driven by benefit mechanism. Therefore, in order to promote the worm to spread in the network, the node will provide false trust. Assume that, at each time step, the infected state node increases its direct trust by increasing the number of successful transactions with its neighbor nodes, as shown in formula (5):

$$Q_\sigma = \frac{\|\sigma - 1\|}{2} \times \mu^2 \tag{5}$$

In formula (5), σ represents the attack propagation risk of the source node, and μ represents the number of successful transactions. Heap overflow based attacks vary greatly depending on the attack technique used, and heap based buffer overflows can be classified into three categories depending on the difficulty of the technique used: simple heap overflows, which are often caused by unintentionally supplying an overlong string and are not intended attacks. However, "routing worms" carry address information to make their own code longer, in this sense will slow down the spread of worms. In dynamic monitoring mode, divide and conquer scanning is a strategy for social network worms to collaborate with each other and search easily infected hosts quickly.

Modify the function pointer, by modifying the function pointer to point to the pre-implanted attack code, so as to achieve the goal of attack. Advanced heap overflow, which makes full use of the data structure and method of heap management in social network, is the most frequently used method of dynamic monitoring. The social network worm sends a portion of the address library to each infected host, and each host scans it for the address it gets. After Host A is infected with Host B, Host A assigns a portion of the address it carries to Host B, and Host B scans that portion. Formatting string attacks are mainly caused by programmers' laziness. String formatting output function is a commonly used means of human-computer interaction in C language. It differs from ordinary functions in that the number of parameters of such functions is variable, and the string formatting parameters guide the social network to complete the corresponding input and output work.

Because dynamic monitoring is user-triggered, it spreads slowly, but these worms do not cause communication anomalies in the discovery of targets, making them inherently more secure. The CRClean worm is a passive worm, waiting for Code Red II to probe the activity. When it detects an infection attempt, it launches a counter-attack back and forth that should infect the attempt. If the counterattack is successful, it removes the CodeRed II worm and installs itself on the machine. When the called function finishes executing, the social network frees up the stack space allocated for the local variable of the function and restores the value that already exists in the register. The return address from the stack is then sent to the instruction register in the social network, and the program starts execution at that address. The disadvantage of this strategy is that the same host may be repeatedly scanned, causing network congestion. The Hit-list scan method is one of the fastest theoretically propagating scans. It first by the worm writer by scanning and other means of detection to collect the worm can attack the network of all the host information.

The worm then attacks the collected host. Heap based buffer vulnerabilities can occur if data is copied to buffers on the heap without checking.

3 Simulation

3.1 Build an Experimental Environment

Simulation experiment is based on Python experimental platform, the scale-free network constructed in the total number of nodes N = 1000, the average length of 6 nodes. The number of nodes in each state of the network at each time was selected as observation object. The simulation experiment needs to be tested on the dataset, and the labeled dataset is usually used. There are two ways to get tagged datasets: one is to use publicly tagged datasets, such as DARPA, KDD 99, and NSL-KDD. The second is to construct a new data set from the actual network traffic. The prototype social network was deployed at the exit of the network, and we used tcpreplay software to replay background traffic and worm traffic on different hosts, respectively. Among them, the background traffic without attack is the network traffic of one month, and the worm traffic is the traffic of lion worm and mscan worm respectively.

Data collection from actual network traffic has problems of classification accuracy and labeling. The outlier-based detection method usually needs to be trained with non-attack data. However, the data collected from the actual network may not meet this requirement. The initial parameters of the network are as follows: 0.01 for infective state nodes, 960 for susceptible state nodes, 10 for infective state nodes, 10 for threatening state nodes, 10 for isolating state nodes, and 20 for immune state nodes. Under these conditions, the experiment was repeated. Shaft K. and Abbass H.A. describe methods for generating tagged datasets. The dataset consists of the background traffic obtained from the real network and the attack traffic obtained by simulation. To tag the background traffic, execute Snort IDS.

3.2 Experimental Results

In order to evaluate the effectiveness of the designed dynamic monitoring method, the data mining -based dynamic monitoring method for worm attacks on social networks and the cloud computing -based dynamic monitoring method for worm attacks on social networks are selected and compared with the designed dynamic monitoring method for worm attacks on social networks, and the number of immune nodes of the three dynamic monitoring methods are tested respectively under different runtime conditions. The experimental results are shown in Tables 1, 2, 3, 4 and 5:

As can be seen from Table 1, when the runtime is 8 h, the average number of immune nodes of the social network worm attack dynamic monitoring method and two other social network worm attack dynamic monitoring methods is: 796, 589, 582, as can be seen from Table 2; when the runtime is 16 h, the average number of immune nodes of the social network worm attack dynamic monitoring method and two other social network worm attack dynamic monitoring methods is: 624, 454, 463, as can be seen from Table 3; when the runtime is 24 h, the average number of immune nodes of the

Table 1. Number of immune nodes with 8 h running time (piece)

Number of experiment	Dynamic monitoring method of social network worm attack based on data mining	Social network worm attack dynamic monitoring method based on cloud computing	The design of the social network worm attack dynamic monitoring method
1	536	594	774
2	520	568	782
3	541	571	790
4	559	529	805
5	629	531	846
6	635	566	773
7	548	519	854
8	537	548	873
9	559	567	828
10	548	582	793
11	622	637	811
12	670	644	764
13	664	698	758
14	612	577	739
15	648	602	745

social network worm attack dynamic monitoring method and two other social network worm attack dynamic monitoring methods is: 514, 328, 362; as can be seen from Table 4, as can be seen from Table 4, when the runtime is 32 h, the average number of immune nodes of the social network worm attack dynamic monitoring method and two other social network worm attack monitoring methods is 344, 2727, 282; as can be seen from Table 5, as can be seen from Table 5, when the social network worm attack dynamic monitoring method is 40 h, the average number of social network worm attack dynamic monitoring methods are designed respectively: 177, 1747, and two other social network worm attack methods are 1747.

According to the above experimental results, this method has certain advantages in dynamically monitoring social network worm attacks. Because this paper improves the decision tree algorithm, which can identify the worm types and extract the worm propagation path according to the worm attack penetration theory and the propagation mechanism. It optimizes the way that worm attacks are dynamically monitored.

4 Conclusion

The dynamic monitoring method of worm attack in social network designed in this paper enriches the research in related fields, and provides a new method for accurately

Table 2. The number of immune nodes with a running time of 16 h (piece)

Number of experiment	Dynamic monitoring method of social network worm attack based on data mining	Social network worm attack dynamic monitoring method based on cloud computing	The design of the social network worm attack dynamic monitoring method
1	512	484	612
2	498	455	609
3	446	432	588
4	413	419	637
5	459	478	612
6	426	455	659
7	487	461	616
8	412	473	602
9	404	481	645
10	415	492	637
11	503	501	625
12	488	522	644
13	467	418	633
14	458	406	617
15	416	469	628

Table 3. The number of immune nodes that run for 24 h (piece)

Number of experiment	Dynamic monitoring method of social network worm attack based on data mining	Social network worm attack dynamic monitoring method based on cloud computing	The design of the social network worm attack dynamic monitoring method
1	331	375	476
2	302	384	555
3	294	396	561
4	316	345	423
5	338	353	565
6	349	377	488

(*continued*)

depicting the model. Aiming at the influence of social network worms on network traffic, an entropy based detection method for social network worms deployed on routers is

Table 3. (*continued*)

Number of experiment	Dynamic monitoring method of social network worm attack based on data mining	Social network worm attack dynamic monitoring method based on cloud computing	The design of the social network worm attack dynamic monitoring method
7	351	349	492
8	311	358	503
9	297	316	521
10	304	372	564
11	322	369	597
12	366	359	496
13	387	346	485
14	341	372	520
15	315	358	466

Table 4. The number of immune nodes with a running time of 32 h (piece)

Number of experiment	Dynamic monitoring method of social network worm attack based on data mining	Social network worm attack dynamic monitoring method based on cloud computing	The design of the social network worm attack dynamic monitoring method
1	265	269	303
2	288	248	312
3	226	267	344
4	293	285	319
5	267	291	306
6	253	288	358
7	249	267	347
8	278	294	369
9	282	311	366
10	264	305	384
11	259	278	375
12	277	267	317
13	301	255	345

(*continued*)

Table 4. (*continued*)

Number of experiment	Dynamic monitoring method of social network worm attack based on data mining	Social network worm attack dynamic monitoring method based on cloud computing	The design of the social network worm attack dynamic monitoring method
14	269	294	366
15	306	314	349

Table 5. Number of immune nodes with 40 h runtime (piece)

Number of experiment	Dynamic monitoring method of social network worm attack based on data mining	Social network worm attack dynamic monitoring method based on cloud computing	The design of the social network worm attack dynamic monitoring method
1	169	174	221
2	188	185	205
3	165	196	226
4	167	173	249
5	182	185	258
6	174	176	237
7	169	182	241
8	153	171	218
9	164	169	209
10	185	179	264
11	172	162	257
12	169	181	282
13	182	163	274
14	175	194	269
15	174	165	288

proposed. This method belongs to the detection method of social network worm based on traffic, that is, analyzing the statistical information of network traffic without analyzing the load of network transmission. The improved decision tree is introduced into the worm propagation process, and the heterogeneous model of network topology can better identify malicious nodes, which provides a new mechanism for preventing and combating worm attacks. At the stage of real-time worm detection, the network data entropy of the current window is calculated and compared with the learning entropy interval. If the current window is not in the learning entropy interval, the worm is warned. It lays the research foundation for the academic circles.

References

1. Kumar, R., Kumar, P., Kumar, V.: Design and implementation of privacy and security system in social media. Int. J. Adv. Networking Appl. **13**(4), 5081–5088 (2022)
2. Li, Y., Song, M.: Trust update model considering worm propagation risk. Oper. Res. Manage. Sci. **29**(10), 163–172 (2020)
3. Li, Y., Zhang, H.: Benign worm propagation model in industrial control networks. Control Eng. China **27**(7), 1286–1292 (2020)
4. Cheng, H., Zhang, R., Zhang, S., et al.: Power failure sensitivity analysis based on improved decision tree. Microcomputer Appl. **36**(3), 144–148 (2020)
5. Li, Y., Li, W., Chen, N., et al.: Battlefield target assistant judgment technology based on improved decision tree. Command Inf. Syst. Technol. **11**(1), 62–67 (2020)
6. Achar, S.J., Baishya, C., Kaabar, M.K.A.: Dynamics of the worm transmission in wireless sensor network in the framework of fractional derivatives. Math. Methods Appl. Sci. **45**(8), 4278–4294 (2022)
7. Panigrahi, R., Borah, S., Bhoi, A.K., et al.: A consolidated decision tree-based intrusion detection system for binary and multiclass imbalanced datasets. Mathematics **9**(7), 751 (2021)
8. Zhou, L., Li, J.: Security situation awareness model of joint network based on decision tree algorithm. Comput. Simul. **38**(5), 264–268 (2021)

A Method for Dynamic Allocation of Wireless Communication Network Resources Based on Social Relations

Changhua Li[(✉)] and Wei Ge

Jiangxi University of Software Professional Technology, Nanchang 330041, China
lichanghua4562@163.com

Abstract. Due to the large uncertainty of signal reception, the transmission speed of the wireless communication network resource allocation method is low. Based on social relations, a dynamic allocation method for wireless communication network resources is designed. Establish a wireless ad hoc network composed of a random number of nodes to determine the upper limit of its security capacity, and the nodes at the edge need to undertake more communication transmission loads on average. According to the similarity between users, a triple closed model is used to measure the social trust relationship in wireless communication networks. Considering the social trust and encounter frequency among users, the optimal relay node is selected for the wireless network communication link. The sum of data transmission rates under each allocation scheme is calculated and sorted in descending order, and resources are dynamically allocated based on the principle of maximum overall network throughput. Compared with other schemes through simulation, the dynamic allocation method of wireless communication network resources based on social relations can increase the network transmission speed and improve the communication performance.

Keywords: Social relationship · Wireless communication network · Resource allocation · Dynamic allocation · Social trust · Relay node

1 Introduction

The surge of mobile data traffic and the popularization of mobile intelligent terminal equipment will undoubtedly bring a very heavy burden to the future wireless mobile communication system. For example, the shortage of spectrum resources, the overload of base stations and the congestion of communication links will greatly hinder the development of future communication networks. Wireless communication networks can be divided into infrastructure-supported networks and wireless ad hoc networks according to topology types. Infrastructure-supported networks, such as wireless cellular communication networks, use wired networks as the backbone network, and infrastructure-supported nodes (such as base stations) provide data forwarding and user service services to other nodes within their coverage. Social information reduces the interaction between them

W. Fu and L. Yun (Eds.): ADHIP 2022, LNICST 469, pp. 431–445, 2023.
https://doi.org/10.1007/978-3-031-28867-8_32

in real life. On the contrary, Internet social software is more and more deeply embedded in people's lives, so social attributes have become a factor that cannot be ignored in wireless communication. Since there is no content of interest or sufficient trust between some users, the establishment of network links may also be hindered in mobile networks while satisfying physical transmission conditions [1].

A wireless ad hoc network is a decentralized transient ad hoc system because it does not depend on pre-existing infrastructure. For example, routers, base stations in wired networks, or wireless switches and access points. In a wireless ad hoc network, each node participates in routing by forwarding data for other nodes, and dynamically determines the node that forwards data based on network connectivity and routing algorithms. In a wireless communication network, each user can not only transmit. It is also possible to receive signals, that is, the user is both a server and a client. And each user has automatic routing function [2]. Each user in the network shares some of the hardware resources they own, such as information storage, information processing, and network connectivity. Users can access these shared resources directly without going through an intermediate entity. This network system breaks through the geographical limitations of traditional wireless networks, and can be deployed more quickly, efficiently, and at low cost, and the failure of any node will not affect the operation of the entire network, and has strong robustness. Device discovery is generally performed by the device sending a known sequence of synchronization or reference signals to detect whether there are other devices around, and requires that the two peer devices sending the beacon have the same sending time and appropriate spatial location. If the devices do not match and there is no redundant matching information, they can only be detected by random beacons, which incurs a loss of time and energy.

In a wireless communication network, the communication probability between nodes is related to the social relationship between nodes. Compared with nodes without social relationship, nodes with social relationship have a higher probability of communication. Obtaining social features from social networks will play an important role in improving the performance of mobile networks. A social network refers to a structure consisting of a set of entities and a series of connections between them, and its main characteristics are community, centrality, and social relations. With social networks, the formation of invalid communication connections can be avoided. From the perspective of content transmission, content popularity reflects the strength of social relationships in the network to a certain extent. When the strength of the social relationship is high, the access and transmission of the most popular part of the content accounts for the vast majority of network traffic. Factors in the social realm.. For example, users' social relations, their willingness to join and their expected data content consistency will also have an impact on the network cluster formation strategy, which will further affect the resource allocation efficiency of the network under multicast communication.

Some scholars have proposed a dynamic resource allocation method for wireless communication networks based on automatic differentiation. The method combines energy harvesting technology with multiary quadrature amplitude modulation of traditional communication. Based on the table format value function reinforcement learning algorithm—Q-learning and SARSA algorithm to find the optimal transmission strategy for each time slot of the automatic differentiation communication system, and finally realize the dynamic resource allocation. However, in the face of a large amount of resources, the transmission rate needs to be improved. Some scholars have proposed a dynamic allocation method of wireless communication network resources based on hierarchical game. This paper proposes two new penalty strategies to promote cooperation among master user networks. By triggering the strategy to promote the formation of cooperation between the main user network to achieve the optimal overall profit, and finally achieve the Nash equilibrium profit. This paper proposes a dynamic allocation method of wireless communication network resources based on social relations to improve the overall performance of the communication network.

2   A Method for Dynamic Allocation of Wireless Communication Network Resources Based on Social Relations

2.1   Determining the Security Capacity of Wireless Communication Networks

This paper considers building a wireless ad hoc network composed of a random number of nodes distributed on a two-dimensional ring with an area equal to the expected number of nodes. The topological structure of the two-dimensional ring can be used to avoid the influence of boundary effects. The positions of the nodes are generated according to a two-step shot noise Cox process. Under the centralized control scheme, the equipment of the cellular network such as the base station will control the users of the wireless network communication, including the establishment of the communication session and other control information. When entering into the process of information exchange between the two parties, the equipment in the cellular network such as the base station will not participate, and all information is sent on the communication link between users [3]. According to the isomorphic Poisson point process of density, a set of cluster heads are randomly distributed on a two-dimensional torus. Each cluster head independently generates nodes of its own cluster, and the density of nodes generated by the cluster head at any position is obtained by the superposition of the node generation process of each cluster head, which can be expressed by the following formula:

$$p(a) = \sum_{i=1}^{m} fh(b, a) \tag{1}$$

In Eq. (1), $p(a)$ represents the node density at position a; m represents the expected number of clusters; i represents the cluster head number; f represents the number of nodes generated by the cluster head; h represents the dispersion density function; b represents the cluster head. The dispersion density function is direction invariant, so the node density at location a is only related to the Euclidean distance between b and a. In the communication mode, a ranking-based model is adopted to describe the wireless network social relationship characteristics in which nodes are more likely to communicate with nearby neighbor nodes. This model is used to estimate the average transmission distance of unicast transmissions. For example, when the user reuses the resources of the cellular network, the base station will send a corresponding signaling message through the control link to notify the device to reduce the transmit power during communication. On the one hand, it reduces the interference to the cellular network users, and on the other hand, it also reduces the energy consumption of the device. When there are resources that can be reused in the network and the user has communication needs, the base station can also implement the function of resource management, that is, the communication user [4].

Legitimate nodes can send packets to legitimate nodes in the same cell or in adjacent cells. According to the definition of cell area, the area of each cell in the inner ring is much smaller than that of each cell in the outer ring. From the perspective of the interference model, the transmit power of the legitimate nodes located in the inner ring should also be smaller than the transmit power of the legitimate nodes located in the outer ring. It is assumed that the channel fading between different users and between users and base stations in the system obeys a large-scale fading model. The noise in all channels is white Gaussian noise with mean 0. Moreover, the distance between the wireless communication network user pair and the cell base station is far greater than the communication link distance between the two. Therefore, the transmit power of a legitimate node is related to the position of the ring where it is located.

The transmission range of a node is related to the density of nodes in the area. During the transmission process, it is necessary to restrain the interference from other nodes in the adjacent area to ensure the successful transmission of the node. The upper limit of the security capacity of the wireless communication network can be calculated by the following formula:

$$c(o) = m \int_0^{2\pi} \int_0^{\sqrt{r}} wp(a)da \tag{2}$$

In Eq. (2), o represents the equivalent wireless resource area; c represents the upper limit of network capacity; r represents the radius of the general cluster; w represents the weight parameter corresponding to the heterogeneous distribution of nodes. The idea of relay communication technology is to use one or more relay nodes for communication between two parties, which can convert a link with a lower communication rate into two or more links with better channel performance. Thus, the system throughput can be improved. When the degree of heterogeneous distribution of nodes is high, the capacity of each node will be reduced due to the reduction of the equivalent wireless resource area. Therefore, the nodes at the edge of each cluster domain need to bear more communication transmission load on average.

2.2 Metrics of Social Trust Relationships in Wireless Communication Networks

In the social domain, mobile users form a social network with stable social relationships, and the connections between users represent the strength of users' social relationships. Whether there is a social path between users has nothing to do with the physical link in the physical domain, but is determined by the social relationship between users. Social modeling among wireless communication network users takes the form of a social graph $W = (P, Q)$, where the set of vertices P represents the set $S = (1, 2, \cdots, s)$ of devices in the network. Vertex $s \in S$ represents a mobile device or a user carrying this device.

Edge $q \in Q$ represents the social connection (distance) between two users x and y with direct interaction. In this network model, it is divided into social level and physical level. The social level reflects the real social relationships of users in real life. Social relationship is a concept that reflects the strength of the relationship between users in a social network, and the lower layer is the physical level. If user x and user y are within the range where they can communicate, direct wireless communication is used between them. Otherwise, a multi-hop wireless network communication path will be established between user x and user y. When the distance is close enough, the communication device can communicate and transmit information through the wireless network. In the physical layer, the physical channel is affected by the communication environment. Users can transmit information through the base station of the cellular network, or use the wireless network communication technology. In this case, it is very important which device the user chooses as a relay to forward his data. Because the selected device must be its trusted device, data forwarding will fail even if there is only one untrusted device [5].

In reality, a user device is likely to have no direct connection with its neighbor devices, so it is necessary to analyze how to calculate indirect trust through direct trust. The social relationship between users is related to many factors. The relationship between relatives and friends, encounter history and geographical location between users can all affect the strength of the social relationship [6]. Different users form different groups according to the strength of their social relationships. In real scenarios, network user pairs are composed of a sender and a receiver. In order to ensure the communication quality between wireless communication network users, the distance between user pairs needs to meet the maximum distance condition for communication. The triple closure model is an effective tool to identify the indirect trust relationship between them. The measurement process of social trust relationship in wireless communication network is shown in Fig. 1.

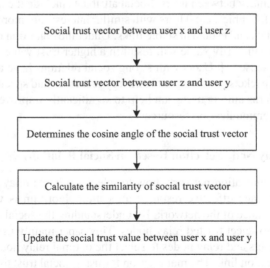

Fig. 1. Social trust relationship measurement process

Suppose there are social connections in the social graph between user z and user y, and between user z and user x. The social trusts between user x and user z, and between user z and user y are ϑ_{xz} and ϑ_{zy}, respectively. During communication, each cellular user will occupy a fixed channel resource allocated by a base station. Users share the channel resources occupied by ordinary cellular users for communication, so the two will interfere with each other during communication. To limit interference, we assume that a user pair can only use one channel. Therefore, user pairs do not interfere with each other. Then, considering the influence of user z, the mutual friend of user x and user y, the social trust between user x and user y can be updated as:

$$\vartheta_{xy} = \sqrt{\vartheta_{xz}^2 + \vartheta_{zy}^2 - 2\vartheta_{xz}\vartheta_{zy}\beta(x, y)} \tag{3}$$

In Eq. (3), ϑ_{xy} represents the social trust between user x and user y; $\beta(x, y)$ represents the similarity between user x and user y. The strength of social relations between users often affects the stability of network connections, and users with close social relations with surrounding users are more willing to establish and maintain wireless communication and share data. The calculation equation of similarity is as follows:

$$\beta(x, y) = \frac{\sum_{j=1}^{u} \left(\varphi_{xj} - \varphi_x'\right)\left(\varphi_{yj} - \varphi_y'\right)}{\sqrt{\sum_{j=1}^{u} \left(\varphi_{xj} - \varphi_x'\right)^2} \sqrt{\sum_{j=1}^{u} \left(\varphi_{yj} - \varphi_y'\right)^2}} \tag{4}$$

In Eq. (4), φ represents the user's interest rating for content j; u represents the total number of contents; and φ' represents the average rating. At the social level, the frequency of communication, the duration of communication, and the interaction activities are used to represent the intimacy between users. Social attributes include the user's gender, job, age, education level, hobbies, etc. Users with similar interests are more likely to interact or request similar data and establish stable links. Under the condition of the same social trust value, a larger similarity value will result in a higher trust value, which is more in line with common sense [7]. Users with strong social relations have a high probability of forwarding data packets, requiring more energy resources and spectrum. Discovering the interaction patterns among users can help to significantly improve the allocation of wireless network communication resources.

2.3 Optimal Relay Node Selection Based on Social Relations

In a wireless communication network, how to select a suitable relay node and how to design a reasonable and effective resource allocation algorithm is very important to improve the performance of the network. By understanding the social relations between users, it is helpful to select trusted relay nodes. This paper mainly uses the social relationship between users and relay nodes to select the optimal relay node for the wireless network communication link. The main reason for using social trust to decide candidate relay nodes is to enable reliable and trustworthy wireless communication between users. Forwarding data through trusted nodes can not only provide better privacy protection, but also increase the chance of data being successfully forwarded. The area division mechanism is used to limit the number of user candidate relay nodes. Two circles are drawn with the source user and the communication user as the center and the communication distance between the source user and the relay nodes in the network as the radius. Only relay nodes located in the overlapping area of the two circles are eligible to be candidate nodes in the network. The schematic diagram of the area division mechanism is shown in Fig. 2.

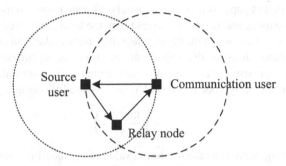

Fig. 2. Schematic diagram of the regional division mechanism

Usually, terminal devices are held by users, and the social relationship between users plays a decisive role in whether the terminal device is willing to become a relay node. Different groups are formed between different users according to the strength of social relations. According to the relationship and needs between users, spectrum resources need to be allocated in different groups. Users who are more popular in the network and interact more with other users can be regarded as relay nodes in the network. In social networks, the willingness of nodes is very important for the formation of social relations and the dissemination of information, and the strength of social relations between two users plays a very important role in whether they can form relay wireless communication. That is, the stronger the social relationship between users, the easier it is to form relay wireless communication.

Considering the social trust and the frequency of encounters among users, a set of candidate relay nodes can be reasonably allocated to users. Candidate relays in the set can encounter frequent encounters with the current user and are willing to provide data forwarding cooperation. Sort users in descending order according to the strength of their social relationships. Then, according to the order of centrality, it is judged in turn whether each user can become a relay node. In addition, distance information should also be considered in the judgment. When the distance between the user to be determined and the relay node that has become a relay node exceeds the specified distance, the user to be determined can be used as the next relay node. According to the above method, all relay nodes will be finally determined. The user's candidate relay set must meet the following conditions:

$$k = \{\vartheta_{lk} \leq \vartheta_0; v_{lk} \geq v_0\} \tag{5}$$

In Eq. (5), k represents the relay node; l represents the target user; ϑ_0 represents the social trust threshold; v_{lk} represents the encounter probability between the target user and the relay node; v_0 represents the encounter probability threshold. Relay nodes, as resource providers, play a dominant role in the cooperative communication process. Obviously, the relay node will consume resources in the process of data forwarding, such as energy and cache resources. With the help of the base station, every mobile user can detect the neighboring devices in his vicinity at any time. Based on this information, the user is able to know the number of encounters he has with other users.

The scenario in this paper is how to reasonably allocate power to maximize the benefits of cooperative users and requesting users when the total amount of power that relay nodes can provide is fixed when multiple requesting users make data requests. Assuming that the encounter duration distribution among users is continuous, the encounter duration distribution is usually centered around the mean. The encounter probability of the target user and the relay node can be calculated by the following equation:

$$v_{lk} = 1 - \left(\frac{\min \delta_{lk}}{\Delta t} \right)^{\varepsilon} \tag{6}$$

In Eq. (6), $\min \delta_{lk}$ represents the minimum value of the meeting interval between the target user and the relay node; Δt represents the communication time; ε represents the probability distribution function parameter. In order to ensure the successful transmission of data and the minimum transmission delay. The user will prefer to choose a device that he can encounter frequently as his relay.

2.4 Establishing the Dynamic Allocation Model of Wireless Communication Network Resources

Mobile devices are usually carried by people, and there is a social relationship between people. This relationship is usually related to a short-term stable social structure and form, and even forms a medium- and long-term social network. Due to the randomness of wireless channels, there is a large uncertainty in signal reception, and users with strong social connections have a higher probability of successfully transmitting data within a certain period of time. In order to find the best assignment result from all feasible mappings for channel assignment, the channel matching process can be abstracted into the maximum weight matching process of the bipartite graph. The objective function can be expressed as the sum of the rates of two users using the same channel plus the rate of a cellular user occupying a channel alone. Each node can only transmit interest data packets or content to nodes in the same cell or adjacent cells, so the node can perceive the transmission of interest data packets or content in adjacent cells.

Users gradually form a stable trust relationship due to their similar interests and the existence of historical interaction information, and jointly optimize the effect of dynamic resource allocation by combining network social characteristics and network physical characteristics. The essence of resource allocation is to determine the correspondence between resource blocks and users on the premise of maximizing throughput and ensuring the quality of service for different types of users. Considering all possible allocation combinations in the network, then calculate the sum of data transmission rates of cellular network users and D2D users under each allocation scheme according to the above calculation method of the sum of data transmission rates, and sort them in descending order. The dynamic allocation model of wireless communication network resources is shown in Fig. 3.

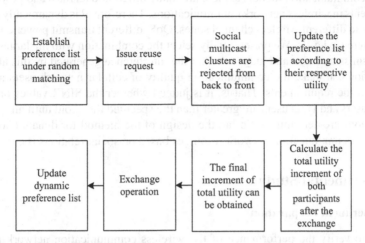

Fig. 3. Dynamic Allocation Model of Wireless Communication Network Resources

Resource allocation allows multiple social multicast clusters to reuse the same cellular user resources to improve spectrum utilization efficiency. From the perspective of channel interference relationship and physical domain model, the mathematical expression of the optimization problem to maximize the overall throughput of the network can be written as:

$$F = \max \mu \eta + \chi \varsigma \tag{7}$$

In Eq. (7), F is the overall throughput of the network; μ is the social matrix reflecting the strength of social relations between users; η is the utility function of rate and social relations; χ is the data rate; ς is the channel gain of the sender and receiver. Define a power control variable, which represents that the user's transmit power on the channel in the joint communication mode cannot exceed the power limit on the channel. It is assumed that a cellular user can only reuse one wireless communication channel, and a wireless communication user can reuse multiple cellular channels. When an interest data packet is sent to a node, if the content requested by the interest data packet exists in the node cache space of the same cell or a neighboring cell of the node, the interest data packet request is satisfied. The node storing the request content of the interest packet will opportunistically (while active) send the request content of the interest packet to the node that received the interest packet. Users establish connections with wireless communication networks whose stability is related to the social strength of multicast clusters. The corresponding utility should be a utility function of the reachability rate and the strength of the link's social relationship. The calculation equation of the utility function can be expressed as:

$$\eta = \chi \theta \log_2\left(1 + \mu^2\right) \tag{8}$$

In Eq. (8), θ represents the power cost, which is used to switch between data rate and power utility. The rate received by the user can be calculated from the channel state. For a fixed channel, it can only be used by one user in either the cellular mode or the D2D communication mode. As there are many different data rates and power utility requirements in wireless network communications. Therefore, θ is dynamically adjusted according to different wireless channel states, QoS, different transmit power constraints and interference levels. We preferentially select the combination with the largest sum of data transmission rates, and judge whether the interference generated by this allocation combination can make the communication quality of cellular network users and D2D users meet the requirements. That is, it is judged whether the SINR values of cellular network users and D2D users are greater than the specified threshold until all allocation combinations are determined. So far, the design of the method for dynamic allocation of wireless communication network resources based on social relations is completed.

3 Experimental Study

3.1 Experiment Preparation

In order to verify the performance of the wireless communication network resource dynamic allocation method based on social relations proposed in this paper, it is verified in the MTALAB simulation platform next. The simulation scenario considers the independent distribution of a single cell, the coverage radius of the cell is 500m, and the cellular users and D2D are randomly distributed in the cell. The base station is located in the center of the cell and is responsible for resource scheduling of the entire system. Each cellular user has the same number of orthogonal resources, and D2D users reuse uplink resources. The maximum transmission distance of D2D communication is set to 20m, and the transmit power of cellular users and D2D users is 25dBm. D2D sending and receiving devices form D2D pairs through a pair matching algorithm according to social and physical constraints. At the same time, it is assumed that cellular users occupy all spectrum resources in the system, and the shadow fading of all communication links in the system obeys a log-normal distribution with a mean value of 0. The simulation parameters are shown in Table 1.

Table 1. Simulation parameter table

Serial number	Simulation parameters	Value	Serial number	Simulation parameters	Value
1	Cell radius (m)	500	5	Noise power (dBm)	−135
2	D2D maximum distance (m)	20	6	Path Fading Index	2.5
3	D2D maximum transmit power (dBm)	25	7	Minimum data rate (bps/Hz)	0.48
4	Base station transmit power (dBm)	40	8	Cellular user transmit power (dBm)	25

In order to better reflect the performance of the method, the dynamic allocation of wireless communication network resources based on social relations is compared with the methods based on automatic differentiation and layered game theory.

3.2 Results and Analysis

Tables 2, 3, 4 and 6 show the comparison results of the network transmission speed of each network resource dynamic allocation method under the test conditions of different numbers of users.

Table 2. Comparison of network transmission speed with 20 users (Mbps)

Number of tests	Dynamic resource allocation of wireless communication network based on social relationship	A dynamic resource allocation method for wireless communication network based on automatic differentiation	Hierarchical game-based wireless communication network resource dynamic allocation method
1	100.15	89.39	93.89
2	102.47	91.08	94.58
3	103.58	92.47	85.67
4	100.96	86.14	94.36
5	104.65	87.05	86.05
6	102.21	88.66	95.14
7	103.54	89.83	93.48
8	105.88	91.55	92.12
9	104.62	90.21	91.23
10	102.36	92.64	90.16

Table 3. Comparison of network transmission speed with 40 users (Mbps)

Number of tests	Dynamic resource allocation of wireless communication network based on social relationship	A dynamic resource allocation method for wireless communication network based on automatic differentiation	Hierarchical game-based wireless communication network resource dynamic allocation method
1	123.07	112.16	108.14
2	128.44	110.53	107.58
3	129.56	110.85	106.82

(continued)

Table 3. (*continued*)

Number of tests	Dynamic resource allocation of wireless communication network based on social relationship	A dynamic resource allocation method for wireless communication network based on automatic differentiation	Hierarchical game-based wireless communication network resource dynamic allocation method
4	126.65	109.68	105.63
5	125.22	113.34	112.26
6	127.81	114.59	103.50
7	128.35	109.26	109.47
8	126.68	108.03	112.79
9	128.06	115.45	114.61
10	127.23	113.74	115.52

Table 4. Comparison of network transmission speed with 60 users (Mbps)

Number of tests	Dynamic resource allocation of wireless communication network based on social relationship	A dynamic resource allocation method for wireless communication network based on automatic differentiation	Hierarchical game-based wireless communication network resource dynamic allocation method
1	142.65	124.63	118.06
2	146.87	126.92	124.44
3	148.54	125.51	119.86
4	147.21	122.72	120.63
5	140.42	123.41	122.35
6	145.16	120.18	118.52
7	139.48	123.29	120.72
8	146.80	124.86	122.12
9	142.52	123.81	125.28
10	143.61	122.65	126.94

When the number of users is 20, the network transmission speed of the dynamic allocation method of wireless communication network resources based on social relations is 103.04 Mbps, which is 13.14 Mbps and 11.37 Mbps higher than the dynamic allocation method of wireless communication network resources based on automatic differentiation and hierarchical game theory.

Table 5. Comparison of network transmission speed with 80 users (Mbps)

Number of tests	Dynamic resource allocation of wireless communication network based on social relationship	A dynamic resource allocation method for wireless communication network based on automatic differentiation	Hierarchical game-based wireless communication network resource dynamic allocation method
1	167.49	141.49	136.43
2	165.84	139.57	142.97
3	164.67	138.25	144.84
4	165.30	136.36	145.61
5	162.56	142.03	139.35
6	163.23	143.85	139.16
7	168.15	140.67	140.78
8	167.47	138.58	138.52
9	169.82	137.78	137.36
10	165.63	142.22	142.26

When the number of users is 40, the network transmission speed of the dynamic allocation method of wireless communication network resources based on social relations is 127.11 Mbps, which is 15.35 Mbps and 17.48 Mbps higher than that of the dynamic allocation method of wireless communication network resources based on automatic differentiation and hierarchical game.

When the number of users is 60, the network transmission speed of the dynamic allocation method of wireless communication network resources based on social relations is 144.33 Mbps, which is 20.53 Mbps and 22.44 Mbps higher than that of the dynamic allocation method of wireless communication network resources based on automatic differentiation and hierarchical game.

When the number of users is 80, the network transmission speed of the dynamic allocation method of wireless communication network resources based on social relations is 166.02 Mbps, which is 25.94 Mbps and 25.29 Mbps higher than that of the dynamic allocation method of wireless communication network resources based on automatic differentiation and hierarchical game.

When the number of users is 100, the network transmission speed of the wireless communication network resource dynamic allocation method based on social relationship is 182.41 Mbps, which is 21.22 Mbps and 17.79 Mbps higher than the automatic differentiation and hierarchical game-based wireless communication network resource dynamic allocation method.. With the increase in the number of D2D users, the network transmission speed is also increasing. The reason for this phenomenon is that the number of users participating in the communication increases, and the throughput of the system also increases. In the same simulation environment, the performance of the

Table 6. Comparison of network transmission speed with 100 users (Mbps)

Number of tests	Dynamic resource allocation of wireless communication network based on social relationship	A dynamic resource allocation method for wireless communication network based on automatic differentiation	Hierarchical game-based wireless communication network resource dynamic allocation method
1	182.92	159.34	164.08
2	180.67	158.41	162.56
3	181.94	159.77	166.62
4	185.81	162.69	165.35
5	182.46	161.06	164.87
6	183.25	162.33	167.04
7	180.38	160.85	166.56
8	181.62	163.57	165.74
9	182.43	164.22	164.13
10	182.57	159.64	159.29

method proposed in this paper is improved in terms of network transmission rate, so the performance of the dynamic allocation method is better than that of the compared resource allocation methods.

4 Conclusion

With the popularization of smart terminals and the rapid growth of data traffic, the emerging demand for local services is growing. Wireless network communication can improve spectrum utilization, reduce the burden on equipment, and increase system capacity. This paper proposes a dynamic allocation method of wireless communication network resources based on social relations, in order to achieve the goal of improving the network transmission rate. We construct a wireless ad hoc network consisting of a random number of nodes to determine an upper limit of its security capacity, with edge nodes on average bear more communication transmission load. Considering the social trust and encounter frequency, the users. The sum of the data transfer rates under each allocation scheme is calculated and sorted in descending order, and the resources are dynamically allocated according to the principle of the maximum overall throughput of the network. Due to the complex and open nature of social networks and the dynamic changes of mobile users, there may be security issues such as malicious nodes stealing, forging or tampering with information in the network, resulting in unreliable information sources in social networks. Therefore, it is necessary to further design the detection mechanism of non-cooperative nodes to realize the safe sharing of data.

References

1. Zheng, B., Sun, Y., Wu, Y., et al.: Ultra-dense network resource allocation based on deep reinforcement learning. Electron. Measure. Technol. **43**(9), 133–138 (2020)
2. She, X., Zhan, Q., Wu, C.: Multi-node information resource allocation recommendation algorithm based on collaborative filtering. Comput. Simul. **38**(6), 419–423 (2021)
3. Wang, Z., Zhang, T., Xu, W., et al.: Dynamic caching placement and resource allocation in UAV emergency communication networks. J. Beijing Univ. Posts Telecommun. **43**(6), 42–50 (2020)
4. Cao, Q., Wang, H.: Spectrum resource allocation in 5G/B5G mobile communication network. Commun. Technol. **53**(8), 1918–1922 (2020)
5. Qian, H.: Research on network resource allocation strategy based on user QoS. Mob. Commun. **45**(2), 124–128 (2021)
6. Feng, L., Xie, K., Zhu, L., et al.: 5G ultra-reliable and low latency communication resource scheduling for power business quality assurance. J. Electron. Inf. Technol. **43**(12), 3418–3426 (2021)
7. Wang, H., Tan, G., Zhou, S.: Dynamic resource allocation of D2D heterogeneous network based on delay constraint. Electron. Measur. Technol. **43**(17), 130–136 (2020)

A Novel Weight Adaptive Multi Factor Authorization Technology

Ruiqi Zeng[2,3]([✉]), Leyu Lin[2,3], and Yue Zhao[1,2,3]

[1] Science and Technology on Communication Security Laboratory, Chengdu 610041, China
[2] No. 30 Research Institute of China Electronics Technology Group Corporation, Chengdu 610041, China
zengruiqi@sina.com
[3] Electronics Technology Cyber Security Co. Ltd., Chengdu 610041, China

Abstract. Facing the increasingly complex computer network system, the importance of network security has become increasingly prominent. Authentication and authorization are important components of computer network security protection. Authentication is the process of verifying the user's identity, and authorization is the process of verifying the user's right to access. In terms of authorization, although there has been a lot of research on related aspects, from the perspective of authority determining factor, it mainly focuses on single-factor authority. For authority control in complex scenarios, single-factor often has great limitations, because it cannot carry out precise control. This paper takes multi-factor authorization as the research object, studies the role of multi-factor in the authorization process, and designs a simple multi-factor authorization algorithm. On this basis, the factor weight adaptive technology is studied, and two adaptive weight algorithms are designed to meet more precise authority control in complex scenarios. Through construction and testing of the actual prototype system, the utility and advantages of multi-factor and weight adaptation in authorization are verified, which expands ideas for subsequent in-depth study of authorization technology in complex scenarios.

Keywords: Authorization · Authority control · Multi-factor · Weight adaptive

1 Introduction

Internet is open to the whole world, and any unit or individual can conveniently transmit and obtain various information on the Internet. The open, shared, and international characteristics of the Internet pose a challenge to computer network security. There are many factors that constitute insecurity to computer information, including human factors, natural factors and accidental factors. Computer network security is to protect hardware, software, and data resources in computer network system from being damaged, modified, or leaked due to accidental reasons or malicious attacks, so that network system can operate continuously and reliably, and network service is normal and orderly [1].

W. Fu and L. Yun (Eds.): ADHIP 2022, LNICST 469, pp. 446–457, 2023.
https://doi.org/10.1007/978-3-031-28867-8_33

In computer security system, identity verification and authorization protection are an important part. Authentication is used to verify user's credentials, such as user name, user ID, etc., to determine user's identity. Common authentication is usually done through username and password, sometimes combined with authentication factors. There may be several authentication factors: single-factor authentication, which is the simplest method of authentication, and usually relies on simple passwords to grant users access to specific systems (such as websites or networks); two-factor authentication, which is a two-step verification process that requires not only a user name and password, but also some sort of user characteristics to ensure a higher level of security; multi-factor authentication, which is the most advanced authentication method, uses two or more security conditions in independent authentication categories to grant users access to the system [2].

Another aspect of system protection is authorization. Authorization is the process of determining whether an authenticated user can access a specific resource. Authorization occurs after system successfully authenticates user's identity, and finally grants user some access to resources (such as information, files, databases, funds, locations, and almost any content). Authorization determines user's ability to access system and extent to which it can reach.

In the process of authorizing a user, it can be determined by a single factor or by using multiple factors. The single-factor authorization system uses only one condition as the authorization basis, such as role authorization. Multi-factor authorization is the process of using two or more conditions to control user access. Thus, access security is effectively improved by combining multiple attribute conditions, such as user characteristics, environmental factors and resource attributes [3].

In multi-factor authorization, in order to better express relationship between each factor and improve accuracy of the final authorization result, it is necessary to assign weight to each factor and establish a multi-factor authorization algorithm. Generally, the weight of the system is pre-configured and will not change arbitrarily, but for complex scenes, a fixed weight value often cannot reflect changes in environment and cannot achieve a very accurate authorization effect. At this time, a multi-factor authorization method with adaptive weights can be considered [4]. Self-adaptive means that in the process of processing and analysis, the processing method, processing sequence, processing parameters, boundary conditions or constraint conditions are automatically adjusted according to characteristics of the processed data, so as to adapt to statistical distribution characteristics and structural characteristics of the processed data, and achieve the best treatment effect. Weight adaptation means that the weight automatically adjusts and adapts to changes in environment to achieve the best weight value.

At present, there are not many related researches on weight adaptive multi-factor authorization control. In order to explore weight adaptive and multi-factor authorization algorithms, this paper first studies the role of multi-factor in authorization, and gives some multi-factor authorization algorithms to verify the advantages of multi-factor authorization in complex scenarios. On the basis of this research, weight adaptation is discussed, some weight adaptation algorithms are proposed, and the design scenarios are verified. Finally, the verification results are summarized, and the follow-up research is prospected.

2 Multi-factor Authorization

In network security, user identification and authorization are very important parts, among which authorization is to grant permissions to user after identification. There are many ways of authorization, such as ABAC [5], RBAC, DAC, MAC, etc. Some authorization systems are based on a single factor. This type of system is often used in simple scenarios. For some complex scenarios, it cannot achieve good results. At this time, a multi-factor authorization method is required. Multi-factors can take multiple factors into consideration, realize multi-dimensional and multi-level authorization, and achieve more precise authority control.

2.1 Multi-factor Combination

Single factor analysis refers to the analysis of a certain variable at a certain moment, such as time factor, spatial factor, and personnel characteristics. Through the analysis of a certain variable, a certain conclusion can be drawn, or a certain operation can be carried out.

Multi-factor analysis is a series of statistical analysis methods that study the relationship between multiple factors and the individuals with these factors. Multi-factor analysis is used to explain the direction and extent of the overall change caused by each factor change when the total change of a phenomenon is affected by three or more factors. Through multi-factor analysis, it is possible to observe the target in multiple dimensions, obtain target characteristics, gain a deep understanding of the system from different angles, and find related relationships, so as to obtain more comprehensive and in-depth results and support subsequent operations more accurately.

In a single-factor authorization system, authorization is often performed based on a certain factor, such as granting a certain authority based on a certain identity. For general simple scenarios, single-factor authorization can meet the target requirements, but in some complex scenarios, the actual conditions are often more complicated, and single-factor conditions cannot meet the needs well. In this case, it is necessary to combine multiple factors to analyze at the same time, so as to make a more comprehensive and accurate judgment, and carry out more precise authority control. At the same time, multiple factors can achieve different combinations, thereby enabling multi-level authorization from different angles and dimensions to meet richer needs.

In the design of our authorization scheme, we adopted a multi-factor analysis method and used multiple factors as the basis for judgment to construct a complex authorization control scenario. When building our prototype system, a scenario with three factors acting simultaneously was simulated, and the accuracy value of each factor was used as a measure to achieve more precise authority determination.

2.2 Multi-factor Weighting Algorithm

When a multi-factor strategy is adopted, multiple factors jointly affect the decision-making, and the analysis result is calculated by the multi-factor weighting algorithm.

We use weighted average method, which is to set a weight for each factor and find their weighted average [6]. The calculation formula is shown in formula 1.

$$score = \sum_{k=1}^{k=n}(V_k * W_k) \tag{1}$$

In the above formula, the measured value of each factor is multiplied by its own weight, and their sum is the final result. In addition to the weighted average method, other algorithms can also be used, or you can customize the algorithm according to the actual situation.

In our prototype system, we simulated four factors, as shown in Fig. 1. The four factors are: facial recognition, gender, coat color, and identification accuracy weighted average. Among them, the weighted average of factor accuracy is a comprehensive value of the previous three factor accuracy, which is obtained by the above weighted average algorithm.

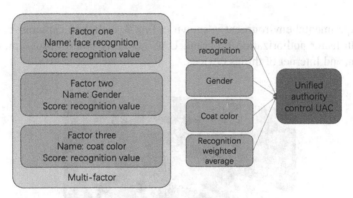

Fig. 1. Authority factors.

2.3 Multi-factor Authorization Prototype System Experimental Results

In order to verify the effect of multi-factor authorization, we designed a prototype system to simulate the role of multiple factors in the authorization determination process.

In our design, we deployed three identification factors, namely: name, gender, and coat color. The system structure diagram is shown in Fig. 2. Three cameras are used to shoot separately, corresponding to the three factors of name, gender, and coat color. Each camera has a monitoring program which records captured image and sends the pictures to the multi-factor authorization program. The multi-factor authorization program contains a multi-factor authorization algorithm to calculate the final recognition value. The final recognition value here is the weighted average of the recognition degrees mentioned above. The multi-factor authorization program also includes weight configuration, and the weight of each factor can be manually set to meet customized requirements. After that, the system will sort out all user attributes, including: name, gender, coat color, and final recognition value, and send these attributes to the unified authority control UAC system connected to the multi-factor authorization program for authority determination. For users who pass the authorization judgment, the program will call the door lock application of the Internet of Things to open the door lock.

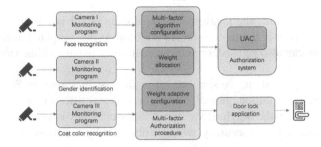

Fig. 2. Weight-adaptive multi-factor authorization prototype system architecture.

The experimental environment is shown in Fig. 3, including cameras, camera program, multi-factor authorization program, UAC authority control program, door lock application, and Internet of things devices.

Fig. 3. Experimental environment.

The overall system configuration interface is shown in Fig. 4. Including weight setting, weight algorithm, and weight adaptive algorithm. The weight adaptive algorithm will be discussed in next section.

Fig. 4. System configuration user interface.

The factor weight can be manually set to meet specific needs. The configuration result is shown in Fig. 5.

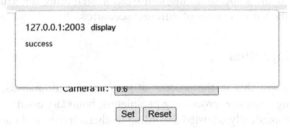

Fig. 5. Weight configuration successful.

The effect of multi-factor authentication is shown in Fig. 6. It can be seen that the determination strategy is defined, multiple user characteristics are obtained through different cameras, and the final score value is calculated, after that all the characteristic value and final score value are passed to authority control system to authorize user to unlock the door. The system message records the entire multi-factor authority process.

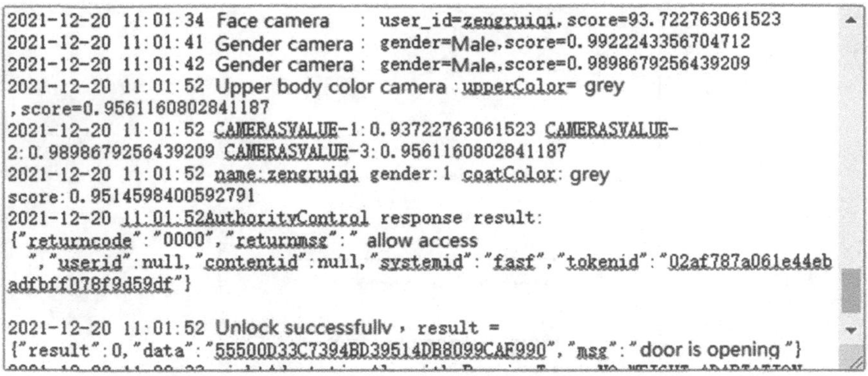

2021-12-20 11:01:34 Face camera   : user_id=zengruiqi, score=93.722763061523
2021-12-20 11:01:41 Gender camera : gender=Male,score=0.9922243356704712
2021-12-20 11:01:42 Gender camera : gender=Male,score=0.9898679256439209
2021-12-20 11:01:52 Upper body color camera : upperColor= grey
, score=0.9561160802841187
2021-12-20 11:01:52 CAMERASVALUE-1:0.93722763061523 CAMERASVALUE-
2:0.9898679256439209 CAMERASVALUE-3:0.9561160802841187
2021-12-20 11:01:52 name:zengruiqi gender:1 coatColor: grey
score:0.9514598400592791
2021-12-20 11:01:52AuthorityControl response result:
{"returncode":"0000","returnmsg":" allow access
", "userid":null,"contentid":null,"systemid":"fasf","tokenid":"02af787a061e44eb
adfbff078f9d59df"}

2021-12-20 11:01:52 Unlock successfully , result =
{"result":0,"data":"55500D33C7394BD39514DB8099CAF990","msg":"door is opening "}

Fig. 6. System running process display.

2.4 Multi-factor Authorization Advantages

In our multi-factor authority determination design, we combine multiple factors for authority control, which can be applied to complex control scenarios. Factors can be adjusted according to the actual situation to monitor in different dimensions, and control more precisely. At the same time, the multi-factor weighting algorithm can also be adjusted to meet the various needs of different scenarios.

3 Weight Adaptation

Self-adaptation means that in the process of processing and analysis, the processing method, processing sequence, processing parameters, boundary conditions or constraint conditions are automatically adjusted according to characteristics of the processed data, so as to adapt to statistical distribution characteristics and structural characteristics of the processed data, and achieve the best treatment effect. Applying adaptive technology to multi-factor authority control can continuously achieve the best control effect as the environment changes.

3.1 Weight Adaptation

In multi-factor algorithm, each factor has a weight, which is used to determine factor proportion, and illustrate its importance. In some simple scenes, the weight is fixed, but in many complex scenes, fixed weight does not reflect real situation, because environment is constantly changing, and the actual proportion and importance of each factor are also constantly changing. At this time, the weight needs to be changed accordingly, automatically adapting to changes in the environment, that is, the weight is adaptive [7].

3.2 Adaptive Algorithm

There are many kinds of adaptive algorithms, the more common ones in industrial applications are LMS, RLS algorithms, etc. [8].

For our multi-factor authorization process, our multi-factor weights have customized some algorithms according to our own scenarios, including fixed weight, time period adaptation and average value adaptation.

Fixed weight means that the weight value of each factor is a fixed value, and it will not change as the environment changes.

The time period-based adaptive algorithm takes time periods as a reference and sets different weight values in different time periods. In the factor setting, three factors (name, gender, and coat color) are set. The weight of each factor is set separately according to the time period. The weight values of name, gender, and coat color in first time period are set as: 0.6, 0.2, 0.2, and the weight values in second time period are set to: 0.2, 0.6, 0.2, and the weight values in third time period are set to 0.2, 0.2, and 0.6 respectively. This simulates the weight change of different time periods (Fig. 7).

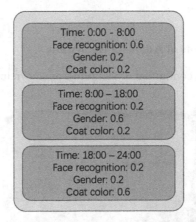

Fig. 7. Time period weight adaptation.

The average value adaptive algorithm is based on the average value of recognition accuracy of each camera. The specific algorithm process is, first record the recognition accuracy of the last 20 photos taken by a single camera (a certain factor), and then calculate their average accuracy value, the calculation formula is shown in formula 2. After that, the accuracy averages of all factors are compared and sorted, and finally a set of weights is defined and these weights are assigned to the ranked factors.

$$averageScore = (\sum\nolimits_{k=1}^{k=n} V_k)/n \qquad (2)$$

Figure 8 is the algorithm flowchart. In our setting, we also define three factors: name, gender, and coat color, and define three weights: 0.45, 0.35, 0.2. Three weight values are assigned to each of the three factors respectively based on the order of the average value of each factor recognition degree. In this way, the weight value of each factor will be adjusted in real time according to recent recognition accuracy status. Here we simulate a scenario, in which the factor with higher recognition accuracy will be assigned a high weight value, and the factor with lower recognition accuracy will be assigned a low weight value.

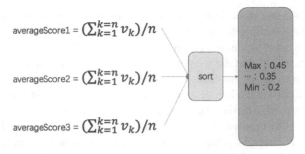

Fig. 8. Average value adaptive algorithm steps.

A variety of weight adaptive algorithms can be configured to use the optimal algorithm at the right time.

3.3 Experimental Results of Adaptive Algorithm

After prototype system is configured, system runs as shown in Fig. 9.

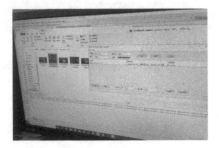

Fig. 9. System operation diagram.

After our prototype system is deployed, no adaptive algorithm is selected in initial stage, so it displays "NO_WEIGHT_ADAPTATION" as shown in Fig. 10. At this time, there is no weight adaptation, and each factor weight is initial value. In our experiment, the default weights of three factors name, gender, and coat color are set to: 0.33, 0.33, 0.33.

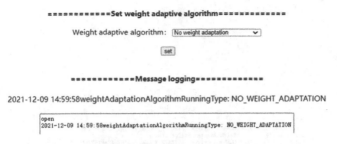

Fig. 10. No weight adaptation.

Set system weight adaptation algorithm to "time period weight adaptive", the output window will show "TIME_PERIOD_WEIGHT_ADAPTATION", and we can see that weight values of the three factors (name, gender, and coat color) are set to: 0.2, 0.6, 0.2, match time period 8:00–18:00 (Fig. 11).

Fig. 11. Time period weight adaptation.

Set system weight adaptation algorithm to "Average weight adaptive", the output window will show "AVERAGE_VALUE_WEIGHT_ADAPTATION". In initial stage, the average recognition value of the three factors (name, gender, and coat color) are set to 0.8, 0.6, and 0.4 by default, so the weight values of the three factors default to: 0.45, 0.35, 0.2. The default weights can be seen in the output window (Fig. 12).

The average value weight adaptation algorithm will count average value of the recognition degree of each factor for the last 20 values to reflect the latest environmental changes. In the subsequent system operation, the weight value changed in real time will be used for calculation, and the result will be used in authority judgment.

Fig. 12. Average value weight adaptation.

3.4 Weight Adaptation Advantages

Through our multi-factor weight adaptive algorithm, the weight can be automatically adjusted with time and environment changes in real-time and will achieve best matching-degree for factor weight. At the same time, the adaptive algorithm can also be configured to meet different needs in different scenarios.

4 Conclusion

This paper studies multi-factor authorization technology with adaptive weights. The paper first explains that with the development of computer network system, authentication and authorization play an important role in computer network security. Then it explains the drawbacks of single-factor authorization and conducts related research on multi-factor authorization, and designs a simple multi-factor authorization algorithm. On this basis, the factor weight adaptive technology is studied, and two adaptive weight algorithms are designed to meet more precise authority control in complex scenarios. Through construction and testing of the actual prototype system, the utility and advantages of multi-factor and weight adaptation in authorization are verified, which expands ideas for in-depth research of authorization technology in complex scenarios.

This research innovatively adds multi-factor authorization algorithms, carries out multi-factor correlation analysis, and can realize multi-level and more accurate authority control, which is impossible in general authorization system. At the same time, the proposed factor weight adaptive algorithm can adapt factor weight to environmental changes, so that the subject analysis is more in-depth, and the authorization process can adapt to more complex application scenarios which general authorization systems cannot be applied to.

In the follow-up research, we will continue to in-depth study of weight adaptation and authorization algorithms in complex scenarios, and introduce some more complex analysis systems, such as big data platforms and artificial intelligence, to further increase the accuracy of authority control.

Acknowledgements. Foundation Item: Supported by Sichuan Science and Technology Program (No. 2021YFG0164).

References

1. Nameless. Cybersecurity Law of the People's Republic of China. Communique of the Standing Committee of the National People's Congress of the people's Republic of China, 2020(3):9
2. Jing, K., Zhang, X., Xu, X.: An overview of multimode biometric recognition technology. In: The 6th International Conference (2018)
3. Ayfaa, B., Apa, C.: LMAAS-IoT: lightweight multi-factor authentication and authorization scheme for real-time data access in IoT cloud-based environment. J. Network Comput. Appl. (2021)
4. Melki, R., Noura, H.N., Chehab, A.: Lightweight multi-factor mutual authentication protocol for IoT devices. Int. J. Inf. Secur. 19(6) (2020)
5. Shen, H.B., Hong, F.: Research on attribute-based authorization and access control. J. Comput. Appl. **87**, 39–45 (2007)
6. Dong, H.L., Park, D.: An efficient algorithm for fuzzy weighted average. Fuzzy Sets Syst. **87**(1), 39–45 (1997)
7. Sun, M., Dou, H., Yan, J.: Efficient Transfer Learning via Joint Adaptation of Network Architecture and Weight. Springer, Cham (2020)
8. Narayan, S.S., Peterson, A.M., Narasimha, M.J.: Transform domain LMS algorithm. IEEE Trans. Acoust. Speech Signal Process. **31**(3), 609–615 (1983)

Multi-source Data Collection Data Security Analysis

Lei Ma[1][(✉)] and Yunwei Li[2]

[1] Beijing Polytechnic, Beijing 100016, China
malei235@tom.com
[2] Beijing Youth Politics College, Beijing 100102, China

Abstract. In order to improve the data effect of multi-source data collection and shorten the time of data collection, this paper proposes a data security analysis method of multi-source data collection. The white noise on the blank data field and knowledge background is removed through data processing, the multi-source data acquisition and access control function is optimized, the encrypted symmetric key is obtained, the filtered data is forwarded to the corresponding trusted exchange agent, the data security exchange characteristics are extracted, and the data security analysis mode is set. Experimental results: the average time consumed by the security analysis method of multi-source data collection data in this paper and the other two security analysis methods of multi-source data collection data are 94.283 s, 129.940 s and 130.121 s respectively, which proves that the performance of the security analysis method of multi-source data collection data in this paper is more perfect.

Keywords: Multi-source data · Data collection · Data security · Data sharing · Data fusion · Semi-structured data

1 Introduction

With the deepening of multi-source data security research, people are no longer satisfied with simply integrating and encapsulating interrelated distributed and heterogeneous data sources. Conventional data sharing and integration can no longer satisfy users' needs in data semantics and knowledge need. Data security analysis is the collaborative processing of heterogeneous data from multiple data sources to achieve the purpose of streamlining data, reducing redundancy, synthesizing complementarity and capturing collaborative information. Data aggregation can combine different statements about the same object scattered in different places to get more complete information about the object. The objective existence of multi-source data and the difficulty of seamless integration of data lead to many problems in the effective management and sharing of information data and files in business processes [1, 2]. Finding an effective data fusion method can handle the intricate relationships between different data sources in a large amount of multi-source data, facilitate the analysis of business processes between network security devices, and make the security analysis operations of related devices simpler and

W. Fu and L. Yun (Eds.): ADHIP 2022, LNICST 469, pp. 458–472, 2023.
https://doi.org/10.1007/978-3-031-28867-8_34

more convenient. At the same time, data aggregation generally involves security issues, because after merging the data and passing some reasoning, some conclusions may be drawn that the data publisher does not expect (may be required by other publishers). Multi-source datasets are large datasets. Compared with traditional datasets, big data is characterized by containing a large amount of unstructured data and semi-structured data. The purpose of multi-source data processing analysis is to discover new and hidden value in data sets, and to efficiently organize and manage large data sets. Compared with traditional data integration, in some cases, people are more concerned about the new semantic meaning exhibited by the aggregated data. Data lays a good foundation for knowledge representation because of its good conceptual hierarchy and support for logical reasoning. Data can be reused, thus avoiding repetitive domain knowledge analysis. We should make good use of existing data, improve the quality of data sources as effectively as possible, and reduce the loss of human, material and related resources in the process of information mining. It has become a basic problem faced by today's computer science and technology to easily and quickly screen out useful data feature information from massive multi-source data or to understand the correlation between data.

To this end, a data security analysis method for multi-source data acquisition is designed to effectively reduce the time required for label generation.

2 Data Security Analysis of Multi-source Data Collection

2.1 Improve Data Processing Flow

The selection and quality of data sets are a crucial condition. A good data set should be evenly distributed, cover a wide range, and have real and effective data, so that the model can correctly learn the parameters required by the trainer. However, the aggregation of data is huge, and the quality of all the data in the original dataset cannot be guaranteed, so many data cleaning algorithms for the original dataset are constantly developing. The principle of data cleaning is to use the existing technical means and methods to clean the "dirty data" by analyzing the causes and existing forms of "dirty data", and convert the "dirty data" into data that meets the data quality or application requirements, thereby Improve the data quality of the dataset. Data cleaning is the detection and screening of outliers in data sets. The data sets used in machine learning training all have certain distribution laws, revealing certain data laws. Therefore, the so-called outliers usually refer to outliers. Due to this feature, before analyzing the security of the data, a model needs to be established in advance. Objects that cannot be fitted with high quality by the model can be regarded as abnormal points. The same is true for other models, but this requires a pre-estimation of the data [3–5]. The means of cleaning are: removing noise data and irrelevant data in the source data set, processing missing data and cleaning "dirty data", removing white noise in blank data fields and knowledge background, considering time sequence and data changes, etc., to complete repeated data processing and the default data processing to complete the conversion of data types. Based on distance cleaning, the distance between data is specified, and the distance from other data objects beyond this distance is regarded as an abnormal point. Data cleaning can be divided into supervised and unsupervised categories. A supervised process, under the guidance of domain experts, analyzes the collected data, removes clearly erroneous noise data and duplicate records, and fills in missing data. The main flow of data processing is shown in Fig. 1:

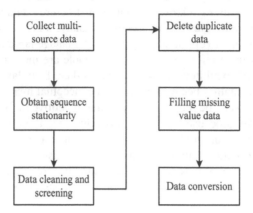

Fig. 1. The main flow of data processing

As can be seen from Fig. 1, the main processes of data processing are: collecting multi-source data, obtaining sequence stationarity, data cleaning and screening, deleting duplicate data, filling missing data, and data transformation. The unsupervised process is to use the established rule base for data cleaning. Generally speaking, data outliers are points with rare attributes in the data set, whose attribute values are very irregular, or relatively low in frequency in the data set, then by calculating the number of occurrences of the values of the corresponding attributes in the data set, we can Effectively assess whether this data is an outlier. Another important aspect of data cleaning is the transformation of data types, usually referring to the discretization of continuous attributes. Generally speaking, the discretization methods that are independent of the class include the equidistant interval method, the equal frequency interval method and the maximum entropy method. According to the above, data set cleaning is a necessary step in machine learning model training, which can help filter out data that is not helpful or even harmful to training. These data do not conform to the overall distribution of the data set, and the model cannot be trained in the data set. It helps, but it will increase the complexity of model training. The methods related to categories include division method and merge method. Through discretization, the size of the data table can be effectively reduced and the classification accuracy can be improved. To use the original data, it must be cleaned, not only to check the storage format of each attribute, but also to check whether its actual content conforms to the specification, such as handling vacancies, identifying and deleting outliers, deleting some duplicate records, and correcting The validity of the attribute value is checked, etc. During the training process, model parameters need to pay more attention to the fitting loss caused by abnormal points, and at the same time, the fitting of normal data will inevitably be affected, which will increase the time and computing resources required for model training, and affect the efficiency of training. In severe cases, it will affect the results of training, resulting in larger training errors and even more serious business losses. This problem is more noticeable in the scenario where joint data sources participate in data cleaning. Fill gaps with the most probable values: The gaps can be identified using regression, Bayesian formal methods tools, or decision tree induction, etc. For example, using the attributes of other customers in the data set, a decision tree can be constructed to predict the vacancy value of income. In the joint data source scenario, because the data of each participant is mixed, in this solution setting, although all parties hold the same type of data. The problem of detecting and eliminating duplicate records is one of the main issues of research in the field of data cleaning and data quality. In the process of merging multi-source heterogeneous data, it is necessary to import a large amount of data from various data sources. Ideally, for an entity in the real world, there should only be one corresponding record in the data source. However, due to the differences in collection methods and storage methods, the data of each participant may have great discrepancies, such as the format of data storage, the dimensions of the data, and the distribution of characteristics and attributes of the data., the above differences between the data may cause a loss of model accuracy. However, when integrating multiple data sources represented by heterogeneous information, due to various problems such as data input errors, differences in format and spelling, etc. in the actual data, it is impossible to correctly identify multiple records that identify the same entity., so that logically pointing to the same real-world entity, there may be multiple different representations in the merged data, that is, the same entity object may correspond to multiple records. Although each participant can perform data cleaning locally, due to the differences in the data itself, the cleaning algorithm may also have different choices, and the cleaning results of each participant may not meet the requirements after data fusion, so the data of all parties are collected together. It is more feasible and can guarantee the cleaning results to deal with them uniformly.

2.2 Optimize Multi-source Data Collection Access Control Function

This model uses smart contracts to implement attribute-based fine-grained access control of multi-source data, and adds access control to the ciphertext acquisition process for dynamic permission judgment, so that the resource owner acts as the only promoter and message in the data sharing process changer. With the exponential growth of multi-source data, coupled with the limited resources of local storage, if a large amount of data is stored locally, it will inevitably bring serious challenges to local storage capacity. Therefore, multi-source data owners upload multi-source data to the cloud to save local storage space. At the same time, the blockchain engine shared general ledger technology effectively ensures the reliability of meta-information storage and the auditability of judgment execution. The main body of the multi-source data sharing process includes the data owner and the multi-source data requester. The direction of multi-source data transmission is from the data owner to the data requester. In order to realize the secure storage of data and the association between on-chain data and off-chain data, we adopt technologies such as blockchain, smart contracts, and IPFS. However, in order to ensure the security of multi-source data, it is necessary to conduct security analysis on multi-source data. One method is to download multi-source data directly from the cloud for security analysis. This method is undoubtedly the best in terms of correctness, but it consumes a lot of resources and time and reduces the efficiency of auditing. Assume that there are several data points in the original space, including clean data points and noisy data points. The filtering operation of noisy data causes the manifold learning algorithm to map the original high-dimensional data points to the low-dimensional space, so that the topological structure of the data is not disturbed by the noise points, or the influence of the noise points is minimized. Then the expression formula of manifold structure data is:

$$D = \sum \frac{E(L - \alpha)}{2} \tag{1}$$

In formula (1), E represents the number of adjacent points, L represents the mapping result, and α represents the translation vector. The two important processes of the multi-source data security sharing process are the data owner uploading resources and the data requester requesting resources. The data owner generates a symmetric key to encrypt the data to be uploaded. The data owner stores the encrypted data in IPFS and obtains the storage address in IPFS. In order to reduce the client auditing overhead, the main method at present is that the multi-source data owner entrusts the auditing task to a third-party auditing agency for auditing. The third-party audit agency adopts the method of random sampling, that is, extracts a part of all multi-source data uploaded by users to the cloud for security analysis [6]. According to the audit results of this part, the integrity of the overall multi-source data is estimated to determine whether the multi-source data is safe. The data owner obtains the encrypted symmetric key through the access control module. The multi-source data owner calls the data management contract to save the data name, storage address, encryption key, dynamic access policy, and author information to the chain state database. The data requester calls the data list method to locate the required data. According to the eigendecomposition, the minimum weighted mean square value of multi-source data is obtained:

$$G_\beta = \frac{H^2 + \varepsilon}{\|1 - \beta\|} \tag{2}$$

In formula (2), H represents the inverse transformation coefficient, ε represents the translation vector, and β represents the weight of the sample point. This method takes both correctness and auditing efficiency into consideration. Compared with the first method, the correctness decreases, but the auditing efficiency is doubled. The multi-source data integrity audit model generally includes three entities: multi-source data owners, cloud service providers, and third-party auditors. The multi-source data requester obtains the storage address information and decryption key through access permission determination. The data requester downloads the encrypted data content and decrypts it with the symmetric key to obtain the metadata content.

2.3 Extract Data Security Exchange Characteristics

From an application-oriented perspective, multi-source data security exchange can be divided into two modes: custom data security exchange and stream security exchange. Data encryption is an effective means of protecting data leakage during transmission or cloud storage. At present, according to whether the keys for data encryption and decryption are the same, it can be divided into symmetric encryption algorithm and asymmetric encryption algorithm. Customized data security exchange is a process of uniformly adapting, converting, filtering, transmitting and loading static heterogeneous data in a specific format based on exchange policies [7]. The basic characteristics of the symmetric encryption algorithm are that it is easy to implement, suitable for encrypting a large amount of data, and the length of the plaintext and the length of the ciphertext are equal, which are the outstanding advantages of the symmetric encryption algorithm. Of course, it will also bring some disadvantages accordingly. It is necessary to build a data transmission channel that both parties can trust in the real environment, which is basically difficult to achieve in an open Internet environment. The characteristics of this data security exchange mode are that it is generally oriented to specific exchange objects, and has strong control over the data exchange process. Exchange, database synchronization, etc. The multi-source data security exchange mode is shown in Fig. 2, and the specific workflow is as follows.

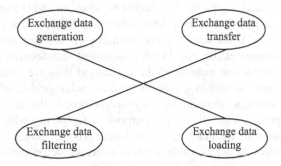

Fig. 2. Multi-source data security exchange mode

As can be seen from Fig. 2, the multi-source data security exchange mode includes: exchange data generation, exchange data transmission, exchange data filtering, and exchange data loading. For exchange data generation, the trusted exchange agent at the sending end is responsible for extracting the data to be exchanged according to the exchange strategy customized by the exchange parties, and then the adapter converts the exchange data into the required format. In addition, once the encrypted data scale increases, the encryption key also increases, and the storage and maintenance of the encryption key becomes a burden for the user. The most fatal disadvantage is that it cannot solve the problems of tampering and denial of messages. Then, the exchanged data is protected and encapsulated by cryptographic technology, and finally the exchanged data is forwarded to the data security exchange server according to the protocol defined by both parties. For exchange data transfer, the data security exchange server, as the controller of the data security exchange, establishes a dedicated secure data channel with both parties of the data exchange, and provides data forwarding for the exchange parties. After the data owner uploads the data to the cloud storage, in order to save the local storage space, the local copy of the data is deleted. When the data owner wants to delete the data copy in the cloud, a deletion order is issued to the cloud service provider. The trusted exchange agent forwards data in a transparent way through "push mode" or "pull mode", without affecting the data exchange between the exchange parties at the application level. Trusted cloud service providers directly delete data, but due to the untrustworthiness of cloud service providers in the current complex network environment, they may only perform logical deletion, and the real data copies are still stored in the cloud. More serious cases are Directly rent this part of the storage space to other tenants, so that other tenants directly obtain copies of your data, resulting in data leakage. Exchange data filtering, the data forwarded to the data exchange server is filtered according to the customized exchange policy, and the data security exchange server forwards the filtered data to the corresponding trusted exchange agent through a dedicated secure data channel according to the customized exchange task.. After the exchange data is loaded, the trusted exchange agent of the receiver verifies the exchange data after receiving the exchange data. After the verification is passed, the exchange data is adapted, converted and loaded into the target system according to the customized exchange strategy. The main idea of the deterministic deletion scheme based on access control is to assign access rights to shared users. When the data is to be deleted, the user's access rights are revoked, so that the data cannot be accessed and indirectly achieves the goal of deterministic deletion of data. This solution is difficult to avoid illegal access technology by hackers. If the access rights technology is not very strong, the security of data still faces serious challenges. In the customized data security exchange mode, since the source, format and content of the exchanged data are relatively fixed, it is easy to protect it, and the exchange process used to exchange data often becomes the main target of the attack, so the main security threats faced in this mode are: The attack on the exchange process can achieve the purpose of tampering with information and spreading malicious code by attacking the exchange process. Based on this, the nature of data security exchange in this mode needs to focus on the protection and control of the exchange process, which can realize the credibility analysis and verification of the exchange process during the exchange execution process, so as to ensure the security of data exchange.

2.4 Set Data Security Analysis Mode

The security analysis is mainly to prove that the timestamp-based signature mechanism is secure in polynomial time, that is, the cloud service provider must store the data owner's files in order to generate valid evidence to respond to the challenge request of the third-party auditor, if the cloud service provider Arbitrary dishonesty will not yield a valid answer. The problem of detecting and eliminating duplicate records is one of the main issues of research in the field of data cleaning and data quality. In the process of merging multi-source heterogeneous data, it is necessary to import a large amount of data from various data sources. Ideally, for an entity in the real world, there should only be one corresponding record in the data source. The main steps of the data security analysis mode are shown in Fig. 3:

As can be seen from Fig. 3, the main steps of the data security analysis mode include: a data interception module, a data encryption/decryption module and a data key module. The function of the data interception module is to intercept the final data processed by the business logic layer before storing it in the distributed storage system of the cloud

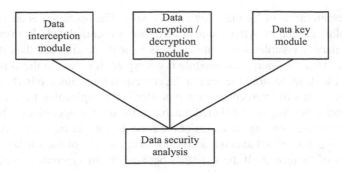

Fig. 3. The main steps of the data security analysis mode

platform, and generate the data ciphertext through the data encryption module. In addition, when the user requests data, the ciphertext data read by the data layer is decrypted and transmitted to the business logic layer for processing. However, when integrating multiple data sources represented by heterogeneous information, due to various problems such as data input errors, differences in format and spelling, etc. in the actual data, it is impossible to correctly identify multiple records that identify the same entity.. The main function of the data encryption/decryption module is to encrypt the stored data and decrypt the read data. The system obtains the secret key generated according to the secret key generation algorithm through the data secret key module, uses the secret key to encrypt the data, and uses the decryption key provided by the user to decrypt the encrypted data when accessing the data. Use the following formula to represent the encrypted data set of the participants:

$$R = \sum\nolimits_{q=1}^{p} \frac{1}{\eta} \times K_{pq} \tag{3}$$

In formula (3), p, q represents two adjacent data nodes, η represents the frequency of occurrence of attribute values, and K represents the weight of attributes. The entities that logically point to the same real world may have multiple different representations in the merged data, that is, the same entity object may correspond to multiple records. Duplicate data can lead to incorrect merge patterns, so it is necessary to deduplicate data in the dataset to improve the accuracy and speed of subsequent merges. The data key module is mainly responsible for managing master keys, generating and distributing data keys. In order to ensure the security of data storage, the form of secondary encryption is adopted. The RSA algorithm master key encrypts the AES algorithm data key. When decrypting, the ASE data key is obtained by decrypting the RSA algorithm private key. The data key is responsible for encrypting the data. Encrypt and decrypt. Each duplicate record detection method needs to determine whether two or more instances represent the same entity. An effective detection method is to compare each instance with other instances to find duplicate instances. In order to detect and eliminate duplicate records from a dataset, the first problem is how to determine whether two records are duplicates. The uploading process of secure data can be divided into: data legitimacy verification, data business logic processing, encryption and storage in the cloud database [8–10]. Data security verification is to ensure that the data format is correct and the content meets the requirements of business logic processing. The operation is checked by the interface before execution. After the data verification is completed, the corresponding functional interface completes the data business logic processing and then the file data is processed. After the encryption module is encrypted, it is sent to the corresponding class of the Dao layer to store the data in the corresponding database in the cloud. The encryption module will intercept and encrypt files before uploading files to HDFS for storage or data uploading to distributed databases for storage operations. This requires comparing the corresponding attributes of the records, calculating their similarity, and then performing a weighted average according to the weight of the attributes to obtain the similarity of the records. If the similarity between the two records exceeds a certain threshold, the two records are considered to be matched., otherwise, it is considered a record pointing to a different entity. The data to be encrypted first determines whether parallel encryption is required according to the size of the data. The data that does not need parallel encryption is directly encrypted by the hybrid encryption algorithm, and then the storage module interface is reflectively called. Large files that need to be encrypted in parallel are uploaded first through MapReduce for parallel hybrid encryption. The sort-merge method is a standard method for detecting exact duplicate records in a database. Its basic idea is to first sort the dataset and then compare adjacent records for equality. This method also provides an idea for detecting duplicate records on the entire dataset, and most of the existing methods for detecting duplicate records are also based on this idea. The data that needs to be encrypted in parallel will be divided into data blocks, and then the data blocks will be allocated to different processors by the MapReduce master node according to the allocation rules. The AES algorithm will encrypt each data block, and the encryption of all data blocks is completed. After that, through the reduce function, the encrypted data blocks are combined and processed to obtain the final ciphertext and the decryption key of the AES algorithm is generated. Each independent file has its own key to ensure that all files will not be caused by the cracking of one key. Encrypted data is broken. After obtaining the AES decryption key, the AES key will be encrypted with the RSA public key of the key management module. After completion, the ciphertext data will be stored in the cloud platform data server.The specific flow of data security analysis of multi-source data collection is shown in Fig. 4.

3   Experimental Tests

3.1   Experiment Preparation

The experiment uses a local virtual machine to load the open source project OpenStack for performance testing, in which the Hadoop sub-project in OpenStack is mainly used to build the required experimental environment. First, the Hadoop cloud environment is built to verify the DPOML algorithm and RFMML algorithm proposed in the paper. According to the existing equipment of the laboratory and the previous scientific research work, the UCI machine learning security data source is selected for the experiment. The hardware configuration of the experimental computer is Dell OptiPlex 3020 Mini Tower desktop, the processor is Inter Core (TM) i7-4590@3.30 GHz quad-core, the memory is 8 GB (Hynix DDR3 1600 MHz), and the main hard disk is GALAXY CX0128ML106-P (128 GB solid state). Hard disk) and Western Digital WDC WD5000AAKX-75U6AA0

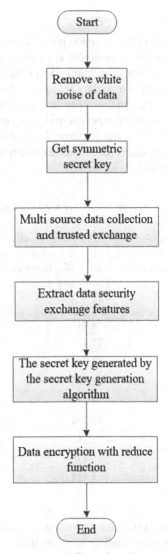

Fig. 4. Data security analysis flow of multi-source data collection

(Blue Disk) (SOOGB mechanical hard disk). The data source is obtained through the Flume component in Hadoop, and then the security data is stored uniformly. This experiment uses four types of security data, namely firewall logs, IDS logs, NetFlow security data, and DNS security data. MultiInputFormat is used to process multi-source data. The deployed Linux system is Centos and Hadoop, and the function is the function library provided by PBC, which is programmed and developed in python language. Finally, the attack test is carried out on the target host. The attack behavior includes application layer, session layer, transport layer, data link layer and network layer, basically covering all layers of the Internet protocol.

3.2 Experimental Results

In order to test the effectiveness of the designed multi-source data acquisition data security analysis method, a comparative experiment is carried out to discuss. The data security analysis method of multi-source data acquisition based on blockchain and the data security analysis method of multi-source data acquisition based on clustering algorithm are selected respectively, and the data security analysis method of multi-source data acquisition in this paper is selected for experimental comparison. The time consumption of security label generation of three multi-source data collection data security analysis methods is tested under different query data volume conditions. The experimental results are shown in Tables 1, 2, 3 and 4.

Table 1. Query data volume 20 GB Security label generation time (s)

Number of experiments	Data security analysis method for multi-source data collection based on blockchain	Data security analysis method for multi-source data collection based on clustering algorithm	Multi-source data collection data security analysis method in this paper
1	72.833	71.009	56.028
2	74.929	74.677	55.993
3	72.091	72.319	53.362
4	71.829	72.062	52.019
5	73.713	71.055	52.044
6	72.640	74.314	53.372
7	72.031	72.298	51.001
8	73.218	72.341	53.083
9	72.090	71.476	51.462
10	74.216	73.318	52.646

According to Table 1, when the amount of query data is 20 GB and the number of experiments is 10, the security label generation time of the blockchain method is 74.216 s, the security label generation time of the clustering algorithm is 73.318 s, and the security label generation time of the method in this paper is 52.646 s; The security analysis method of multi-source data collection data in this paper, compared with the other two security analysis methods of multi-source data collection data, consumes an average of 53.101 s, 72.959 s and 72.487 s for security label generation respectively.

According to Table 2, when the amount of query data is 40 Gb and the number of experiments is 8, the security label generation time of the blockchain method is 98.717 s, the security label generation time of the clustering algorithm is 96.590 s, and the security label generation time of the method in this paper is 62.546 s; The security analysis method of multi-source data acquisition data in this paper, compared with the other two security analysis methods of multi-source data acquisition data, consumes an average of 64.013 s, 95.036 s and 94.254 s for the generation of security labels, respectively.

Table 2. Query data volume 40GB security label generation time (s)

Number of experiments	Data security analysis method for multi-source data collection based on blockchain	Data security analysis method for multi-source data collection based on clustering algorithm	Multi-source data collection data security analysis method in this paper
1	92.736	89.636	67.973
2	98.090	91.017	65.611
3	89.567	92.367	63.544
4	92.381	96.873	62.628
5	91.884	94.091	61.710
6	98.980	98.563	63.008
7	96.862	93.488	64.767
8	98.717	96.590	62.546
9	95.099	95.607	65.330
10	96.045	94.312	63.012

Table 3. Query data volume 60GB security label generation time (s)

Number of experiments	Data security analysis method for multi-source data collection based on blockchain	Data security analysis method for multi-source data collection based on clustering algorithm	Multi-source data collection data security analysis method in this paper
1	120.678	118.937	98.673
2	122.087	121.663	96.556
3	121.990	119.579	95.474
4	118.664	122.367	97.329
5	122.491	123.291	99.09
6	118.368	120.398	105.089
7	121.276	118.467	103.673
8	119.564	122.094	102.182
9	123.820	122.678	96.321
10	124.093	120.022	98.334

According to Table 3, when the amount of query data is 60 GB and the number of experiments is 9, the security label generation time of the blockchain method is 98.717 s, the security label generation time of the clustering algorithm is 122.678 s, and the security label generation time of the method in this paper is 96.321 s; The security analysis method of multi-source data collection data in this paper, compared with the other two security analysis methods of multi-source data collection data, consumes an average of 99.272 s, 121.303 s and 120.950 s for generating security labels, respectively.

Table 4. Query data volume 80 GB security label generation time (s)

Number of experiments	Data security analysis method for multi-source data collection based on blockchain	Data security analysis method for multi-source data collection based on clustering algorithm	Multi-source data collection data security analysis method in this paper
1	163.323	162.563	112.676
2	162.093	163.643	104.442
3	158.248	161.334	108.110
4	161.654	158.873	113.533
5	160.765	162.232	107.708
6	163.238	163.112	112.699
7	158.548	158.111	109.457
8	159.235	163.090	108.245
9	160.345	161.765	110.220
10	163.896	160.874	111.433

According to Table 4, when the amount of query data is 80 GB and the number of experiments is 10, the security label generation time of the blockchain method is 163.896 s, the security label generation time of the clustering algorithm is 160.874 s, and the security label generation time of the method in this paper is 111.433 s; The security analysis method of multi-source data collection data in this paper, compared with the other two security analysis methods of multi-source data collection data, consumes an average of 109.852 s, 161.135 s and 161.560 s for security label generation respectively.

According to Table 5, when the amount of query data is 100 GB and the number of experiments is 5, the security label generation time of the blockchain method is 207.637 s, the security label generation time of the clustering algorithm is 207.771 s, and the security label generation time of the method in this paper is 138.789 s; The security analysis method of multi-source data collection data in this paper, compared with the other two security analysis methods of multi-source data collection data, consumes 145.175 s, 199.265 s and 201.353 s respectively for the generation of security labels.

The data in Tables 1, 2, 3, 4 and 5 show that our party has high multi-source data collection efficiency under different data volumes. This is because the method in this paper removes white noise on the blank data domain through data processing, and shortens the time-consuming of generating security labels by using trusted switching technology.

In order to verify the security of multi-source data collection data of different methods, the blockchain method, clustering algorithm and the method in this paper are used to verify the integrity of multi-source data collection data, and the results are shown in Fig. 5.

Table 5. Query data volume 100 GB security label generation time (s)

Number of experiments	Data security analysis method for multi-source data collection based on blockchain	Data security analysis method for multi-source data collection based on clustering algorithm	Multi-source data collection data security analysis method in this paper
1	201.883	207.119	145.121
2	202.729	193.391	153.362
3	189.647	189.093	142.004
4	193.289	202.088	143.737
5	207.637	207.771	138.789
6	201.562	203.627	142.088
7	193.322	188.028	153.489
8	188.976	201.672	139.421
9	212.389	214.091	152.355
10	201.220	206.646	141.387

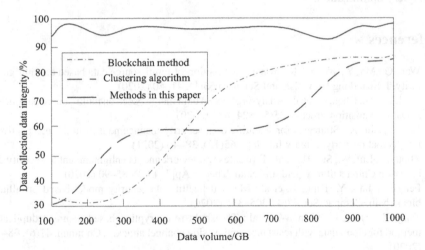

Fig. 5. Data integrity of multi-source data acquisition

It can be seen from Fig. 5 that when the amount of multi-source data collection is 200 GB, the integrity of multi-source data collection data of the blockchain method is 31%, the integrity of multi-source data collection data of the clustering algorithm is 42.5%, and the integrity of multi-source data collection data of the method in this paper is 95.1%; When the amount of multi-source data collection data is 600 gb, the integrity of multi-source data collection data of blockchain method is 78.1%, the integrity of multi-source data collection data of clustering algorithm is 58.9%, and the integrity of multi-source data collection data of this method is 96.8%; This method always has a high integrity of multi-source data collection, which indicates that the multi-source data collection data security of this method is higher.

4   Concluding Remarks

The data security analysis method of multi-source data collection in this paper, in terms of data security analysis, mainly solves the problem of dynamic operation of multi-copy data, and prevents forgery and forgery between distributed storage nodes of a single cloud service provider during the data integrity audit process. Substitution and collusion attacks while reducing data leakage to third-party auditors. At the same time, in the merging and preprocessing of multi-source data, the methods used in data cleaning are discussed, such as dealing with missing values, removing outliers, removing duplicate records, etc. In addition, the methods of data transformation, such as normalizing the data, have also been improved. The method of data merging is to collect data from multiple data sources and store them in a consistent data store. In terms of data deterministic deletion, it mainly solves the fine-grained operation of data by users and adds a trusted verification mechanism after the deletion operation is completed. The future research direction is mainly to continuously improve the subject at the level of multi-dimensional data query optimization.

References

1. Wan, Q., Ma, Y., Wei, L.: Knowledge acquisition of multi-source data based on multigranu-larity. J, Shandong Univ, (Natural Science) **55**(1), 41–50 (2020)
2. Yu, L., Li, S., Chen, C., et al.: Analysis of ocean data merge based on multi-source parameters. J. Data Acquisition Process. **35**(5), 824–833 (2020)
3. Luo, J., Liu, X.: Strategies for scientific data security management from the perspective of intellectual property. Library Inf. Serv. **65**(12), 38–46 (2021)
4. Zhou, X., Liu, W., Sui, H., et al.: Five safes framework and its enlightenment to security data access in China's library field. Inf. Stud. Theory Appl. **43**(3), 85–90 (2020)
5. Feng, T., Jiao, Y., Fang, J., et al.: Medical health data security model based on alliance blockchain. Comput. Sci. **47**(4), 305–311 (2020)
6. Tang, X., Zhou, L., Shan, W., et al.: Threshold re-encryption based secure deduplication method for cloud data with resistance against side channel attack. J. Commun. **41**(6), 98–111 (2020)
7. Lv, G., Chen, L., Xiao, R., et al.: Simulation of quantitative assessment method for data security situation of wireless network communication. Comput. Simul. **37**(7), 337–340,372 (2020)
8. Jiang, L., Tang, Z.: Multi source data acquisition system based on Flume, Kafka and HDFS. Inf. Technol. Informatization **06**, 115–117 (2021)
9. Xu, H., Xu, Z., Chen, M.: Multi source and multi dimensional reading data collection and digital portrait based on xAPI. Educ. Commun. Technol. **16**(04), 59–63 (2020)
10. Wang, J., Guo, Y., Wen, X., Wan, F.: Multi source data acquisition and comprehensive evaluation system for smart business district. Comput. Eng. **45**(01), 284–291 (2019)

Detection Method of Fake News Spread in Social Network Based on Deep Learning

Yandan Lu[1(✉)] and Hongmei Ye[2]

[1] School of Literature and Media, Guangxi Normal University for Nationalities, Chongzuo 532200, China
gxmzsfxy20010023@163.com
[2] Department of Chinese, Changji University, Changji 831100, China

Abstract. The current detection of fake news spread in social networks does not consider the correlation between news text and images, resulting in inaccurate detection results. A detection method for fake news spread in social networks based on deep learning is devised. The size of the time period is dynamically adjusted according to the number of news in the time period, and features are extracted uniformly for comments/retweets in the same time period. Preprocess social network news data to ensure that the vast majority of text is covered and controlled within the range of machine computing power. Multi-modal features are mined and constructed from images, texts and user-side information. Modal fusion does not use the addition of residuals, but splices the residuals and attention matrices, and then sends them to the fully connected layer to convert the dimension size, and then Update the modal. The fused feature vector is input into the feedforward network for classification, and the prediction result is obtained. The experimental results show that the design method can improve the detection accuracy, and the deep features it contains can more effectively detect fake media content in social networks.

Keywords: Deep learning · Social network · Fake news · Communication detection · Detection method · News communication

1 Introduction

Mobile information flows at a high speed in today's highly developed Internet, and people are completely accustomed to obtaining required information from major Internet platforms, especially popular domestic social network platforms such as Weibo and Zhihu. The social network platform represented by Sina Weibo attracts hundreds of millions of users to share, interact and disseminate information on its platform by virtue of its openness, flexibility, free, instant and many other features, which accelerates the speed of information exchange between people and depth. While the explosive information and rich content of social network news brings convenience to people, its shortcomings are gradually exposed. Internet users can speak freely on the Internet in real time, and can also adapt news through forwarding or secondary creation. It is an increasingly

W. Fu and L. Yun (Eds.): ADHIP 2022, LNICST 469, pp. 473–488, 2023.
https://doi.org/10.1007/978-3-031-28867-8_35

serious problem to spread information at almost zero cost and to flood the Internet with spam and even fake news. The spread of false rumors will have large-scale negative effects on society and cause social unrest. Online rumors are defined as unsubstantiated explanations that have played a key role in the life cycle of a social networking platform and have caused a certain social impact. At the same time, the diversity of information, the freedom of expression, and the fission of dissemination speed have made false media content and false speeches an excellent opportunity to come to power, and have also made social platforms such as Weibo become the current flood of false information in China. The main source and transmission medium of my country, and the scope of influence is unprecedentedly huge. At the same time, activities such as forwarding and rubbing heat have increased the intensity of short-term outbreaks of false information, and the scope of influence has further expanded. Compared with traditional text information, information with images or videos can not only provide a richer plot to attract more readers, but also often increase the credibility of news, which is often exploited by false rumors. False rumors often use false or fake images and inflammatory language to mislead readers and spread quickly. Such false information not only confuses the public, but also easily arouses bad public sentiment, deepens social conflicts, and affects the prosperity and stability of the country. Fake news that is intentionally or unintentionally reposted not only affects the direction of public opinion, but also violates the people's right to know, and the country's credibility on the Internet will be weakened in the long run. Therefore, research on fake news detection technology is of great significance to help countries and network platforms curb the generation and spread of fake news and maintain social stability [1, 2]. The time period of rumors can be divided into incubation period, outbreak period and settlement period, corresponding to the stage of just release, extensive discussion, verification or no longer having social impact. Generally speaking, rumors should be detected as soon as possible during the incubation period of rumors. Once malicious false rumors reach the outbreak period, they may have a very large negative impact on society. Due to the complexity and variety of media content and its highly confusing nature, it is difficult for humans to achieve a high recognition rate based on experience, which is not conducive to the rapid detection and blocking of false media content. Artificial intelligence technology has been well used with the continuous progress of science and technology and the in-depth exploration of the ancestors, and has achieved very good detection results in text classification tasks. The detection of online rumors can not only purify the ecological environment of the network in today's information overload environment, but also help the public identify effective information, and also improve the credibility and credibility of the platform. Computer technology optimizes the human and material cost of replacing manual review and improves the efficiency of news review. Fake news has the characteristics of frequent occurrence and rapid spread, and it can spread all over the Internet in ten minutes. It is difficult to accurately predict and detect fake news from massive information in a timely manner only by manual screening. This paper proposes a detection method of fake news spread in social networks based on deep learning, which helps to advance the research process of text mining and semantic understanding. It can timely respond to the detection of massive Internet fake news through the more accurate classification algorithm model of fake news prediction, improve the efficiency of news review, optimize the quality of news, and improve the reading experience of users.

2 Detection Method of Fake News Spread in Social Network Based on Deep Learning

2.1 Divide the Social Network News Dissemination Cycle

More and more people begin to use online media to obtain news information with the increasing popularity and simplification of mobile applications [3]. Compared with traditional media, online media has the advantages of quick access to information and easy sharing and communication. Spreading rapidly, these messages may provide ground for reactionary activities, shake people's confidence and cause panic, or increase the unnecessary workload of staff [4]. In the detection of fake news dissemination in social networks, since social networks are a platform with continuous dynamics, if only the performance of features at a certain moment is modeled, the characteristics that the features will change over time are ignored [5]. Since the trend of information dissemination is affected by many aspects, in order to reflect the important influence of the actual characteristics of the network on the dissemination, the method based on the local structural characteristics of the network is firstly proposed, followed by some global roaming counting methods and random block walk models. This paper analyzes the time series propagation mode of information events in order to mine the change mode of features. The practice of dividing the event propagation cycle at equal intervals will cause most of the data to be allocated in the early time interval, and there is little or no available information in the later time interval. For a regular network, since the connections between nodes are regular links based on a known strategy, the path length between any two nodes is longer, but the clustering coefficient between them will be high. Contrary to regular networks, nodes are randomly linked with a certain probability for random networks, so the path length between any two nodes will be shorter, but the clustering coefficient will also be lower. When the propagation cycle is divided at equal intervals because the popularity of events is generally in the early stage, the number of comments/numbers of information events in the middle and late time intervals decreases sharply, and even in many middle and late time intervals there is not a single comment/repost, which leads to the follow-up model. There is no effective information to exploit to extract the temporal variation of features. The social network has the advantages of both the regular network and the random network. The length of the characteristic path between any nodes is small, but the clustering coefficient is quite high. It is more inclined to a certain type (regular or random) network and is controlled by the parameter γ. $\gamma = 0$ represents a regular network, and $\gamma = 1$ represents a random network. The process of converting a regular network into a random network is to reconnect the edges existing in the regular network with γ probability. The dissemination period of information events is no longer divided at equal intervals in this paper, but the size of the time period is dynamically adjusted according to the number of news in the time period, and the features are extracted uniformly for comments/reposts in the same time period. The division process of social network news dissemination cycle is shown in Fig. 1.

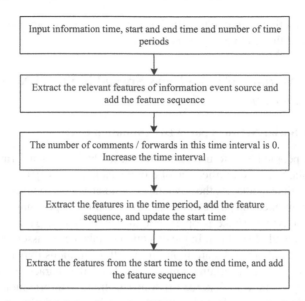

Fig. 1. The division process of social network news dissemination cycle

For information events, the information dissemination period is set, and this time period is divided into M time periods. The information event source is directly divided into a time period in order to highlight the text content characteristics of the information event source in particular, and the number of microblogs and the time interval in this period are initialized to 1. False information on social media seems to be endless, but in fact, a considerable number of events have been judged as rumors. Rumors from many years ago can cause a commotion with a small modification of the place and time. A large amount of similar information often appears in a short period of time after the same false information occurs. If these messages are put into the model indiscriminately, a lot of time will be wasted. Comments/retweets are divided into M-1 time periods. First, the time interval is 1 h. If the number of comments is 0, the time interval is increased, and the time interval is divided into 2 h, and so on. If the number of comments in the current time period is not 0, extract all comment-related features in the current time period. For the last time period, the time interval is no longer cumulative from 1 h, but is set directly to the end time. Information dissemination in online networks is often characterized by high outbreaks, rapid demise, and high anonymity. When tracking the dissemination trajectory of a piece of information, due to various objective reasons, we usually lack information about the underlying dissemination path network, so based on the observed It is a great challenge to reconstruct the hidden propagation paths through the information diffusion process. On social media, the content of fake news is updated and spread quickly, but manual review has the problem of lag and inefficiency. Therefore, it is of great significance to carry out automatic detection of fake news.

2.2 Social Network News Data Preprocessing

We found that there are a lot of invalid data in the data set by observing the original features of the data set, mainly reflected in data duplication and missing attribute values. There are more than 100 occurrences in a concentration, which will undoubtedly affect the subsequent feature extraction and model construction. Therefore, this paper preprocesses the original data to facilitate the subsequent process [6]. Chinese word segmentation is the basis for the text classification of the model constructed in this paper. After successfully segmenting an input Chinese text, the efficiency of computer recognition of words can be greatly improved. Improving the accuracy of Chinese word segmentation can often improve the accuracy of text classification results. Different word segmentation algorithms and thesaurus will affect the final detection effect from the qualitative analysis for text classification tasks. Text preprocessing is required for data that is too long or too short, including removing stop words and data filtering. After the stop words are removed, the overall length of the data is shortened, and words such as "you", "de", and "ba" that are common and have no obvious effect on text feature extraction are removed from the content. Similarly, URLs, user nicknames, etc. existing in the data are replaced with spaces through regular expressions. The statistical-based word segmentation method borrows mathematical theory, the most common and relatively mature are Hidden Markov, Maximum Entropy, Conditional Random Field, etc. Given a large number of texts that have been divided into words, divide a sentence, and then formulate different division methods, and calculate the probability of the division results respectively, and take the word segmentation method with the largest division probability. The text content of the news and the user's personal introduction each have a strong impact on the authenticity of the news, and the connection between the two can also be used as an effective feature for discrimination, such as the possibility of true news when the two are related. Will be significantly improved, and the probability of true news will be greatly reduced when the two are completely unrelated. Since there are many possible combinations of the context of a word, the matrix-based distribution representation usually produces the problem of combinatorial explosion. Furthermore, the sequence of words in the test dataset is likely to be different from the sequence of all words seen in the training set, resulting in poor generalization. Therefore, this article splices the news text with the blogger's self-introduction, and also splices the blogger's geographical information and news section categories into the blogger's introduction. The spliced text is as follows.

$$T = N_T + D_T + C_T + L_T \tag{1}$$

In formula (1), T represents the spliced text; N_T, D_T, C_T, L_T represents news text, user description text, news section category and user geographic location, respectively. Since the statistical based word segmentation method was proposed, the speed of text segmentation and the accuracy of word segmentation have been significantly improved. It is no longer necessary to build a very complete dictionary, and it does not require a deep linguistic knowledge to identify new words and eliminate ambiguity. Using computers to process text classification tasks must represent the word or word after word segmentation into a numerical form that can be recognized by the computer, that is, convert the text into word vectors. It can often be processed manually for data with few missing values, while a large number of missing values will change the distribution of the data, so it is necessary to delete or fill in the missing values. The handling of missing values is shown in Fig. 2.

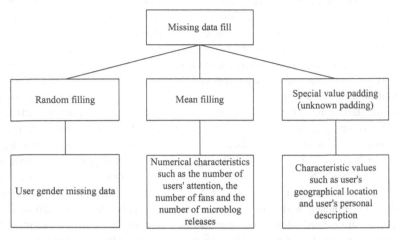

Fig. 2. How to handle missing values

The average special symbols in the text account for 8.3% according to the analysis of the news text content in the dataset, and the text content mainly includes various information such as the body, title, interactive topic, contact information and so on of the news. The neural network-based representation converts sparse word vectors into denser vectors through neural networks rather than statistical matrices. It avoids the curse of dimensionality and can model more complex contextual content. Word2vec generates word vectors by learning the context information of the words, which belongs to the neural network distribution representation. This paper extracts them in a specific format and fixes them at the beginning of the text for the titles and interactive topics in the text, and removes special symbols, http URLs, HTML tags and garbled characters from the news text by regular matching. There is also a type of data in social media, which is characterized by extremely short text length, the main information is reflected in pictures or videos, and the text only contains words such as "view and forwarded video". This type of data text does not contain valid information, temporarily. It is not considered in this article, so it is discarded. Taking into account the problem of machine computing power, the max length of this article is set to 128, which can ensure that most of the text is covered and controlled within the range of machine computing power.

2.3 Fusion of Multimodal News Features

Existing fake news detection often uses single-modal data, such as text, propagation mode or feature engineering, but with the development of the Internet, the ways of presenting news are more and more diverse, and a piece of news usually includes text, images, comments, etc. [7]. Multiple sources of information. This paper will construct multimodal features from images, texts and user-side information to improve fake news detection models. Since the tweets of social networks are mostly short texts, this paper uses LSTM and Text-CNN to extract text features to extract the time-series semantic features and n-gram local features of the text. The pre-trained word vector is used as the input of the model, which is input to LSTM and Text-CNN respectively. The text features are obtained through LSTM, and the hidden layer representation vector is $Z \in SW$, where S is the number of words and W is the feature dimension. The local features of the text are obtained through the Text-CNN module. Images in fake news tend to be of lower quality, clarity and resolution than real news images. The communicator leads to continuous compression of news images through direct copying, multiple forwarding or tampering. Periodic features will be displayed in the frequency domain features for the secondary compression and tampering of images. The DCT algorithm is used to extract the frequency domain features of news images to capture the image structure tampering and compression information in this paper. The extraction process of frequency domain features by discrete cosine transform is shown in Fig. 3.

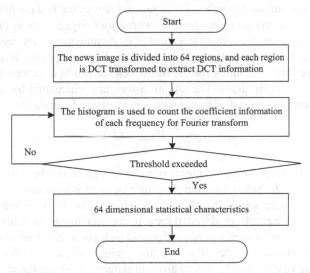

Fig. 3. Extraction process of frequency domain features of news images

User status information includes the user's gender, number of followers, number of followers, location, user description, number of microblogs, news category and other fields. This paper constructs effective features by mining the hidden information of these fields. Taking the number of user followers as an example, fake news is distributed mainly by users with less than 10,000 followers. At the same time, as the number of followers increases, the proportion of fake news in the total number of news releases gradually decreases. The follower growth rate and the follower growth rate in the feature table are the ratio of the number of followers and the number of followers to the registration time, respectively, and the friendly intimacy and the friendly responsiveness are the ratios of the number of mutual followers to the number of followers and the number of fans. Posting activity is the ratio of the number of microblogs to the registration time, and the user reputation value is the ratio of the number of followers to the sum of the number of followers and followers. The features obtained by the above calculations are all continuous values. In light of the actual situation, those with a large number of fans are often certified by Weibo or the official platforms of various media, and the news released is usually true, while those with a small number of fans are often personal bloggers, who publish or forward news to the end. More casual, resulting in the spread of fake news. Modal fusion does not use the addition of residuals, but splices the residuals and attention matrices, and then sends them to the fully connected layer to convert the dimension size, and then update the modality. Convert the text features generated by LSTM into key and value, namely K and V, the calculation process is as follows:

$$\begin{cases} K = f(Z; \varphi_1) \\ V = f(Z; \varphi_2) \end{cases} \tag{2}$$

In formula (2), f represents a fully connected layer without activation function; φ_1 and φ_2 are training parameters. A multi-head attention mechanism is adopted to obtain multi-dimensional attention weights. Propagation features include the number of likes, comments, retweets, and user engagement, where user engagement is the ratio of the number of comments to the sum of the number of comments and retweets. In general, the higher the user engagement, the more attention it can attract, the faster the spread. The above values are continuous values. Multi-head attention is to repeat the attention calculation process many times. The attention weight is calculated by scaling the dot product each time, and the calculation process is shown in formula (3).

$$A = h\left(\frac{f(Z; \varphi_2)\varphi_1}{\sqrt{I}}\right) \tag{3}$$

In formula (3), A represents the attention weight; h represents the activation function; I is the transformed feature dimension. Multiple types of information can complement each other, and it is unreasonable to rely only on a single modality for fake news detection. Statistical features are constructed from three aspects: user information, text statistics and image statistics. These features will be used as user-side contextual features to participate in the construction of the model. The attention weights of each updated modality are computed through a two-layer feed-forward neural network. The updated image features and text features are fused through attention weights.

2.4 Establishment of a Fake News Spread Detection Model in Social Networks Based on Deep Learning

The detection model of fake news dissemination in social networks established in this paper based on deep learning [8] consists of a multimodal feature extraction module, a multimodal feature fusion module [9] and an output module. Traditional machine learning methods require us to construct features by ourselves, but manually constructed features often fail to represent deeper connections between features. Neural networks can automatically capture potential connections in the data, reducing the overhead of manually constructing features for researchers [10]. Among them, the multimodal feature extraction module includes three sub-modules: the text feature extraction module, which uses the BERT model to extract features such as news text semantics and style; the image feature extraction module, which includes the content information and frequency domain information of the image; the contextual feature extraction module, It includes the extraction of user portrait features and statistical features. The event discriminator consists of two fully connected layers. Blog posts in a data set are first labeled with the event to which they belong. Assuming there are k types, the goal of the event discriminator is to correctly assign incoming data to the event to which it belongs. A convolutional filter with window size h takes as input a contiguous sequence of h words in a sentence and outputs a feature. The output vector of the Text-CNN layer is used as the input of the multi-head self-attention layer. By setting the multi-head self-attention, the features between messages in multiple sub-representation spaces can be effectively captured. During training, the feature extractor and false information detector achieve the purpose of improving the detection ability of fake news by reducing the detection loss as much as possible. At the same time, the feature extractor also tries to maximize the loss of event discrimination to achieve the event The discriminator learns the purpose of potentially identical representations between events. The feature vector of text based on the pretrained model can be represented as:

$$\alpha = \tanh[\theta_1 \tanh(\theta_2 + \varepsilon_2) + \varepsilon_1] \tag{4}$$

In formula (4), α is the text feature vector finally obtained in this paper; θ_1 and θ_2 are the weight matrices of the fully connected layer; ε_1 and ε_2 are the corresponding offsets. The idea of the MLM module is to randomly cover the input sentences during training, some words are masked out, and then let the model predict those words by learning the input context. This self-supervised task is similar to cloze. Since Transformer processes a sentence with a distance of 1 for each word, the word prediction mask can take into account the entire sentence. For image data, this paper adds a Batch Normalization layer between the fully connected layer and the activation function in order to speed up network convergence and prevent over-fitting. Finally, after Droupout, the obtained vector is the image content feature vector. Calculated as follows:

$$\begin{cases} \beta = \mathrm{Dr}\tanh\big[B(\theta_2\beta' + \varepsilon_2)\big] \\ \beta' = \tanh[B(\theta_2\chi + \varepsilon_2)] \end{cases} \tag{5}$$

In formula (5), β is the image content feature vector; β' is the frequency domain feature; Dr is the Dropout processing; B is the Batch Normalization layer; χ is the input image. The task of NSP is to determine whether two sentences are contextual, and to learn the features of sentences through this task. The data used for training are sentence pairs extracted from the corpus. 50% of these sentence pairs are contextually coherent, the other 50% are incoherent, and the second sentence is randomly extracted. This paper uses multiple attentions to calculate the attention weight, and then adds the weighted value to the query value to obtain the potential interaction features between messages. The message features of N intervals can be obtained through the above steps. Single domain features only focus on intra-domain differences, while ignoring the relationship between feature domains. The combination of features also has significance for news judgment. In the above, we obtained the image and text features respectively. The correlation between the text and the image is considered, the consistency between the news text and the image can also be used as the basis for fake news detection. This paper combines text features and image features in order to enhance the feature intersection between image and text. The input required by GraphSAGE is the node feature and the node adjacency matrix. In this model, the node is each blog post, and the feature is the text feature plus the event feature. The text features are extracted by BERT, and the event features are extracted manually. The adjacency matrix is represented by the text similarity between blog posts. The discrete features input from the user side are spliced with numerical features after passing through Dense Embedding, and then 256-dimensional user context features are obtained through two fully connected layers [10]. The multimodal feature vector is obtained by concat splicing, which is expressed as:

$$F = [\alpha, \beta, \alpha \otimes \beta, \eta] \tag{6}$$

In formula (6), F represents the multi-modal feature vector; $\otimes$ is the point multiplication operation; η represents the feature set of the user modality. The feature vector is input into the feedforward network for classification, and the prediction result is obtained. The specific calculation process is as follows:

$$P = \text{sig}[\theta_1 \tanh(F\theta_2 + \varepsilon_2) + \varepsilon_1] \tag{7}$$

In formula (7), P represents the prediction result; sig represents the sigmod activation function. The BLSTM layer uses two unidirectional LSTMs in different directions to memorize the forward propagation information and the backward propagation information respectively. Parent object. The hidden state of forward propagation and the hidden state of backward propagation are concatenated, thereby extending the learning ability of the model for the forward and backward directions. So far, the design of the detection method of fake news spread in social network based on deep learning is completed.

3 Experimental Study

3.1 Experimental Scheme

A comparative analysis experiment is designed In order to verify the overall effectiveness of the detection method of social network false news dissemination based on deep learning. The experimental scheme is as follows:

1) Before the experiment, we prepared for the experiment. The experiment took Weibo as the research object, used the Scrapy Redis framework for parallel crawling, counted the collected data sets, and gave detailed information to explain the experimental environment.
2) Determine the experimental performance indicators, and analyze the performance of the detection method through specific performance indicators, including accuracy, accuracy, recall and F1.
3) After determining the experimental performance indicators, carry out comparative analysis. The comparative methods are detection method of fake news spread in

social network based on deep learning, detection method of fake news spread in social network based on support vector machine and A detection method of fake news spread in social network based on Naive Bayes, Verify the effectiveness and feasibility of this method.

3.2 Experiment Preparation

The training data set used in the experiment in this paper is Weibo blog posts, including text and images. Each blog post has a unique corresponding id and topic field. There are three types of tags: news without judgment, real news and fake news. Due to the limitations of the Weibo API, it is impossible to crawl a large number of unauthorized users' data, and crawling through the API cannot meet the data requirements. Therefore, this article uses website page parsing to obtain Weibo information. The specific method is described as follows. The Weibo page needs to be logged in to view, so this article uses Selenium to simulate login, and saves the cookie information after login to the database to prepare for subsequent crawling. According to the topic area, it is divided into seven major categories: "social life", "medicine and health", "sports and entertainment", "technology", "financial business", "military politics" and "educational examination". In addition to text information, data on social networks also includes external URL links, message prompts (in the form of @ plus username), hashtag tags (indicating the topic to which the text belongs), and emoji pictures. We need to preprocess the original data first for these unconventional data information. This article uses the Scrapy-redis framework for parallel crawling, that is, multiple Scrapy-redis processes run at the same time to crawl, and finally save the data to MongoDB. Hashtag tags and expressions have certain meanings, and no special treatment is performed here. For the URL, replace it with a special link/web page link, and replace it with mention someone/mention someone for the @ plus username. The text is then further processed using word segmentation tools. The quartiles represent the values of the points at the 25th, 50th, and 75th percentiles after sorting, respectively. The maximum length of 1982 is too long, which is an abnormal point. In this experiment, the truncation method is adopted, that is, for all data, the content after the length exceeds a certain closed value max length is deleted. The results of the original data statistics show that 75% of the data length does not exceed 140. The statistical information of the microblog dataset used in the experiment is shown in Table 1.

The data is randomly divided into a training set, a validation set and a test set, to verify the application effect of the detection method of fake news spread in social networks. The deep learning experiments involved in this chapter are all based on the deep learning framework TensorFlow, which has built-in multiple open source software libraries for numerical computing, and supports CPU computing and GPU acceleration.

3.3 Experimental Performance Index

In order to effectively verify the performance of the design method, this paper selects the accuracy rate, precision rate, recall rate and F1 as the evaluation indicators The accuracy rate represents the proportion of correctly classified samples (true TP and true negative TN) to the total number of samples. Precision represents the proportion of predicted

Table 1. Statistics of Weibo dataset

Category	Numerical value	Category	Numerical value
Number of events	40658	Event maximum number of comments	516
False rumor	23612	Event minimum number of comments	3
Real information	22837	Average number of comments on events	22.6
Number of comments	803695	User number	720186

positive samples (true TP) that are predicted to be positive (true TP and false positive FP). The recall rate represents the proportion of predicted positive samples (true TP) to actual positive samples (true TP and false negative FN). F1 is related to its precision and recall, and is equal to the sum of the inverse of precision and the inverse of recall. The larger the F1 value, the more robust the model. The calculation formula is as follows.

The accuracy rate formula:

$$M1 = \frac{TP + TN}{TP + TN + FP + FN} \tag{8}$$

The precision rate formula is:

$$M2 = \frac{TP}{TP + FP} \tag{9}$$

Recall rate formula:

$$M3 = \frac{TP}{TP + FN} \tag{10}$$

F1 formula:

$$F1 = \frac{1}{M2} + \frac{1}{M3} \tag{11}$$

The larger the above experimental performance index value, the better.

3.4 Results and Analysis

The detection method of fake news dissemination in social networks mainly involves classification tasks. The evaluation of the classification effect can be done through a confusion matrix, which is used to compare the predicted results with the actual values. The rows represent the predicted values and the columns represent the actual values. The test results of deep learning-based detection methods for fake news spread in social networks are compared with support vector machine-based and Naive Bayes-based detection methods. Taking all the features involved in this paper as input, the base classifier is trained with 3 detection methods. This paper selects the accuracy rate, precision rate, recall rate and F1 as the evaluation indicators, and uses the pros and cons of the detection effect to verify the effectiveness of the deep learning-based method for the dissemination of fake news in social networks. The experimental comparison results are shown in Table 2–Table 5, respectively.

Table 2. Accuracy comparison

Testing frequency	Detection method of fake news spread in social network based on deep learning	Detection method of fake news spread in social network based on support vector machine	A detection method of fake news spread in social network based on Naive Bayes
1	0.8916	0.8134	0.8307
2	0.8858	0.8168	0.8344
3	0.8925	0.8255	0.8281
4	0.8832	0.8122	0.8325
5	0.8963	0.8086	0.8266
6	0.8858	0.8264	0.8438
7	0.8920	0.8137	0.8352
8	0.9044	0.8322	0.8286
9	0.9027	0.8289	0.8364
10	0.8986	0.8165	0.8251

According to the test results in Table 2, the accuracy of the deep learning-based social network fake news dissemination detection method is 0.8933, which is 0.0739 and 0.0612 higher than that of the support vector machine-based and naive Bayes-based comparative detection methods.

According to the test results in Table 3, the accuracy of the deep learning-based social network fake news dissemination detection method is 0.9177, which is 0.0788 and 0.0669 higher than that of the support vector machine-based and Naive Bayes-based comparative detection methods.

According to the test results in Table 4, the recall rate of the deep learning-based social network fake news dissemination detection method is 0.8767, which is 0.0423 and 0.0576 higher than that of the support vector machine-based and naive Bayes-based comparative detection methods.

According to the test results in Table 5, the F1 value of the deep learning-based social network fake news dissemination detection method is 0.8967, which is 0.0601 and 0.0621 higher than that of the support vector machine-based and naive Bayes-based comparative detection methods. It can be seen from the above classification results that the detection method of fake news dissemination in social networks proposed in this paper integrates various features, which helps to improve the accuracy. The design

Table 3. Accuracy comparison

Testing frequency	Detection method of fake news spread in social network based on deep learning	Detection method of fake news spread in social network based on support vector machine	A detection method of fake news spread in social network based on Naive Bayes
1	0.9208	0.8242	0.8509
2	0.9114	0.8485	0.8658
3	0.9257	0.8268	0.8527
4	0.9225	0.8436	0.8366
5	0.9336	0.8353	0.8433
6	0.9262	0.8525	0.8655
7	0.9152	0.8412	0.8582
8	0.9026	0.8244	0.8378
9	0.9119	0.8377	0.8446
10	0.9074	0.8552	0.8522

Table 4. Comparison of recall rates

Testing frequency	Detection method of fake news spread in social network based on deep learning	Detection method of fake news spread in social network based on support vector machine	A detection method of fake news spread in social network based on Naive Bayes
1	0.8656	0.8309	0.8137
2	0.8828	0.8255	0.8248
3	0.8705	0.8468	0.8019
4	0.8633	0.8236	0.8262
5	0.8966	0.8353	0.8354
6	0.8682	0.8285	0.8185
7	0.8825	0.8296	0.8263
8	0.8757	0.8441	0.8027
9	0.8942	0.8374	0.8135
10	0.8676	0.8422	0.8284

Table 5. Comparison of F1 values

Testing frequency	Detection method of fake news spread in social network based on deep learning	Detection method of fake news spread in social network based on support vector machine	A detection method of fake news spread in social network based on Naive Bayes
1	0.8923	0.8275	0.8319
2	0.8969	0.8368	0.8448
3	0.8973	0.8367	0.8265
4	0.8919	0.8335	0.8314
5	0.9147	0.8353	0.8393
6	0.8963	0.8403	0.8413
7	0.8986	0.8354	0.8419
8	0.8889	0.8341	0.8199
9	0.9030	0.8376	0.8288
10	0.8871	0.8487	0.8401

method has higher mining potential than the comparison method, and the deep features it contains can more effectively detect fake media content in social networks.

4 Concluding Remarks

The advent of the self-media era has given Internet news a richer form of expression, but it has also increased the difficulty of fake news detection. This paper proposes a detection method for fake news dissemination in social networks based on deep learning. This method can effectively improve the detection effect and has effectiveness and advantages. The model in this paper is offline learning, that is the model is trained with the existing data set, and then the model is predicted. If the model is directly used online, the effect may not be very ideal. Everyday information is changing rapidly, how to build a data-driven incremental model is also one of the future research directions.

References

1. Lou, J.: Detection methods of fake news for social networks. J. Zhejiang Inst. Commun. 21(2), 106–110 (2020)
2. Qiu, G., Li, X., Han, K.: Simulation of information interception model of rumor spreading power in social network. Comput. Simul. 38(4), 209–212, 217 (2021)
3. Bhari, P.L.: Use of machine learning and detect fake profiles in a social media network. ECS Trans. 107(1), 11905–11920 (2022)
4. Lopez-Vizcaino, M.F., Novoa Francisco, J., Carneiro Victor, et al.: Early detection of cyberbullying on social media networks. Future Generation Comput. Syst. 118(2), 219–229 (2021)

5. Xu, M., Zhang, Z., Xu, X.: Research on spreading mechanism of false information in social networks by motif degree. J. Comput. Res. Dev. **58**(7), 1425–1435 (2021)
6. Li, L., Liu, Y., Hou, L.: Detection of fake news on emergency public health events based on adversarial neural network. Chinese J. Med. Libr. Inf. Sci. **30**(7), 1–9 (2021)
7. Zhang, G., Li, J.: Detecting social media fake news with semantic consistency between multi-model contents. Data Anal. Knowl. Discovery **5**(5), 21–29 (2021)
8. Mukherjee, D., Chajraborty, S., Ghosh, S.: Deep learning-based multilabel classification for locational detection of false data injection attack in smart grids. Electr. Eng. **104**(1), 259–282 (2022)
9. Zhang, G., Li, J., Hu, X.: Fake news detection based on multimodal feature fusion on social media. Inf. Sci. **39**(10), 126–132 (2021)
10. Sang, C., Xu, W., Jia, C., et al.: Prediction of evolution trend of online public opinion events based on attention mechanism in social networks. Comput. Sci. **48**(7), 118–123 (2021)

Tracing Method of False News Based on Python Web Crawler Technology

Hongmei Ye[1(✉)], Yandan Lu[2], and Gang Qiu[3]

[1] Department of Chinese, Changji University, Changji 831100, China
Yehm1025@126.com
[2] School of Literature and Media, Guangxi Normal University for Nationalities,
Chongzuo 532200, China
[3] Department of Computer Engineering, Changji University, Changji 831100, China

Abstract. At this stage, false news is rampant in the new media environment. Due to the wide dissemination channels of false news, the large amount of information and data, and the difficult governance of false news in the news communication industry, this paper proposes a false news Traceability Method Based on Python web crawler technology, builds a false news traceability management mechanism, and combines Python web crawler technology to build a false news traceability evaluation system to achieve the goal of false news traceability, Finally, through experiments, it is proved that the method of tracing the source of false news based on Python web crawler technology has high practicability and accuracy in practical application, and fully meets the research requirements.

Keywords: Python network · Reptile technology · Fake news · Information traceability

1 Introduction

In the new media environment, with the development of Internet technology and the upgrading of mobile terminals, great changes have taken place in people's way of obtaining information and reading behavior. Under the background of massive information, users tend to follow blindly to obtain information, making false news shops. Under the Internet technology, the social media platform is the main platform for information release and transmission. Under the social media platform, everyone can act as the publisher of news and release information in real time [1]. Under the diversified social platforms, the communication modes and purposes of large amount of information have also become diverse. Based on the communication mechanism of false news under the background of new media and combined with the relevant theories of management, this paper analyzes 110 samples of false news in the past 11 years, studies the development characteristics and corresponding governance means of false news in the past 11 years, changes the previous governance at the macro level, and explores based on tracing the source, Make false news on the basis of evidence-based governance. The proliferation

W. Fu and L. Yun (Eds.): ADHIP 2022, LNICST 469, pp. 489–502, 2023.
https://doi.org/10.1007/978-3-031-28867-8_36

of false news in the era of self media makes the responsibility unclear in the process of governance, early warning and accountability of false content become difficult problems. Putting forward the false news Traceability Method Based on Python web crawler technology is the fundamental means to solve the false news. Starting from the definition of the concept of traceability, learn from the successful experience of traceability in other fields, and explore the traceability mechanism suitable for false news [2].

According to the distribution and development trend of social public opinion, it is predicted whether changes in public opinion will be leveled after the source of false news dissemination is analyzed. At present, there is no technical breakthrough in the supervision of false news at home and abroad. A "decentralized traceability database" based on traceability is proposed. By using the "public chain" and "alliance chain" under Python technology and the python network crawler technology under government supervision, the unification from macro to micro is realized, and the double protection of the authenticity of information is realized. False news has existed for a long time, but its communication mode and characteristics are also changing with the passage of time. Therefore, under different circumstances, different management measures must be taken to effectively curb the spread of false news [3]. At present, with the development of network technology and the diversification of social networks, the propagation speed, coverage and far-reaching impact of false news in the new media environment must be traced from the source to achieve comprehensive management if we want to solve the problem at the source.

2   Fake News Information Feature Collection Based on Python Web Crawler Technology

2.1   Fake News Traceability Management Mechanism

The current Traceability Technology is mainly based on Python web crawler, and realizes systematic traceability with the help of the decentralized characteristics of Python web crawler technology. As a comprehensive method and means to solve problems, "mechanism" needs to be tested and proved by practice. Traceability has been proved to be an effective method to solve problems through practice in other fields, and even corresponding mechanisms have been formed in other fields for unified and coordinated operation [4]. At the same time, the mechanism itself contains institutional factors and requires relevant personnel to abide by them. In the above chapters, this paper has carried out a detailed analysis and summary on the release and dissemination process of false news in the new media environment. On this basis, by establishing a set of false news traceability mechanism for overall cooperative operation, this traceability mechanism involves all information receivers, communicators, platform builders, news supervision departments and other parts in the context of new media. On the premise of all parts' participation, analyze the traceability mechanism of false news to make it more systematic and theoretical, so as to guide practice more effectively [5]. At present, the tracing of false news needs to learn more from the application of sources in other fields, and find a set of experience mechanism that can guide its own development in combination with the characteristics of false news communication., As the quality level of journalists is becoming more and more uneven, many news editors have not received professional training and obtained relevant qualifications, resulting in frequent news communication events that ignore professional ethics and violate the principle of news authenticity, and false news is prohibited repeatedly.

With the explosive development of the self-media industry, users publish information anytime and anywhere, and attract the attention of other self-media users through the combination of pictures, texts and videos. Demand has led to the emergence of information islands. There is information asymmetry between communicators and receivers. In the new media environment, due to the competition between interests, there are also problems such as information asymmetry between communicators. In the face of a large amount of information, it is difficult for media operators to centrally share information. Even if false news occurs, error correction and clarification cannot be unified, and the rumor-refuting platforms are fighting each other. It is difficult to centralize information, and supervision and tracking become complicated and difficult [6]. Finally, some practitioners neglected their duties, copied and pasted content at will and uploaded unverified fake news. After being exposed, they evaded responsibility and shied each other's responsibilities, resulting in the ambiguity of the responsible subjects, and retrospective accountability became a necessary means. Under the massive amount of information, users are highly satisfied with the convenience and timeliness of Internet news, but they are not very recognized in terms of credibility and seriousness. On different media news clients, the same Events are reported in various forms and in huge quantities. Therefore, on this basis, false news reports are also displayed on major platforms in multiple forms and channels. Media platforms urgently need to integrate massive content resources to improve their authority [7]. Based on this, the dissemination model of Chinese online media false information is displayed, as shown in Fig. 1:

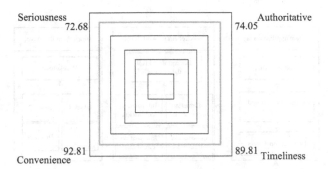

Fig. 1. The dissemination model of disinformation in Chinese online media

The definition of "tracing the source" in the online entry is: looking for the origin upstream, which is compared to tracing the source against the current, and later extended to the pursuit of the source. This term was first proposed by the European Union as a title of a relevant system for food safety. There are roughly three traditional traceability technologies. One is RFID radio frequency technology, which attaches signals to specific items and records relevant data on chips. When you need to know the detailed information of items, you can identify and read out the flow direction of relevant data products through radio signals: the second is QR code or bar code, Record the batch, place of origin and other information of relevant products through QR code or bar code [8]. In the new media environment, the dissemination of false news presents fission and data network dissemination, and the transmission path is difficult to record and the volume is large. At present, the main means to deal with false news is to refute rumors. However, new media information takes the form of mesh geometry quantitative diffusion, which needs to be realized in combination with the decentralization technology of Python web crawler: decentralization means that the center is no longer the content of information, but the whole information dissemination process from generation, release and forwarding, that is, to control the information life cycle, Thus, a set of accounting mechanism starting from traceability is realized, and on this basis, a set of "decentralized traceability database" based on information is formed. At the same time, decentralization is not absolute "decentralization", but having multiple centers to jointly maintain and supervise the whole link. In the case of multi-party maintenance, the content cannot be easily changed and stored more safely. By providing a decentralized content market, news interviews, content publishing, text copyright and distribution of press releases can be carried out at the same time within this scope, and even a set of agreements can be established through the authenticity of content audit to give certain rewards to publishers. After tracing the source of information through decentralized technology, each news received by users can find the original publisher and the whole communication path. In addition to being forced to trace, the communication of the whole information is also determined in an organizational way. After the "decentralized traceability database" formed under the background of Python web crawler technology is determined, it still needs a set of laws and regulations and credit investigation mechanism to improve and restrict. Based on this, the technical architecture of news management based on Python web crawler is optimized, as shown in Fig. 2.

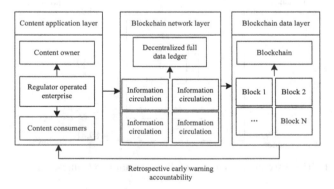

Fig. 2. News management technology architecture based on Python web crawler

In the infection graph G(V2E, P) obtained based on the IC model, due to the large scale of the real social network, it is inefficient to find the source node from all the propagation nodes. Therefore, in order to quickly and efficiently find the initial propagation node, we screen all the propagation nodes, eliminate those propagation nodes that cannot be source nodes, and narrow down the candidate node set. Due to the propagation of false information in the network, it will definitely cause a certain influence, which means that the source node must have a certain degree of propagation. The dissemination ability is not too low, because if the dissemination ability of the source point is too low, the false information will not be able to spread well. Therefore, this paper defines the propagation capability () of a single propagation node u as the ratio of the number of neighbor nodes that accept the false information propagated by node u to become a propagator among the neighbor nodes of node M to the total number of neighbor nodes of node M, the following formula:

$$I(u) = Ea - \frac{Mn_u}{AN_u} \tag{1}$$

In the formula, n is the number of infected neighbor nodes of node M: N is the total number of neighbor nodes of node u. The propagation capacity of a single node, we can filter all propagation nodes, therefore, this paper takes the propagation capacity (a) The set A composed of all nodes greater than a certain threshold E is called the potential source node set. The value of E is based on the maximum probability that the initial propagation node can be selected into the set A as the standard. This paper believes that if the propagation capability (u) of a node in the infection graph is less than the threshold E, it is considered that the node cannot be used as a network If the dissemination source node of false information exists, the false information will not be able to spread out. Therefore, the setting of the threshold is reasonable without considering the special circumstances, and the specific expression of the node set A of the propagation source of potential false information is shown in the following formula:

$$\Lambda_u = \left\{ I(u) | \frac{n_u}{N_u} > \varepsilon \right\}, u \in V \tag{2}$$

Before the data selected in the above data mining process is formally used as the index of model construction, it is necessary to check whether the data in the table contains noise value or missing value. If so, it is necessary to clean the data, otherwise it will have a serious impact on the data mining effect, which will seriously affect the accuracy of the prediction results of the prediction model. Data cleaning refers to the last procedure to find and correct the identifiable errors in the data file, which is responsible for filling the missing parts of the data, identifying the outliers and removing the impact of noise. Due to the huge amount of data, data cleaning can be completed quickly with the help of the powerful operation ability of the computer. The data cleaning process is shown in Fig. 3.

After the above data mining and data cleaning process, the selected model construction indicators are shown in Table 1 below.

Python web crawler is an artificial network with a wide range of properties, which is composed of multiple neurons. The network is a parallel computing model designed for data analysis and summary based on the structure and operation characteristics of human brain Python web crawler. The biggest feature of the model is that it has good self-learning ability, self-organization, superior fault tolerance and nonlinear mapping ability. In the past, news ratings were affected by various factors such as broadcasting mode, content, time period and other channel programs. However, since the popularity

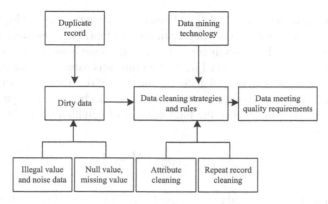

Fig. 3. Data cleaning process

Table 1. Model input indicators

Month	Network click data parameters
1	2.5% (x1)
2	2.3% (x2)
3	2.3% (x3)
4	2.7% (x4)
5	3.5% (x5)
6	2.5% (x6)
7	2.4% (x7)
8	2.4% (x8)
9	2.5% (x9)
10	2.6% (x10)

of various electronic devices, especially smart phones, news ratings are more affected by network click data. The Internet click through rate is affected by various other factors, which will show a certain disorder and nonlinearity. Therefore, it is impossible to accurately predict the fluctuation of news ratings by using linear methods such as regression prediction.

2.2 Fake News Information Management of Python Web Crawler Technology

In the Internet era, the amount of information is large, and the data collection of fake news has become more difficult. Currently, information is mainly disseminated through social media platforms, and the information content of the operating platform is collected and counted to establish a complete resource database. It is more urgent to use the decentralization characteristics of Python web crawler technology to achieve information path restoration [9]. The resource sharing is completed through the "centralized traceability database". In the database, the background of the database can be trusted. Data such as the number of transmissions are tracked. At the same time, the system contains a large-scale corpus and data statistics of social platform users. Users in the database have an evaluation level, which is divided according to the user's usual communication behavior. The decentralized traceability database uses the corpus and user behavior database. The published content is collected, and at the same time, it will automatically analyze whether there are sensitive words or topics, and automatically generate review results based on the publisher's credit evaluation. After the first stage of verification, the decentralized traceability database will send the content and review evaluation report. To the operator platform, the platform will conduct a second review, the content is completed and passed the review, which confirms the authenticity of the information and then disseminates it to the users. In this way, the content and the disseminator start to check at the source of dissemination, ensuring the authenticity of the source of the content. Specifically as shown in Fig. 4:

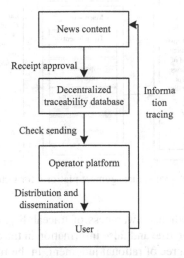

Fig. 4. Decentralized source measurement database information traceability process

In view of the large volume and numerous sources of false news, it is difficult to overcome the problem of tracing the source by relying on the strength of one party alone. It needs the joint cooperation and collaborative governance of macro and micro levels. However, from the perspective of national macro laws and policies, the binding force of law may not be able to take into account and reach all aspects of false news communication. With the development of Internet technology, false news under social media is the product of irrational competition in the communication market, while China's market economy lags behind. At the same time, the media industry belongs to the field of ideology and the forms of media are diverse, Government regulation sometimes fails to grasp the attributes of such industries: as a profit-making organization, the operation platform is committed to meeting the needs of users, often ignoring social benefits. False news is the product of the media's excessive pursuit of individual interests. Regulating users' illegal behavior on social media requires the formation of industry alliances among industries and the formulation and observance of common industry rules in addition to the management of joint supervision platform, With the wide application of Python web crawler technology, the news industry also needs to keep pace with the times, improve the sense of responsibility, combine the third-party alliance based on "alliance chain" technology built by the operation platform with the "public chain" platform based on users, and jointly complete the source tracing of content, communication path monitoring and audience influence control under the technical means of "decentralized traceability database", as shown in Fig. 5.

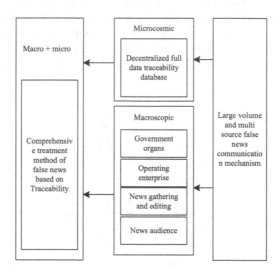

Fig. 5. Integrated management platform for virtual news

Cultivation of news audience awareness of traceability: For news recipients, it is more difficult to identify the true and false information in the context of new media, but it still requires a certain degree of rational judgment. In the future when information is traceable, news audiences It is very important to cooperate with them, not to blindly spread or believe rumors, and resolutely resist news extortion. Macroscopically, abide by national laws and regulations, correctly use various search engines, portal websites and other intelligent terminal platforms, and at the same time be good at using their own supervision power, and have a certain binding effect on news editors. Government agencies, enterprise platforms, news editors, and audience interaction, in order to realize the promotion and development of traceability technology.

2.3 Realization of Fake News Information Traceability Based on Python Web Crawler Technology

The data patterns of different data sources are different in semantics and syntax. When crawlers crawl the data of online social networks or load data from existing data sets, they will inevitably face the problems caused by heterogeneity. Considering the structure of data from a more abstract level can effectively reduce the required workload. Firstly, the source schema or source attribute name will be mapped to the classes and attributes in the semantic model. An automatic mapping algorithm is usually used to generate candidate mapping rules, and these mappings will be further checked. Then, data conversion is carried out based on these rules. Most of them are one to many and many to many relationships. The identification, mapping and transformation of these relationships require some additional manual operations [10]. However, after the mapping rules are established, the ETL process is automatically executed through the program. Some common patterns used to establish relationships between individuals are summarized in Table 2:

Table 2. Common models of object properties

Pattern	Type	Example
Mode A	Embedded	Microblog posts are returned in JSON format and directly embedded into posts. The relationship between posts and user accounts is embedded
Mode B	External & amp; explicit	In some data sets, posts are stored in a directory named after the platform, so the connection between the online platform and posts is external and displayed
Mode C	External & amp; implicit	An organization has established accounts on different social networking platforms, which are relational, external and implicit

Python web crawler technology based on input and output parameters mainly includes three core components: preprocessor, service filter and service matcher. The overall framework of Python web crawler technology is shown in Fig. 5. Among them, service filter and service matcher involve using the relationship in ontology for calculation. When there are many candidate services, the calculation takes a long time. Therefore, in the preprocessor, the search space is reduced according to the ontology referenced in the semantic description of service and request. After that, the service filter is responsible for further filtering the service, checking the consistency between the request and the input and output of the service, excluding completely inconsistent candidate services, and further reducing the search space. Finally, the service matcher outputs the sorted candidate service list according to the semantic similarity. The overall framework of Python web crawler technology is shown in Fig. 6:

When the information is propagated from the source node to other nodes, the longer the propagation time interval on the propagation path, the later the node that the path reaches will receive the information. The propagation time interval is the forwarding event interval 7684 on each edge on the propagation path. Therefore, if the distance of nodes on the network is properly defined, nodes that are further away from the source node will be activated later. Replacing the source node with another node would violate this fact. Using the above characteristics, the traceability estimation function can be designed. To take advantage of this property, the required traceability function needs to

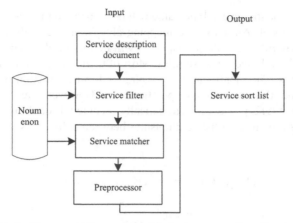

Fig. 6. The overall framework of Python web crawler technology

be able to reflect how close a node is to an early-activating node and far from a late-activating node. The overall framework of the information traceability method is shown in Fig. 7.

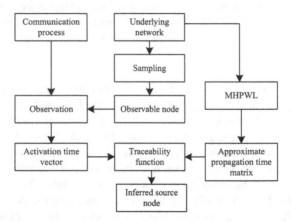

Fig. 7. The overall framework of the information traceability method

Information traceability mainly includes three key parts: (1) using sampling method to select observable nodes and observe them to obtain the activation time vector of information transmission process; (2) The approximate expression of information propagation time between nodes is obtained according to the underlying network; (3) According to the approximate propagation time and activation time vector from each node to observable node, the possibility of each node may be the source node is estimated by using the information traceability function, and the inferred source node s is obtained.

3 Analysis of Experimental Results

In order to better evaluate the method proposed in this paper, this paper sets the edge propagation probability to obey the uniform distribution on (0, 1). The comparison standard is based on the above algorithm idea, using the average error distance and The error rate is used to prove the accuracy of the experiment, that is, the shortest distance between nodes (Hops) to represent the error distance. If the shortest distance is 0, the node evaluated by the experiment coincides with the actual node. And the error rate represents the accuracy of identifying the source point in all infection maps, that is, the average error distance. In order to evaluate the TSRA algorithm proposed in this paper, it is compared with the existing algorithms for the single-point traceability problem.

3.1 Experimental Result

In this paper, 200 experimental simulations are performed on three real-world datasets, and 200 node information C = (V, E) is obtained, and the error distance (Hops) and error rate of the positioning source point are compared. The experimental results are recorded in Fig. 8 in:

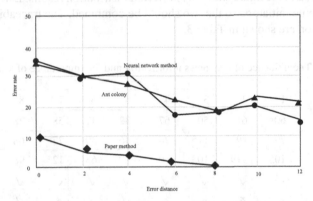

Fig. 8. Comparison of error distances in network information traceability

The traceability time factor is particularly important, and early important participants refer to users with a high comprehensive influence index r who participated in the topic earlier. Define the release time of G as T = Earliest(T: CG}, T = atest(IcCG], and the time T when v ∈ G participates in the cascade, then M = T–1. In order to distinguish the early period when M is small and r is large Important participants, because the scale and duration of different thematic events are different, the method of grading and quantification is adopted for comparison. The Mt level is the total time T-T is divided into 10 segments, and then the high-influence nodes with r > 2 are classified according to the participation level. The time of association is projected into different Δt levels. For example, 14% of the nodes of false information 1 are involved, Fig. 9 shows the comprehensive influence distribution of different news information.

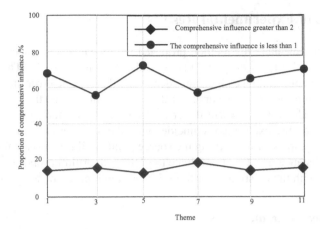

Fig. 9. Distribution of comprehensive influence of different news information

The news information nodes are projected into different Δt levels according to the time of participating in the cascade. For example, false information 1 has 14% of the nodes involved, and the figure shows its projected distribution results. Early participants may include some initiators, so the two should be combined. The traceability results of false information are shown in Table 3:

Table 3. The influence of fake news information and the importance of traceability

Serial number	1	2	3	4	5	6	7	8	9	10
Initiator	60	61	50	67	32	47	58	28	35	58
Key participants	79	49	115	78	46	39	77	68	68	29
Event source	105	119	103	103	89	63	125	98	62	78
A	√	√	√	√	√	√	√	√	√	√
B	×	×	×	×	√	×	×	√	√	√
C	√	×	√	×	√	√	√	×	√	√
D	√	√	√	√	√	×	√	√	√	√
E	×	√	×	√	√	√	√	√	√	√

Where "√" means reaching the index" ×" Indicates that the indicator is not met. It can be seen that the event source is basically locked within 129 D, and the excavation effect is achieved. Compare information traceability methods from an overall perspective. The table shows the success rate of the combination of different traceability methods (Table 4).

This success rate is calculated separately for all propagation processes and observations on each network. As shown in the table, the overall traceability effect of the traceability estimation method in this paper is significantly better than the traditional

Table 4. Success rates of different traceability methods

Information content (GB)	500	1000	1500	2000
Paper method	0.8809	0.9596	0.8785	0.8972
Neural network	0.3026	0.6006	0.5199	0.6581
Ant colony	0.4513	0.3088	0.3285	0.3785

two methods, and the traceability accuracy and traceability time are significantly higher, which can obtain a higher success rate and fully satisfy the research.

3.2 Experimental Analysis

From the experimental results, it can be concluded that the false news Traceability Method Based on Python web crawler technology designed in this paper can basically keep the success rate of traceability above 87% when tracing the source of massive news data, up to 96%, which can be highly practical and accurate, and can be used as a technical means for the news industry to control false news.

4 Conclusion

Based on Python web crawler technology, this paper designs a false news traceability method, aiming at the propagation of false news in massive new media news information, and implements false news traceability through Python web crawler technology. Through experiments, it is verified that the designed false news traceability method can be completed in the actual application process, has high practicality and accuracy, and can effectively maintain the authenticity of network news in the new media environment, Provide technical support for the governance of false news in the news communication industry.

References

1. Fang, Q., Cheng, Y.: Design and implementation of distributed crawler based on Docker container. Electron. Design Eng. **28**(08), 61–65 (2020)
2. Yuan, J., Wang, X.: Art blockchain certificate traceability model based on three chains. Appl. Res. Comput. **38**(10), 2915–2918+2925 (2021)
3. Wang, J.: Tracing and en route filtering analysis of false data based on Python crawler technology. Henan Sci. Technol. **40**(22), 27–30 (2021)
4. Jing, L., Siyu, F., Yafu, Z.: Model predictive control of the fuel cell cathode system based on state quantity estimation. Comput. Simul. **37**(07), 119–122 (2020)
5. Zhu, Q.: Design of public opinion analysis and early warning system based on Web crawler. Electron. Des. Eng. **28**(22), 56–60 (2020)
6. Wang, Y., Zhu, S., Hou, S., Wei, Z.: Application of Python based crawler technology in big data environment. Inf. Commun. **08**, 189–190 (2020)

7. Chun, L.: Design of big data acquisition system based on web crawler technology. Modern Electron. Tech. **44**(16), 115–119 (2021)

8. Yu, H., Zhang, S., Liu, Z., et al.: Propagation source tracing algorithm based on priori estimation. Pattern Recogn. Artif. Intell. **33**(01), 86–92 (2020)

9. Chen, C., Zhou, L.: Tracing and filtering of false data based on Python crawler technology. Comput. Simul. **38**(03), 346–350 (2021)

10. Wang, L., Wang, S.: Dilemma traceability and model innovation: research on personal information cooperative governance based on the blockchain. Chinese Public Administration **12**, 56–61 (2020)

Design of Mobile Monitoring System for Natural Resources Audit Considering Risk Control

Huang Meng[1](✉) and Xuejing Du[2]

[1] Jiangsu College of Finance and Accounting, Lianyungang 222000, China
huangmeng2022@126.com

[2] Shanghai University of Political Science and Law, Shanghai 201701, China

Abstract. In order to better ensure the health of natural resources and environment and avoid the risk of environmental pollution, this paper puts forward the design method of natural resources audit mobile monitoring system considering risk control, optimizes the hardware structure of natural resources audit mobile monitoring system, further optimizes the system software function and operation process, optimizes the natural resources risk identification and control method, and constructs the management index of natural resources audit mobile monitoring. Finally, the experiment proves that the mobile monitoring system of natural resources audit considering risk control has high practicability in the process of practical application and fully meets the research requirements.

Keywords: Risk control · Natural resources · Mobile monitoring

1 Introduction

China has a huge land area and complex natural environment. Under this background, many natural resources are not well and effectively protected, which is very easy to cause resource pollution, damage the ecological balance, and even threaten the monitoring of the masses, resulting in natural resource risk problems. Based on this, it is necessary to monitor the state of natural resources in real time, but the previous human monitoring consumes a lot and is difficult to obtain the state information of natural resources in real time and effectively, It is necessary to combine risk control and modern monitoring equipment for resource monitoring. As a new computing mode, natural resource risk control is the focus of the new generation of information technology, which is of self-evident importance and will bring significant changes to the whole industry and even the whole society [1]. Therefore, a design method of natural resources audit mobile monitoring system considering risk control is proposed. Determine the unified monitoring indicators of the three databases. The system can be divided into five functional modules: monitoring data collection module, monitoring data transmission module, monitoring data management module, early warning module and system management module. The system adopts a distributed architecture and is divided into monitoring server and monitoring agent. In the application process of domestic industry, a massive data cloud storage technology is developed to solve the problems of high-speed and massive digital content storage and subsequent operation at a low cost.

© ICST Institute for Computer Sciences, Social Informatics and Telecommunications Engineering 2023
Published by Springer Nature Switzerland AG 2023. All Rights Reserved
W. Fu and L. Yun (Eds.): ADHIP 2022, LNICST 469, pp. 503–518, 2023.
https://doi.org/10.1007/978-3-031-28867-8_37

2 Natural Resources Audit Mobile Monitoring System

2.1 Configuration of System Hardware Equipment Structure

The monitoring system requirements of natural resource risk control platform can be described from two aspects: physical machine monitoring requirements and virtual machine monitoring requirements. The physical machine monitoring is similar to the traditional cluster monitoring, while the virtual machine monitoring needs some new features. From the perspective of monitoring, the physical machine infrastructure of the risk control platform is not much different from the traditional cluster system. The infrastructure is all physical machines, and the number will not change too frequently compared with virtual machines [2]. Just mark the area to which the physical machine belongs. The natural resource risk data storage management system can be simplified as shown in Fig. 1. The figure shows all monitored physical machines, and their number, configuration and IP settings are basically unchanged.

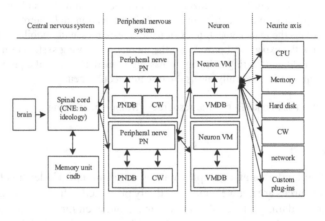

Fig. 1. Natural resource risk data storage management system

The monitoring data storage module receives the monitoring data and stores it structurally. On the one hand, the monitoring system provides users with monitoring view, on the other hand, it provides monitoring data API for other application modules. The system shall also provide early warning function to notify the user when there is an error in the system [3]. At the same time, the monitoring module shall have strong stability and will not collapse after long-time operation. It can run on common versions of Linux and windows systems. Monitoring items can be added and deleted flexibly. If you want to add another special monitoring item, you can also change the system configuration. The monitoring agent on all nodes reads the system performance data every 5min on average, and the impact on the system performance and network bandwidth cannot exceed 5%. This monitoring index can enable the user to understand the current running state of the CPU, including the percentage of time the CPU is in user mode, the percentage of time in system mode, the percentage of idle time, the percentage of soft interrupt and hard terminal time, and the number of interrupts received by the CPU in no second. The data requirements of the operating system are shown in Table 1:

Table 1. Data requirements of operating system

Domestic system software	Monitoring indicators
Domestic operating system	Basic system information
	CPU Information
	I/O information
	Virtual memory information
	Network traffic information
	Process information

In order to enable users to clearly understand the reading and writing status of the system disk, this part of the monitoring indicators include the number of merges per second, the reading and writing times of IO devices, the reading and writing times of sectors, the average IO data size of each device, queue length, waiting time, service time and other information. The database includes Dameng, Shentong and Jincang databases, although they are three different databases, However, in order to enable users to compare and analyze the three databases on the basis of understanding the status of all aspects of the database, this topic will determine a unified monitoring index for the three databases. The monitoring indicators of domestic database are shown in Table 2:

Table 2. Database data requirements

Domestic system software	Monitoring indicators
Domestic database	Basic information of database
	Other resource information
	Monitoring interface information

The scalability of the system is essential in the design process of the system. In the way of obtaining data, the method of analyzing data, and even in the design of the interface mode of the system to provide external data, it needs to be considered emphatically [3]. The business layer mainly includes various functional modules in the system, which is the core part of the whole architecture. This layer is mainly responsible for processing various requests accepted by the presentation layer and returning the request processing results to the presentation layer [4]. In addition, this layer is also responsible for the management of monitoring tasks and monitoring data in the monitoring system. The specific system monitoring task control and management framework is shown in Fig. 2:

Monitoring task control module, which is the main monitoring program of the system. In the monitoring system, because different monitoring objects have different methods to obtain data, different monitoring objects have different data acquisition procedures. The monitoring task control module is responsible for the initialization of the task when the monitoring system is started and calling the corresponding data acquisition program

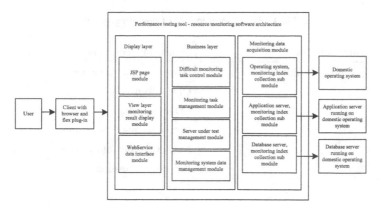

Fig. 2. System monitoring task control management framework

when the monitoring task starts [5]. When the user adds a new monitoring task to the resource monitoring system, the main process of the resource monitoring system will allocate multiple threads according to the newly added task monitoring indicators. If the monitoring task added by the user includes two monitoring indicators of CPU and IO, the system process will allocate one thread for CPU and IO respectively. The corresponding thread will be responsible for timing the monitoring tasks of the corresponding monitoring indicators and calling the monitoring data acquisition module of the monitoring indicators. When the monitoring task reaches the start time, the threads with different monitoring indicators will call the data collection method of the corresponding monitoring indicators. When the monitoring task ends, the corresponding threads will close the collection of monitoring information and close their own threads [6]. In terms of data acquisition method, the design process should not only consider the accuracy of the selected data acquisition method in data acquisition, but also consider the applicability of the acquisition method when facing different data sources. For example, in different versions of linuix environment, whether the same data acquisition command will obtain the same monitoring data is also required to be scalable in the method of analyzing data. When the data source of the monitoring system changes, That is, when the collected data format is likely to change, whether the old data analysis method will adapt to the new data format, or whether the old sentence analysis method is easy to expand on the basis of the new data format, it is also necessary to consider the scalability of the external data interface when designing the external data interface of the system [7]. When the data source or data format changes, and when the system adds monitoring items, whether the data interface service provided by the system can adapt to the changes of the provided data format is also extremely important.

2.2 System Software Function Optimization

The system can be divided into five functional modules: monitoring data acquisition module, monitoring data transmission module, monitoring data management module, early warning module and system management module, as shown in Fig. 3:

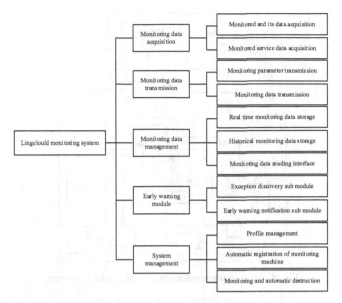

Fig. 3. Optimization of system software function structure

As can be seen from Fig. 3, the monitoring data collection module includes monitoring and its data collection and monitoring service data collection; The monitoring data transmission module includes monitoring parameter transmission and monitoring data transmission; The monitoring data management module includes real-time monitoring data storage, historical monitoring data storage and monitoring data reading interface; The early warning module includes an abnormality finding sub module and an early warning notification sub module; The system management module includes file management, automatic registration and monitoring of monitoring machine and automatic destruction. The research on risk management of regional natural resource shortage should include risk overview of natural resource system, risk factor identification of natural resource shortage, risk analysis and quantification of natural resource shortage, economic loss assessment of natural resource shortage, risk evaluation of natural resource shortage, research on risk management measures of natural resource shortage and risk decision-making of natural resource shortage, We can also study the control standard of regional natural resource shortage risk [8]. The research on natural resource shortage risk management involves resources, environment, society, economy and other systems. It is a complex large-scale system decision-making and management problem. The relevant contents involved in the research include natural resources macroeconomic model, natural resources water demand prediction model and natural resources system simulation model. The specific contents of natural resources risk management research include natural resources risk analysis model, natural resources economic loss model, natural resources risk decision-making model and natural resources risk management model [9]. The relationship between the components of the natural resources system risk management research and the models involved is shown in Fig. 4:

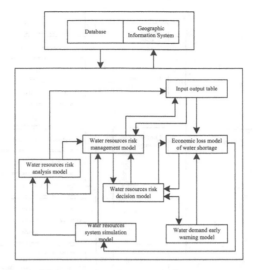

Fig. 4. Risk management content of resource system

After the risks are identified, estimated and evaluated, and several feasible risk treatment schemes are put forward, the decision-makers need to analyze and make decisions on the possible risk consequences of various treatment schemes, that is, decide which risk treatment policies and schemes to adopt. According to their progressive relationship, the research on risk decision-making can be divided into single objective risk decision-making (in $_t$) and multi-objective risk decision-making (fist | in $_i$), Considering the utility of decision-makers, the concept of utility can also be introduced to make decisions on natural resource hedging schemes.Risk estimation and evaluation is often the core content of risk analysis. Risk estimation, also known as resource risk measurement, refers to the quantitative estimation of the probability p of risk occurrence and its loss degree D by analyzing a large number of accident data collected on the basis of risk identification and using the methods of probability theory and mathematical statistics. The general risk estimation methods include subjective estimation, objective estimation Extrapolation methods include forward extrapolation and Monte Carlo digital simulation methods. The overall risk estimation of the system is carried out through the analysis of comprehensive risk events and the resulting loss analysis. Generally speaking, the probability description of system risk loss should be established. The general form of risk calculation of the system is:

$$p(\text{loss}_k) = \sum_i D \sum_m p(\text{in}_t) \times p(\text{fist} | \text{in}_i) \tag{1}$$

The dynamic resource discovery mechanism thread of risk control realizes the response mechanism of the self-organizing function of the monitoring system on the server side, and its flow is shown in the Fig. 5.

During initialization, create a listening socket and enter the listening state. If there is a request information, parse the request information, and query whether there are records of the corresponding resource domain in the database according to the parsed

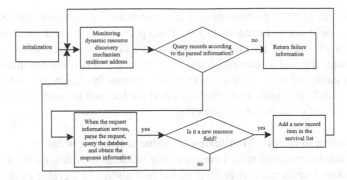

Fig. 5. Server side process of dynamic resource discovery mechanism

information. The database is the database maintained by the management program of the cloud platform. The information when the resource is online is stored in the database, including MAC address, host relationship, resource domain, etc., but the information in the database is static, That is, it will not change dynamically after writing [10]. If the corresponding record cannot be found, the query failure information is returned. If the corresponding record can be found, you should also judge whether the resource sending the request is the first resource in the new resource domain. If so, you should also add a new resource domain record item to the resource domain survival list structure. After judgment, the queried information includes the resource type, host relationship, name of the resource domain The contact information of the spokesperson in the resource domain is returned to the resource sending the request [11, 12]. Cloud platform overview is used to display a series of resource information of the platform; The status monitoring real-time curve displays the performance information of resources, and can display the resource performance of each resource domain; The centralized control provides an instruction issuing interface, which is responsible for transmitting the instructions entered by the user to the background processing program. The decision-making process of independent review and analysis of natural resources risk is shown in the Fig. 6:

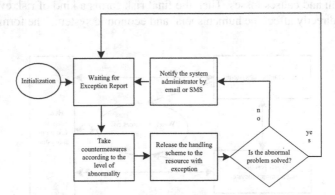

Fig. 6. Decision making process of independent audit and analysis of natural resources risk

The principle of status monitoring real-time curve display is to generate PNG pictures according to the stored data and display the PNG pictures on the web interface. In order to display the changes of the curve in real time, PNG pictures need to be generated at a certain frequency and displayed on the web interface. When the monitoring page is opened, the page will request data from CN at a certain frequency, and a PNG picture will be generated and displayed on the web interface for each request. When the request frequency is small enough, a real-time effect can be achieved.

Implementation of Mobile Monitoring of Natural Resources Audit

Like the general concept of risk, there are many definitions of natural resource system risk. In random hydrology, it is defined as the probability of a crash; In the economic analysis and evaluation of natural resources engineering, risk refers to the possibility or probability that a decision-making index is less than or greater than a specified value 1 in the whole application time of the project when considering the randomness of characteristic indexes. The risk in resource operation is defined as "the possibility or probability of accident during operation and operation of the reservoir and the degree of deviation from the normal state or expected goal". To sum up, natural resource system risk generally refers to the unexpected events and their probability and the resulting loss degree in the natural resource system under specific space-time environmental conditions. Specifically, the research on the risk of natural resource system includes the research on the reliability of the operation of natural resource system itself (the opposite of risk relative to risk), and its research object is the causes of risk events and the probability of risk events; And the potential adverse impact or harm of natural resource system crash on human property, health, psychology and ecological environment, that is, the probability distribution of loss caused by the crash in currency. Natural resource risk is actually a system involving "man nature society". If the accident has no impact on human and society, it can not be called a risk event. As a natural resource, there are several modes for the occurrence of risk events: one is because the uncertainty of risk factors acts on the natural resource engineering system. The natural resource engineering regulates this uncertainty to a certain extent. If it exceeds the engineering regulation ability, it may form a risk event. After the risk event is formed, it acts on the human social and economic system and causes losses, Then the final risk forms a kind of risk event that the risk factors directly affect the human social and economic system. The formation mode

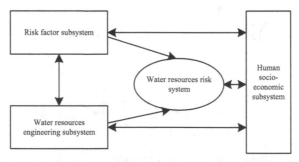

Fig. 7. Schematic diagram of natural resource risk system

of any kind of risk event is inseparable from the risk factor system and human social and economic system (Fig. 7).

There are various risks in the natural resource system. The methods of risk analysis and evaluation are different, but their basic modes are the same. They can be summarized into defining system problems, risk identification, risk analysis (risk estimation and evaluation), risk treatment and risk decision-making, as shown in Fig. 8.

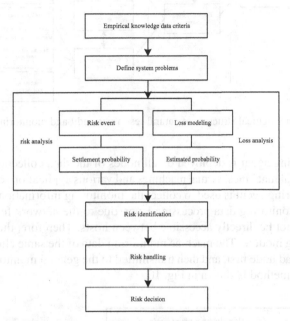

Fig. 8. General process of natural resources system risk assessment

The system is a distributed architecture, which is divided into monitoring server and monitoring agent. The monitoring agent is used to collect local information and communicate with the monitoring server. The monitoring server is used to receive the data sent by the monitoring agent, store and manage the data, implement the early warning strategy, and provide the monitoring data access interface. The monitoring agent is deployed on each monitored physical machine and virtual machine. One is designated as the monitoring server in each monitored domain, and the monitoring server program is deployed on it. Deploy monitoring agents on all service nodes, storage nodes and other devices in the computing device layer and common computing resource layer of the monitored system. Specifically, collect virtual machine monitoring information, host monitoring information, virtual network monitoring information, virtual storage monitoring information, public service monitoring information, etc. on the device. Collect the information from the monitoring agent on the server side, and finally present it to the administrator. Processing layer, data collection layer and monitoring application layer. The figure shows the system hierarchy of the risk control platform (Fig. 9).

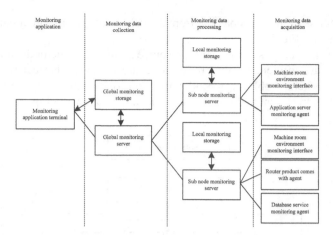

Fig. 9. Hierarchical structure of natural resources audit and monitoring system

The monitoring agent is deployed on all nodes of the data collection layer system, including physical machines, virtual machines and various application servers of the system. The monitoring agent is used to collect the monitoring information of the machine locally. In the monitoring data processing layer, due to the network firewall and other reasons, it may not be directly accessible between hosts. Therefore, this layer is added to the monitoring module. The machine monitoring data of the same cluster is collected on the cluster head node first, and then transmitted to the general monitoring server. The data acquisition method is shown in Fig. 10.

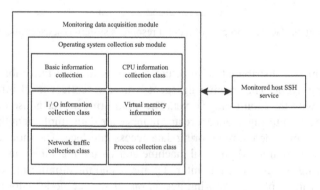

Fig. 10. Resource audit risk data collection method

The monitoring application layer includes monitoring display interface and monitoring data API. The monitoring display interface provides three monitoring interfaces: area view, physical machine view and virtual machine view. Display real-time monitoring information and historical monitoring information to users. Use the corresponding Java classes defined in the operating system acquisition sub module in the monitoring system, remotely connect to the monitored host through the risk control protocol, send the data acquisition command to the monitored host at the same time, control the data acquisition of the monitored target host, and then transmit the collected data to the data processing class of the data acquisition layer (Fig. 11).

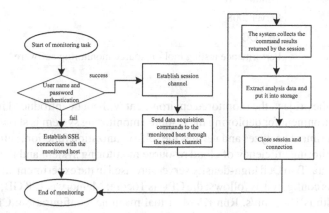

Fig. 11. Review and identification process of natural resource risk data party

In this data acquisition mode, when the system monitoring task starts, the main monitoring program calls the corresponding monitoring data acquisition class according to the monitoring index selected by the user. The monitoring class of each monitoring item first establishes a remote connection with the monitored host through the login user name and password of the risk control protocol. The establishment of the connection adopts the verification method of the risk control standard. If the verification fails, the connection class will throw an exception to the monitoring task and end the monitoring task; If the verification is successful, the monitoring system will establish a session channel with the monitored host through the risk control protocol, and send the data acquisition command to the monitored host through the channel. After the monitored host executes the data acquisition command, it will also return to the monitoring system in time through the session channel. After receiving the original data, the acquisition program at the system end will extract and analyze it, and store the extracted monitoring data into the system database. When the monitoring end time expires, the system will control to close the connection between the session channel and the monitoring host, so that the monitoring ends.

3 Analysis of Experimental Results

3.1 Experimental Environment

The software environment used in the test includes the software environment for deploying the resource monitoring system and the software environment of the monitored host. The software environment of the monitored host is a domestic operating system equipped with domestic database and domestic application server. Here, due to the confidentiality

Table 3. Server side software environment

Software category	Software name	Edition
Operating system	Windows 10	SP10
Database	Server2008	V5.6
Browser	360, QQ	V9
Tested sample	Performance testing tool - resource monitoring software	V1

problems of the project, the monitored environment will not be introduced in detail. The software environment for deploying the resource monitoring system is shown in Table 3:

This experiment analyzes and verifies the self-organization function, self repair function, the monitoring efficiency of cloud resource monitoring system and the accuracy of monitoring data. Two Dell high-density servers are used in the experiment, and each computing node is configured as follows: the CPU is 16core; Memory is 128 GB; Disk is 3TB and bandwidth is 1000 mb/s. Run KVM virtual machine, configuration: CPU is 2core; Memory is 4GB; Disk is 50 GB; Bandwidth is 1000 mb/s. The server adopts Debian 6 0.5, the virtual machine adopts windows ver2008 operating system. The distribution and configuration of computing nodes are shown in Table 4.

Table 4. Distribution and configuration of computing nodes

Node	Number of VMS	Is it a control node CNE	Server
Q1	15	Yes	A
Q2	35	No	B
Q3	35	Yes	A
Q4	35	No	B

Each node acts as a separate domain, in which all virtual machines belong to form an independent peripheral nervous system pn. The experiment is as follows: firstly, N1 is configured and started as the control node. After all the virtual machines are started and the network data transmission volume of the system is stable: start N2, N3 and N4 nodes at the same time, and designate the No. 1 virtual machine of each node as the management node MPN of the peripheral nerve. After the virtual machines of the three nodes are established and the network data transmission curve tends to be stable, the experiment ends.

3.2 Experimental Results and Analysis

The main purpose of system function test is to check the integrity and availability of system functions. This section tests the functions of the system through system test cases to test whether the system meets the needs of users. This section describes the functional test results of the system through two modules: monitoring server management function and user-defined monitoring task. The management function test results of monitoring server are shown in Table 5.

Table 5. Server management module test table

Functional module	Server management function			
Use case description	Users can manage and monitor the server on the server page			
Function description	Prerequisite	Testing procedure	Expected results	Pass or not
Display host information	There is a monitored host in the system	Enter the server interface	You can view the monitored host information	Yes
Add monitored host	This host does not exist in the monitoring system	Enter the server management interface, enter the added host information, and click the Add button	Pop up the prompt of adding successfully	Yes
	A host with the same information has been added	Enter the server management interface, enter the same information as the added host, and click the Add button	Pop up the prompt box of adding failure and prompt that the server already exists	Yes
	This host does not exist in the monitoring system	Enter the server management interface, enter the added host information, and click the Add button	Pop up the prompt box of adding failure, and prompt the wrong format of the input information	Yes

The non functional test of the system mainly tests the data accuracy of the resource monitoring system. Through the functional test of the system, it is known that each functional module in the system operates normally, but the normal operation of the system function does not mean that the monitoring data of the system is accurate. This section will test the accuracy of monitoring data. The data acquisition objects in the monitoring system include system basic information, CPU information, IO information, virtual memory information, network traffic information and process information of domestic operating system; The application server information of manufacturer a and manufacturer B of domestic application servers are compared with the monitoring data collected by the above data object monitoring data collection program with the monitoring results collected by some authoritative third-party monitoring tools. The following are the monitoring data obtained by the CPU monitoring data acquisition program and the monitoring data results obtained by other monitoring tools. Table 66 shows the eight groups of monitoring data obtained by the CPU acquisition program of the resource monitoring system every five seconds, and table shows the CPU monitoring data collected by other third-party monitoring tools at the same time. See Table 6 and Table 7 for details:

Table 6. CPU sampling data of resource monitoring system

Time	User	Nice	Sys	Iowait	Irq	Soft	Steal	Idle
10:15:13	0.65	0.01	1.45	0.00	0.00	0.00	0.00	98.2
10:15:20	0.65	0.01	1.45	0.00	0.00	0.00	0.00	98.2
10:15:25	0.65	0.01	1.45	0.00	0.00	0.00	0.00	98.2
10:15:28	0.65	0.01	1.45	0.00	0.00	0.00	0.00	98.2
10:15:33	0.65	0.01	1.45	0.00	0.00	0.00	0.00	98.2
10:15:38	0.65	0.01	1.45	0.00	0.00	0.00	0.00	98.2
10:15:45	0.65	0.01	1.45	0.00	0.00	0.00	0.00	98.2
10:15:49	0.65	0.01	1.45	0.00	0.00	0.00	0.00	98.2

Table 7. CPU sampling data of other monitoring tools

Time	User	Nice	Sys	Iowait	Irq	Soft	Steal	Idle
10:15:13	0.00	0.00	0.00	0.00	0.00	0.00	0.00	100.0
10:15:20	0.00	0.00	0.00	0.00	0.00	0.00	0.00	100.0
10:15:25	0.00	0.00	1.00	0.00	0.00	0.00	0.00	99.0
10:15:28	0.00	0.00	0.00	0.00	0.00	0.00	0.00	100.0
10:15:33	0.00	0.00	0.00	0.00	0.00	0.00	0.00	100.0
10:15:38	0.00	0.00	0.00	0.00	0.00	0.00	0.00	100.0
10:15:45	0.00	0.00	5.00	0.00	0.00	0.00	0.00	95.0
10:15:49	0.00	0.00	1.00	0.00	0.00	0.00	0.00	99.0

It is found that the data obtained by the third-party monitoring tool is different from the data obtained by the third-party monitoring system. Through data comparison, it is found that in addition to CPU monitoring information, IO virtual memory and network traffic information have such phenomena. The monitoring data collected by other modules of the system has no such problems and is consistent with the monitoring results of other monitoring tools. Through the non functional test of the system, it is determined that the system has problems in the accuracy of monitoring data of CPU, IO virtual memory and network traffic, which is deviated from the monitoring results obtained by other monitoring tools. Through the functional test of the monitoring system, the results show that each functional module of the system runs normally, the system is relatively stable, and meets the needs of users. The selection of test cases in functional testing is more in line with the requirements of test case writing, so the test process and test results are more objective and can reflect the real state of the system. In the process of data quasi determination test in the non functional test of the system, the deviation problem in the monitoring data of individual monitoring items is found. This chapter deeply analyzes the problem through the data acquisition method of the system, and gives specific solutions. Overall, after the system test, the stability and availability of the system are determined to meet the needs of users, and the data acquisition method is further improved in the test process.

4 Conclusion

Resource monitoring system is a distributed system based on wide area environment, which can monitor and audit all risk data in the environment. Its goal is to provide a unified, centralized and visual management tool for local and remote administrators, and provide reference for the decision-making of administrators or other grid users by evaluating the performance of the system. Starting with the system design, this paper expounds the functional requirements of the system and the design ideas of system architecture, information model and communication, and finally gives the reference implementation of a system prototype.

Fund Project. Jiangsu Educational Science "14th Five-Year Plan" project: Innovation and practice of training mode of rural auditors under the background of full audit coverage, NO.D/2021/03/24.

References

1. You, Y., Wang, H., Ren, T.A., et al.: Storage design of tracing-logs for application performance management system. J. Softw. **5**, 1302–1321 (2021)
2. Zhang, G., Hu, Y.Y., Han, X.L., et al.: Design of distributed water quality monitoring system under circulating water aquaculture mode of freshwater pearl mussels. Trans. Chinese Soc. Agric. Eng. **5**, 1302–1321 (2021)
3. Lian, J., Fang, S.Y., Zhou, Y.F.: Model predictive control of the fuel cell cathode system based on state quantity estimation. Comput. Simul. **7**, 119–122 (2020)
4. Sun, X.M., Shen, C.G.: Design and research of multi-parameter PLC monitoring system for greenhouse environment. Tech. Autom. Appl. **33**, 78–81 (2020)
5. Du, Z.M., Bai, P.R., Han, X.: Design of Training platform for oilfield water injection hardware control and remote monitoring system. Comput. Measur. Control **4**, 53–56 (2020)
6. Qiu, S.l., Pang, J., Jin, L.S.: Value realization mechanism of ecological goods in natural resources: an analytical framework of the regime complex. China Land Sci. **1**, 10–17+25 (2021)
7. Lan, M., Lin, A.W., Jin, T., et al.: Quantitative analysis of knowledge maps of natural resources accounting and assessment research in China based on CiteSpace. Resources Sci. **4**, 621–635 (2020)

8. Trysnyuk, T.V.: Mobile environmental monitoring system of the Dniester: modeling of technical system of hydro resources and extreme floods. Environ. Saf. Natural Resources **2**, 121–128 (2021)

9. Gebrehiwot, S.G., Bewket, W., Mengistu, T., Nuredin, H., Ferrari, C.A., Bishop, K.: Monitoring and assessment of environmental resources in the changing landscape of Ethiopia: a focus on forests and water. Environ. Monit. Assess. **193**(10), 1–13 (2021). https://doi.org/10.1007/s10661-021-09421-3

10. Xu, W., Zhang, Z., Wang, H., et al.: Optimization of monitoring network system for Eco safety on Internet of Things platform and environmental food supply chain. Comput. Commun. **151**, 320–330 (2020)

11. Lee, S., Gandla, S., Naqi, M., et al.: All-day mobile healthcare monitoring system based on heterogeneous stretchable sensors for medical emergency. IEEE Trans. Industr. Electron. **10**, 8808–8816 (2020)

12. Teo, T.W., Choy, B.H.: in. In: Tan, O.S., Low, E.L., Tay, E.G., Yan, Y.K. (eds.) Singapore Math and Science Education Innovation. ETLPPSIP, vol. 1, pp. 43–59. Springer, Singapore (2021). https://doi.org/10.1007/978-981-16-1357-9_3

Psychological Motivation Model of Employee Turnover Risk Management Based on ERP System

Han Yin[✉]

School of Traffic and Transportation, Xi'an Traffic Engineering Institute, Xi'an 710300, China
xinqiba001@sina.com

Abstract. In the analysis of employee turnover motivation, there is a lack of quantitative feedback on turnover risk characteristics, resulting in a high turnover rate of employees. Which is based on ERP system to build employee turnover risk management psychological motivation model. From the salary, promotion mechanism, work pressure, career planning and staff personal five aspects, analysis of the impact of turnover risk management factors. From the management content of each module in the ERP system, extract and decompose the enterprise risk factors that affect the turnover intention of core employees, identify the characteristics of resignation risk, and achieve the purpose of data-driven operation. The psychological motivation model of employee turnover risk management is established to reveal the internal mechanism of employee turnover. Based on the reasons for leaving, the implementation path of risk management is put forward to motivate employees to work through various channels. Integrate the above solutions into the ERP system to realize the psychological motivation analysis of employee turnover risk management. The test results show that the psychological motivation model based on ERP system can make full use of employee information, reduce the turnover risk index and the turnover rate of employees, provide effective decision-making feedback for the management, and avoid the loss of core employees.

Keywords: ERP system · Employee turnover · Risk management · Psychological motivation · Turnover risk · Model construction

1 Introduction

At present, the economic development of our country has entered a new normal, the supply-side structural reform is deepening, and the pace of industrial transformation and upgrading is also accelerating in an all-round way. Under the circumstances of the new normal, the economic growth of our country has shown a trend of accelerated conversion of old and new growth drivers. Meanwhile, the deepening of the market-oriented reform of interest rates and the acceleration of financial disintermediation, as well as the rise of financial technologies under the tide of innovation and entrepreneurship have promoted the vigorous development of the socialist market economy. The competition among enterprises has gradually shifted from product competition to technology and talent competition. The competition for talents is becoming more and more fierce. The war of seizing is a war without gunpowder smoke. It not only provides the opportunity for the staff to flow, but also puts forward new requirements and challenges to the human resource management of enterprises. Under the background of the market economy of our country, however, talent flow is an important phenomenon that is increasing and becoming more and more normal. Of course, such flow has a positive effect on society and enterprises. For example, it can make social resources more reasonably relocated and allocated. Good competition among talents can form a social ethos advocating learning. From an economic perspective, talent flow can stimulate consumption, stimulate the economy, and enliven the operation and development of society. Mobility is one of the important characteristics of human resources. Mobility of personnel within an organization can promote the optimization of resource allocation and efficient operation of the organization. However, the voluntary termination of labor contract will bring more losses to employers and indirectly increase operating costs, especially the loss of core talents, which will bring a heavy blow to enterprises and jeopardize the operation and survival of enterprises [1]. In recent years, the theory and practice of human resource management tend to regard the individual employee as the research unit. Turnover motivation, turnover impact factors, turnover motivation model and turnover prevention measures are hot issues in recent years. The turnover of employees should be treated dialectically. The opportunity for employees to realize their self-value has been increased. At the same time, the high turnover rate has become a thorny problem for enterprises. The impact of employee turnover on a company can be seen in many ways. The direct impact of employee turnover on the organization is to increase the cost of recruitment. The vacancy needs to be filled by new employees. Therefore, the human resources department of the company has to expend manpower and material resources to complete the new recruitment. In order to make the enterprise win in the market economy, the enterprise must grasp the basic characteristics of employee turnover, analyze the existing human resources management methods, and solve the problem of high turnover rate. When an employee leaves his post and moves to another company, his salary is raised, his working environment is improved and his career development is better. When other employees in the original unit know the situation, their will to stay in the same post is easy to be shaken. Then it is easy to trigger the chain effect, resulting in a greater degree of staff turnover phenomenon. How to reduce the turnover rate of employees and avoid the loss of core employees has become a difficult problem for most companies.

Based on this background, this paper constructs the psychological motivation model of employee turnover risk management based on ERP system, from the five aspects of salary, promotion mechanism, work pressure, career planning and individual employees, this paper analyzes the influencing factors of turnover risk management. From the management content of each module in the ERP system, extract and decompose the enterprise risk factors that affect the turnover intention of core employees, and identify the characteristics of resignation risk. Build the psychological motivation model of employee turnover risk management, propose the implementation path of risk management, and realize the psychological motivation analysis of employee turnover risk management. This model can make full use of employee information, reduce the turnover risk index and the turnover rate of employees, provide effective decision feedback for the management, and thus avoid the loss of core employees.

2 Establishment of Psychological Motivation Model for Employee Turnover Risk Management Based on ERP System

2.1 Analysis of Influencing Factors of Employee Turnover Risk Management

In this paper, the factors affecting employee turnover behavior are classified into 5 categories, as shown in Fig. 1.

Whether the specific position salary level is superior to the average salary level of the industry is embodied in the competition of salary level, which has a direct impact on the use and turnover of knowledge workers. The existing compensation system is complex, including salary, bonus, incentive and welfare, but different types of projects can not be shared by everyone, need to be determined according to the actual work ability of employees. Employees attach great importance to the value of their own, the sense of achievement as an important pursuit of their work. Value is often reflected in the level of treatment provided by the organization. Moreover, these pay levels to a large extent affect the performance of employees, and then related to the pursuit and realization of their own values. The motivation of young employees is easily dampened because the salary is related to the length of the employee's service. But for the old employees, the salary is not changed after reaching a certain length of service. In addition, different employees have different bonuses and benefits in the process of work, resulting in psychological imbalance of employees, increase the probability of leaving. The level of compensation and benefits reflects the return of the costs they pay for the accumulation of knowledge and their personal abilities. They measure their ability and value by the level of their wages and benefits, thus showing their own value. The internal posts and responsibilities are clear, and different posts and titles directly affect the salary and welfare of employees, as well as the right to say and make decisions on internal affairs of the company.

The promotion mechanism of enterprises is influenced by many rules. Lack of staff capacity, performance evaluation and recognition, when there is no wage increase, are often deductible items, such as absence from work, security work did not do a good job deducting wages, etc., employees in the hopeless situation of promotion will seek stability, that no error is good, will produce a large number of "Buddhist" employees, seriously affecting the development of the company. Good young employees are not rewarded, and managers who move into middle management have a long way to go. Compensation and benefits, as well as the importance attached to them, develop slowly, leading to lay-offs and job-hopping. In addition to their salary, employees place great importance on the realization of their own value in the hope of gaining promotion through hard work.

If not objective performance appraisal affects their own promotion, it will make them lose confidence in their own professional premise, and thus choose to leave the enterprise. More than 30% of employees feel they are overburdened. Some managers suffer from physical and mental overdrafts, and if given the opportunity, they all choose to look for new development opportunities. If the performance evaluation index is not up to the standard, the promotion will be slow and the salary will be reduced, which will make the outgoing staff feel that the salary does not meet their expectations.

At present, there is no perfect medium and long-term talent demand and training plan, and the medium and long-term talent demand and training plan is essential for a growing enterprise. When young employees first enter the workforce, they are not very clear about their development goals and the positions they are suitable for. When they officially enter the workforce, they may find that they are not suitable for the job, or that their style of doing things does not match the overall style of the company, resulting in a willingness to leave. Human resources strategic planning and the group's overall strategic planning is not seamless, the lack of overall planning directly leads to the uncertainty of enterprise future development, the group's overall strategy can not be effectively implemented [2]. Any standard scientific system, is to rely on people to implement, when the enterprise's demand for talent did not consider ahead of time, there is no overall planning.

Good talent matching needs to achieve two points: one is that the level of knowledge and skills among members should be close to form a resultant force; Employees are less familiar with each other, and their busy schedules make them less likely to spend time outside of work. Good team cooperation in the work, to enhance the collective sense of membership, resulting in a sense of belonging, prefer to stay in the enterprise to work, to avoid brain drain.

2.2 Identification of Exit Risk Characteristics Based on ERP System

2.2.1 ERP System

Based on the analysis of the influencing factors of employee turnover behavior, ERP system is used to identify the characteristics of turnover risk, and the qualitative description is transformed into quantitative representation. ERP system is an enterprise resource management system which integrates enterprise management idea, business process, basic data, manpower, material resources, computer hardware and software. Through the use of ERP, enterprises can use computers to carry out automated management of enterprise personnel and information resources. The implementation of ERP will not fundamentally change the company's governance structure and rules of procedure, but through the clear division of position-responsibility-authority and strict establishment of authority, the responsibilities and authority in terms of decision-making, implementation and supervision can be further clarified, and a scientific and effective division of labor and balance mechanism can be formed. Through ERP to achieve the strategic objectives and business planning decomposition, to unify the plan to guide the ERP system in all aspects of staff work, to achieve a planned, controlled management. From the perspective of internal structure, job responsibilities and business processes, the implementation of ERP makes the organizational structure become flat, the vertical perspective of hierarchical relationship is greatly reduced, the information transmission is more smooth, and the speed and accuracy of transmission are enhanced; through the matching of responsibilities and powers, the power and responsibilities are implemented to individuals. ERP systems work only when they are combined with people who fully understand their functions and work hard. For employees, the implementation of ERP, through the correspondiir own post responsibilities and operating manuals, can clearly define their own responsibilities and authority, the correct exercise of authority. Therefore, the application of ERP system not only improves the management level of enterprises, but also improves the quality level of employees, makes full use of human resources to maximize the role of people.

2.2.2 Identify the Characteristics of Turnover Risk

The employee turnover risk management indicator system is established in the enterprise ERP system. The data dimension has a and the initial number of indicators is b. The corresponding employee information system is established, which is expressed as:

$$w = \langle \alpha, \beta, \chi, \gamma \rangle \tag{1}$$

In formula (1), w refers to the employee information system; α refers to the object set, including a data dimension; β refers to the set of α features, including b indicators; χ refers to the set of α values, where the number of data is $a \times b$; γ refers to the function that forms the information system, describing the function that forms the information system, describing the characteristic value of the object. Based on the management content of each module in the human resources management system structure, extract and decompose the enterprise risk factors that affect the turnover intention of core employees, and combine the macro factors and the personal factors identified by the employee portrait data to select the risk indicators hierarchically [3]. According to the entropy weight method to select indicators, combined with the stochastic analytic hierarchy process to determine the comprehensive weight of indicators to complete the index system. According to the above description, we get the data that represent the object and the original index, then we process the data in relation, get the form of two-dimensional table, and then store the information in relational database [4]. The transformed 2-D table can be represented as follows:

$$p = (\gamma, z) \tag{2}$$

In formula (2), p stands for the transformed two-dimensional table; z is a collection of indicators for talent resignation, including all information. The actual data after making operation decision can be used to describe the business of each human resource management department and to quantitatively feedback the actual effect of management decision. When the value of a description property is a language literal or several numeric values, it is discontinuous and needs to be discretized. FCM algorithm can improve the diversity of objects and different clusters in the same cluster to the greatest extent. The value functions that construct FCM are as follows:

$$f = \sum_{x=1} \sum_{y=1} c_{xy} \vartheta_{xy}^2 \tag{3}$$

In formula (3), f represents the value function of FCM; c_{xy} represents the range to which the x th and y data points attributes belong; ϑ_{xy} represents the distance between the x th and y data points. The formula for calculating the c_{xy} is as follows:

$$c_{xy} = \frac{1}{\sum\limits_{g=1} \left(\frac{\vartheta_{xy}}{\vartheta_{gy}}\right)^2} \tag{4}$$

In formula (4), g represents the fuzzy cluster center grouped by all indicators. Through the results of exit risk forecast and identification, assist the decision-making and operation management. The risk level of active brain drain of an enterprise is the basis of business prediction, which indicates the severity of the brain drain of core employees. Determining the risk level is conducive to enterprise managers to clearly understand the current situation of employee turnover and timely control and adjust the loss of core employees.

2.2.3 Establishment of Psychological Motivation Model for Employee Turnover Risk Management

Employee turnover is a process, turnover is a part of it. To study the problem of employee turnover, it is necessary to analyze the inherent law of employee turnover and explore the psychological motivation of employee turnover risk management. Through the modeling of the psychological motivation of risk management of employee turnover, this paper reveals the internal mechanism of employee turnover more clearly. The psychological motivation model of employee turnover risk management is shown in Fig. 2.

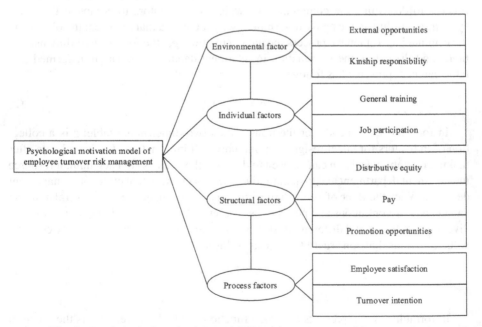

Fig. 2. Psychological motivation model of employee turnover risk management

Environmental factors include employment opportunities and family responsibilities. Both positive and negative reference factors were used in the model. Keeping the employee on the payroll will help the employee achieve the kinship responsibility, so the kinship responsibility is related to leaving the company. The separation rate of married women is higher than that of unmarried women. That's because women's relative responsibilities are greater than men's, forcing them to exit the labor market or relocate to suitable jobs. The employment opportunities in the labor market are positively correlated with employee turnover. The proportion of technical post turnover is higher than that of administrative post, personnel post and logistic post. One of the main reasons is that technical post employees have more external opportunities.

Personal factors include general training, job participation, and positive/negative emotions. Universal training refers to the extent of staff training, the more generalized the training, the higher the turnover intention, showing a positive correlation trend. Most of the training organised by companies is general training that helps employees acquire job-related generic skills that can be applied to other companies as well. Job participation, as its name implies, refers to the degree of contribution to job content and employee effort, which has a direct impact on job satisfaction and organizational commitment and has a negative correlation with resignation [5]. There is a close relationship between job participation and employee turnover, and the employee with high job participation has a relatively low turnover rate. The turnover rate of relatively busy departments is lower than that of the departments with less workload. Positive/negative emotions refer to an employee's personal perception of emotions, and those with more negative emotions are more likely to leave. Enterprise human resources management is limited to the traditional personnel management, there is no perfect performance appraisal and salary system, such as pay no clear implementation program. On the one hand, the salary increase is arbitrary, on the other hand, the salary increase depends on the decision of the supervisor and the general manager. Unfair distribution will lead to high turnover.

Structural factors include remuneration and promotion opportunities. Structural variables further affect turnover intention by influencing organizational commitment, and then indirectly affect turnover behavior. Compared with enterprises in the same industry, salary level is not strong, salary, housing provident fund, social security and other remuneration and benefits are not small gap with large enterprises in the same industry. Department leaders and ordinary staff pay gap is too large, department leaders are three or four times the salary of ordinary staff, unequal pay for equal work occurs from time to time. The study found that salary is the most important factor in deciding to leave. The relationship between employees and the organization is an exchange, the employees provide labor output, the organization gives corresponding return. Employees pursue maximum self-interest. If there are multiple benefits and costs juxtaposed in an organization, the employee trade-offs to maximize benefits. The company has not established the professional promotion channel, has not made clear the system and the promotion condition, but the middle-level cadre's selection mostly is the leader intentionality decision, has not made the request clearly, the ordinary staff often in a post many years, these will cause the staff to leave the post psychology [6].

Process factors include job satisfaction, job search behavior and turnover intention. Process factors are indirect factors that affect turnover behavior. Employee's job satisfaction is an important factor influencing employee's turnover, which is influenced by many factors, such as high pressure and dissatisfaction with salary. Employees have certain expectations of the organization. If the organization satisfies such expectations, it will strengthen the recognition and satisfaction of employees to the organization, thus generating attachment feelings to the organization and maintaining the membership of the organization for a long time [7]. Satisfaction with the company was lower among those who left. The positivity and intention to leave are positively correlated with the possibility of leaving. Through the above analysis, the psychological motivation model of employee turnover risk management is constructed.

2.3 Implementation Path of Employee Turnover Risk Management

According to the psychological motivation model of employee turnover risk management, in order to alleviate the problem of employee turnover, the following implementation path is proposed. First, formulate and improve human resources strategic planning. Human resources planning plays a vital role in the implementation of the overall development strategic planning of enterprises, and the two plans should keep the same frequency resonance. The design process of human resource strategic planning is shown in Fig. 3.

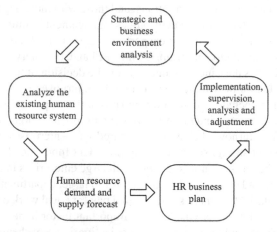

Fig. 3. Design process for human resources strategic planning

Scientific human resources strategic planning can closely link the human resources management of enterprises with the overall development strategy of enterprises. At the beginning of the recruitment should be discussed with the candidates on future career development planning and other issues, grasp the new employee's self-career planning and development needs. As for the work arrangement of employees, they may be given some freedom in a moderate manner, and in terms of the work contents with strict procedures, they shall not deviate from the standards and complete the work step by step, while for the work with room for play, they shall respect the different opinions of employees, allow employees to express different opinions, so as to create a better communication atmosphere among colleagues, colleagues and leaders and attract employees to work harder. Forecast the demand and supply of human resources, subdivide the supply gap, find out the truth, and then provide a scientific basis for rational deployment of human resources strategy.

Second, establish a reasonable salary system. Managers keep abreast of the local labour market profile during the implementation of the plan and adjust the relevant coefficients and the unit compensation management plan in a timely and flexible manner to ensure the external competitiveness of compensation. Enterprises shall take into account strategic, competitive, incentive and nature of the remuneration design, adjust the remuneration structure ratio, ensure stable growth of basic wage, fully adjust the floating wage ratio in accordance with the nature of job positions, and optimise the incentivisation role in a targeted manner. Allow employees to meet basic needs at the same time more effectively play an incentive role. On the basis of a comprehensive consideration of the external competitiveness of the unit pay, to re-adjust and design the pay level. In a reasonable range, plan the budget for the whole year, reduce unnecessary spending, so that the end of the year can retain sufficient funds as a year-end bonus to the staff.

Third, strengthen the performance system and optimize the assessment mechanism. The contents of assessment should be distinguished and each has its own emphasis. For example, the technical post assessment shall focus on product quality, operation process proficiency, process innovation, energy consumption reduction, etc., the research and development post shall focus on the progress of new product research and development, project development, patent application, etc., and the sales department shall mainly assess product sales, market expansion, channel planning, customer relationship and maintenance, etc. In addition to the heavy tasks can be detailed division of labor, each task clear to the individual, after the completion of the task signed confirmation, paper version retained, as the basis for assessment management archive. The assessment should pay attention to the result, give priority to the contribution, highlight the merit pay and widen the income gap. Perfect assessment management can make more objective evaluation of employees, and at the same time in the study of work efficiency and staff capacity to play a role. According to the different circumstances, distribute the weight reasonably.

Fourth, the establishment of multi-ladder promotion mechanism. Salary, rank and position should be parallel among different steps. Employees can choose different promotion paths according to their own career development plans. Reasonable career promotion mechanism can effectively prevent employees from changing their jobs because of unclear career development, as well as starting their own business and so on. In order to change this situation, we should start with the promotion of professional and technical posts, classify the posts, supplement the promotion channels for professional and technical personnel, and motivate the staff through diversified promotion channels. At the beginning of each year, the enterprise shall provide an opportunity to transfer and adjust the three categories of employees, namely, management, sales and technology. The application shall be submitted by the employees, examined by the human resources department and the business department in charge and approved by the leader in charge. The operation may be carried out. The three categories of positions at the same level may be freely converted according to the needs to provide more space and choices for the employees to develop freely.

3 Experimental Study

Based on the risk index, this paper measures the application effect of the psychological motivation model of employee turnover risk management based on ERP system. Risk index is essentially a qualitative method for grading and comparing risks. Risk index is used to measure the potential turnover willingness of employees, including the overall satisfaction of internal and external environment and their own conditions. Set the critical value of the risk index to 1, beyond which the employee's overall satisfaction is low and there is potential turnover risk. Risk index = probability of risk occurrence × Risk influence and risk controllability. The analysis results of the psychological motivation model of employee turnover risk management based on ERP system are compared with those based on BP neural network and clustering algorithm. The employees of an enterprise shall be classified into four groups according to their service years, namely, working for less than 3 years, 3–5 years, 5–10 years and more than 10 years, and the risk index shall be calculated in each group. The experimental results are shown in Tables 1, 2, 3 and 4.

Table 1. Risk index of employees working for less than 3 years

Number of tests	Psychological Motivation Model of Employee Turnover Risk Management Based on ERP System	Psychological Motivation Model of Employee Turnover Risk Management Based on BP Neural Network	Psychological Motivation Model of Employee Turnover Risk Management Based on Clustering Algorithm
1	0.823	1.307	1.164
2	0.954	1.218	1.267
3	0.881	1.659	1.371
4	0.868	1.426	1.558
5	0.956	1.563	1.485
6	0.923	1.532	1.526
7	0.935	1.685	1.653
8	0.892	1.222	1.585
9	0.904	1.304	1.792
10	0.857	1.241	1.561

According to the results in Table 1, the risk index of the psychological motivation model of employee turnover risk management based on ERP system is 0.899, which is 0.517 and 0.597 lower than that based on BP neural network and clustering algorithm.

Table 2. Employee risk indicators for jobs 3–5.

Number of tests	Psychological Motivation Model of Employee Turnover Risk Management Based on ERP System	Psychological Motivation Model of Employee Turnover Risk Management Based on BP Neural Network	Psychological Motivation Model of Employee Turnover Risk Management Based on Clustering Algorithm
1	0.810	1.126	1.247
2	0.854	1.256	1.385
3	0.881	1.289	1.256
4	0.728	1.168	1.260
5	0.735	1.035	1.332
6	0.762	1.122	1.255
7	0.823	1.213	1.227
8	0.856	1.244	1.374
9	0.812	1.281	1.245
10	0.813	1.027	1.113

According to the results in Table 2, the risk index of ERP-based psychological motivation model is 0.807, which is 0.369 and 0.462 lower than that of BP neural network and clustering algorithm.

Table 3. Employee risk indicators for 5–10 years of work

Number of tests	Psychological Motivation Model of Employee Turnover Risk Management Based on ERP System	Psychological Motivation Model of Employee Turnover Risk Management Based on BP Neural Network	Psychological Motivation Model of Employee Turnover Risk Management Based on Clustering Algorithm
1	0.622	0.869	0.806
2	0.583	0.981	0.853
3	0.657	0.850	0.988
4	0.668	0.823	0.967
5	0.744	0.836	0.734
6	0.514	0.968	0.751

(continued)

Table 3. (*continued*)

Number of tests	Psychological Motivation Model of Employee Turnover Risk Management Based on ERP System	Psychological Motivation Model of Employee Turnover Risk Management Based on BP Neural Network	Psychological Motivation Model of Employee Turnover Risk Management Based on Clustering Algorithm
7	0.552	0.897	0.825
8	0.663	0.856	0.902
9	0.638	0.933	1.043
10	0.655	0.862	1.084

According to the results in Table 3, the risk index of ERP-based psychological motivation model is 0.630, which is 0.258 and 0.265 lower than that of BP neural network model and clustering model.

Table 4. Risk indices of employees working for 10 years or more

Number of tests	Psychological Motivation Model of Employee Turnover Risk Management Based on ERP System	Psychological Motivation Model of Employee Turnover Risk Management Based on BP Neural Network	Psychological Motivation Model of Employee Turnover Risk Management Based on Clustering Algorithm
1	0.549	0.846	0.876
2	0.405	0.779	0.845
3	0.556	0.781	0.712
4	0.428	0.762	0.763
5	0.562	0.630	0.893
6	0.586	0.752	0.702
7	0.429	0.624	0.638
8	0.467	0.787	0.654
9	0.434	0.742	0.781
10	0.423	0.613	0.727

According to the results in Table 4, the risk index of the psychological motivation model of employee turnover risk management based on ERP system is 0.484, which is 0.248 and 0.275 lower than that based on BP neural network and clustering algorithm.

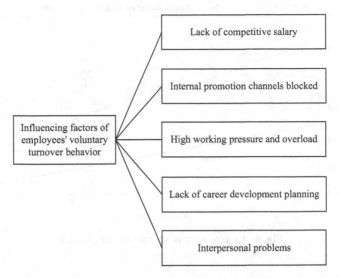

Fig. 1. Influencing factors of employee voluntary turnover

In order to verify the turnover rate of enterprise employees based on the psychological motivation model of Employee Turnover Risk Management of ERP system, the turnover rates of enterprise employees of the three models are compared, and the comparison results are shown in Fig. 4.

It can be seen from Fig. 4 that with the increase of working years, the turnover rate of employees in the three models will decrease. When the service life is 10 years, the turnover rate of enterprise employees based on the psychological motivation model of Employee Turnover Risk Management of ERP system is 0.48%, the turnover rate of enterprise employees based on the psychological motivation model of Employee Turnover Risk Management of BP neural network is 0.65%, and the turnover rate of enterprise employees based on the psychological motivation model of Employee Turnover Risk Management of clustering algorithm is 0.71%. It can be seen that the turnover rate of enterprise employees based on the psychological motivation model of Employee Turnover Risk Management of ERP system is low.

From the above results, we can see that the risk index obtained by the two comparison methods is too high to provide effective decision-making reference for the management. The model can make full use of the employee information in ERP system, understand the employee's thought and provide effective decision-making feedback for the management, so as to avoid the loss of core employees.

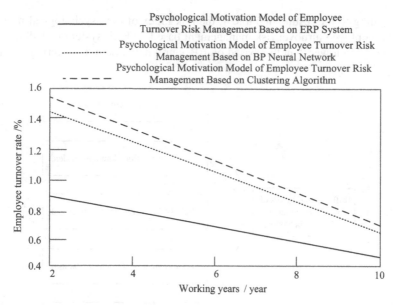

Fig. 4. Employee turnover rate of three models

4 Conclusion

ERP system avoids the disadvantages of single data analysis scenario, considers the role of data and analysis results in the overall operation management and the impact on operation decision. Based on the ERP system, this paper constructs the psychological motivation model of employee turnover risk management, and provides management decision strategies and suggestions to reduce employee turnover risk. The model will be improved in the future research, such as expanding the sampling range, increasing the sample capacity and improving its representativeness. In addition, stratified sampling can also be used to carry out comparative analysis in order to fully highlight the personality differences of the samples. Provide more effective guidance to enterprises to reduce employee turnover rate.

References

1. Song, H., Cheng, Y.: Compulsory citizenship behavior and new generation employees' turnover intention. Econ. Manage. J. **43**(4), 108–121 (2021)
2. Liu, L., Yan, Y.: Study on influential mechanism of career adaptability on employee entrepreneur-ship intention. J. Zhengzhou Univ. Light Ind. (Soc. Sci. Edition) **22**(1), 57–64 (2021)
3. Li, P., Zhou, Y., Wang, Z., et al.: Why do service employees engage in customer-oriented deviant behaviors: a motivational perspective account. Chinese J. Manage. **18**(9), 1325–1334 (2021)
4. Zhao, Y.: Method of information system damage data isolation based on CTMO model. Comput. Simul. **37**(12), 479–483 (2020)

5. Li, Z., Lin, M., Li, Y.: A study on the employees turnover behaviors and causes based on grounded theory. Hum. Resour. Dev. China **37**(7), 21–33 (2020)
6. Bu, W., Zhao, M., Qi, P., et al.: The influences of employee engagement on the turnover intention of the new-generation employees —based on the research on organizational identity. J. Jiangxi Inst. Educ. **41**(5), 19–22 (2021)
7. Yao, Z., Zhang, X.: Research on the influence of congruence in time pressure on employees' turnover intention —emotional exhaustion as a mediator. Soft Sci. **34**(9), 109–115 (2020)

Prediction Method of Consumption Behaviour on Social Network Oriented to User Mental Model

Han Yin[✉]

School of Traffic and Transportation, Xi'an Traffic Engineering Institute, Xi'an 710300, China
xinqiba001@sina.com

Abstract. The current consumption behavior prediction method is mainly to use historical data modeling, through finding the laws of the data, to predict the user's consumption behavior. But the consumption psychology of users in social network will be impacted by the information in social network. The current usage methods ignore the impact of users' psychology on consumption behavior. In order to improve the above defects, the prediction method of social network consumption behavior oriented to user mental model is studied. Psychological characteristics refer to the stable characteristics of psychological activities. After understanding the psychological characteristics of users, the user's social network consumption psychological model is established. Dynamic identification is performed according to the user's personal preferences. According to the fit between user preferences and commodity characteristics, the utility value of commodities is obtained, and the social network consumption behavior is predicted using the differentiation ensemble learning model. The experimental results show that the average prediction accuracy is up to 90.52%, about 15% higher than the original method, and the proposed method has good stability for different conditions.

Keywords: User mental model · Social network · Consumption behavior · Behavior prediction · Subdivision ensemble learning model

1 Introduction

With the way of shopping from offline to online migration of a large scale, the market of online stores is huge, but the same fierce competition. In an era of booming big data and the Internet, the interaction between shopping platforms and social networking platforms has led to a proliferation of valuable data. Relevant businesses and departments make full use of these valuable data and information, target consumers' behaviors in a small way to form targeted operational strategies, and predict the future short-term and long-term economic situation to carry out regulation and maintenance in a higher dimension. Consumer behavior theory is a theory based on a series of factors, such as consumer psychology, desire, priority choice, etc. When consumers have a strong desire to buy goods, the utility of goods is very large; conversely, the utility of goods

W. Fu and L. Yun (Eds.): ADHIP 2022, LNICST 469, pp. 534–545, 2023.
https://doi.org/10.1007/978-3-031-28867-8_39

is small. Businesses can attract a large number of new consumers at low cost if they follow up the data information in real time, on the one hand, to understand consumers, on the other hand, to monitor competitors' trends, and on the other hand, to mobilize commodity prices to launch a variety of preferential activities. After all, in a competitive market, commodity prices matter. But note here that information contains value, but data is cluttered, worthless, even misleading and burdensome if not analyzed in a decent and appropriate way. In addition to understanding how consumers make choices about products and services based on information, the study of consumer behavior can also understand the relevant experience and satisfaction of consumers after making behavioral decisions. It is inseparable from the marketing activities of the enterprise market and is the basis of marketing decision-making [1]. In addition to online consumer buying data reflecting business information, prophase behavior data are also of great value. When businesses accurately grasp consumption habits and preferences, they can develop personalized recommendation and marketing methods to reduce operating costs and increase profits. At present, the prediction of consumer behavior is mainly based on the analysis of consumer online search behavior, comparison and selection information before consumption and historical consumption, and the prediction result of whether the user will buy a certain kind of goods or not is obtained by using the feature engineering of selection model and residual grey model and the method of model fusion, classifier and residual grey model. But also faces many problems, such as cold start, do not consider the content of the product label information.

User's behavior is determined by their psychological state. Studying user's consumption psychology will help to understand user's needs and provide multi-intelligentized services. The traditional research method of consumer psychology based on questionnaire is not only time-consuming and labor-intensive, but also subjective. With the coming of big data era, it is possible to use historical behavioral data to model user psychology. Mental model is the result of the interaction between the user's knowledge system and the current environment, which is similar to the user experience theory. It has qualitative characteristics as well as uncertain characteristics. Psychological model embodies a way of thinking, and becomes a visual angle for people to know things consciously or unconsciously. A mental model is a known experience in which the user knows something new. When users are faced with a product, they tend to understand the new product from their own experience, from the stored knowledge structure and past experience [2]. Using the user's mental model can understand the market of the product from the user's point of view, and get a better user experience. According to the above analysis, the consumer behavior of social network users contains a lot of information that can assist the business to make business decisions. Considering the advantages of user mental model in analyzing user behavior, this paper will study the prediction method of social network consumption behavior oriented to user mental model. The K-means clustering algorithm is used to construct the consumer psychology model of users' social network. The user preference is expressed according to the user emotion calculation in the construction of the user psychological model, so as to improve the accuracy of prediction. Predict the consumption behavior of social network users through commodity utility evaluation, and improve the overall prediction effect.

2 Prediction Method of Consumption Behavior on Social Network Oriented to User Mental Model

2.1 Establishing a Consumer Psychological Model of User Social Network

2.1.1 Predict Consumer Psychology of Users

In order to use the user's mental model, we need to master the user's mental state and the help of cognitive psychology, i. e., the psychology of information processing. Social network users will have different consumption psychology when they spend money. And different consumption psychology may lead to the consumer behavior decision. The psychology of consumers can be divided into quirks, beauty, convenience, reality, difference, habits psychology, economic psychology, different consumer psychology of packaging preferences and needs are different. Social networks provide users with a variety of services, users can enjoy the social network brought about by a variety of services, such as making friends, access to information, games, instant messaging. Every day, users use the Web, especially social networks, to generate huge amounts of data, such as tweets, comments, ratings of products, articles or news likes, check-ins at places of interest, clicking/browsing favorite products or news, favorites and purchases of products of interest, uploading travel pictures to social networks, uploading videos of their own filming to video sites, and so on. User-centered concept has been continuously infiltrated in various industries and fields, and the behavior of users in social networks is the reflection of their true thoughts and thoughts, which plays an important role in analyzing and understanding users' preferences [3]. Therefore, by analyzing the consumption data of user social network, we can build a universal consumer psychological model of user social network, which can help to predict consumer behavior more accurately.

The user behavior data set is processed and the number of occurrences of each behavior is counted. In general, the most frequent click-and-view behavior occurs when users browse social networks. Therefore, click-and-view behavior is the main behavior in data sets. Further, the number of collections and the number of last purchases are similar, and the number of purchases occurred the lowest. This shows that users will be after the collection of goods, the probability of purchase will be greatly increased. To the user clicks to look at the behavior and the frequency of the same item to buy statistics, for the same item, only once to buy the majority of users. The frequency decreases with the increase of the number of purchases. In addition, the analysis of online consumer behavior, data types can be collected complex, data types, including the consumer ID of the integer data, long integer data that time stamps. Data dimensions vary, including the conversion rates of goods obtained through purchase and browsing. The data angle is also very broad, including indicates consumer sex, the age and so on personal information; Represents the product traffic, the sales volume and so on merchant information. Screening This example is intended to determine whether a purchase will occur based on a large amount of previous data. The final data of this study includes consumer ID, commodity category ID, consumer behavior and timestamp of each behavior.

2.1.2 Using k-Means Clustering Algorithm to Build the Consumer Psychology Model of Social Network Users

This paper uses k-means clustering algorithm to cluster the consumption data and consumption psychology of users, and establishes the consumption psychology model of social network users. Consumer psychology is divided into four categories. Consumers in the first category mainly pursue the singularity of the goods, and pursue new packaging, new technology and new technology. Consumers in the second category have higher requirements for the beauty brought by packaging, and attach great importance to the craft, color and shape of the goods, while attaching importance to the economy and uniqueness of packaging. The third category of consumers in the consumption of the main focus on habitual, higher brand loyalty, and pay attention to the practicality of goods. The fourth category of consumers in the consumer's main psychological reality, followed by convenience, in the purchase of convenient, convenient packaging on the basis of more attention to the cost-effective goods.

According to the current types of users' consumption psychology, the cluster is classified by the clustering algorithm, and the user's consumption-related data and consumption preferences are taken as the feature vectors of the clustering algorithm. The initial cluster centers of each fern cluster are randomly selected, and the distance between the cluster centers and the samples to be clustered is calculated according to the following formula:

$$d\left(c_i, x_j\right) = \sqrt{\sum_{j=1}^{n}\left(c_i - x_j\right)^2} \tag{1}$$

Among them, c_i is the cluster center of each consumer psychology cluster, and x_j is the data of consumer behavior to be clustered. The parameters of k-means algorithm are trained by building feature training set manually. After clustering all the consumption data, the distance center of each cluster is updated. The data with the shortest distance from other data in each cluster is calculated as a new cluster, and the clustering process is repeated again until no new cluster center is generated. After the mapping relationship between user's psychology and user's social network consumption behavior is established, the user's consumption psychological model can be obtained.

2.2 Dynamic Identification of Users' Social Network Consumption Preferences

Static personalized recommendations have the ability to identify the user's past preferences, but if the user's preferences change, recommendations based on past preferences do not always provide the user with the greatest utility. As user preferences evolve, users tend to change their behavior over time. The ever- changing preferences of users have a great influence on the accuracy of recommendations. However, user preferences may not necessarily evolve gradually, but can also evolve rapidly between two consecutive periods of time, which means that user preferences may be similar to or completely different from those of the previous period of time in a continuous period of time. Therefore, it has been studied that the user's preference is expressed by a series of behaviors before the purchase. Such as buying, viewing, like points. Generally speaking, user behavior can be divided into goal feedback and assistant feedback. Therefore, considering the preferences of future users can provide more accurate recommendations [4].

In the field of service application, the situation involved is different according to the different application needs, and the more common one is to classify the situation into service requester situation, service provider situation and service situation. In the field of pervasive computing, because pervasive computing is the fusion of information space and physical space, people can get digital services at any time, anywhere and transparently in this fusion space, so situational information plays a vital role in it, and contextual and situational awareness computing has been attached great importance.

The factor of user preference is expressed as the user's preference for product features, including the feature dimension of user preference and the degree of user preference. The reason why users buy goods is that the information reflected by the content of goods accords with the user's preference, and because the user's preference is variable, this paper selects two time intervals of half a month and a month to obtain the historical purchase data of users in these two time periods, and expresses the user's preference through the calculation of user's emotion in the user's psychological model. The specific steps are as follows:

(1) Computation method for user preference of a single user t moment: at any time as t moment, obtain historical purchase data of the user for the previous week up to the t moment, compute the emotional value q of the purchasers' b based on all user behavior data sets of the same category of goods purchased in the history, use the product label as the dimension of user preference, and use the emotional value q as the degree of user preference. If there are k dimensions of user preference, then the individual user t moment preference can be expressed as follows [5]:

$$P_t = \{(b_1 : q_1), (b_2 : q_2), \cdots, (b_k : q_k)\} \tag{2}$$

(2) Computation method for user preference for $t + 1$ moment of a single user: take half a month as time span, half a month after commencement of t moment as $t + 1$ moment, obtain historical purchase data from the t moment to the $t + 1$ moment of the user, take the historical purchase behavior of users of the same type of commodities purchased in the past as the data set, compute the emotional value s of the b of the purchaser, the commodity label as the dimension of user preference, and the emotional value q as the degree of user preference. During this period of time, several uncontrollable factors may change the dimension of user preference or the degree of user preference. Therefore, the updating rules for user preference are as follows: if a new commodity buyer arises, the corresponding emotional value shall be added directly; if several commodities of the same type are purchased, the emotional value of the purchaser $q' = q \times w + q' \times w'$ shall be updated as the new user preference in accordance with the method of weighted proportional accumulation. If the updated user preferences have k' dimensions, the $t + 1$ moment preference of a single user can be expressed in the following formula.。

$$P_{t+1} = \{(b_1 : q_1'), (b_2 : q_2'), \cdots, (b_{k'} : q_{k'}')\} \tag{3}$$

Based on the dynamic identification of consumers' social network consumption preference, this paper predicts consumers' social network consumption behavior by using the subdivision ensemble learning model.

2.3 Predicting Consumption Behavior of Social Network Users

In the actual user social network purchase process, the user's choice depends on his assessment of the utility of the commodity in the transaction process $w_{s,u,i}$, which is a combination of the user's preference and the characteristics of the commodity. Denote the latent factor vectors of users and products as r_u and s_i, respectively, there are:

$$w_{s,u,i} = r_u^T s_i \tag{4}$$

The expression above indicates that the more consistent the user's preferences are with the characteristics of the product, the more the product is in line with the consumption habits of the user, and the higher the utility value of the product $w_{s,u,i}$; on the contrary, the more deviation the characteristics of the product deviate from the user's preferences, and the more inconsistent the user's consumption habits are, the lower the utility value of the product $w_{s,u,i}$ is. For rational users, the probability of choosing a certain item depends on the relative utility of the item in the whole set, and there is a positive correlation between them. Specifically, in the online consumption forecasting problem, users will consider the final opportunity cost and benefit in each transaction process, and choose the goods with utility value not less than $w_{s,u,ioc}$ to purchase. That is, in an arbitrary purchase, the utility of the commodity k purchased by the user shall not be less than that of the best alternative:

$$w_{s,u,k} \geq \max w_{s,u,i} = w_{s,u,ioc} \tag{5}$$

In a user's historical purchase, the final purchase k is logged and therefore always known; the best alternative is not, and the platform does not directly know and record the user's preferences, so any product that appears in the purchase is likely to be the best alternative. To obtain the best alternative to each purchase, the utility value of each purchase is calculated in terms of formula (4) for each application of the product in question. In fact, although the user's preference is not directly reflected in the product sequence of the purchase behavior, it can still be used to analyze the user's preference in the purchase behavior. For example, the number of clicks a user makes on different items in a sequence reflects the user's preferences to some extent. Based on this, this paper proposes a behavior sequence utility function $F(s, i)$, using $F(s, i)$ to analyze the product sequence can be used to estimate and judge the utility of the product to the user. Form of definition: The sequence of clicks made by a user in a historical purchase is recorded as cq in order, the length of cq is N, and the position in which the commodity k constitutes a set $P(s, i)$, the sequence utility of the commodity k in a historical purchase is:

$$F(s, i) = \sum_{k \in P(s,i)} e^{-(N-k)/N} \tag{6}$$

The sequential utility of each commodity in all purchasing behavior can be calculated by the above expression. Based on the sequential utility of goods and the dynamic identification of consumer psychology and consumer preference, a subdivision ensemble learning model is used to predict consumer behavior.

The category of a consumer is related not only to his initial state, but also to the state of other consumers in the consumer's social network who are closely related to him. Ordinary consumers will listen to the advice of the buyer, the purchase impulse, become the buyer. In order to facilitate the research, it is assumed that the information that consumers receive is positive, that is, after the general consumers obtain the information, they will have the desire to buy and become potential consumers or buy and become buyers directly. Set rules for the conversion of conduct between various types of consumers as follows [6]:

R1: Interaction between general consumers and purchasers, general consumers may obtain certain product information and then transform into potential consumers;
R2: A potential consumer may become a buyer after interacting with the buyer;
R3: After a period of time, buyers no longer have the desire to disseminate information on the conversion of immunity to consumers;
R4: Immunized consumers will no longer accept information from purchasers, but may continue to purchase products over time:
R5: Buyers will spread consumer information to the general consumer, potential consumers and immune consumers do not have the ability to spread.
R6: General consumer contact with the buyer may lead to direct purchase behavior and thus into the buyer.

The process is as follows:

(1) Amount of information exchanged: The amount of information exchanged when a consumer comes into contact with other types of consumers. At the moment t, agent i and agent j contact, agent i and agent j will change the amount of information in i.
(2) Amount of updated information: the total amount of information on changes in Agent i shall be calculated and updated within the t time limit.
(3) Comparative information content: the conversion of consumer categories shall be carried out according to the information content of agent i and the threshold of information content for each category of consumers.

According to the change of consumers' consumption psychology, historical consumption data and social interaction information are collected, and DSEM is used to predict the consumers' consumption behavior.

DSEM is a nested ensemble learning model, which subdivides the dataset by sample filtering. The model is built from multiple Bagging modules, each of which contains several base classifiers. The training set of each Bagging module in the DSEM is obtained by sample filtering. DSEM can be independently trained to n Bagging modules by sample filtering layer by layer, and then the final learning device can be obtained by set strategy for these n Bagging modules. To implement sample filtering, DSEM sets a strong rule binding policy in each of the Bagging modules, giving the sample that cannot be filtered through the rules to the next Bagging module. Strong rule-combining strategy means that when Bagging model predicts a sample, more than 70% of the base learner must predict the sample into the same class before the model can give the prediction results. For example, in a binary classification problem, the positive sample is labeled as 1, the negative sample is labeled as 0, the base learner is $b(x)$, and the strong rule classifier consisting of the base learner is $B(x)$. The strong rule-binding strategy of $b(x)$ is [7]:

$$B(x) = \begin{cases} 1, \sum b(x) \geq r \times E \\ 0, \sum b(x) \leq (1 - r) \times E \\ np, eles \end{cases} \tag{7}$$

The E represents the number of base learners contained in the Bagging module; the r is the strong regular coefficient, $r \in [0.7, 1]$. By setting the combination strategy of strong rules, the model can avoid the random guessing of uncertain samples, so that each layer of Bagging model can only predict the sure samples and filter the uncertain samples to the next layer. Through filtering layer by layer, DSEM can subdivide the original dataset into several subsets, and use different Bagging models to fit each subset to improve the accuracy of prediction.

Each layer of the Bagging module is trained based on SFTraing, giving misclassification and samples that have not passed the rules to the next layer of the model. The training framework based on sample filtering is shown in Fig. 1.

When a test sample enters the DSEM model, it is first predicted by the first Bagging module. If the sample passes through the strong rules of the Bagging module, the classification of the test sample by the Bagging module is used as the prediction result. So far, we have completed the research on the prediction method of consumption behavior of social network oriented to user mental model.

3 Test Study

3.1 Test Preparation and Process Design

This section will test the predictive effect of this method, so as not to fail to achieve the expected research goal in the practical application of this method.

Firstly, the data of social network consumption behavior is preprocessed according to the following contents, using userbehavior.csv data set for experiments, and the actual consumption behavior corresponding to the consumption behavior data is used as a reference for the prediction accuracy.

Data preprocessing mainly includes four operations: duplicated data de-duplicating, missing data processing, unreasonable data processing and data format conversion. Data de-duplication refers to the processing of the user order table. All the fields in the user order table have the same values, so it is necessary to de-reprocess them. There are still some fields in the user order table that have different numbers of items but the values of other fields are the same. This kind of data is de-duplication and the first article of duplicate data is kept. Missing data processing is mainly for the user's age and the

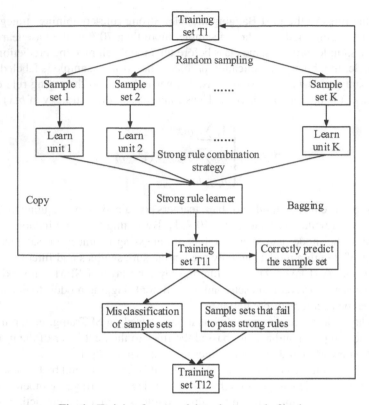

Fig. 1. Training framework based on sample filtering

parameters of the product field, delete the missing data for too much, the missing data for the average filling.

Based on this dataset, several experiments were designed to evaluate the performance of behavioral prediction methods by using some common indexes such as precision, recall, F1 value.

The formula for calculating the accuracy and recall rates is as follows:

$$P = \frac{TP}{TP + FP} \tag{8}$$

$$R = \frac{TP}{TP + FN} \tag{9}$$

Among them, *TP* indicates that the forecast is positive, and the actual forecast is correct. The *FP* indicates that the forecast is positive and the actual forecast is wrong. *TN* indicates that the forecasts are negative and that the forecasts are actually wrong. *FN* indicates that the forecast is negative and that the actual forecast is correct.

The following is the formula for calculating the F1 value:

$$F\_s = \frac{(1 + \beta^2)R \times P}{\beta^2(R + P)} \tag{10}$$

The F1 value is a tradeoff between accuracy and recall, and it measures the effectiveness of the classification based on the ratio of the recall to the weighted importance of accuracy as determined by the manually set coefficient β. F1 value can reflect the prediction performance of positive samples. Comparing the prediction method proposed in this paper with the behavior prediction method based on support vector machine and the prediction method based on Bayesian model, the performance of consumption behavior prediction method is evaluated by comparing the values of three indexes.

3.2 Test Results

The accuracy and recall of the three methods for predicting the consumption behavior of social network users on the same historical consumption behavior dataset are shown in Table 1.

Table 1. Comparison of accuracy and recall rates of consumer behavior forecasts

Serial number	Method in this paper		SVM based forecasting method		Prediction method based on bayes	
	Accuracy/%	Recall rate/%	Accuracy/%	Recall rate/%	Accuracy/%	Recall rate/%
1	94.28	88.43	82.18	75.66	75.71	73.57
2	90.85	87.95	70.53	76.08	86.14	69.46
3	87.43	84.09	78.75	80.14	63.85	73.76
4	88.17	86.43	84.07	81.55	74.08	64.41
5	91.25	88.47	77.83	77.08	74.91	57.65
6	90.59	85.19	80.62	77.32	60.57	79.34
7	94.27	87.94	72.39	76.83	79.73	63.42
8	87.68	86.26	74.24	79.18	63.64	74.83
9	91.38	87.87	81.64	80.71	84.76	73.49
10	89.26	83.25	73.05	77.46	85.38	72.99

Analysis of the data in Table 1 shows that the accuracy and accuracy of this method in predicting consumer behavior in social networks are significantly higher than those in the other two consumer behaviors. The average prediction precision and accuracy of this method are 90.52% and 86.59% respectively, 77.53% and 78.20% respectively based on SVM, 74.88% and 79.29% respectively based on Bayesian method. From the numerical point of view, the prediction effect of this method is better. Among the different forecasting methods, the difference between the accuracy and the accuracy of the prediction method based on the Bayesian model is small, and the difference between the prediction accuracy and the accuracy of the prediction method based on the support vector machine is obvious. It shows that this method is more stable for different social network users to predict consumer behavior.

The F1 values of the three methods under different coefficients β are compared as shown in Fig. 2.

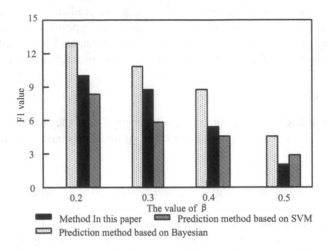

Fig. 2. Comparison of F1 values under different adjustment coefficients

Analysis of the information in Fig. 2 shows that under different coefficients β, the predicted F1 value of the method is higher than that of the other two comparison methods. According to the definition of F1 value, the prediction performance of this method is more comprehensive, and the accuracy and recall are more balanced. To sum up, the consumer behavior prediction method based on user mental model proposed in this paper is effective and can predict consumer behavior accurately, which provides technical support for marketing strategy.

4 Conclusion

In the field of modern consumer behavior research, scholars believe that the ultimate goal of consumer behavior is to make themselves satisfied. With the development of network economy, it can not only affect the offline operation decision of market economy, but also affect the online consumption decision of consumers. Since the vast majority of users do not give explicit feedback (grading and commenting) on the goods they have consumed, the implicit feedback of users plays an important role in grasping users' needs and describing users' psychology. User psychological model can analyze the user's consumption data, can help businesses more accurately analyze the user's psychology and thus accurate sales. This paper studies the consumer behavior prediction method of social network based on user mental model, and uses this method to predict the consumer behavior and consumer propensity in social network. The accuracy of the method is verified by experiments. The method can accurately predict the consumption behavior of users in the social network from their consumption psychology. At present, the research on social networks is mainly comprehensive, and the professional research on social networks is not mature. In future research, professional social networks need to be the focus of research, and the prediction of user consumption behavior needs to be constantly updated to better grasp the needs of users.

References

1. Zhang, F., Ni, R.: Research on online users consumption prediction model based on SOMTE. Math. Practice Theory **50**(15), 49–59 (2020)
2. Zhou, L.: Empirical analysis on the prediction of tourism consumption price index construction for tourism consumption of urban and rural residents—Test based on VAR model. Prices Monthly **06**, 1–5 (2019)
3. Ye, Z., Chen, E.: Study on the prediction of natural gas consumption based on residual grey model. J. Chongqing Inst. Technol. **32**(09), 99–102 (2018)
4. Hu, X., Zhang, H., Dong, J., et al.: Prediction model of user buying behavior based on CNN-LSTM. Comput. Appl. Softw. **37**(06), 59–64 (2020)
5. Han, L., Jiang, G., Wang, H.: Fast mining simulation of online behavioral information characteristics under big data analysis. Comput. Simul. **36**(6), 346–349 (2019)
6. Hu, X., Zhang, H., Dong, J., et al.: Prediction of ensemble learning-based new users' repurchase behavior on e-commerce platform. Modern Electron. Tech. **43**(11), 115–119+124 (2020)
7. Ge, S., Ye, J., He, M.: Prediction model of user purchase behavior based on deep forest. Comput. Sci. **46**(09), 190–194 (2019)

Construction Safety Monitoring System Based on UAV Image

E. Yang[✉], Jing Yang, Kangyan Zeng, and Yan Zheng

Department of Architectural Engineering, Chongqing College of Architecture and Technology,
Chongqing 401331, China
eyangYang26785@163.com

Abstract. In the process of construction safety monitoring, the construction image is not corrected, resulting in low monitoring accuracy and high false alarm rate. Therefore, a construction safety monitoring system based on UAV image is designed. The hardware framework of the system is designed by the camera mounted UAV module and stm32f103c8t6 as the system control core through wireless network communication technology. Based on the hardware design, the UAV image is corrected, and the image features are extracted by the brisk algorithm according to the preprocessing results. The kernel function SVM is introduced to classify the UAV images and realize the safety monitoring of engineering construction. The system test results show that the monitoring accuracy of the designed monitoring system is higher than 90%, and the false alarm rate is low. It can monitor the construction safety behavior in real time, and is more flexible and convenient in actual use.

Keywords: UAV image · Architectural engineering · Construction safety · Monitoring system · Brisk algorithm · Support vector machine

1 Introduction

The number of construction companies has grown rapidly, and the number of construction workers is also increasing. The construction area and investment amount have repeatedly hit new highs. The development of the construction industry has gradually diversified, and the design aspect has become more and more broad. Due to the characteristics of the construction industry, the diversity and complexity of employees in this industry, the cultural level is different, and the professional quality is also uneven [1]. The construction industry is characterized by complex production activity systems, long construction periods, and constantly changing processes, which make risks and potential safety hazards run through the project, with great danger and many unsafe factors. Urban and rural construction and development are inseparable from project construction and management. Whether it is survey and design before construction, or dynamic data monitoring and quality inspection during and after construction, measurement work runs through project construction and management. There are many hazardous areas in the

W. Fu and L. Yun (Eds.): ADHIP 2022, LNICST 469, pp. 546–556, 2023.
https://doi.org/10.1007/978-3-031-28867-8_40

modern building production process, which are difficult to achieve by human monitoring. Taking strict measures for construction site safety supervision is the main policy of construction site supervision. Focusing on the core of engineering quality and engineering safety issues, taking effective supervision is a compulsory course in the construction industry [2].

In recent years, it has become more and more difficult for on-site supervision to adapt to the increasingly complex construction environment. The main reason is that the number and energy of supervisors are limited, and they cannot cover every corner of the construction site. Therefore, the use of automated monitoring systems can comprehensively and quickly It is necessary to detect the unsafe behavior of workers to ensure the safety of engineering construction. In the current research on building construction safety management, there are various technical solutions. The early application of security surveillance technology was mainly based on video surveillance technology. The surveillance cameras used analog cameras, which could only control the safety of the construction site from the perspective of video surveillance. However, monitoring cannot be implemented for hazardous environments such as fire protection, lightning in high-rise buildings, and strong winds. With the popularization and use of sensor technology, sensors are gradually introduced into the field of construction engineering [3]. Such as fire monitoring. At this time, most of the sensors are wired sensors, and the sensors are both analog and digital. The use of wired sensors increases the type of on-site information monitoring, but it has the problem of deployment and wiring. In the field use, construction bumps also occur, resulting in the line being cut off, and even the sensor cannot be recovered in the end. The construction safety monitoring using wireless communication technologies such as RFID can realize multi-point distributed monitoring, but this monitoring system leads to a lot of controller nodes on site, and the deployment and wiring workload on site is huge, which is very inconvenient to use [4]. In the current construction project, the scale of the project is getting bigger and bigger, in order to ensure the construction period, a lot of manpower and material resources are invested at the same time. In addition, the height of building construction projects is more and more forecast, and the safety monitoring of high-rise buildings is more difficult. The performance requirements for security monitoring systems are further improved.

UAV technology is a multidisciplinary comprehensive technology born from the development of science and technology. It integrates GPS global positioning technology, remote sensing technology, communication technology and UAV driving technology. It can not only carry out precise remote control, but also confirm and record spatial geographic information. As UAV technology becomes more and more mature, and considering its advantages such as safety, high cost performance and mobility, more professional application projects have begun to use UAVs to complete [5]. The image files of the construction site captured by the drone can be processed to obtain relevant information on the construction site such as area, volume, slope, height, etc., so that the supervisor can supervise the construction site in the monitoring room. UAV images are relatively unrestricted by time and space, and the imaging effect is good. At present, there are many successful cases of using UAVs in aerial surveillance.

Based on the above analysis, this paper will design a construction safety monitoring system based on UAV images. Through the UAV module, the image data transmission and communication module, the operation control module and the surrounding circuit design system hardware part, the system software design will be completed by preprocessing the construction image data, extracting the feature points in the UAV images and identifying the dangerous behaviors in construction, According to the system hardware and software design, the construction safety monitoring system is designed and applied to the actual construction project management.

2   Design of Hardware Part of Construction Safety Monitoring System Based on UAV Image

The hardware part of the construction safety monitoring system designed in this paper is mainly composed of the UAV equipped with the UAV image acquisition device, the core control module of the monitoring system, the UAV data transmission communication module and the data processing module. The hardware framework of the engineering construction safety monitoring system is shown in Fig. 1 [6].

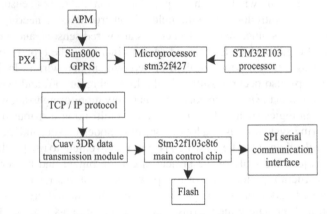

Fig. 1. Schematic diagram of the hardware framework of the engineering construction safety monitoring system

The flight control unit of sim800c GPRS is pixhawk flight control, which has APM and Px4 firmware and corresponding ground station connection. GPRS is connected to 32-bit microprocessor stm32f427 for flight control, and STM32F103 processor is used as additional fault protection standby controller. The GPRS of the UAV is embedded with TCP/IP protocol and connected with cuav 3DR data transmission module. Stm32f103c8t6 is selected as the main control chip for control. The design process of each hardware module in Fig. 1 will be described in detail below.

2.1   UAV Module

In order to ensure the safe and reliable operation of this system, it is first necessary to select a UAV that can fly stably and has strong wind resistance. Compared with fixed-wing UAVs, rotary-wing UAVs are smaller and simpler in structure. At the same time, they have lower requirements for flight space, can fly in different attitudes, and can hover in the air for a long time. In addition, the multi-rotor UAV offsets the additional rotational moment due to the different steering between adjacent rotors, and does not need to use the tail alone to achieve body balance, so it has better maneuverability and a simpler and more reliable structure. To sum up, this design chooses multi-rotor UAV to realize the safety monitoring of construction engineering.

Compared with the load capacity, this design has higher requirements for the endurance of the UAV, because the quadrotor UAV is selected. The flight control unit of the quadrotor UAV adopts PIXHawk flight control, which has two sets of firmware and corresponding ground station software, APM and PX4, and has an operating frequency of 168 MHz. It has a 32-bit microprocessor STM32F427 as the main controller, and a 32-bit STM32F103 processor with independent power supply as an additional fault-protected backup controller, and has a large storage space and various sensors [7].

The UAV remote control unit consists of two parts: the transmitter and the receiver. The transmitter is installed on the ground end, and the receiver is mounted on the UAV end and connected to the flight control. The transmission process of the control signal sends instructions to the receiver through the transmitter of the remote control, and the receiver decodes and transmits it to the flight control system, and the UAV makes corresponding actions according to the signal instructions.

This design chooses SIM800C GPRS to provide direction guidance for UAV flight. When the power supply voltage of the drone is under-voltage or the onboard monitoring terminal is powered off, the monitoring terminal cuts off the external power supply and enables the backup battery. When the external power supply restores the power supply, the backup battery can be charged.

2.2 Image Data Transmission and Communication Module

The communication between the UAV and the ground station, the communication between the ground station and the monitoring terminal adopts wireless data transmission mode, the data transmission communication adopts UART serial communication, and the baud rate is set to 115200. The communication process is to send one byte at a time, and send 30 consecutively, among which there are data flag bits and various data. The UAV digital radio is a professional data transmission radio with the help of digital signal processing and software radio technology. The data transmission station used in this article is the CUAV 3DR data transmission module, with a maximum transmission distance of 5000 m. The UAV image transmission module adopts channel coding, video compression, modulation and demodulation and signal processing technology to transmit the video obtained by the camera in the quadrotor UAV to the ground station in real time by wireless transmission. This design uses 5.8 GHz analog image transmission to transmit analog video, transcodes the high-definition video obtained by the camera on the UAV into 1080P, and finally converts it into a digital signal and transmits it to the video display interface of the ground station.

The GPRS of the drone is embedded with the TCP/IP protocol, and the GPRS communicates with the ground station and the drone flight control through the TCP protocol. The data sent by the drone to the ground monitoring mainly includes flight status information, flight control setting information, running status information, etc. The total data volume is relatively large, about 300 bytes. In actual communication, the above data information is divided into multiple data packets according to the requirements, and the corresponding data packets are returned according to the request information of ground monitoring. The communication protocol data frame format of downlink data is shown in Table 1.

Table 1. Data frame format of downlink data link communication protocol

Project	Numerical value	Illustrate
STX	2 Bytes	Data frame header, indicating the beginning of the data packet sent to ground monitoring
Length	1 Byte	Total packet length
CmdID	1 Byte	Feedback command ID to ensure that the command is sent successfully
PacketID	1 Byte	Packet ID, used to distinguish different packets
DataPacket	n Bytes	Valid data, used to represent data information,
CheckSum	2 Bytes	Sum check to ensure the correctness of the data

The communication protocol format of the uplink data sent by the ground station to the UAV is similar to the downlink data protocol format.

2.3 Operation Control Module and Surrounding Circuit Design

According to the design requirements of the control system, the proposed cost, and the future scalability, the concentrator circuit selects the main control chip model STM32F103C8T6, which is a 32-bit microcontroller based on the ARM Cortex-M3 core of the STM32 series, 2 V~3.6 Wide voltage supply range of V, the maximum operating frequency is 72 MHz. It also features single-cycle multiply instructions and hardware divides, as well as a priority programmable interrupt system. The chip also has 64 KB of Flash memory and 20 KB of SRAM memory. In addition, it also integrates a wealth of on-chip peripherals, multiple timers, DMA controllers, serial ports, ADC, SPI interfaces, etc., with low cost, high speed Fast and cost-effective.

The external Flash chip W25Q128 is a 128 M-bit large-capacity serial flash memory device. It uses the SPI serial communication interface and connects to the STM32 on-chip SPI interface to realize the control read and write of the W25Q128. It has 8 pins, only four signal lines are needed for communication, and the design requirements can be met by configuring these four signal lines accordingly. The W25Q128 connection circuit is shown in Fig. 2.

The power supply voltage of the main control chip is 2–3.6 V, and 3.3 V is selected as the power supply voltage; the working voltage of the GPRS module is required to be in the range of 3.3–4.8 V, and 4 V is designed for power supply; in the TTL to USB interface circuit, the CH340G chip needs 5 V power supply; The storage module requires a 3.3 V supply voltage. According to the voltage requirements of the above modules,

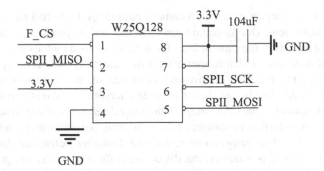

Fig. 2. W25Q128 connection circuit

the concentrator circuit is designed to output 3 V values of the power supply to ensure the normal operation of the circuit. The specific design is as follows:

The power module adopts 12 V input, and the on-off of the concentrator circuit is controlled by the module power switch. First, the DC step-down conversion chip MP2359 is stepped down to obtain the 5 V voltage required by the TTL to USB module. The 5 V voltage passes through two forward low dropout voltage regulator AMS1117 chips respectively, and obtains a 3.3 V voltage for the STM32F103C8T6 main control chip. The design power supply voltage of GPRS module SIM800C is 4 V, and its working peak current is close to 2 A, so DC-DC step-down chip MP2303 is selected for step-down, and the required 4 V power supply voltage is obtained.

With the hardware support of the safety monitoring system designed above, the image analysis and processing of the construction work images collected by the UAV is carried out to realize the safety monitoring function of the entire construction project.

3 Software Part Design of Construction Safety Monitoring System Based on UAV Images

On the basis of the above system hardware design, this paper designs the software part of the system. First, it corrects and stitches the distortion difference of the building engineering construction image data collected by the UAV, and then uses the classifier to identify the dangerous behaviors of engineering construction in the UAV image, and distinguishes the construction scene with dangerous factors from the normal scene, so as to complete the safety monitoring of building engineering construction.

3.1 UAV Image Processing

UAVs are easily affected by airflow, wind speed and wind direction. Therefore, the influencing factors of its own structure on image quality can be divided into three points: poor camera objective lens distortion, pixel size and resolution, and flight stability. The data processing function is mainly to correct and splicing the image data collected by the UAV, and then generate digital orthophoto images, digital elevation models, etc. deal with.

Camera lens correction is based on camera calibration to correct the error between the camera and the lens. Due to certain errors in the design, production, assembly and other technical aspects of the camera lens, the image obtained by the camera will have a certain optical distortion, which further affects the reliability and accuracy of the aerial triangulation results. Therefore, eliminating the optical distortion error of the lens to the greatest extent is a key link before the air-to-three encryption, and it is also a prerequisite for ensuring the quality of the UAV image data. After determining the internal orientation elements, optical distortion parameters and area array deformation parameters of the digital camera, photogrammetry can be performed. However, before the obtained digital image is used in actual production, the distortion difference of the image needs to be corrected according to the camera calibration parameters. The total correction model for camera distortion is [8]:

$$\Delta X = \Delta x \left(q_1 r^2 + q_2 r^4 \right) + 2p_1 \left[r^2 + 2\Delta x^2 \right] + 2p_2 \Delta x \Delta y + \xi \Delta x + \zeta \Delta y$$
$$\Delta Y = \Delta y \left(q_1 r^2 + q_2 r^4 \right) + 2p_1 \left[r^2 + 2\Delta y^2 \right] + 2p_1 \Delta x \Delta y \tag{1}$$

Among them, ΔX, ΔY are the correction values of the image point; Δx, Δy are the coordinate differences between the image point coordinates and the image principal point coordinates in the image coordinate system; r is the radial distance; q_1, q_2 are the radial distortion coefficients; p_1, p_2 are the tangential distortion coefficients; ξ is the CCD non-square scale coefficient, and ζ is the CCD non-orthogonality distortion coefficient.

The main task of radiation correction is to reduce the influence of image distortion and noise as much as possible by eliminating the chromatic aberration between images, restore the most original state of the image, improve the contrast between the images, and better match and stitch the subsequent UAV images. to prepare. Histogram equalization and histogram matching are the most commonly used image enhancement methods.

Evenly distributing the histogram of the original image is the basic mechanism of histogram equalization, so as to achieve the purpose of expanding the dynamic range of the gray value of the image, thereby improving the overall contrast of the image. The gray value of the UAV image can be regarded as a random variable in the interval [0, 1]. Let the gray value be $g (0 \leq g \leq 1)$, the gray value of the transformed pixel is g', and P_g and $P_{g'}$ are the probability of random variables g and g'. Density function, the transformation function is T, then there is the following formula [9]:

$$\begin{cases} g' = T(g), (0 \leq g \leq 1) \\ P_{g'} = P_g |dg / dg'| \end{cases} \tag{2}$$

The gray value g corresponds to the gray value g' one-to-one, and the transformation function T must meet the following conditions:

(1) T is uniquely determined in the interval [0,1] and increases monotonically;
(2) $0 \leq T \leq 1$ ensures that the pixel gray value after mapping is within the valid range.

When the image gray value is K order, the transformation result can express the formula:

$$g'_k = T(g_k) = \sum_{K=1}^{K} P_g^j = \sum_{K=1}^{K} \frac{N_j}{N} \tag{3}$$

In the formula, k represents the gray level of the image; N represents the total number of pixels; N_j represents the number of pixels on the j gray layer; P_g^j represents the probability density of the j gray layer; $T(g_k)$ represents the transformation of the pixels in the k gray layer function; g'_k is the final transformation result. UAV images can not only get the grayscale histogram equalization result of one image, but want to get the color image equalization result, then it is necessary to first divide the color image into RGB components, and realize the histogram equalization of each component separately. Then combine the three components.

Histogram equalization can only automatically enhance the contrast of the entire image to obtain a globally equalized histogram, but the calculation process cannot adjust the parameters and is not easy to control. The histogram matching can select an image with excellent effect as the matching object, and correct the histogram of the original image to make it into a specified shape. After processing the images collected by the UAV, the classifier is used to identify the images, so as to realize the function of engineering construction safety monitoring.

3.2 Engineering Construction Hazard Identification

Before using the classifier to identify the dangerous behaviors or factors of engineering construction in the UAV image, the features in the image are extracted as recognition vectors. In this design, the brisk algorithm is used to extract the feature points in the UAV image, and the expression is:

$$S(I) = \frac{g'_k(\Delta X, \Delta Y)}{T(g_k)} \tag{4}$$

According to the feature extraction results, identify the construction risks of the project. The specific process is as follows [10]:

1) Construct n octave layers (denoted by c_i) and n intra-octave layers (denoted by d_i). Assuming that there is an image I, the octave layer is generated: the c_0 layer is the original image, the c_1 layer is 2 times downsampling of the c_0 layer, the c_2 layer is 2 times the downsampling of the c_1 layer, and so on. The generation of the intra-octave layer: the d_0 layer is 1.5 times the downsampling of I, the d_2 layer is 2 times the downsampling of the d_1 layer (that is, the $2 * 1.5$ times downsampling of I), the d_3 layer is 2 times the downsampling of the d_2 layer, And so on.
2) Perform FAST9-16 corner detection on the image to obtain an image with corner information, and perform FAST5-8 corner detection on the original image I once
3) According to the above steps, the position and scale of the image feature points are obtained, and the two-dimensional quadratic function interpolation is performed on

the FAST score value at the position corresponding to the layer where the extreme point is located and its upper and lower layers, and the real score extreme value is obtained. The point and its precise coordinate position are used as the feature point position; then one-dimensional interpolation is performed on the scale direction to obtain the scale corresponding to the extreme point as the feature point scale.

After extracting the image features from the UAV image, the support vector machine model is used to classify the image features. The essence of SVM is a model for classifying two types of samples, and the optimal classification hyperplane of the samples is calculated to maximize the distance between the classification boundaries between the two classes. The basic expression of SVM can be written as:

$$\min \ 0.5\|\delta\|^2$$
$$s.t. \ y_i\left(\delta^T x_i + \eta\right) \geq 1 \tag{5}$$

Among them, x_i is the sample; y_i is the classification label of the sample; δ is the slope of the expression of the classification hyperplane; η is the intercept of the expression of the classification hyperplane. The optimization objective of the above equation is quadratic and the constraints are linear, which is easier to solve when it is converted into a Lagrangian dual problem. The kernel function is introduced to reduce the influence of image dimension on classification. The discriminant of SVM is as follows:

$$f(x) = K(x_i * x)\mathrm{sgn}\{a_i y_i(x_i * x) + b\} \tag{6}$$

Among them, $K(x_i * x)$ is the kernel function of SVM; a_i is the Lagrange multiplier; b is the mapping function. Using hyperplane UAV image classification to distinguish construction scenes with risk factors from normal scenes. When images containing construction risk factors are classified in the classifier, an alarm is issued to facilitate timely supervision by construction managers. The above software program for realizing construction safety monitoring function is transplanted into the hardware module of the UAV remote monitoring system, and the design and research of the construction safety monitoring system based on the UAV image is completed.

4   System Performance Test Experiment

The construction safety monitoring system based on drone images is designed above. Before applying the system to actual project management work, it is necessary to accurately test the operation of each module of the system and the performance of the system to ensure that the system can be used in practical applications. to achieve the expected design targets.

4.1   Experimental Content

Run test and performance test on the monitoring system designed above. In the operation test, the data transmission between the UAV and the ground in the system, the connection of hardware modules, etc. are mainly tested, so as to verify whether the system can operate normally after assembly. In the system performance test, the monitoring accuracy rate, false alarm rate and algorithm running time are counted, and the alarm accuracy rate and alarm real-time performance of the system are tested, so as to judge whether the system meets the reliability and real-time performance of construction safety monitoring.

4.2 Experimental Results

Power on each hardware module of the system separately to make sure that the hardware modules are connected properly. After the control core loads the test program, the indicator light flashes according to the program requirements, indicating that the system is running normally.

In the system performance test, the drone is used to shoot the simulated construction scene in the simulated scene. After adding the interference image to the image collected by the drone, the system recognizes and counts the alarm correct rate, false alarm rate and response time of the system. Table 2 below shows the system performance test results.

Table 2. System performance test results

Interference image ratio/%	Correct rate/%	False alarm rate/%	Response time/s
5	98.5	0.02	0.8
10	98.2	0.02	1.1
15	97.6	0.04	1.2
20	97.1	0.08	1.2
25	95.8	0.06	1.4
30	93.7	0.07	1.3
50	93.3	0.09	1.4

According to the above system performance test results, the construction safety monitoring system based on UAV imagery designed in this paper not only improves the operating performance of the system significantly, but also has an accuracy rate higher than 90% for monitoring dangerous events during the construction process, and false alarms The rate is low, and the construction danger alarm interval is short. It can give early warning of dangerous events in time, and has the convenience of use. It has a good auxiliary role in the construction safety monitoring work.

5 Concluding Remarks

The comprehensive quality of construction practitioners is not high, the management level of managers is low, the safety identification ability is low, and the safety operation ability is low. Many hidden dangers cannot be sent information and dealt with in a timely manner, and they often focus on the progress of the project in order to pursue interests. To maximize, in order to be able to deliver work on time, we will not hesitate to take risks and fail to meet the requirements of safe operation. In addition, there are many manual operations, high-altitude operations, and cross-operations in construction projects, resulting in many hidden dangers and accidents. Frequent occurrence. In recent years, the application scope of UAVs has been continuously expanded, and currently three major research fields of military, civilian and scientific research have been formed. The application of UAV technology in construction quality management is one of the effective auxiliary means to reduce the unsafe behavior of workers and ensure the safe production of construction projects. Therefore, this paper designs a construction safety monitoring system based on UAV images, and tests the performance of the system to verify that the performance of the system can meet the needs of practical applications.

References

1. Xu, L., Fu, M., Wang, C.: Research on the application of BIM technology in safety management of prefabricated buildings. Constr. Econ. **42**(04), 53–56 (2021)
2. Hu, Q., Tian, X., He, Z.: Risk assessment for the green building construction safety based on the five-element connection number set analy-sis model. J. Saf. Environ. **21**(05), 1880–1888 (2021)
3. Yang, J., Chen, D.: Dual channel remote video synchronization monitoring method based on embedded web. Comput. Simul. **38**(02), 477–481 (2021)
4. Yang, K., Ahn, C.R.: Inferring workplace safety hazards from the spatial patterns of workers' wearable data. Adv. Eng. Inform. **41**(AUG.), 100924.1–100924.11 (2019)
5. Ye, X., Wang, T., Zhao, Z., et al.: Upgrading design and function innovation of safety risk monitoring system for rail transit construction. Urban Rapid Rail Transit. **34**(01), 110–114 (2021)
6. Rodrigues, S., Costa, D.B.: Integrating resilience engineering and UAS technology into construction safety planning and control. Eng. Constr. Archit. Manag. **26**(11), 2705–2722 (2019)
7. Liu, L.: Design of agricultural UAV remote monitoring system based on DER communication technology. J. Agric. Mech. Res. **43**(07), 230–234 (2021)
8. Sun, L., Wang, Q., Shi, K., et al.: Overview of computer vision research in construction safety field based on knowledge graph. Saf. Environ. Eng. **28**(02), 44–49 (2021)
9. Xu, Q., Chong, H.Y., Liao, P.C.: Collaborative information integration for construction safety monitoring. Autom. Constr. **102**(JUN.), 120–134 (2019)
10. Li, J., Zhang, J., Yu, Y.: Intelligent ground monitoring system of UAV inspection based on power grid industry. Inf. Technol. **44**(06), 134–138 (2020)

Research on Construction Risk Monitoring Method Based on Mobile Terminal

Jing Yang$^{(\boxtimes)}$, Kangyan Zeng, Caixia Zuo, and E. Yang

Department of Architectural Engineering, Chongqing College of Architecture and Technology, Chongqing 401331, China
jane451301513@163.com

Abstract. In building construction, due to the large number of staff and complex construction operation process, there will be a large number of unsafe factors, which will bring potential safety hazards to the construction project. Therefore, these factors affecting the construction safety must be monitored and managed to prevent the occurrence of safety accidents. In order to better ensure the safety of construction and avoid construction risks, a construction risk monitoring method based on mobile terminal is proposed. Combine mobile terminal technology to collect and identify construction characteristic data, build a construction risk management framework based on mobile terminal, build corresponding evaluation systems for different construction risks, optimize evaluation algorithms, and improve the construction risk monitoring system. Finally, it is confirmed by experiments. The construction risk monitoring method based on mobile terminal has high practicability in the process of practical application, and fully meets the actual requirements.

Keywords: Mobile terminal · Building construction · Risk monitoring · Evaluation system

1 Introduction

The establishment, analysis, research, design and plan of construction risk monitoring are based on the prediction of the future situation and on the normal and ideal technology, management and organization. These factors may affect the realization of the project objectives and the actual operation process of the project. The internal and external interference factors that cannot be determined in advance in these construction projects are called the risk of construction projects [1]. There are mainly the following five common safety accidents in construction, namely, falling from height, object strike, mechanical injury, electric shock and collapse. According to relevant survey data, the incidence of these five safety accidents accounts for more than 80% in construction. There are many reasons for the occurrence of these five safety accidents, mainly including the following points: first, the lack of construction ability of construction personnel leads to the occurrence of safety accidents, and the standardized operation is not carried out in accordance with the standard workflow. Secondly, inadequate safety protection measures are also

W. Fu and L. Yun (Eds.): ADHIP 2022, LNICST 469, pp. 557–570, 2023.
https://doi.org/10.1007/978-3-031-28867-8_41

one of the main factors causing safety accidents. The third is the management factor. The lack of awareness of safe production management and improper management of construction personnel are also one of the factors causing construction safety accidents. In the construction site, due to the lack of safe and reasonable supervision and management, it is very easy to cause construction safety accidents [2]. In the construction, the construction needs a variety of construction raw materials and a large number of construction equipment, which will affect the safety of construction operation. The work quality and construction ability level of construction personnel will have a direct impact on the construction quality.

In view of the above problems, relevant scholars have proposed some construction monitoring methods, such as building construction safety monitoring methods based on computer vision. This method first analyzes the application scope of computer vision technology in construction safety management, and summarizes the potential application of computer vision technology in the safety management of construction elements such as man-machine, material, method and environment by combing the provisions of the specifications, Two human related safety management, construction worker position safety management and occupational health and safety management, are selected as the main research contents. Aiming at the position safety management of construction workers, this paper analyzes the functional requirements of the position safety management system, summarizes the information needed to identify the position, trajectory and unsafe area of workers using computer vision technology, proposes the calculation method of the risk degree of workers' behavior in breaking into unsafe areas, and builds a recognition module based on the position and trajectory of workers. The unsafe area identification module and the danger degree calculation and warning module constitute an integrated computer vision construction worker position safety management system. In addition, some scholars have proposed a method of monitoring the construction process of prefabricated buildings based on feature extraction of point cloud models. They have completed the collection of point cloud data on the construction site by using a three-dimensional laser scanner, and obtained the point cloud data on the construction site of prefabricated buildings based on PCL platform, the point cloud data is preprocessed, and the prefabricated component segmentation, geometric feature extraction and measurement are completed Determine the site design elevation by 3D laser scanning technology; Simulate the assembly construction of prefabricated components and realize the monitoring of building construction. Although the above method can realize the monitoring of building construction, there is a problem of large risk error in the monitoring. Therefore, this paper proposes a method of building construction risk monitoring based on mobile terminals. The main research routes of this method are as follows:

(1) Collect and identify the characteristic data of building construction through mobile terminal technology.
(2) Build a construction risk management framework, build a corresponding evaluation system for different construction risks, and improve the construction risk monitoring system.
(3) The experimental results show that the proposed method is practical and fully meets the practical requirements.

2 Construction Risk Monitoring Method Based on Mobile Terminal

2.1 Construction Risk Management Framework Based on Mobile Terminal

2.1.1 Functional Framework of Safety Risk Management Module

Risk management is the top priority of the daily safety management of construction. The level of risk management directly determines the level of on-site safety production. However, mass casualties will always occur in the actual work, which may be due to the inaction of the safety management personnel on the construction site, or due to centralized resumption of work, rush time of some enterprises, rush for tasks, etc. In short, the main reason is that the construction safety risk management is not standardized, or the lack of a certain link of the management process [3]. Therefore, in order to ensure that the whole process of project construction safety and construction management is controllable and realize the systematization of risk management, the research of risk management module is a necessary choice for construction safety management. The security risk management module mainly includes four sub modules: risk source database, risk source upload, risk source audit and risk patrol. The functional framework of safety risk management module is shown in Fig. 1.

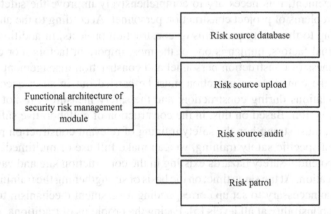

Fig. 1. Functional framework of safety risk management module

Construction risk collaborative management based on mobile terminal mainly refers to the cooperation of two or more different resources and management subjects to jointly complete the safety management of the construction site in the safety management of the construction site. Specifically, the safety collaborative management of construction engineering mainly includes the all-round coordination between different organizations, different departments, different application environments, people and people, people and machines, science and technology and tradition. On the basis of information technology, the safety collaborative management of construction site is to make full use of a variety of information technologies in the safety management of construction engineering, take the construction unit as the main body, and multiple organizations related to construction safety participate in the safety management and supervision of construction site, so as to ensure the smooth progress of construction engineering construction, and eliminate its unsafe factors to the greatest extent [4]. Under the condition of mobile terminal technology, the safety collaborative management level of construction site mainly includes three subsystems: safety training multimedia and assessment, construction site video monitoring, intelligent sensing and Internet of things technology monitoring, as shown in Fig. 2.

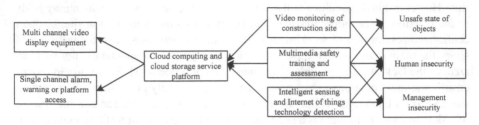

Fig. 2. Construction site safety system management framework

Under the condition of mobile terminal, in the construction of construction project safety management, it is necessary to comprehensively improve the safety awareness and safety problems of project construction personnel. According to the analysis of the causes leading to the safety problems of construction projects, in addition to the irresistible external factors, human is one of the most important factors. For example, the safety awareness of construction personnel and construction management personnel is insufficient, the construction operation must be carried out in strict accordance with relevant regulations during construction, and the safety measures are not set in place during construction. Based on this, in the construction of collaborative safety management system, must strengthen the safety training of relevant construction personnel. In the process of specific safety training, we can make full use of multimedia to visually display the potential safety hazards existing in the construction site and carry out early warning education. At the same time, on the basis of strengthening the training of relevant staff, it is also necessary to set up corresponding assessment mechanism to implement the safety responsibility at all levels [5]. Facing the problems of traditional construction methods, such as high energy consumption, serious waste of resources, large pollution emissions and insufficient labor force, the advantages of building friendly to the environment, high construction efficiency and large reduction of construction personnel are gradually emerging. At the same time, the performance of architecture in engineering quality is also better than that of traditional architecture.

2.1.2 Safety Monitoring and Early Warning Process

According to the theoretical research on the causes of risk accidents, the causes of safety accidents in engineering projects can be attributed to track crossing, that is, when there is overlap between dangerous sources such as human unsafe behavior, material unsafe state and unsafe environment, it is easy to cause safety accidents. Therefore, the fundamental goal of safety monitoring and early warning is to effectively find the hazard sources in the overlapping state or the upcoming track crossing events. All construction activities of the building take place in a fixed space. In order to facilitate safety monitoring, this space can be divided into several fixed size space unit cubes [6, 7]. In each space unit, various hazard information sources can be extracted and feature vectors can be constructed to reflect the safety risk of the space unit. There is a difference between the cell space feature vector with higher security risk and the cell space feature vector with lower security risk. The model constructed by this method has the following advantages: the model has learning ability, and the accuracy of the model continues to improve with the increase of training sample size; The model has good universality and can be applied to different engineering projects: it greatly reduces the system development cost and has better economy [8]. To sum up, the safety monitoring and early warning process based on space unit is shown in Fig. 3.

Fig. 3. Analysis, identification and early warning process of space unit construction

There are two key problems in constructing the safety monitoring and risk early warning model. The accurate feature vector is the basis for constructing the safety risk classification and prediction model of the space unit. Therefore, it is necessary to determine how to construct the feature vector of the space unit first, so as to fully describe the various safety States of the space unit. The feature vector classification model of spatial unit based on mobile terminal is established by using mobile terminal technology.

2.2 Construction Risk Evaluation Algorithm

In recent years, in order to facilitate the communication and coordination between the project participants and improve the management and control ability of the project management of the participating units, mobile terminal technology has been introduced into the building continuously. With the maturity of technology, mobile terminal technology has become an indispensable technology in the management and control process of construction projects [9, 10]. Figure 4 shows the mobile terminal construction risk monitoring and management system.

A large number of prefabricated components need to be used in building construction. Component stacking and storage may cause collapse and overturning accidents. See Table 1 for relevant hazard sources.

The discriminant matrix is the relative importance relationship between the child attributes under the same parent attribute. A series of importance relationships are compared in the same matrix to obtain the importance contribution of the child attribute

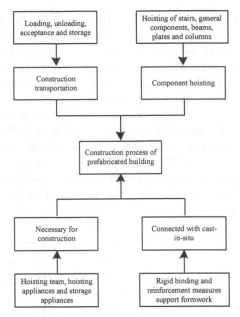

Fig. 4. Mobile terminal construction risk monitoring and management system

Table 1. Hazards in components and storage

Number	Hazard source	Occurrence link
A	The number of stacking layers exceeds the specified load pressure	Stack
B	Improper stacking form of components	Storage
C	Insufficient stacking stiffness and bearing capacity	Storage
D	No reinforcement measures are taken when components are stacked	Stack
E	Triangular channel steel support is not placed symmetrically	Stack
F	Component stacking without base plate	Stack
G	The stacking area of components is not within the effective range	Construction preparation
H	The ground of component stacking area is not hardened	Construction preparation

to the parent attribute. Using the discriminant matrix, the relative weight relationship between different levels and parent-child level attributes is gradually realized, and finally integrated into the absolute weight of each level element in the complete system. The specific operation of establishing the discrimination matrix is to establish the discrimination matrix table according to the hierarchical relationship of the index system, and then judge the relative importance of elements according to the important scale table. The discrimination matrix is shown in Table 2.

Table 2. Discrimination matrix

Target layer	S1	S2	S3
S1	1	Q12	Q13
S2	Q11	1	Q13
S3	Q21	Q22	1

The method of using discriminant matrix to solve the index weight method is mainly divided into two steps: the first step is to calculate the relative weight of the elements of a single decision matrix to the W-level elements, and the second step is to calculate the absolute weight of the elements of each layer combined with the hierarchical relationship and relative weight. Calculate relative weights. First normalize the discrimination matrix in columns:

$$w_{ij} = \frac{1}{\sum\limits_{i=1}^{n} w_i}, \quad (i = 1, 2, \cdots, n, \ j = 1, 2, \cdots, n) \tag{1}$$

Further, sum the discrimination matrix row by row:

$$w_k = \sum_{j=1}^{n} w_{ij} - n \tag{2}$$

Then, the column direction w_k is normalized to obtain the weight λ_{max} of the risk factor, which is the relative weight of each element. Calculate the discriminant matrix and bring λ_{max} into the formula for testing, and only if the ratio of CI to RI is less than 0.1 can be satisfied:

$$CI = \frac{(\lambda_{max} - w_k)}{n - 1}, \quad (n = 1, 2, \cdots) \tag{3}$$

By referring to the definition of physics, it is judged that the dispersion state of each risk index is recorded as $(r_{ij})_{m \times n}$, so as to evaluate the scheme. Due to its objectivity, accuracy and recognition are high. The calculation process is as follows: the construction matrix A contains m evaluation indicators and n objects, namely:

$$A = w_k (r_{ij})_{m \times n} - 1 \tag{4}$$

Since the dimension difference of each indicator $f_{t'}$ is relatively large, in order to eliminate its influence, it is necessary to normalize the data:

$$b_y = A \frac{r_{iy}}{r_{max} - r_{min}} \tag{5}$$

$$e_i = -\frac{1}{\ln n} \sum_{t=1}^{n} A(b_y - \ln f_{t'}) \tag{6}$$

Calculate the indicator weight according to the above steps, and the entropy value of the i indicator under (m, n):

$$w_{ej} = \frac{1 - e_1}{m - \sum\limits_{i=1}^{m} e_i} \tag{7}$$

During construction, due to the large quality of prefabricated components on the construction site, it is common for multiple cranes to operate at the same time, which may easily lead to hoisting accidents such as falling, impact, and overturning. See Table 3 for related hazards.

Table 3. Hazardous sources in component hoisting

Number	Hazard source	Occurrence link
A	Unreasonable setting of lifting points	Hoisting process
B	Improper selection of hoisting equipment	Construction preparation
C	Long term overload operation of equipment	Hoisting process
D	Operation error of hoisting operators	Hoisting process
E	Cross interference of tower crane parallel operation	Hoisting process
F	The construction personnel are in the blind area of the tower crane driver's vision	Setup script
G	Worker position conflicts with lifting route	Hoisting process
H	Construction workers did not wear safety belts	Setup script
I	Heavy snow, fog, heavy rain and other bad weather	Hoisting process

The source of danger in Table 3 occurs in the hoisting process. Because the reserved steel bars are pulled out during the hoisting process or the location of the hoisting point is unreasonably selected, it is easy to cause overturning accidents. If there is a danger source A, it is easy to cause overturning accidents. In addition, in the construction site of group tower operation, when multiple tower cranes operate in parallel, it is easy to generate C danger source, which leads to tower crane collision accidents. In addition, when installing prefabricated components, D hazard source may cause a high-altitude fall accident; E hazard source will cause construction difficulties such as reduced visibility at the construction site and prefabricated components swaying with the wind, which can easily lead to accidents.

2.3 Realization of Construction Risk Monitoring Based on Mobile Terminal Technology

Based on the real-time monitoring and early warning mechanism of construction safety, the real-time monitoring and early warning model of building construction safety is constructed by using real-time monitoring data collection and safety risk classification and identification model, as shown in Fig. 5.

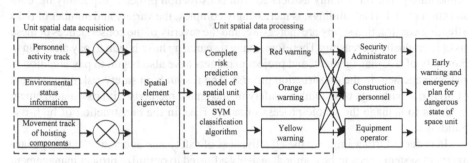

Fig. 5. Model for real-time monitoring and early warning of space unit safety

Construction risk factors involve people, materials, machines, etc., and have a high degree of dispersion. In order to ensure the accuracy of the analysis results of construction risk factors, this paper has studied a large number of relevant domestic and foreign literature, and fully combined with expert interview records (Appendix B), divide these 20 risk factors into five categories according to their essential attributes: personnel, materials, machinery, technology, and management. Finally, the risk evaluation index system of the building construction stage is obtained, as shown in Fig. 6.

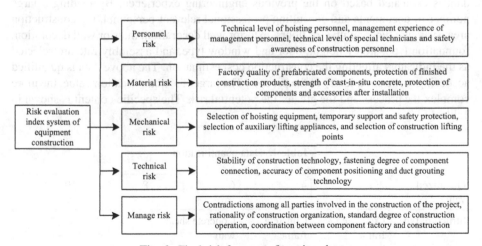

Fig. 6. Final risk factor confirmation chart

The real-time monitoring space unit feature vector is used as the input variable of the space unit security risk classification model, and the output result is the risk identification result. Movement trajectories of workers and hoisting equipment can be monitored through GPS data combined with RFID technology. The risk identification results include: red warning, orange warning, yellow warning and no security risk. Depending on the type of risk warning, different response measures should be taken at the construction site. For the red warning, emergency intervention measures should be taken immediately. The uncertainty associated with construction projects, especially the construction of high-rise buildings, is increasingly complex, the variety and number of risks affecting construction safety are increasing, and the severity of the consequences of these risks is gradually increasing. Therefore, many researchers have begun to pay attention to the study of risk management, and project managers have also begun to pay attention to risk management in the construction process. The research on safety risk management of high-rise buildings is to combine safety management methods with risk management methods to manage the whole process of safety risks in the construction of high-rise buildings.

In order to further improve the effectiveness of safety management, the safety management system needs to be comprehensively adjusted to optimize project management centered on safety technology management and the actual construction needs of high-rise buildings. The optimization of the safety management system needs to be combined with the overall coordination of skyscraper construction project management, supervision and management methods and application of the management system, in order to achieve an overall improvement in the effectiveness of safety management.

3   Analysis of Results

The sample is described by the characteristics of the construction project, and the risk data is estimated based on the previous engineering experience. By reading a large number of documents and consulting professional relevant personnel, the construction area, structure type, number of floors, exterior wall decoration, interior wall decoration, foundation type, storey height, door and window type, and assembly rate are selected as the indicators affecting the construction cost estimation. The above data is quantified according to the complexity of the process. The larger the quantification value, the more complex the process and the greater the potential risk. The specific content is shown in Table 4.

Table 4. Data quantification

Quantized value	2	4	6	8	10
Structure type	Concrete frame	Concrete frame cast-in-situ shear wall	Concrete shear wall	——	——

(*continued*)

Table 4. (*continued*)

Quantized value	2	4	6	8	10
Foundation type	Independent foundation	Strip foundation	Raft foundation	Pile raft foundation	Box foundation
Exterior wall decoration	Face brick	Imitation stone brick	Common coating	Lacquer	——
Interior wall decoration	Cement mortar	Mixed mortar	Latex paint	Coating	——
Assembly area	——	——	By actual data (percentage)	——	——
Number of layers	——	—	According to actual data	——	——

The sample data of 35 completed construction projects were selected, and the MAY-LAB2014 software was used for relevant programming. The first 30 sample data were selected as training samples, and the last 5 sample data were used as testing samples. According to the data quantification table and the sample data of the building, the data quantification processing is performed on the training samples and the testing samples. The processing results are shown in Table 5:

Table 5. Quantification of training data

Category	Input vector									Output vector
Serial number	C1	C2	C3	C4	C5	C6	C7	C8	C9	0 (yuan/m^3)
A	14.83	3	1	7	1	2.9	2	3	55	1765
B	16.8	2	1	8	2	2.9	3	3	55	1925.6
C	23.15	3	1	13	3	3	1	3	55	2192.6
D	16.95	2	1	7	3	3	2	1	55	1741.6
E	26.92	1	2	10	3	2.8	3	1	55	2182
F	28.36	1	2	21	3	2.7	3	1	55	2512
G	27.65	3	3	8	4	2.9	2	3	65	2295.6
H	31.3	2	2	13	3	2.9	3	2	75	2689.5

Identify the main influencing factors that affect project risk. Through the calculation, it is concluded that the ability of relevant personnel, the risk of capital supply, the unreasonable division of rights, responsibilities and interests, and the mistakes in production allocation decision-making are the main influencing factors. We should focus on it and take measures to reduce the corresponding risks. Table 6 shows the calculation results of the risk factor of unsafe factors.

Table 6. Calculation results of the risk factor of unsafe factors

Sort	Unsafe factors	Risk coefficient
A	Unreasonable division of rights, responsibilities and interests	1.322
B	Production allocation decision error	0.789
C	Risk of capital supply	0.628
D	Ability of relevant personnel	0.487

Taking concrete structure as an example, the application advantages of building risk monitoring are analyzed, as shown in Table 7.

Table 7. Application advantages of building risk monitoring

Primary coverage	Fabricated concrete structure	Cast in situ concrete structure
Construction progress	Production and construction are separated, and both are carried out at the same time. The construction efficiency is greatly improved, one floor in 3–4 days, and the labor demand is reduced by 50%	The construction efficiency is low and the speed is slow. It takes 6–7 days for one floor, and requires a lot of manual cooperation
Construction quality	The large-scale assembly line production in the factory greatly reduces the quality problems. At the same time, the construction accuracy control is high and the project quality is greatly improved	The error and precision control are low, the spatial size deformation is large, the inspection batches are uneven, and the quality is difficult to guarantee
Resource utilization	Based on meeting the requirements of skill emission reduction and environment-friendly advocated by the state	Large water consumption, large power consumption, serious material loss and large amount of construction waste abandoned in construction
Environmental protection	Due to the special construction process of prefabricated buildings, prefabricated components are not processed on site, avoiding all kinds of pollution caused by concrete pouring on the construction site	The construction site has high dust and noise, and a lot of waste water and garbage, which not only wastes resources, but also needs human treatment

To verify the construction risk monitoring error, the building construction safety monitoring method based on computer vision and the prefabricated building construction process monitoring method based on point cloud model feature extraction are compared with the methods in this paper. The results are shown in Fig. 7.

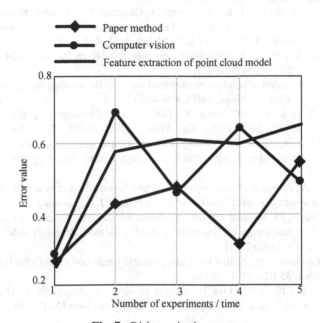

Fig. 7. Risk monitoring error

It can be seen from the results in Fig. 7 that the construction risk monitoring method for mobile terminals constructed in this paper has an error within 5%, so the estimation accuracy is relatively high. Once the sample data and the trained network are obtained, the estimation process is very simple, which saves the tedious work of applying the quota to calculate the engineering volume, which can greatly improve the monitoring and management speed.

4 Concluding Remarks

In construction site construction, safety production management is an important part of construction management. In specific construction management, safety management belongs to a systematic and comprehensive management process. Based on information technology, construction projects must fully combine the current status of safety production management, take effective measures, actively build an information collaborative management system, and comprehensively strengthen the safety management level of construction projects.

References

1. Baudrit, C., Taillandier, F., Tran, T., Breysse, D.: Uncertainty processing and risk monitoring in construction projects using hierarchical probabilistic relational models. Comput.-Aided Civil Infrastruct. Eng. **34**(2), 97–115 (2019)
2. Lee, P.C., Wei, J., Ting, H.I., Lo, T.P., Long, D., Chang, L.M.: Dynamic analysis of construction safety risk and visual tracking of key factors based on behavior-based safety and building information modeling. KSCE J. Civ. Eng. **23**(10), 4155–4167 (2019)
3. Li, M., Wang, W., Zhang, Z.: Study on construction risk factors based on ISM and MICMAC. J. Saf. Environ. **22**(1), 22–28 (2022)
4. Zhong, X.: Simulation of building construction safety early warning model based on support vector machine. Comput. Simul. **36**(8), 459–462 (2019)
5. Duan, Y., Zhou, S., Guo, Y., Wang, X.: Safety risk and strategy of prefabricated building construction based on SEM. J. Civil Eng. Manag. **37**(2), 70–75, 121 (2020)
6. Chen, R., Jiang, A., Dong, Y., Xiong, Q., Lu, Y.: Construction and application of risk assessment model for prefabricated construction quality. J. Rail. Sci. Eng. **18**(10), 2788–2796 (2021)
7. Hua, Y., Li, Z., Zhang, S.: The comparison of SSGF construction system with prefabricated construction system and traditional construction system. J. Eng. Manag. **34**(03), 23–27 (2020)
8. Alipour-Bashary, M., Ravanshadnia, M., Abbasianjahromi, H., Asnaashari, E.: A hybrid fuzzy risk assessment framework for determining building demolition safety index. KSCE J. Civ. Eng. **25**(4), 1144–1162 (2021)
9. Yan, Y.: BIM-based construction key chain progress prediction model after typhoon disaster. J. Catastrophol. **35**(02), 39–43 (2020)
10. Wei, R., Li, C., Jiang, W., Luo, H.: Design of safety management system for access of construction work area. J. Civil Eng. Manag. **37**(02), 136–141+150 (2020)

Construction Risk Monitoring Method of Subway Foundation Pit Engineering Based on Simulated Annealing Neural Network

Tengfei Ma[1], Mingxian Zhang[1], Zhuoli Ji[1(✉)], Shuibing Zhang[2], and Yi Zhang[3]

[1] College of Civil Engineering and Planning, Liupanshui Normal University, Liupanshui 55300, China
jizhuoli920122@163.com

[2] Shaoxing Construction Project Quality and Safety Supervision Station, Shaoxing 312000, China

[3] Xi'an Siyuan University, Xi'an 710038, China

Abstract. Traditional subway foundation pit engineering construction risk monitoring methods have slow convergence speed when solving large-scale practical problems, which affects the accuracy of monitoring. Therefore, a subway foundation pit engineering construction risk monitoring method based on simulated annealing neural network is designed. By identifying the accident risk sources of foundation pit engineering, understanding the accident causes and prevention mechanism among human, machine and environment, the classification of risk sources of foundation pit engineering is obtained, and the safety risk monitoring index system is constructed. The monitoring indicators are analyzed in detail, and the annealing neural network is optimized, and the process of double-layer simulated annealing algorithm is designed to realize risk monitoring. In the case simulation experiment, the designed monitoring method and the traditional method are used to monitor the project. The monitoring experimental results show that the proposed method can accurately predict the deformation of the subway tunnel through the monitoring data of the deep foundation pit construction adjacent to the existing subway tunnel.

Keywords: Simulated annealing neural network · Subway foundation pit · Construction risk monitoring

1 Introduction

With the rapid development of cities and the growth of urban population, the development and utilization of underground space is a more effective way to open up the living space of human beings. Urban underground railways, underground factories, underground garages, underground commercial streets, underground substations, underground shopping malls, underground warehouses, underground civil air defense projects, municipal underground engineering, and military underground engineering have increased [1].

W. Fu and L. Yun (Eds.): ADHIP 2022, LNICST 469, pp. 571–587, 2023.
https://doi.org/10.1007/978-3-031-28867-8_42

Whether it is the construction of the subway project of municipal engineering or the construction of the deep foundation of high-rise buildings, all of them involve the excavation of the foundation pit project, of which the foundation pit project of the subway station is one of the main manifestations. In view of the frequent occurrence of subway construction accidents, it is one of the effective measures to ensure the smooth progress of subway construction to use information-based monitoring equipment to grasp construction data and take corresponding measures through analysis. Therefore, in the process of subway construction, it is very necessary to analyze the uncertain factors that cause construction accidents by using risk management tools and implement key risk monitoring, and take preventive measures in advance to avoid losses. The research on construction risk monitoring includes the risk assessment model of deep foundation pit construction of metro station based on dea-ahp and BP neural network [2]. WBS-RBS is used to identify the risk factors in the construction process, and a two-level risk evaluation index system is established by AHP to determine the evaluation matrix. On the basis of the above, data envelopment is used to calculate the index weight, and the index weight value is used as the input of BP neural network to predict the construction risk of deep foundation pit of subway station and finally determine the risk level. However, in the process of construction risk monitoring, the traditional method often has the problem of slow convergence when solving large-scale practical problems, which affects the accuracy of monitoring. Therefore, a method of construction risk monitoring of Metro Foundation Pit Based on simulated annealing neural network is designed. By identifying the accident risk sources of foundation pit engineering and understanding the accident causes and prevention mechanisms among people, machines and environment, the classification of the risk sources of foundation pit engineering is obtained. Based on this, the safety risk monitoring index system is constructed, and each monitoring index is analyzed in detail. The annealing neural network is optimized, and the double-layer simulated annealing algorithm flow is designed to realize the project construction risk monitoring. The experimental results show that the absolute error and relative error of this method are kept below 0.01% and 0.14% respectively. It shows that this method can accurately predict the deformation of the subway tunnel through the construction monitoring data of the deep foundation pit adjacent to the existing subway tunnel.

2   Research on Risk Monitoring Method of Metro Foundation Pit Construction Based on Simulated Annealing Neural Network

2.1   Identify Risk Sources of Foundation Pit Engineering Accidents

In the research of risk source identification of foundation pit engineering accident, the occurrence of construction engineering quality accident has both technical and management reasons. There are four conditions for building accident: object, objective aspect, subject and subjective aspect. The man-made sources of risks in building deep foundation pit projects can be divided into the limitations of survey and design, the temporary nature of the foundation pit support project, the dynamic nature of the implementation process and the complexity of the organization, environmental sensitivity and the social hazard and the project There are five categories of manager's blindness [3]. The environmental sources of foundation pit engineering risks are mainly divided into four categories: geological factors, hydrological factors, surrounding environmental factors and construction factors. Classification by man, machine, and environment is a rational model for examining the cause of accidents and accident prevention mechanisms. The relationship between the cause of accidents and prevention mechanisms between man, machine, and the environment with management as the boundary is as follows:

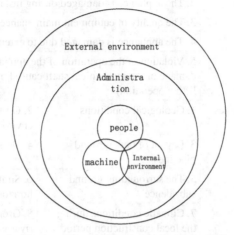

Fig. 1. Man machine environment management diagram

This section first finds out the accident classification standards and related literature summary according to relevant regulations. Based on the risk occurrence mechanism of subway station foundation pit engineering, as shown in the figure above, summarizes the previous team's research on risk sources and summarizes the foundation pit engineering from "The 37 risk sources" under the man-machine-environment-management model are as follows (Fig. 1 and Table 1):

Table 1. Risk sources of foundation pit engineering under the model of "man, machine, environment and management"

Risk source classification	A detailed description	
Actor risk source	1. Professional skills	2. Operation violation
	3. Work experience	4. Improper emergency handling
	5. Physical condition	6. On-site monitoring errors
	7. Safety awareness	8. Insufficient personnel
	9. Construction project supervision and construction contractors have insufficient experience or human error	

(*continued*)

Table 1. (*continued*)

Risk source classification	A detailed description	
	10. Malicious damage to engineering buildings or facilities by contractors or third parties	
Risk sources of construction tools	1. The quality of the equipment is unqualified	
	2. The appliance is damaged during installation	
	3. The quality of equipment maintenance is not up to standard	
	4. The appliance is damaged due to external force	
	5. Violation of the operation of the appliance (such as the damage of the mechanical shaft caused by the error of the lifting operation)	
Environmental risk sources	1. Geological conditions	2. On-site production environment
	3. Layout of underground pipelines	4. On-site space layout
	5. The surrounding ground subsidence	6. Stratum subsidence and horizontal displacement
	7. Climatic conditions during the local construction period	8. Groundwater and adjacent river water level
	9. Ground transportation	10. Rebound of foundation pit
	11. On-site security measures setting	12. Surrounding ground buildings
Organize and manage risk sources	1. Purchase of safety equipment	2. Drain of key talents
	3. The group has no cohesion	4. Unreasonable organizational structure
	5. Reward for safe work	6. Distorted information communication
	7. Safety awareness of managers	8. Managerial skills
	9. Management makes employees feel unfair and dissatisfied	
	10. -8 Management safety training and education (corporate management, project management personnel)	

Reasonable and comprehensive use of risk source identification methods provides method support for finding the risk sources of foundation pit engineering. Through the introduction of the scope of application of risk source identification methods and the advantages and disadvantages, the expert investigation method, safety checklist and risk breakdown structure (RBS) The method has been further applied in identifying the risk sources of foundation pit engineering accidents and determining the types of foundation pit engineering construction accidents [4].

2.2 Establish a Safety Risk Monitoring Index System

According to a large number of literatures, this paper summarizes the key problems in the construction process of deep foundation pit project, which are foundation pit support safety, foundation pit excavation safety and groundwater control [5, 6]. The construction of deep foundation pit adjacent to the existing subway is often more complicated than that of the general deep foundation pit construction. The above three problems affect the deformation of adjacent existing subway tunnel. (1) The impact of foundation pit support on adjacent existing subways. The supporting project of deep foundation pit plays a pivotal role in the construction of deep foundation pit. The supporting construction of deep foundation pit can play a role in retaining soil and water, ensuring the balance of stress on the overall structure of the entire foundation pit. Stable and safe. The type, thickness, insertion depth and type of supporting system, spacing, pre-loading size and back pressure soil reservation of the foundation pit retaining structure will all affect the ability of the foundation pit retaining system to resist deformation. In addition, the interaction between retaining wall and soil will also affect the displacement of retaining wall and soil outside the pit. (2) The influence of excavation of foundation pit on adjacent existing subway. In the construction process of deep foundation pit adjacent to the existing subway, the selection of excavation scheme will also affect the deformation of subway. Different geographical location, adjacent to the existing subway deep foundation pit in the excavation process will have different impact on the deformation of the subway. Earthwork excavation of foundation pit engineering is a unloading process. In this process, the change of ground load will affect the deformation of subway. (3) The impact of groundwater on neighboring existing subways. During the construction of a deep foundation pit, the change of the groundwater level will affect the deformation of the foundation pit, as well as the structural deformation of the adjacent deep foundation pit subway. The construction process of the foundation pit needs to control the height of the groundwater level, which requires precipitation of the foundation pit or recharge of groundwater. During the precipitation process of the foundation pit, the pore water in the stratum will dissipate, causing the subway to undergo vertical displacement and settlement. Similarly, the recharge of groundwater will cause subway.

The rise of the structure. This kind of influence of groundwater on the subway will seriously threaten the normal and safe operation of the subway. Therefore, the change of groundwater is one of the non-negligible factors that cause the deformation of subway tunnels. (4) The index of the influence of the construction of deep foundation pit adjacent to the existing subway on the subway deformation is established. According to the influence of the deep foundation pit construction on the deformation of the adjacent existing subway and the influence of each construction process on the adjacent existing subway deformation in the construction process [7, 8], combined with the research method adopted in this paper, the data obtained from the monitoring of the foundation pit in the construction process and the data obtained from the subway monitoring are analyzed, so as to analyze the future of the foundation pit Therefore, the safety risk control measures can be formulated in advance to control the deformation of foundation pit and subway.

According to the analysis results and monitoring conditions, a safety risk technical monitoring index system is constructed for the safety risks of deep foundation pit construction, as shown in Table 2.

Table 2. Safety risk monitoring index system of metro tunnel deep foundation pit

Primary indicators	Secondary indicators	Level three indicators
Safety risk monitoring index system for deep foundation pit of metro tunnel	Foundation pit safety	Maintain the horizontal displacement of the pile (wall) body
		Maintenance pile (wall) top horizontal displacement
		Vertical displacement of maintenance pile (wall) top
		Vertical displacement of ground wall
		Steel support axial force
		Anchor cable tension
		Uplift at the bottom of the foundation pit
		Excavation depth of foundation pit
		Distance between foundation pit excavation and subway tunnel
		Groundwater level
	Subway tunnel safety	Horizontal displacement of tunnel structure
		Vertical displacement of tunnel structure
		Tunnel section convergence

The support system of the deep foundation pit can maintain the force balance of the overall structure of the entire foundation pit, can effectively ensure the stability and safety of the foundation pit during the construction process, and prevent the occurrence of safety accidents. (1) Pile (wall) horizontal displacement: the deformation of the diaphragm wall can most intuitively reflect the safety status of the foundation pit, which is a problem that needs to be paid attention to during the construction of the foundation pit. The diaphragm wall is under pressure from groundwater and soil [9, 10]. As the construction progresses, the underground conditions of the foundation pit are constantly changing, and the underground diaphragm wall and various piles are also in a constantly changing dynamic environment. Therefore, we should always pay attention to the horizontal displacement of the pile (wall) and monitor it from time to time to prevent the excessive horizontal displacement of the pile (wall) from being damaged, which may cause risks or safety accidents. The horizontal displacement of the pile (wall) is generally monitored and measured with a total station and an inclinometer. (2) Vertical displacement of pile (wall): in addition to the influence of groundwater and soil on the displacement of ground connected pile (wall), the friction resistance at the wall side and the bearing capacity at the bottom of wall also have certain influence on the vertical displacement of pile (wall). As a part of the foundation, diaphragm wall should bear not only horizontal lateral load but also vertical bearing capacity. The vertical displacement of pile (wall) in retaining structure will also affect the safety of deep foundation pit project. The vertical displacement of pile (wall) is generally monitored by geometric level or hydrostatic level. (3) Axial force of steel support: the change of axial force of internal support always reflects the state of supporting structure of deep foundation pit. The setting of axial force of steel support should ensure the stability of supporting structure, so as to ensure the stability of the whole deep foundation pit. It is necessary to pay attention to the change of axial force of steel support and find out the abnormal situation in time, so as to take measures to prevent the occurrence of risk accidents of deep foundation pit. The axial force of steel support is usually monitored by axial force meter or reading instrument. (4) Uplifting of the bottom of the foundation pit: During the excavation of a deep foundation pit project, the unloading of the soil in the pit will change the pressure at the bottom of the pit, causing the soil to rebound and deform, which will lead to the uplift of the bottom of the foundation pit. If the bulge at the bottom of the foundation pit is too large, it may cause risks during the construction of the deep foundation pit project. Under normal circumstances, as the excavation of the foundation pit continues to advance, the rebound of the bottom of the foundation pit will continue to increase, resulting in an increase in the uplift value of the bottom of the foundation pit. Therefore, the monitoring of the uplift at the bottom of the foundation pit is also an important risk indicator. Generally, a level gauge is used to monitor the uplift at the bottom of the foundation pit, and it should be performed immediately after each excavation is completed. (5) Anchor cable tension: the setting of anchor cable plays a role in strengthening the surrounding strata of deep foundation pit. The setting of anchor cable can control the deformation of diaphragm wall, reduce the displacement of supporting structure and ensure the stability of foundation pit. (6) Excavation depth of foundation pit: during the excavation of foundation pit, the balance of stratum will be destroyed, resulting in upward uplift of soil at the bottom of pit, lateral deformation of retaining structure of foundation pit and changes of surrounding stratum of foundation pit, which will lead to ground settlement and deformation of adjacent subway tunnel. Therefore, the influence of excavation depth of deep foundation pit construction on subway deformation is very important. (7) The distance between the excavation of the foundation pit and the subway tunnel: the impact of the construction of a deep foundation pit adjacent to the existing subway tunnel on the deformation of the subway tunnel is related to its relative position. The different positional relationship between the two makes the degree of impact on the existing subway tunnel different, which is related to the distance between the excavation position of the foundation pit and the subway tunnel. Generally speaking, the closer to the subway tunnel, the greater the impact on the deformation of the subway tunnel. Therefore, different distances have different effects on subway tunnels, and different distances require different protection measures. (8) Groundwater level: in the construction process of deep foundation pit, the treatment of groundwater is a very important link. The groundwater level will affect the stress change of foundation pit and the vertical displacement of adjacent existing subway tunnel. However, in the construction process of deep foundation pit, the groundwater level should be reduced reasonably to make the construction smooth. The influence of the change of groundwater level on the deformation of subway tunnel needs to be monitored according to the change of water level and the change of subway tunnel. At the same time, reasonable subway protection measures should be formulated to control the safety of subway tunnel and the safety construction of foundation pit. (9) Vertical displacement of the tunnel structure (including differential settlement): Due to the complexity of hydrogeology in the stratum, unloading of foundation pit excavation, groundwater control, and other construction processes will affect the vertical displacement of the subway tunnel structure. The sinking or floating of subway tunnels will have a certain impact on the safe operation of the subway. To monitor the vertical displacement of the existing subway tunnel structure, a level, a total station or an automated monitoring method is generally used. (10) Horizontal displacement of tunnel structure (including differential horizontal displacement): in the process of deep foundation pit construction and reinforcement of subway surrounding, the horizontal displacement of subway tunnel will change due to the change of soil lateral pressure. The change of horizontal displacement of existing subway tunnel structure is also an important factor affecting the safe operation of subway. In the process of deep foundation pit construction, it is necessary to strengthen the frequency of monitoring and find out the law between it and the construction process of deep foundation pit, so as to predict the subsequent safety situation of subway, and formulate relevant safety control measures in advance. (11) Tunnel clearance deformation: Tunnel clearance deformation, also known as clearance convergence, refers to the phenomenon that the rock and soil near the subway tunnel rush into the subway tunnel space after the excavation of the subway tunnel. Generally, it refers to the relative position between two points on the side of the tunnel. The changes that have occurred. Excessive tunnel clearance will affect the operation safety of subway trains, which is an important indicator in subway safety monitoring.

2.3 Optimize Annealing Neural Network

Although the traditional simulated annealing algorithm can not solve the problem of slow simulated annealing based on the idea of the traditional neural network simulated annealing, it can not solve the problem of optimal solution based on the idea of neural network simulated annealing Quasi annealing optimization is called network algorithm. The first layer of simple annealing is a similar simulated annealing idea. It first sets a larger learning rate $\eta(0.5 - 0.9)$ and momentum term coefficient $a(0.5 - 0.9)$, and gradually reduces the learning rate and momentum term coefficients, so that the error accuracy quickly drops to a smaller value. When the error accuracy drops to a preset target accuracy value, you can jump out of the training neural network. The second layer of deep annealing, after the simple annealing of the first layer reaches a small error, on this basis, the simulated annealing algorithm is used to find the global optimal solution. The use of double-layer simulated annealing to optimize the BP neural network reduces the training time of the BP neural network, improves the training speed of the neural network, and solves the problem that the training of the BP neural network cannot obtain the global optimal solution. On the basis of the traditional BP algorithm, adding learning rate η and momentum coefficient a, using the idea of simulated annealing, a similar simulated annealing algorithm is proposed, called the first-level simple annealing. The algorithm starts to set larger $\eta(0.5 - 0.9)$ and a, and then after training each batch of samples, η and a are gradually reduced in the following way:

$$\begin{cases} \eta = \eta - \dfrac{0.01}{10^q} \\ a = a - \dfrac{0.01}{10^q} \end{cases} \tag{1}$$

The initial value of q is set to 0, and after each batch of sample training, q becomes $q + 1$ until the calculated error accuracy reaches the preset target error accuracy or the maximum number of iterations reaches the preset value.

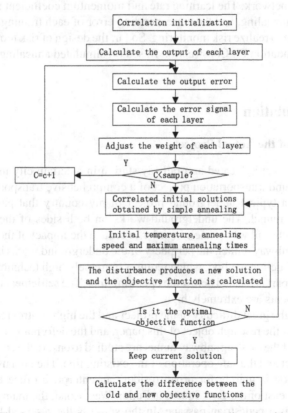

Fig. 2. Flow chart of two-layer simulated annealing algorithm

The main purpose of the first layer of simple annealing is to shorten the training time of BP neural network, improve the training speed, and make the error accuracy rapidly decline. If the error accuracy of the preset target can be achieved, the calculation will jump out. Otherwise, it will go into the second layer deep annealing. According to the first layer of simple annealing and the second layer of annealing, combined with the standard BP algorithm, a two-layer simulated annealing optimization algorithm is obtained (Fig. 2).

Deep annealing is modified on the basis of the traditional simulated annealing algorithm. Its main purpose is to search for the global optimal value. Based on the traditional simulated annealing algorithm, the main improvements of deep annealing are as follows:

(1) The error accuracy of each sample in the last error precision value obtained by the first layer simple annealing is taken as the initial value.

Finally, the total error accuracy value is used as the initial objective function value.

Disturbance neural network weights V (input layer to hidden layer weight matrix) and w (hidden layer to output layer weight matrix).

The objective function is the sum of the error precision values of each training sample.

(4) When recalculating the error accuracy value of each sample, the weight adjustment formula of traditional BP algorithm is still used to modify the weight matrix V and W of neural network. The learning rate and momentum coefficient used are the first layer of simple annealing, and the corresponding error of each training sample is taken as a new partition to realize risk monitoring. So far, the design of risk monitoring method for Subway Foundation Pit Construction Based on simulated annealing neural network is completed.

3   Case Simulation

3.1   Overview of the Actual Project

This example project is located at a railway station in a certain city in Central Plains. It is an underground transportation project of a comprehensive transportation hub (East Square), and is a typical underground project in my country that passes through an existing subway tunnel. The underground works on both sides of the subway tunnel are very close to the subway tunnel. In order to reduce the impact of the construction in progress on the subway tunnel, the periphery of the underground works is constructed by the reverse construction method, which requires extremely high technical requirements, and the crossing part is located below the operating line. The settlement and deformation control requirements are extremely high.

Considering the great difficulty of the project and the high control requirements, this project is taken as the research object of this paper, and the deformation control standard and technology of the existing subway tunnel are studied to ensure the construction safety of the new project and the safe operation of the existing line. The land use planning near the railway station is mainly for urban residential, enterprise office and commercial activities. To the east of the East Square is Putian West Road, the main road of the city. Under the road is a pedestrian passage. In the south is the reserved land for the bus terminal station. On the west side is the railway station. Underground in the south side are the transfer stations of Metro Line 1, line 5 and city line 3. The north side is the highway passenger station. There is a national road passing through the underground next to the project, and the two station section of Metro Line 1 in the city passes through the underground in the middle of the project.

The north-south structure of the square is 175 m east-west and 121 m north-south. Shunzuo District of the central island is 109 m east-west and 56 m north-south. The remaining part is the reverse cropping area; the structure of the south area is 176 m east-west and 117 m north-south. Shunzuo District of the central island is 118 m east-west and 66 m north-south. The remaining area is a reversed work area. The height of the commercial area on the first underground level is 6.1 m, and the ceiling height is 5 m after the decoration is completed; the second and third floor parking garages are 4.2 m in height, and the pipeline height is 2.4 mm after the garage is decorated; The three connecting passages connecting the north and south areas are 48 m long, with cross-sectional dimensions of 14.2 m × 3.557 m, 18.5 m × 3.557, and 14.2 m × 3.557 m. The passage connecting the north side with the bus station is 45.5 m long and 16.3 m wide. The positional relationship diagram between the civil air defense project and the section tunnel section is as follows (Fig. 3):

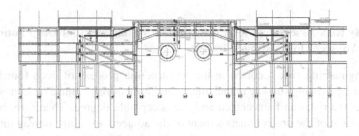

Fig. 3. The relationship between the air defense project and the section of the tunnel section

During the site survey of the East Square project, it was discovered that the interval between the station-Boxue Road station of the first phase of the Metro Line 1 project had been completed, and there was a large amount of soil above the tunnel. At the same time, the soil was within the scope of the East Square project. According to the measurement results, the length of the mound is 220 m along the longitudinal tunnel of the section, and the height of the mound is 14 m at the highest point. The engineering features of this project are as follows:

(1) The foundation pit has deep excavation, large excavation area and many retaining structures.

The foundation pit is rectangular, 269 m long and 176 m wide. The underground water level of the plot is high, and the requirements for foundation pit excavation, precipitation and protection are relatively strict; moreover, there is a bus station around the foundation pit which is about to be constructed, and there may be cross construction. At the same time, due to the limitation of construction site conditions, a large number of enclosure structures need to be constructed. Because of the complexity of the foundation pit project, SMW support method, diaphragm wall support method and deep mixing pile support method are used in the retaining structure of the foundation pit.

(2) It is difficult to protect the safety of operating subways.

The shield section of Metro Line 1 passes under the deep foundation pit project to connect the three connecting passages connecting the north and south foundation pits. During construction, it is necessary to ensure that Metro Line 1 can operate safely and normally. Reinforce the body, which makes the situation faced during the construction process more complicated and the difficulty of the construction greatly increases.

(3) Complex surrounding environment

There are railway stations, underpasses of National Highway 107, and long-distance passenger stations in the immediate vicinity of the project. The impact of construction on the surrounding environment must be considered, and protective measures must be taken; The passages cross; and the underground passages in the core area of the transportation hub to be built on the south side of the square are also likely to be constructed in parallel with the project. The consideration of various factors adds to the difficulty of the project. Therefore, the construction environment is complicated.

3.2   Safety Risk Assessment for Construction Management of Deep Foundation Pits Adjacent to Existing Subways

For the engineering risk occurred in the construction process of deep foundation pit adjacent to the existing subway tunnel, the formulation of risk control measures depends on whether the risk can be accepted and the acceptable degree. Therefore, before the risk assessment of the project management of the adjacent existing subway tunnel deep foundation pit project, it is necessary to formulate a clear risk occurrence probability level and risk acceptance criteria. In this paper, according to the situation of the project, the occurrence probability of risk factors is divided into five levels, as shown in the following Table 3:

Table 3. Probability level of risk factors

Grade	Description	Probability	Valuation
A	Rarely	<0.0003	0.1
B	Rarely	[0.0003, 0.003]	0.3
C	Occasionally	[0.003, 0.03]	0.5
D	Probably	[0.03, 0.3]	0.7
E	Frequently	≥0.3	0.9

The interval of each risk value corresponds to a specific level of safety risk level, as shown in the following Table 4:

Table 4. Safety risk classification

Grade	Value at risk	Acceptance	Security risk description
First level	0–0.25	Can be overlooked	The risk situation is extremely low, the safety status of the project is very good, and no treatment is required
Second level	0.25–0.5	Acceptable	The risk is low, the safety status is good, it needs attention, the possibility of serious injury is small, but there is the possibility of general injury accidents, and routine management review
Three levels	0.5–0.75	Acceptable after treatment	The risk is medium, the safety status is average, the general injury accident is more likely to occur, and rectification is required
Level 4	0.75–1.0	Refuse to accept	The risk probability of the project is high, and there is a large potential risk It is difficult to deal with the consequences of the risk, so we must adjust and pay attention to it continuously

The safety risk level is divided according to the probability and severity of safety risk events. Therefore, the safety risk interval is a comprehensive manifestation of the consequences of safety risk accidents. After determining the safety risk level of the project, the safety risk manager can determine the risk acceptance standard according to the maximum risk that the enterprise and the project can bear, and at the same time determine the safety risk control plan and measures according to the risk acceptance standard, so as to improve the project Carry out effective security risk control. The layout of the foundation pit water level monitoring is as follows (Fig. 4):

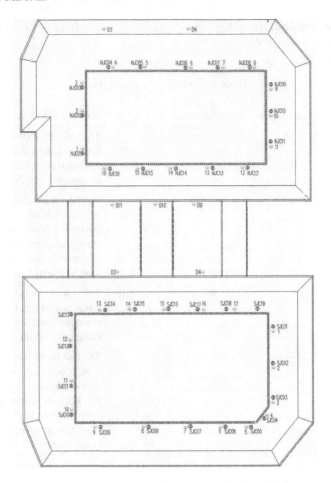

Fig. 4. Layout of foundation pit water level monitoring

In this data collection process, due to the large number of foundation pit monitoring points and subway monitoring layout points, the monitoring data at NJC 16 and NJC 21 were selected for analysis of the monitoring data at NJC 16 and NJC 21 during the initial model establishment. The monitoring data source of the body selects No. 13 monitoring point, the monitoring data of anchor cable tension selects the anchor cable No. 6 monitoring point; the subway monitoring selects the monitoring data of the monitoring point within the scope of channel 1#. According to the actual monitoring situation of the project, the interval of each set of data is 4 days. The monitoring data of the deep foundation pit during the construction process is used to fit and train the various indicators of the subway, and then a dynamic method is used to predict. That is, with the continuous advancement of construction, the data information obtained by monitoring gradually increases. These new data are continuously added to the training samples during the network model prediction process, and the neural network is dynamically trained, adjusted and updated, and the monitoring error is counted..

3.3 Experimental Results and Analysis

On the basis of the above experimental preparation, the detection model based on simulated annealing neural network is established. The NNTool toolbox in MATLAB software is used to replace the data into the compiled program and run it. The results obtained for the first time are as follows (Fig. 5):

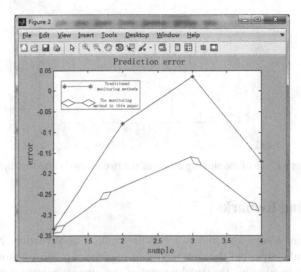

Fig. 5. Comparison of monitoring errors of the first training of the two methods

As shown in the figure above, although the simulated annealing neural network has a high degree of fitting ability, there are still certain errors in the prediction results of the network verification, and some sample points have large errors, so the dynamic method is used to predict, and the new collection is organized. The data is added to the training sample to dynamically train, adjust and update the network model.

Taking 16 groups of sample data as training data and 6 groups of sample data as test data, the results obtained are as follows (Fig. 6):

As can be seen from the above figure, with the increase of the number of training samples, the prediction accuracy of the two detection methods is improved. In the training and testing of traditional monitoring methods, the maximum absolute error is 0.45, the maximum relative error is 7.35%, but the rest of the absolute error is kept below 0.05, the relative error is below 0.7%. In the training and inspection of the monitoring method designed in this paper, the maximum absolute error is 0.14, the maximum relative error is 2.08%, the remaining absolute error is kept below 0.01, and the relative error is kept below 0.14%. Therefore, it can be concluded that the subway foundation pit construction risk monitoring method based on the simulated annealing neural network designed in this paper can accurately pass the adjacent existing The deformation of the subway tunnel can be predicted by the monitoring data of the deep foundation pit construction of the subway tunnel, so that the safety risk control measures of the subway tunnel can be formulated in advance to ensure the safety of the subway in the construction process.

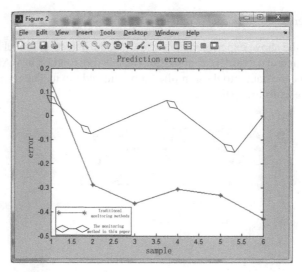

Fig. 6. Comparison of monitoring errors of the two methods after multiple training

4　Concluding Remarks

At present, China's subway projects are at the peak of construction. Not only is the construction scale large, but the complexity of engineering construction is far greater than in the past. In addition, the development of the subway construction market is not perfect, the behavior of market subjects is not standardized, and engineering quality and construction casualty accidents occur from time to time, which greatly increases the risk factors of the project. Because there are many factors affecting the project risk, and some factors affect each other, it is difficult to evaluate the project risk, This paper carries out construction risk monitoring based on simulated annealing neural network. By identifying accident risk sources, understanding accident causes and prevention mechanisms, classifying risk sources, building a safety risk monitoring index system, and analyzing the indicators, optimizing annealing neural network to achieve project construction risk monitoring. However, in the process of project construction risk monitoring, due to the complexity of the algorithm, the monitoring time did not achieve the expected effect and the monitoring efficiency was reduced. In the next study, the algorithm will be improved to shorten the calculation time and improve the monitoring efficiency of subway foundation pit construction risk.

Fund Project. The Guizhou Provincial Department of Education's Youth Science and Technology Talent Growth Funding Project "Study on the Road Performance and Permeable Function Evaluation of Coal Gangue Improvement of Permeable Asphalt Pavement in Liupanshui Mining Area" (Project No.: Qianjiaohe KY Zi [2020] 119).

References

1. Zhong, X.: Simulation of building construction safety early warning model based on support vector machine. Comput. Simul. **36**(08), 459–462 (2019)
2. Cao, Y.: Optimization of adaptive signal control using simulated annealing algorithm. J. Transport. Eng. Inf. **59**(01), 53–59+64 (2018)
3. Jiang, M.: A predication model of bp neural network based on genetic simulated annealing algorithms. Softw. Eng. **021**(007), 36–38 (2018)
4. Hu, Z., Li, W., Qiao, J.: Frequency conversion sinusoidal chaotic neural network based on self-adaptive simulated annealing. Acta Electron. Sin. **47**(03), 613–622 (2019)
5. Gong, S., Liu, J., Mei, L., et al.: Loss optimization model for multi-hop HF radio transmission on sea surface based on BP simulated annealing algorithm. New Industr. Strareg. **4**, 17–21 (2019)
6. Yao, J., Hu, H., Bai, Y., et al.: DOA estimation method based on SAPSO-BP neural network. J. Chongqing Inst. Technol. **380**(05), 189–194 (2018)
7. Li, L., Zheng, H., Ding, C., et al.: Terrorist attack classification based on improved bp neural network. Softw. Guide **18**(05), 21–26 (2019)
8. Luo, Z., Zeng, L., Pan, H., et al.: Research on construction safety risk assessment of new subway station close-attached undercrossing the existing operating station. Math. Probl. Eng. **2019**(1), 1–20 (2019)
9. Srinivasan, G., Venkatesh, M.P.: Risk assessment of construction project in India. IOP Conf. Ser. Mater. Sci. Eng. **993**, 012044 (2020)
10. Park, K., Lee, H.W., Choi, K., et al.: Project risk factors facing construction management firms. Int. J. Civil Eng. Trans. A Civil Eng. **17**(3), 305–321 (2019)
11. Liu, W.: Risk source analysis and identification method of shallow buried bias tunnel construction. Sichuan Archit. **40**(3), 320–322 (2020)
12. Lv, H.: Research on construction safety risk management of new subway projects adjacent to existing subway lines. Doors Windows **011**, 152–153 (2021)

Design of Mobile Monitoring System for Tower Crane in Assembly Construction Based on Internet of Things Technology

Dongwei Zhang, Shun Li$^{(\boxtimes)}$, and Hongxu Zhao

Hunan High Speed Railway Vocational and Technical College, Hengyang 421002, China
lishun521@126.com

Abstract. High data loss rate exists in the mobile monitoring system of assembly tower crane in construction. Therefore, a mobile monitoring system of assembly tower crane based on Internet of things technology is designed. Hardware part: adopt 32-bit data bus, integrate common high-definition multimedia interface; Software part: make use of space geometry principle to construct anti-collision model of tower group, transmit terminal parameters of tower crane safety monitoring system, optimize remote communication protocol of assembly building construction by using internet of things technology, and set up function of mobile monitoring system of tower crane. The experimental results show that the average loss rate of the two systems is 27.871%, 37.807% and 37.452% respectively, which shows that the higher loss rate is improved after the combination of IOT technology.

Keywords: Internet of Things technology · Assembly building · Tower crane · Mobile monitoring · Anti-collision model · Construction quality

1 Introduction

Prefabricated building is a new type of building structure, which responds to the call of national energy saving and green construction, and makes the building industry develop from field work to factory and industrial production. With the rapid development of science, technology and economy, tower cranes are more and more widely used in modern buildings, and the safety monitoring system of tower cranes is also developing continuously [1–3]. The outline of the 13th Five-Year Plan for Housing and Urban-rural Development to improve the preparation and implementation level of urban and rural planning puts forward clear requirements for promoting green buildings, promoting green and low-carbon construction, developing fabricated buildings and accelerating the transformation of the production mode of the construction industry, and at the same time gives high support to the applicability and advancement of advanced fabricated building technologies and product systems, and the construction of national fabricated building production bases. Now the country has more than 400 regular manufacturers of tower cranes, and the annual number of tower cranes by 10% or so growth rate. It can be said

© ICST Institute for Computer Sciences, Social Informatics and Telecommunications Engineering 2023
Published by Springer Nature Switzerland AG 2023. All Rights Reserved
W. Fu and L. Yun (Eds.): ADHIP 2022, LNICST 469, pp. 588–603, 2023.
https://doi.org/10.1007/978-3-031-28867-8_43

that the country is a tower crane production country. Not only to ensure the quality of the building, but also to improve labor efficiency, shorten the construction period, reduce pollution, do our best to meet customer needs, with good prospects for development. The assembled building conforms to the concept of green building in the whole life course and can realize the real recycling. However, with the continuous increase in the number of tower cranes, the quality of the various manufacturers, leading to tower cranes in the construction process there are a variety of risk factors. The annual average number of accidents caused by the global safety of the tower crane on personal safety and material and financial losses. At the same time, every province and city promotes the development of the green structure of the future construction industry from the height of sustainable development of social economy. Therefore, strengthening the real-time monitoring of tower crane construction, improve the safety factor of tower crane, is an urgent and important task at present. However, it is unavoidable to find the fault and find out the cause of the fault through real-time monitoring, so as to reduce the incidence of tower crane accidents and improve the safety factor and work efficiency.

Therefore, the mobile monitoring system for prefabricated tower crane based on Internet of things technology is designed. Tiny210 is used as the main circuit board, and ARM920T core and 32amba bus interface are used to construct the hardware part of the mobile monitoring system of the prefabricated tower crane for building construction, which improves the overall operating efficiency of the system. By studying the anti-collision model, we can judge the distance between tower cranes and improve the monitoring effect of the system. Realize the connection of things based on Internet of things technology, and enhance the connection ability of monitoring system. Through the main functional modules of the tower crane mobile monitoring system, the mobile monitoring server of the tower crane is debugged, and the mobile monitoring system design of the prefabricated construction tower crane based on the Internet of things technology is completed.

2   Hardware Design of Mobile Monitoring System for Assembly Tower Crane in Construction

TINY210 is used as the main circuit board in the mobile monitoring system of assembly tower crane in construction. TINY210 is an ARM Cortex-A8 framework from the core circuit board, its performance is very high. It is an 8-layer circuit board, its integration is very high, reasonable use of gold deposit technology. S3C2440 is a Samsung chip solution designed to provide high performance and low cost for consumer electronics and handheld terminal devices. It uses ARM920T as its core and has a five-stage pipeline. TINY210 core board application of the instruction set for the ARM v7, its main processor is Samsung S5PV210 chip, which can run up to 1 GHza S5PV210 chip application of the PowerVR SGX540 engine, whether 2D or 3D images can be smoothly accelerated, and can also play 1080P video, format can be MPEG4, H.263, H. 264 and so on. The 32AMBA bus interface has the advantages of low power consumption and simple operation, so S3C2440A can be used to develop handheld devices and practical solutions for developers. It provides a relatively perfect solution in cost, power consumption and performance. At the bottom of the TINY210 core board is a double row of pins, spaced two millimeters apart, that pull out most of the functional pins normally used by the main processor. The core board also integrates a common high-definition multimedia interface to make it easy for users to extend the functionality available. TINY210 standard with 512 Mb DDRZ memory, using 32-bit data bus, single-channel mode, running frequency of up to 200 MHz. S3C2440A is designed for image processing and other multimedia applications by Samsung. The main frequency is up to 400 MHz. It supports the network very well, and provides rich on-chip devices, such as 8-channel 10-bit ADC, touch-screen interface, I2C bus interface, USB host, USB device, SD host card, MMC card interface, 2-channel SPI and internal PLL clock frequency doubling. As the first step of automatic monitoring and control, sensor technology will be applied to this design. Sensor is a device to realize energy conversion. In other words, sensor is a device that can sense external information, such as light, force, sound, heat, and so on, and convert these external information into voltage or current signal which is convenient to operate according to reasonable methods. S3C2440A provides a complete set of general-purpose system peripherals, and makes the whole system consumption is minimal, because it has many commonly used function modules, so it also saves the trouble of adding additional equipment. It uses professionally stable CPU core power chips and reset chips to ensure system runtime stability, and supports booting from Nand Flash and Nor Flash. According to the working principle of the sensor to classify, common capacitance sensor, resistance sensor, inductance sensor, Hall sensor, photoelectric sensor, thermocouple sensor, etc. According to the physical quantity of the sensor to classify, common have weighing sensor, displacement sensor, temperature sensor, pressure sensor, gas sensor, etc. The system hardware architecture is shown in Fig. 1:

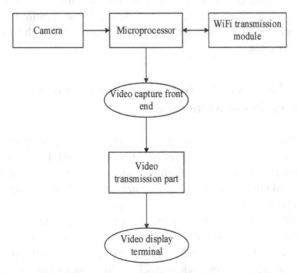

Fig. 1. Schematic diagram of the system hardware architecture

As can be seen from Fig. 1, the system hardware includes: camera, microprocessor, WIFI transmission module, video transmission part, video display terminal and video capture front end. Core board with 512 Mb of memory, and can be equipped with

1GB flash memory, can smoothly run a variety of operating systems, including Android, Linux, WINCE system. SDRAM bytes are 64M, NorFlash and NANDFlash are 2M bytes and 32M bytes, SDRAM chip model is HY57V561620FTP/MT48LC16M16A2, 32M bytes is the capacity of each chip, 16-bit is the width of data, and 32-bit is made up of two pieces. According to the nature of the sensor output signal to classify, the common on-off sensor, its output is 1 or 0. The output signal of analog signal sensor is analog signal. A digital sensor whose output signal is a pulse signal or binary code. The history of sensors can be traced back a long time, and it can be said that sensors have been around since the invention of devices with detection capabilities. However, in the past, the status of sensor technology in contemporary science and technology is not high, leading to the slow development of sensor technology. In NAND Flash, pages and chunks are used as the basic read-write unit, which can greatly improve the efficiency and reduce the cost of production. In NAND Flash, pages and chunks are used to store the necessary programs, such as bootstrap Uboot, Linux kernel and file system image. At present, the development of sensor technology has been closely linked with the development of science and technology. Sensor technology, communication technology and PC technology have become the pillars of information industry. The relationship between the input and output signals of the sensor is called the basic characteristic of the sensor, i. e. the input-output characteristic, which can be divided into static and dynamic characteristics.

3 Software Design of Mobile Monitoring System for Assembly Tower Crane in Construction

3.1 Construction of Tower Group Collision Prevention Model

Tower group anti-collision function Tower crane safety monitoring system is one of the core functions. Before programming the anti-collision software module, we need to study the anti-collision model. The aim of the study is to find an efficient and accurate way to calculate the minimum distance between the tower crane and the tower crane in real time, and make different responses when the minimum distance reaches the warning threshold. Before the tower crane safety monitoring system is put into use, its function must be tested. As the tested object, the safety monitoring system of tower crane has not only the common characteristics of the system, but also its own uniqueness. Therefore, it is necessary to analyze the overall structure and functions of the system under test before testing, and summarize the requirements of the test software according to the specific function points. Besides judging the distance between the tower crane and the tower crane, the model should be able to judge the type, direction and position of the collision, and help the operator to avoid the danger. The anti-collision model provides warning and response when the distance between the tower crane and the tower crane or between the tower crane and the obstacles is less than the safe distance. Regarding the tower crane as a space, suppose that under the premise of the known three-dimensional coordinates of the tower crane, the α is a point on the line and the α coordinate is $(K_\alpha, G_\alpha, L_\alpha)$, and the distance between the two points with the shortest distance is calculated as follows:

$$\begin{cases} K_\alpha = K_1 + \frac{\alpha - K_1}{(K_2 - K_1)} \\ G_\alpha = G_1 + \frac{\alpha - G_1}{(G_2 - G_1)} \\ L_\alpha = L_1 + \frac{\alpha - L_1}{(L_2 - L_1)} \end{cases} \quad (1)$$

At this point, the distance between two lines in space becomes the distance between a point and a line in space. In geometric mathematics, the process of finding the distance between a point and a line in space is as follows:

$$D = \sum \frac{\alpha + \phi}{\sqrt{(K_\alpha, G_\alpha, L_\alpha)}} \quad (2)$$

In formula (2), ϕ denotes a relational parameter. The working environment of tower crane is complex and its motion state is various, so it is necessary to analyze all the possible collision conditions and the possible collision types of the conditions, and sort out the relationship between them to form a systematic and specific procedure to judge the collision conditions. So as to achieve targeted and comprehensive and effective. Safety monitoring systems for tower cranes involve the Internet of Things technology, software engineering technology, wireless sensor network and remote data communication technology and other related fields [4, 5]. The monitoring terminal of tower crane (hereinafter referred to as RTU) is installed in the operating room of the tower crane. Its functions include: collecting sensor data and communicating between tower cranes through ZigBee wireless transmission network. The wireless transmission module is used to transmit the position, displacement, rotation angle, speed and other information of the tower crane, and the space geometry principle is used to calculate and detect the distance between the two tower cranes and between the tower and the protection zone in real time, and to provide early warning and alarm alarm when the alarm and warning threshold set by the scheme is reached [6]. In order to facilitate the subsequent maintenance and expansion, the monitoring terminal (RTU) adopts modular design. Roughly divided into 3 layers: interface display layer, software processing layer and hardware processing layer. The hardware layer collects parameters by sensor, receives and transmits data by communication module, and the software layer is responsible for data processing, including anti-collision calculation by parameters and running data. The interface display layer is responsible for data and image display, and has good man-machine interaction.

3.2   IOT Technology Optimized Assembly Building Construction Remote Communication Protocol

The Internet of Things (IOT) is an extended and extended network based on the Internet, whose core and foundation remains the Internet [7]. At the same time, its client extends and expands to any item and between items for information exchange and communication. Society has become smarter, more convenient and safer by connecting everything around it with the Internet of Things. The overall flow chart of the mobile monitoring system of the assembled tower crane based on the Internet of things technology is shown in Fig. 2:

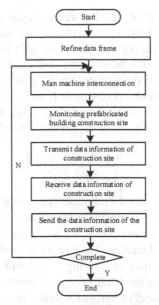

Fig. 2. Overall flow chart of mobile monitoring system of assembled tower crane based on Internet of things technology

Assembly building wireless communication network is characterized by dynamic topology, self-organizing network and multi-routing, which can overcome the obstacles of special geographical and climatic environment and is easy to expand. Applying wireless communication network technology to tower crane security monitoring system has more flexibility than traditional wired monitoring system. Because the system not only simulates the configuration center to complete parameter configuration, but also simulates the collision scene of tower crane terminals, the communication protocol is designed into two kinds: communication and configuration module communication protocol and communication protocol between towers. According to the design principle of communication protocol, the similarities of data frames are refined, and the format requirements of uplink and downlink frames are designed [8]. The protocol version is for the extensibility of the protocol, compatibility between the versions, to facilitate the subsequent upgrade of this article used in the protocol version of 0x04, length of 4 Bits. So in the system, the communication between tower crane and tower crane, the communication between tower crane and server, the communication between tower crane and configuration software are wireless communication, improve the flexibility and convenience. In the system, the communication between the tower crane and the tower crane is through the ZigBee wireless communication module for data transmission, the model used is the Shun Zhou SZ205 series. The functionality of the Fabricated Building Network indicates that the Internet of Things (IOT), at any time, in any environment, in any article, in any person, enterprise, commerce, by any means of communication, meets any service requirements provided, achieves the interconnection of goods to goods, people to goods, and people to people, elevates the Internet to a new dimension and is the trend of future network communication [9, 10]. Before communication, we need to understand the characteristics of ZigBee module, do relevant test, according to its characteristics and the requirements of the design algorithm, its main parameters are set as follows: network type is peer-to-peer, that is, any data sent out by a tower crane, other tower cranes can receive, node type is the terminal node. On the basis of formulas (1) and (2), the partial derivative of the model is obtained. If the extreme value of the equation is required to be obtained, the partial derivative of the unknown number shall be obtained and the partial derivative shall be 0. The expression formula is as follows:

$$\frac{\eta(\alpha - 1)^2}{\eta\phi} = 0 \tag{3}$$

In formula (3), the η represents the minimum value of ϕ. Frame type is to distinguish between different functions of the data frame, length of 5 Bits. The length of the information segment saves the specific number of bytes of communication information, which is convenient for positioning and reading. The equipment code is the unique identification of the object under test. The safety monitoring terminal of the tower crane of the object under test in this paper selects the IIVIEI number, which is unique and consists of 15 Arabic numerals with the number of bytes of 3 Bytes. In an assembly building, different ZigBee communication modules must have the same channel and ID, baud rate, and in the case of a star network, the address of the module must not be the same. On a large site, if there is more than one tower group, the interference between different towers needs to be taken into account, so that different IDs and channels can be used to avoid the interference between different towers. Therefore, a program that can change the channel and ID through software is added to the program. The information section is the specific information of the communication. Check bit is to ensure the accuracy of the data, the information segment data after CRC check, save the check result value, length is 2 Bits. The configuration information transmission frame is applied to the configuration function module, which is used to transmit the parameters that need to be set by the terminal of the crane safety monitoring system. Because of the upper limit of the length of the information segment of data frame, exceeding the upper limit may lead to the increase of packet loss rate and bit error rate. In addition to the online service platform can be real-time view of the running status of the tower crane, the software designed in this paper allows the use of PC monitoring software, site monitoring of the status of the tower crane. Through the monitoring software installed on the site computer and connected with ZigBee wireless communication module, users can monitor the lifting, luffing, wind speed, moment, alarm, violation and other information of the tower crane on the spot. The communication between the tower crane and the field monitoring software is based on the communication protocol with the online platform, sending out a frame of state data every second. The basic information includes the coordinate information of tower crane, the length of tower arm and so on, which can be encapsulated into one frame. But the reserve information is many, therefore encapsulates each reserve the data the frame. The test software simulates one or more monitoring terminals after configuring the tower group, obstacles and other information for the tower crane under test. The communication protocol is the same as that between towers. The system under test needs to simulate the tower crane safety monitoring terminal (RTU) to transmit and receive data frames.

After the installation of the safety monitoring system, it is necessary to configure the terminal with the initial data, such as the coordinate, height and moment curve of the tower crane, and correct the data of amplitude, height, hoisting weight and 0° angle, so that the system can work normally. In the system designed in this paper, these data are still configured in wireless way. Configuration data personnel do not need to climb on the tower crane, only on the ground to operate the computer configuration software, you can achieve the terminal configuration parameters. Among them, the tower group identification code marks different tower groups, each tower machine of the same tower group has the same field value, the role of the same frequency tower group identification code in the same city is different, thus avoiding the same frequency interference, the length of 6 bits. Tower crane identification code marks the same tower group different tower cranes, to facilitate the communication with each other to distinguish each other, the length of 6 bits. The information section is the interactive information section between the towers, and due to the operation of the anti-collision algorithm, the real-time operation data of each tower crane of the same tower group (such as rotation, amplitude variation, lifting height, etc.) are required. The configuration software can read the current parameters of the terminal and send the set parameters to the terminal wirelessly. After receiving the information, the terminal can set the parameters according to the content.

3.3 Setting Up the Function of Mobile Monitoring System of Tower Crane

The safety monitoring system of tower crane can detect the lifting height, luffing, lifting weight, inclination and wind speed of the tower crane, and present them to the driver through the screen with intuitive images and values, and give an alarm when the height, lifting weight, inclination, wind speed, torque and turning angle exceed the limits. The communication protocol between the Internet of Things and the mobile monitoring server of the tower crane shall comply with the communication protocol standards of the mobile monitoring server of the tower crane. The monitoring of obile monitoring server for tower crane can be divided into two categories: monitoring and control. Real-time monitoring tower crane mobile monitoring server running state, and to show the curve, namely oscilloscope function. Generally speaking, the working environment of tower crane is complex, not only there are adjacent tower cranes, but also there are sidewalks, buildings and other sensitive areas. The main functional modules of the mobile monitoring system for tower cranes are shown in Fig. 3:

From Fig. 3, it can be seen that the main function modules of mobile monitoring system for tower crane include wireless network module, communication protocol module, site configuration monitoring module and tower crane group communication module. The control of the monitoring server includes checking the parameters, status and alarm information, modifying the parameters and debugging on line. The interface must show the parameter information of the monitoring server for the convenience of the user. The software needs to be suitable for different types of tower crane mobile monitoring server, so the parameters on the interface need to match the type of tower crane mobile monitoring server. In order to achieve this goal, the sensing layer of the crane safety monitoring system collects and transmits the real-time operating data of the tower crane, such as amplitude change, rotation and height of the tower crane. These data can be presented to

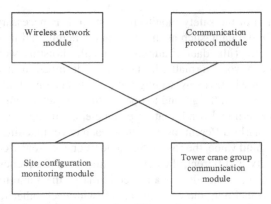

Fig. 3. Main function modules of mobile monitoring system for tower crane

the operators through the LED interface, and the alarm can be given before the collision occurs through the communication between adjacent tower cranes and the anti-collision algorithm. In terms of performance, the real-time monitoring of the tower crane mobile monitoring server is mainly to monitor the current, speed, voltage and other operating state, which changes quickly, so the sampling frequency should be up to 4 kHz. According to the functional requirements of the software, the Android terminal sends control commands to the monitoring server, which can check the parameters of the driver, display the status and alarm information, modify the parameters, and debug the driver. At the same time, in view of the demand of supervision and management of tower cranes on various construction sites, the safety monitoring system of tower cranes uploads the operation data and illegal operation data of the tower cranes at this moment to the platform of the management center so as to achieve effective supervision and management. Based on the practical application of tower crane safety monitoring system, the function table of the system is summarized. On the one hand, because of the difficulty of developing test tools and the consumption of time and manpower, if the object under test changes slightly, the test software will change greatly or even need to be redeveloped, which leads to the low utilization of test tools. Oscilloscope realization of the function: Android receive sampling data, according to the value of sampling data plotted curve, display monitoring objects and data. Different parameters have different range of values. While displaying the data, the graph needs to be translated and scaled so as to be easy to observe. In order to view the parameter configuration and monitoring data of mobile monitoring server of tower crane later, configuration parameters and monitoring data need to be stored. On the other hand, some automation requirements may be difficult to achieve or even impossible to achieve, the development process may cost more than the object under test. Therefore, it is necessary to analyze the specific situation, in the trade-off between the two, in the current conditions as much as possible scalable, reusable. Software architecture is defined as: according to the design principles, the use of different modes of thinking on the system structure to implement a reasonable arrangement. The software also has file-sharing capabilities for remote collaboration and troubleshooting. Of course, the terminal software should be suitable for different models of tower crane mobile monitoring server. In terms of performance, monitoring frequency is required to reach 4000 Hz. If the development of software analogy to build a house, then first of all to lay a good foundation, put up a good structure of concrete columns, if the house is not reasonable, then the follow-up construction work is not good, even if the completion may not be solid. By the same token, architecture plays a fundamental and decisive role in software development.

4 Application Testing

4.1 Test Readiness

Site Environment: This project is mainly to ensure the safety of construction, so the site testing was carried out in a construction site. Firstly, the tower crane is calibrated according to the calibration process. After calibration, the displayed value of height is compared with the actual value, and the difference is corrected according to the calculation formula. The construction site includes 15 residential buildings, in each residential corner of the outer wall and the site gate and logistics management office and other key parts of the installation of a total of 35 video surveillance equipment. Secondly, under the tower crane, the data of the tower crane are read and configured behind a building to verify the feasibility of wireless configuration and the reliability of long-distance communication. CPU: Intel (R) Core (TM) 2Auad CPU Q9500 @ 2.83 GHz, memory: 1.96 GB, operating system: Windows, database: MySQL, web server: Tomcat. When testing the logic of the system's internal code, XUnit is used to do white-box testing, write unit test code, and try to ensure that all test units are covered. Finally, let the tower crane hook down, every other period of distance, with the configuration software to read a set of sample value data, fill in the corresponding actual representative height, read to more than 3 groups of data, with the software to calculate the correction value. At the same time, the internal logic of the software system is verified through code walking and checking. Open ZLG-CANBUSTest debug software, ANBUS communication baud rate is set to 1 Mbps, using the Andro ID terminal to read parameters configuration interface motor parameters.

4.2 Test Results

The application effect of the monitoring system is tested, and the monitoring system based on BIM and the monitoring system based on deep learning are selected to carry out the comparative test. The packet dropout rates of the three systems were tested under different data frame intervals. The experimental results are shown in Tables 1, 2, 3, 4 and 5:

From the Table 1, the packet loss rate of the BIM based mobile monitoring system for prefabricated construction tower crane is 82.006%–85.477%; the packet loss rate of the deep learning based mobile monitoring system for prefabricated construction tower crane is 82.004%–87.412%; and the packet loss rate of the mobile monitoring system for prefabricated construction tower crane in this paper is 69.745%–73.660%. We can see that the average loss rate of the two systems is 71.972%, 83.996% and 84.695%, respectively. The low packet loss rate of the designed system proves that the monitoring effect of the designed system is better.

Table 1. Data frame time interval 10 ms packet loss rate (Mbps)

Number of experiments	Mobile monitoring system of assembled tower crane based on BIM	Mobile monitoring system of tower crane in assembly construction based on deep learning	Mobile monitoring system of assembled tower crane in construction
1	85.316	86.004	71.066
2	84.915	85.113	72.469
3	82.006	87.412	70.135
4	83.774	85.025	69.745
5	82.919	86.747	73.448
6	85.014	85.316	72.148
7	84.223	84.117	73.645
8	85.477	85.933	72.018
9	85.009	83.645	73.162
10	84.131	82.419	72.459
11	83.799	83.112	71.455
12	82.545	84.794	70.040
13	83.746	85.316	71.645
14	82.945	82.004	72.484
15	84.121	83.475	73.660

From the Table 2, the packet loss rate of BIM based mobile monitoring system for prefabricated tower crane in construction is between 40.236% and 51.615%; The packet loss rate of the mobile monitoring system of the prefabricated tower crane based on deep learning is between 44.055% and 46.919%; The packet loss rate of the mobile monitoring system of the prefabricated tower crane for building construction in this paper is between 35.122% and 37.004%. We can see that the average loss rate of the two systems is 36.042%, 44.439% and 45.705%, respectively. The low packet loss rate of the designed system proves that the monitoring effect of the designed system is better.

From the Table 3, the packet loss rate of BIM based mobile monitoring system for prefabricated tower crane in construction is between 26.314% and 38.512%; The packet loss rate of the mobile monitoring system of the prefabricated tower crane based on deep learning is between 25.890% and 33.784%; The packet loss rate of the mobile monitoring system of the prefabricated tower crane in construction is between 14.024% and 19.845%. We can see that the average loss rate of the two systems is 15.802%, 33.649% and 28.777%, respectively. The low packet loss rate of the designed system proves that the monitoring effect of the designed system is better.

As can be seen from the Table 4, the packet loss rate of the BIM based mobile monitoring system for prefabricated tower cranes in construction is between 17.949% and 21.647%; The packet loss rate of the mobile monitoring system of the prefabricated

Table 2. Data frame time interval 50 ms packet loss rate (Mbps)

Number of experiments	Mobile monitoring system of assembled tower crane based on BIM	Mobile monitoring system of tower crane in assembly construction based on deep learning	Mobile monitoring system of assembled tower crane in construction
1	40.236	48.445	36.121
2	51.615	47.022	36.748
3	43.33	46.919	35.122
4	47.116	45.388	36.077
5	45.849	45.117	35.948
6	44.120	46.023	36.745
7	43.649	46.879	35.915
8	42.116	45.011	37.004
9	45.467	46.822	36.129
10	42.008	45.188	35.202
11	43.615	44.919	36.749
12	44.898	43.722	35.124
13	43.164	44.055	36.112
14	44.202	45.199	35.134
15	45.199	44.869	36.499

tower crane based on deep learning is between 18.636% and 22.114%; The packet loss rate of the mobile monitoring system of the prefabricated tower crane in construction is between 9.064% and 13.155%. The average loss rate of the two systems is 11.509%, 19.932% and 20.174%, respectively. The low packet loss rate of the designed system proves that the monitoring effect of the designed system is better.

From the Table 5, the packet loss rate of the BIM based mobile monitoring system for prefabricated tower cranes in construction is between 5.222% and 8.125%; The packet loss rate of the mobile monitoring system of the prefabricated tower crane based on deep learning is between 7.064% and 8.692%; The packet loss rate of the mobile monitoring system of the prefabricated tower crane in construction is between 2.495% and 5.162%. We can see that the average loss rate of the two systems is 4.033%, 7.012% and 7.90%, respectively. The low packet loss rate of the designed system proves that the monitoring effect of the designed system is better.

According to Table 1, Table 2, Table 3, Table 4 and Table 5, the average packet loss rates of the three methods are compared, and the comparison results are shown in Fig. 4:

It can be seen from Fig. 4 that the average packet loss rate of the BIM based mobile monitoring system for prefabricated construction tower cranes is 7.012%, the average packet loss rate of the deep learning based mobile monitoring system for prefabricated construction tower cranes is 7.910%, and the average packet loss rate of the mobile

Table 3. Data frame time interval 100 ms packet loss rate (Mbps)

Number of experiments	Mobile monitoring system of assembled tower crane based on BIM	Mobile monitoring system of tower crane in assembly construction based on deep learning	Mobile monitoring system of assembled tower crane in construction
1	26.314	28.447	16.316
2	31.156	29.613	15.942
3	29.548	28.451	16.945
4	27.825	27.005	19.845
5	26.447	28.391	15.642
6	27.363	27.544	18.001
7	36.845	28.612	16.495
8	37.442	27.331	15.848
9	36.915	25.890	14.349
10	37.122	26.745	15.006
11	38.205	27.288	14.317
12	37.466	28.311	15.212
13	38.512	31.548	14.036
14	37.555	33.784	15.045
15	36.018	32.693	14.024

Table 4. Data frame time interval 500 ms packet loss rate (Mbps)

Number of experiments	Mobile monitoring system of assembled tower crane based on BIM	Mobile monitoring system of tower crane in assembly construction based on deep learning	Mobile monitoring system of assembled tower crane in construction
1	21.166	20.105	13.155
2	19.648	19.877	11.048
3	18.546	18.636	10.978
4	17.949	21.495	11.666
5	20.316	19.645	12.648
6	22.084	18.374	11.745
7	20.784	19.462	10.979
8	19.346	21.007	11.546

(*continued*)

Table 4. (*continued*)

Number of experiments	Mobile monitoring system of assembled tower crane based on BIM	Mobile monitoring system of tower crane in assembly construction based on deep learning	Mobile monitoring system of assembled tower crane in construction
9	18.455	18.655	12.485
10	19.752	19.064	11.616
11	20.613	21.003	12.748
12	21.647	22.114	11.036
13	19.860	21.549	9.064
14	20.499	22.077	10.515
15	18.312	19.552	11.404

Table 5. Data frame time interval 1000 ms packet loss rate (Mbps)

Number of experiments	Mobile monitoring system of assembled tower crane based on BIM	Mobile monitoring system of tower crane in assembly construction based on deep learning	Mobile monitoring system of assembled tower crane in construction
1	6.946	7.064	3.421
2	5.648	8.122	2.495
3	7.565	7.422	3.884
4	8.125	7.136	4.100
5	6.315	8.595	3.676
6	5.222	8.166	4.019
7	6.718	8.122	3.677
8	7.998	8.636	4.552
9	6.349	7.161	3.134
10	7.009	8.495	4.815
11	8.115	8.692	5.162
12	6.348	7.446	4.888
13	7.495	8.055	3.469
14	8.006	7.226	4.181
15	7.414	8.316	5.021

monitoring system for prefabricated construction tower cranes in this paper is 4.033%. The average packet loss rate of the designed system in this paper is lower than that of the comparison method, which proves that the effect of the mobile monitoring system

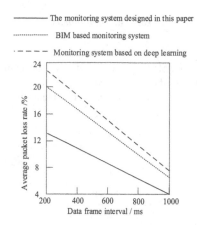

Fig. 4. Comparison results of average packet loss rate of different methods

of the prefabricated tower crane in building construction based on the Internet of things technology is better.

5 Conclusion

Based on this background, this paper designs and realizes the test software, which can replace the traditional manual test mode, save manpower and material resources, shorten the test cycle, and improve the production efficiency. At the same time, in order to solve the problems of the driver's vision blind zone and inconvenient communication with the signal flag bearer, a mobile monitoring system based on Internet of Things is designed. The system combines video compression technology, Wi-Fi wireless transmission technology, image tracking processing technology, Linux technology and S3C6410 microprocessing technology. The test results show that the mobile monitoring system designed in this paper has a good application effect.

Fund Project. In 2021, the "14th Five-Year Plan" of Hunan Province Educational Science "Planning for College Students' Employment and Entrepreneurship Research Special Project "Research on the "Golden Course" Construction of the Integration of Innovation and Entrepreneurship Courses and Ideological and Political Colleges in Higher Vocational Colleges-Taking the High-speed Railway Whole Industry Chain Plan New Entrepreneurship Education as the Example example. Project approval number: XJK21BXJ020, number: XJ211082.

References

1. Song, X.: Design of tower crane real-time interactive safety monitoring platform based on internet of things and BIM technology. Build. Constr. **42**(5), 833–835 (2020)
2. Tong, X., Zhang, F., Zhang, X.: Analysis on the significance of remote monitoring of construction site based on orbit determination transmission and cloud control technology. Intell. Build. City Inf. **11**, 79–81 (2020)

3. Liang, L., Zhang, Z., Lu, L., et al.: Subsection monitoring of tower crane working process based on internet of things. Build. Constr. **44**(1), 156–159 (2022)
4. Liu, S., Liu, D., Muhammad, K., Ding, W.: Effective template update mechanism in visual tracking with background clutter. Neurocomputing **458**, 615–625 (2021)
5. Wang, S., Liu, X., Liu, S., et al.: Human short-long term cognitive memory mechanism for visual monitoring in IoT-assisted smart cities. IEEE Internet Things J. **9**, 7128–7139 (2022). https://doi.org/10.1109/JIOT.2021.3077600
6. Wang, J., Hao, W., Tao, Z., et al.: Safety risk assessment of tower crane operation based on fuzzy Bayesian network. Saf. Environ. Eng. **28**(4), 15–20 (2021)
7. Ding, E., Yu, X., Liao, Y., et al.: Key technology of mine equipment state perception and online diagnosis under internet of things. J. China Coal Soc. **45**(6), 2308–2319 (2020)
8. Liu, S., Wang, S., Liu, X., et al.: Fuzzy detection aided real-time and robust visual tracking under complex environments. IEEE Trans. Fuzzy Syst. **29**(1), 90–102 (2021)
9. Zhang, D., Wang, J., Ji, H., et al.: Research and application of micropower safety monitoring IoT system for mine. J. Commun. **41**(2), 44–57 (2020)
10. Zhou, Q., Huang, S., Cheng, H.: Research on multiple access protocol of internet of things nodes based on probability detection. Comput. Simul. **37**(12), 148–152 (2020)

Personalized Recommendation Method of Innovation and Entrepreneurship Education Resources Based on Social Network Platform

Shun Li, Dongwei Zhang$^{(\boxtimes)}$, and Xiaolin Zhou

Hunan High Speed Railway Vocational and Technical College, Hengyang 421002, China
asa080801@163.com

Abstract. Innovation and entrepreneurship education is an education aimed at training creative thinking and entrepreneurial ability. Due to the increasing types of educational resources, the accuracy of recommendation of educational resources is low, which can not meet the personalized learning needs of learners. Therefore, a personalized recommendation method of innovation and entrepreneurship education resources based on social network platform is proposed. Analyze the relationship between learners and innovation and entrepreneurship education resources, and use information such as resource evaluation, download, and viewing time to continuously update the learner's interest model. Based on the social network platform, the characteristics of users' knowledge needs are extracted, and the knowledge achievements are displayed through the knowledge relationship network. For learners with similar resource preferences and interactive discussion forums, use social relationships to build educational resource learning groups. Find out user groups with similar interests to the target user or project sets with similar characteristics to the target project, and recommend the top 10 innovation and entrepreneurship education resources that are similar to the learner's interest vector model to the learner. The test results show that the personalized recommendation method of innovation and entrepreneurship education resources based on social network platform has high accuracy and can be applied to the personalized recommendation system of innovation and entrepreneurship education resources to meet the personalized learning needs of learners.

Keywords: Social network platform · Innovation and entrepreneurship · Educational resources · Personalized recommendation · Recommendation method · Social relationship

1 Introduction

As a specific form of educational practice in human society, entrepreneurship education aims to cultivate the innovative consciousness, innovative spirit and innovative ability of educational objects, and its emergence and development have reasonable purpose and regularity. Learning resources are one of the important factors that determine the effect of teaching. With the development of learning methods, the requirements for learning

W. Fu and L. Yun (Eds.): ADHIP 2022, LNICST 469, pp. 604–619, 2023.
https://doi.org/10.1007/978-3-031-28867-8_44

resources are also increasing. The quantity and quality of learning resources have been paid more and more attention by national education authorities and researchers. As an important part of higher education, entrepreneurship education involves the repositioning of higher education personnel training functions, the reform of education forms, the adjustment of education content, etc., reflecting the innovative development of education reform. Online educational resources are organized and provided centered on educational platforms. It is impossible for learners to traverse all platforms to retrieve educational resources. On the one hand, learners' personalized learning needs cannot be matched with the most suitable educational resources. On the other hand, high-quality education resources are also not fully utilized. While adhering to its basic theoretical basis, the disciplinary development of entrepreneurship education needs to inject new and beneficial "elements". We should learn from the theories of other disciplines to enrich and deepen our own theories and meet the needs of practice [1]. Equate entrepreneurship education with employment education. Among the majority of teachers and students, the marginalization of entrepreneurship education is difficult to change. While a large number of learning resources bring more optional space to learners, the problem of "information overload" also arises. Because in the process of online learning, teachers and learners are relatively separated in time and space, learners can not get teachers' guidance in time, which is easy to be confused in a large number of information resources. Judging from the various aspects of the current entrepreneurship education, there is a general lack of attention to the internal mechanism of entrepreneurship education. The personalized push of learning resources for innovation and entrepreneurship education will greatly improve the learning quality of online learning, and personalized education has become the development trend of future education. From the perspective of management research, innovation and entrepreneurship education needs to strengthen all aspects, such as increasing the proportion of practical courses, launching various innovation competitions, increasing basic courses in entrepreneurship education, and setting corresponding innovation credit awards [2]. It is very difficult for students to find the resources they are interested in in the massive information. Facing the fields they are not familiar with, it is difficult to identify what content is worth learning, which in turn affects the quality of learning. The basic courses of entrepreneurship education are inseparable from the management of teaching resources. Through personalized recommendation of resources, the coupling between students and professional courses, practical courses and other teaching contents can be improved to meet the practical needs of innovative and entrepreneurial talents training. In the era of mobile Internet, each user occupies a network node, and multiple nodes are connected to form a social network. The relationship and structure are the "trunk" of the whole social network, while knowledge is the "blood" flowing in the social network. The relationship chain formed between nodes, that is, the social network relationship, is the channel to ensure the rapid flow of knowledge. According to the user's historical data and behavior data, the personalized recommendation model of educational resources is established, and the resources that users are really interested in are selected from the huge information resource database and recommended to users. Whether in real life or in the network environment, the strength of a user's social relationship will directly affect the user's knowledge behavior. Users with strong connections in the network are usually groups with the same interests or similar viewpoints, and it is easier for them to establish trust and reciprocal relationships, and to transmit information more efficiently. Based on the social network platform, this paper proposes a personalized recommendation method for innovation and entrepreneurship education resources, which provides certain theoretical support and practical basis for the practice of entrepreneurship education in colleges and universities.

Therefore, a personalized recommendation method of innovation and entrepreneurship education resources based on social network platform is proposed. Through the behavior of learners, analyze the types of innovation and entrepreneurship education of learners. Based on the communication function of the social platform, extract the characteristics of users' knowledge needs. Using the interactivity and similarity between learners, construct learning groups of educational resources. Recommend relevant resources to learners through their evaluation of educational resources.

2 Personalized Recommendation Method of Innovation and Entrepreneurship Education Resources Based on Social Network Platform

2.1 Correlation Analysis Between Learners and Innovation and Entrepreneurship Education Resources

The learner's behavior is mainly divided into two types: explicit feedback behavior and implicit feedback behavior, to a certain extent, it determines the level of the score. Explicit feedback behaviors are mainly learners' evaluation and scoring of learning resources. Implicit feedback behaviors mainly include the number of times the learner clicks on each link, the browsing time of a certain web page, the learner's download of resources, and the learner's operations on a certain part of the document, such as copying and marking. Learners' repeated learning of a certain learning resource or some learning resources is of great help to understand learners' behavior and interests. The traditional collaborative filtering recommendation algorithm mainly depends on whether learners have learned some learning resources and their preferences for learners. It is lack of considering the learning times of different learners. As long as the learned resources are simple, it is considered that learners are interested in the resources. Innovation and entrepreneurship education resources are stored in the resource base of colleges and universities, and have the characteristics of digital processing, network transmission, large storage capacity, diversified retrieval, autonomous learning, and diversified presentation.

The resources of the resource library are rich and diverse. The resources involved in this article include documents, pictures, texts, videos and other types of resources generated in innovation and entrepreneurship projects. Usually, the number of learner's learning of the same resource or the same type of resource represents different degrees of preference, so the number of learning of the resource is a good measure of the degree of preference of the learner. Unlike search engines that require precise keywords, personalized recommendation methods aim to discover a user's underlying preferences and provide a personalized recommendation list based on a user's past activities and profiles. Based on the above analysis, this paper proposes a learner-innovation and entrepreneurship education resource association model, and defines the learner set as: $A = \{a_1, a_2, \cdots, a_x\}$, and the innovation and entrepreneurship education resource set as $B = \{b_1, b_2, \cdots, b_y\}$. Therefore, a binary relationship matrix consisting of the learner set and the innovation and entrepreneurship education resource set can be obtained, as shown below:

$$W_{xy} = \begin{bmatrix} w_{11} & w_{21} & \cdots & w_{x1} \\ w_{12} & w_{22} & \cdots & w_{x2} \\ \vdots & \vdots & \cdots & \vdots \\ w_{1y} & w_{2y} & \cdots & w_{xy} \end{bmatrix} \tag{1}$$

In formula (1), W_{xy} is the binary relationship matrix composed of the learner set and the innovation and entrepreneurship education resource set, the row vector represents the learner, and the column vector represents the learner learning the resource. w_{xy} represents the matrix element. These behaviors can reflect the interests of learners. In order to improve the accuracy of recommendation, this paper adopts a combination of explicit and implicit methods to construct a learner interest model. In the registration interface, let the learner choose the interest label that interests him, and then build the initial interest model according to the learner's choice. The frequency of learning a resource cannot be ignored. The learning frequency of learners to learning resources can reflect different degrees of preference, and the average frequency of resource learning can be defined as:

$$f'(a_x) = \frac{\sum\limits_{y} f(a_x, b_y)}{c(a_x)} \tag{2}$$

In formula (2), $f(a_x, b_y)$ represents the total number of times that learner a_x has learned innovation and entrepreneurship education resources; $f'(a_x)$ represents the average number of times that a_x uses innovation and entrepreneurship education resources; c represents the number of innovation and entrepreneurship education resources that a_x has learned. In the learning process of learners, the explicit and implicit feedback information of learners, such as: evaluation of resources, downloads, viewing time, etc., are collected, and the learner's interest model is continuously updated according to this information. The resources of the resource library are metadata generated according to unified standards and specifications, which facilitates the sharing, management and retrieval of resources. If a learner has studied a certain learning resource more often, it can be considered that the learner has a high degree of attention to the resource or this type of resource. The resources of the resource library can be constantly updated and changed dynamically. Learners can search for learning resources in the resource library, and at the same time upload and share their useful resources to the resource library, so as to ensure the continuous development of the resource library [3]. Taking the average use times $f'(a_x)$ of innovation and entrepreneurship education resources as the critical value is the classification analysis of the recommendation model. When the learning frequency exceeds $f'(a_x)$, learners have a high degree of attention to this type of innovation and entrepreneurship education resources, and tend to recommend this type of resources.

2.2 Extracting User Knowledge Demand Characteristics Based on Social Network Platform

Social networking platforms provide users with the functions of acquiring and exchanging knowledge resources. The social network platform has the integration of virtual and real, fairness and pertinence. It uses Internet technology to create a huge place for knowledge exchange, and provides professional knowledge resources to specific users, thereby avoiding communication barriers caused by differences in identity and status. A social network refers to a collection of social actors and their interactions. A social network is often composed of multiple social actors and their relatively stable relationships due to their interactions and connections. In this process, the social network platform obtains user information and user search information, which is the user's knowledge demand information, which can be used to meet and predict user needs. After the service is over, users evaluate the social networking platform according to their own satisfaction and user experience [4]. A social network platform includes both an individual network of actors and a group network of actors. Among them, the individual network focuses on the network position and role of the individual user, as well as the status of interaction with others; the subgroup network studies how groups connect users through "bridges", and understands the interaction within each subgroup and between subgroups. Therefore, in the process of providing services on social network platforms, users directly express their knowledge needs, and also express their satisfaction with knowledge services.

The services provided by social network platforms are generally knowledge browsing, knowledge query and knowledge push. The interaction of innovation and entrepreneurship education resources on social network platforms is essentially a relationship based on information sharing, mutual assistance and collaboration. Knowledge demand is a sense of dissatisfaction with knowledge generated by social network platform users in a specific situation, and this sense of dissatisfaction or lack can help users achieve a certain goal [5]. Weak ties in social networking platforms expose users to different communities and organizations, provide users with more opportunities to acquire heterogeneous resources, and lay the foundation for interaction and cooperation between

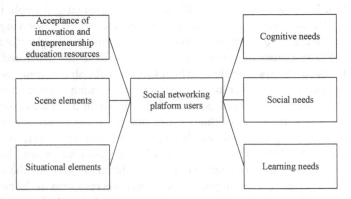

Fig. 1. Types of social network platform requirements

users. At different times and in different places, the knowledge needs of users are constantly changing. The types of requirements for social networking platforms are shown in Fig. 1.

Different from traditional knowledge community, user relationship is not only to maintain social relationship, but to further realize knowledge sharing, exchange and knowledge interaction through social relationship. In the social network platform, users can choose users who have the same interests, hobbies, and goals as their communication objects to communicate and interact with them. The strong relationship in the social network platform is a social relationship based on acquaintances composed of friends, relatives and colleagues. They have more homogeneity. Mutual trust is the premise of participating in social interaction. This trust relationship is embedded into the social interaction network structure, making the relationship between users more stable. The social demand comes from users seeking users with the same knowledge interests and knowledge preferences in order to meet the purpose of knowledge exchange, thus forming a knowledge relationship network, and through the knowledge relationship network to display knowledge achievements, publish knowledge and maintain the knowledge relationship network. User social demand motivation and results are shown in Fig. 2.

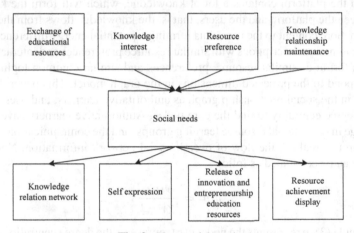

Fig. 2. User social needs

Information interaction is reflected in the interaction between learners and teaching elements, and is a crucial link in the interaction process. It mainly includes three forms of interaction between learners and learning resources, learners and learners, and learners and teachers. The openness of knowledge exchange also enables continuous updating of professional knowledge resources to ensure that the dynamic needs of user groups can be met at all times. Knowledge services provided by social networking platforms are driven by user needs. From the request of the service, the user browses through the knowledge provided by the social network platform or uses a search tool to retrieve it, and the user directly interacts with the social network platform.

2.3 Building Educational Resource Learning Groups Based on Social Relationships

In order to enhance the accuracy of innovation and entrepreneurship education resource recommendation, this paper considers the interaction and similarity between learners and constructs a multi-dimensional social relationship network graph model. For social network platforms, the knowledge needs of knowledge subjects and the realization of the value of knowledge subjects have become the driving force for the development of recommendation services on social network platforms. Social network platforms need to improve the recommendation function and provide users with innovative knowledge service models. Construct a weighted undirected graph, convert the learner model into points in the space, and at the same time divide and combine them according to different dimensions to form different subspaces, so as to describe the relationship between learners in a deeper and multi-dimensional way. At the same time, it can be projected to the specified dimension as required to obtain the learner relationship information in a specific low-dimensional space. Educational resource learning groups are composed of common users with the same views and interests on a topic, or a collaborative community based on a specific task. The users in the social network platform have knowledge needs, and the platform contains a lot of knowledge, which will form the knowledge flow between the platform and the users, that is, the knowledge flows from the platform group with more content to the user, thus forming a potential energy difference [6]. The feature combination descriptions with similar resource preferences and interaction in the discussion forum or similar resource preferences and similar common login locations can correspond to the projected dimensions in the graph model. This paper defines the key nodes in the social relationship graph as authoritative learners, and uses the calculation of degree centrality to find the core nodes. Authoritative learners have influence and prestige in educational resource learning groups, and the communication within and between groups realizes the flow of overall social network information. Node degree centrality can be calculated as follows.

$$u(p) = \frac{v(p)}{|\chi| - 1} \tag{3}$$

In formula (3), p represents the node; $u(p)$ represents the degree centrality value of p; $v(p)$ represents the number of nodes directly connected to the node p through the edge; χ represents the projection result of the point set under the specified dimension. The final degree centrality value is obtained by normalizing the number of nodes in the graph except for the point p itself. Authoritative learners are willing to share innovation and entrepreneurship education resources, and social network platforms are highly recognized. In the process, authoritative learners' sense of achievement is enhanced, and they are willing to contribute resources to social network platforms, thus forming a virtuous circle. Knowledge published by knowledge publishers on social networking platforms is recommended by the platform, which can enhance the knowledge publisher's sense of achievement and enhance others' trust in the publisher. After obtaining the learner feature representation in the two-dimensional space, perform the KMeans algorithm to cluster similar learners according to the spatial distance. In the iterative process, each sample point is firstly assigned to the nearest center point according to the distance, and the Gaussian kernel function is used as the distance to define the edge weight. The specific calculation method is shown in formula (4).

$$h(a_1, a_2) = e^{-\eta \|a_1 - a_2\|^2} \tag{4}$$

In formula (4), $h(a_1, a_2)$ represents any two points in the learning user sample set; e is a natural constant; η represents the parameters of the Gaussian function used in constructing the adjacency matrix. Calculate the average coordinate value in the space according to the newly divided category, update the center point corresponding to each category, and output the final clustering result after the classification result is stable. The platform understands the information matching degree, system function deficiencies and service defects through user feedback information, and more accurately understands user needs. Users can avoid possible wrong operations during interaction by browsing other people's feedback. For learners, attributes such as their professional background, learning resource preferences, forum responses, and common login locations can be used. Learners with more common or similar attributes can be considered to be more closely related. User-generated information is generated from the interaction process between users, and a large amount of content containing users' opinions and ideas generated by users during the interaction process not only enriches the information resources of the academic community, but also enhances the community's knowledge innovation ability. For learners with similar resource preferences and interaction in the discussion area, there is a high probability that they have similar learning and research directions and needs for the same type of resources, and such characteristics can be used to recommend learning partners to each other.

2.4 Design a Personalized Recommendation Model for Innovation and Entrepreneurship Education Resources

In the personalized recommendation of innovation and entrepreneurship education resources, this paper mainly updates the learner's interest vector model through the learners' evaluation of text resources. In this paper, there are five types of evaluation: not interested, not very interested, general, more interested, very interested. The update process of the learner interest vector model is as follows. The learner behavior, that is, the preference information representation part, collectively represents each learner's overall choice information about the educational resources he chooses. The collaborative attribute representation part mainly includes two parts: learner clustering and educational resource category information representation, that is, the number of learning groups divided by learners and the total number of educational resource categories. In the multi-combination structure of learning resource characteristics, it mainly includes innovation and entrepreneurship education resources, knowledge objectives or learner concepts corresponding to the resources, the difficulty level of the resources being set, and the learning time for which the resources are set [7]. Establish a text collection of innovation and entrepreneurship education resources that learners have browsed and evaluated, and extract the text vector model of each text from the resource library. The coding in this part adopts the one-hot coding method after converting the numerical features into continuous features, and the numbers are represented by the corresponding group labels and preset course categories. Traverse all the texts in the text set, and merge the feature word information of the text with the current learner interest vector model. The calculation formula of the feature word weight value is as follows:

$$\varpi = \frac{\ln(t_1 - t_2)}{b} + \vartheta \varepsilon \mu \tag{5}$$

In formula (5), ϖ represents the weight of the feature word; t_1 is the current date of the social network platform; t_2 is the date when the learner's interest vector model last modified the keyword; b represents the half-life, which is set to 7 in this article, which means that people's interest will decay to half of the original after seven days; ϑ represents the learning rate, which is 0.5 in this article; μ represents the keywords of interest points; ε represents the learner The scores corresponding to the evaluation of innovation and entrepreneurship education resources are $-2, -1, 0, 1,$ and 2, respectively. In the multi-group structure of learner's characteristics, it mainly includes the learner, the desired learning goal, the level of difficulty of learning expectations, the upper and lower limits of the expected learning time, and the type of media and content. Since learners are used to recommend resources to each other, it is necessary to consider the diversity of resources accepted by learners, the weight of each learner in the resource recommendation process, whether learners participate in the recommendation process and the proportion of participation. Corresponding constraints can be added to ensure that enough learners participate in resource sharing while maximizing the interests of resource acquirers. Sort the n-dimensional vectors of the learner's interest vector model from large to small according to the size of the feature word weight value, and take the top 20 items with the largest weight value and normalize them. Find user groups with similar interests to the target user or item sets with similar characteristics to the target item. The cosine similarity formula is used to calculate the similarity between the text vector model and the learner's interest vector model. The calculation formula is as follows:

$$\sigma(g_1, g_2) = \frac{\sum\limits_{x} g_1 g_2}{\sqrt{\sum\limits_{x} g_1^2} \sqrt{\sum\limits_{x} g_2^2}} \tag{6}$$

In formula (6), $\sigma(g_1, g_2)$ represents the similarity between the text vector model and the learner interest vector model; g_1 is the weight value of keywords in the text vector model of innovation and entrepreneurship education resources; g_2 is the weight value of keywords in the learner interest vector model. Since different users have different interest characteristics, their preferences for resources have strong subjective factors, resulting in different scoring scales. By improving the cosine similarity algorithm, the user's average score for all resources can be subtracted when calculating the user's score for an item, so that the similarity calculation can be more accurate. The text vector model of innovation and entrepreneurship education resources is compared with the learner's interest vector model, and the top 10 learning resources that are more similar to the learner's interest vector model are recommended to the learners. Based on the above process, the design of the personalized recommendation method for innovation and entrepreneurship education resources based on the social network platform is completed.

3 Experimental Studies

3.1 Experiment Preparation

In order to verify that the personalized learning resources of innovation and entrepreneurship education resources recommended by this method meet the needs of learners, an experimental test is carried out. Experimental data includes not only learning resource data, but also historical data of learners' learning. In existing public datasets, such as edX, World UC and other datasets, they provide dozens of attributes, including course data, learning learner information and learner behavior data. This paper extracts some online learning data information from a university student innovation and entrepreneurship resource database system. At present, the number of users who have successfully registered in the innovation and entrepreneurship resource database system is 268. Users with low downloads are removed through user filtering. There are 237 valid users, 692 project resources, and 1068 project ratings. Combined with the actual situation of learners and learning resources, some data are completed to form the experimental data set of this paper.

3.2 Results and Analysis

In the experimental data set, with the accuracy of recommendation as the evaluation standard, the higher the accuracy, the better the recommendation effect. This paper randomly selects 50–250 users in increments of 50 each time, and the length of the fixed recommendation list is 10. A total of 5 rounds of experiments were conducted, with 10 tests in each round, and the final results were averaged. In this way, with the increase of the number of users, the performance of the personalized recommendation method of innovation and entrepreneurship education resources based on the social network platform is verified. In this experiment, the accuracy is selected as the evaluation standard to measure the overall effect of the recommended method. The accuracy index often uses F1 score value to reflect the quality of the index. It is the harmonic average calculated by accuracy rate and recall rate. Both the former two are taken into account. At present, this index has been widely used in recommendation performance verification. Therefore, this study also uses the F1-score value to evaluate the recommendation accuracy. The verification results of the design method in this paper on the data set are compared with the personalized recommendation method of innovation and entrepreneurship education resources based on deep learning and multi-source data fusion, so as to make an accurate evaluation of the recommendation method. The personalized recommendation results of innovation and entrepreneurship education resources for 50–250 users are shown in Table 1, Table 2, Table 3, Table 4 and Table 5.

In the test of 50 users, the average accuracy of the personalized recommendation method of innovation and entrepreneurship education resources based on deep learning is 0.8005; The average accuracy of the personalized recommendation method of innovation and entrepreneurship education resources based on multi-source data fusion is 0.7853; While the average accuracy of the personalized recommendation method of innovation and entrepreneurship education resources based on the social network platform is 0.8328,

Table 1. Recommendation accuracy for 50 users

Testing frequency	Personalized recommendation method of innovation and entrepreneurship education resources based on social network platform	A personalized recommendation method for innovation and entrepreneurship education resources based on deep learning	A personalized recommendation method for innovation and entrepreneurship education resources based on multi-source data fusion
1	0.8245	0.8079	0.7840
2	0.8387	0.7983	0.7785
3	0.8264	0.8096	0.7867
4	0.8151	0.8065	0.7931
5	0.8322	0.7957	0.8058
6	0.8418	0.7846	0.7926
7	0.8356	0.8013	0.7815
8	0.8325	0.8028	0.7772
9	0.8493	0.8009	0.7823
10	0.8322	0.7982	0.7720

Table 2. Recommendation accuracy for 100 users

Testing frequency	Personalized recommendation method of innovation and entrepreneurship education resources based on social network platform	A personalized recommendation method for innovation and entrepreneurship education resources based on deep learning	A personalized recommendation method for innovation and entrepreneurship education resources based on multi-source data fusion
1	0.7856	0.7549	0.7333
2	0.7823	0.7448	0.7416
3	0.7946	0.7569	0.7265
4	0.7888	0.7481	0.7387
5	0.7967	0.7642	0.7274
6	0.7924	0.7576	0.7301
7	0.7775	0.7417	0.7251
8	0.7893	0.7525	0.7292
9	0.7952	0.7403	0.7268
10	0.7815	0.7545	0.7136

which is 0.0322 and 0.0474 higher than that of deep learning and the personalized recommendation method based on multi-source data fusion.

In the test of 100 users, the average accuracy of the personalized recommendation method of innovation and entrepreneurship education resources based on deep learning is 0.7515; The average accuracy of the personalized recommendation method of innovation and entrepreneurship education resources based on multi-source data fusion is 0.7292; While the average accuracy of the personalized recommendation method of innovation and entrepreneurship education resources based on the social network platform is 0.7884, which is 0.0368 and 0.0592 higher than that of deep learning and personalized recommendation method based on multi-source data fusion.

Table 3. Recommendation accuracy for 150 users

Testing frequency	Personalized recommendation method of innovation and entrepreneurship education resources based on social network platform	A personalized recommendation method for innovation and entrepreneurship education resources based on deep learning	A personalized recommendation method for innovation and entrepreneurship education resources based on multi-source data fusion
1	0.7540	0.7134	0.7089
2	0.7587	0.7268	0.7167
3	0.7468	0.7186	0.7036
4	0.7536	0.7243	0.7023
5	0.7555	0.7115	0.6905
6	0.7511	0.7301	0.6818
7	0.7410	0.7227	0.7144
8	0.7472	0.7155	0.7071
9	0.7525	0.7292	0.6952
10	0.7469	0.7166	0.6858

In the test of 150 users, the average accuracy of the personalized recommendation method of innovation and entrepreneurship education resources based on deep learning is 0.7208; The average accuracy of the personalized recommendation method of innovation and entrepreneurship education resources based on multi-source data fusion is 0.7006; While the average accuracy of the personalized recommendation method of innovation and entrepreneurship education resources based on the social network platform is 0.7507, which is 0.0298 and 0.0501 higher than that of deep learning and personalized recommendation method based on multi-source data fusion.

In the test of 200 users, the average accuracy of the personalized recommendation method of innovation and entrepreneurship education resources based on deep learning is 0.6775; The average accuracy of the personalized recommendation method of

Table 4. Recommendation accuracy for 200 users

Testing frequency	Personalized recommendation method of innovation and entrepreneurship education resources based on social network platform	A personalized recommendation method for innovation and entrepreneurship education resources based on deep learning	A personalized recommendation method for innovation and entrepreneurship education resources based on multi-source data fusion
1	0.7208	0.6845	0.6548
2	0.7146	0.6787	0.6415
3	0.7285	0.6764	0.6577
4	0.7367	0.6851	0.6686
5	0.7231	0.6815	0.6752
6	0.7255	0.6646	0.6663
7	0.7322	0.6573	0.6521
8	0.7213	0.6725	0.6532
9	0.7179	0.6858	0.6475
10	0.7122	0.6892	0.6506

innovation and entrepreneurship education resources based on multi-source data fusion is 0.6567; While the average accuracy of the personalized recommendation method of innovation and entrepreneurship education resources based on the social network platform is 0.7233, which is 0.0457 and 0.0665 higher than that of deep learning and personalized recommendation method based on multi-source data fusion.

In the test of 250 users, the average accuracy of the personalized recommendation method of innovation and entrepreneurship education resources based on deep learning is 0.6287; The average accuracy of the personalized recommendation method of innovation and entrepreneurship education resources based on multi-source data fusion is 0.6130; While the average accuracy of the personalized recommendation method of innovation and entrepreneurship education resources based on the social network platform is 0.6868, which is 0.0581 and 0.0738 higher than that of deep learning and personalized recommendation method based on multi-source data fusion.

According to Table 1, Table 2, Table 3, Table 4 and Table 5, the average accuracy of the three methods is compared, and the comparison results are shown in Fig. 3.

It can be seen from Fig. 3 that the average accuracy of the personalized recommendation method of innovation and entrepreneurship education resources based on deep learning is between 0.6287 and 0.8005; The average accuracy of the personalized recommendation method of innovation and entrepreneurship education resources based on multi-source data fusion is between 0.6130 and 0.7853; The average accuracy of the personalized recommendation method of innovation and entrepreneurship education resources based on the social network platform is between 0.6868 and 0.8328, which is higher than the average accuracy of the comparison method, which proves that

Table 5. Recommendation accuracy for 250 users

Testing frequency	Personalized recommendation method of innovation and entrepreneurship education resources based on social network platform	A personalized recommendation method for innovation and entrepreneurship education resources based on deep learning	A personalized recommendation method for innovation and entrepreneurship education resources based on multi-source data fusion
1	0.6946	0.6244	0.6047
2	0.6882	0.6387	0.6287
3	0.6765	0.6465	0.6165
4	0.6822	0.6236	0.6122
5	0.6853	0.6355	0.6051
6	0.6916	0.6424	0.6090
7	0.6949	0.6312	0.6166
8	0.6875	0.6248	0.6233
9	0.6754	0.6176	0.6028
10	0.6921	0.6023	0.6114

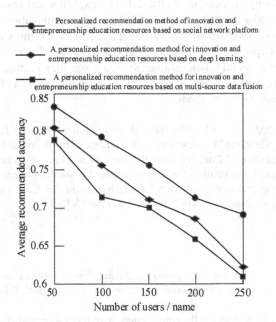

Fig. 3. Average recommendation accuracy of three methods

the personalized recommendation method of innovation and entrepreneurship education resources based on the social network platform is better.

It can be seen that the personalized recommendation method for innovation and entrepreneurship education resources designed in this paper is better than other recommendation methods in terms of F1-score value, which verifies the adaptability of this method to different scale data sets. The experimental results show that the method proposed in this paper is consistent and pertinent with the actual innovation and entrepreneurship education resource recommendation problem.

4 Conclusion

Innovation and entrepreneurship education is a practical education to consciously cultivate the entrepreneurship consciousness of college students and the innovative ideas of entrepreneurs. The rapid development of educational big data enriches learners' learning methods, but it also makes it difficult for learners to obtain suitable learning resources from massive online learning resources for learning. It is necessary to introduce personalized recommendation technology to provide learners with appropriate learning resources. Taking innovation and entrepreneurship education as the research object, this paper proposes a personalized recommendation method of innovation and entrepreneurship education resources based on social network platform. The combination of traditional platform recommendation and Resource Recommendation and sharing among learners makes the learning process more efficient. The experimental results show that this method is feasible and accurate to recommend personalized innovation and entrepreneurship education resources to learners, it can be applied to the recommendation system of innovation and entrepreneurship education. There is still much room for improvement in the work related to social relationship analysis and learning partner recommendation in this study. In future work, learners' negative feedback behavior can be considered and applied to recommendation methods.

Fund Project. In 2021, the "14th Five-Year Plan" of Hunan Province Educational Science "Planning for College Students' Employment and Entrepreneurship Research Special Project "Research on the "Golden Course" Construction of the Integration of Innovation and Entrepreneurship Courses and Ideological and Political Colleges in Higher Vocational Colleges-Taking the High-speed Railway Whole Industry Chain Plan New Entrepreneurship Education as the Example Example subject approval number: XJK21BXJ020, number: XJ211082.

References

1. Shi, L., Li, J.: Innovation and entrepreneurship education in colleges and universities: connotation, dilemma and path optimization. Heilongjiang Res. High. Educ. **39**(2), 100–104 (2021)
2. Wang, Z., Wang, T.: A study on the construction of an innovation and entrepreneurship education ecosystem: the role and function of China's higher education. J. Hangzhou Teach. Coll. (Soc. Sci. Ed.) **43**(5), 63–69 (2021)
3. Li, X., Liang, H., Feng, J., et al.: Design of personalized learning resource recommendation system for online education platform. Comput. Technol. Dev. **31**(2), 143–149 (2021)

4. She, X., Zhan, Q., Wu, C.: Multi-node information resource allocation recommendation algorithm based on collaborative filtering. Comput. Simul. **38**(6), 419–423 (2021)
5. He, Y., Xu, W.: Research and application of personalized education resource recommendation algorithm based on high-dimensional tensor decomposition. WuxianHulian Keji **18**(10), 114–115 (2021)
6. Yang, X.: Research on the application of user portrait in the personalized recommendation of educational resources. Fujian Comput. **37**(1), 52–53 (2021)
7. Wu, Y., Cai, Q., Liu, Z., et al.: Digital resource recommendation based on multi-source data and scene similarity calculation. Data Anal. Knowl. Discov. **5**(11), 114–123 (2021)

Digital Management System of Library Books Based on Web Platform

Xing Zhang[✉]

The Tourism College of Changchun University, Changchun 130000, China
xingxing11852@163.com

Abstract. In view of the slow response speed and single function of the library's book management system, this paper designs a library's digital management system based on the network platform. In the hardware part, t91sam9263 chip is used as the core, and the digital management system is established based on the web platform. In the software part, on the basis of dividing the grid space of digital management of books, the management mode of virtual space of books is designed by using virtualization technology. Introduce user interest indicators to recommend book resources personalized. The system test results show that in the face of high concurrent requests, the response time of the designed system is less than 300 ms, the error rate of book recommendation is less than 5%, and the performance is obviously improved.

Keywords: Web platform · Library · Book management · Digital system · Virtual space · User interest

1 Introduction

University library, as an auxiliary teaching unit in colleges and universities, mainly serves students, teachers and other administrative personnel working in colleges and universities, and is a comprehensive institution providing academic research services. Therefore, no matter in terms of collection or reader groups, it has obviously different characteristics from general social libraries [1]. Readers in colleges and universities have different needs for libraries due to various reasons, such as students' different majors, grades, curriculum arrangement, research directions and hobbies, teachers' different teaching courses, research fields and further studies, etc., and the requirements for library data management system are constantly improving. Today, with the rapid development of information technology, how to finish the work more efficiently and quickly has become people's primary goal. Traditional libraries occupy a large area, and books are widely distributed, so ordinary unfamiliar borrowers can't quickly find the books they need, which reduces readers' enthusiasm for acquiring knowledge to a certain extent and greatly reduces learning efficiency [2].

The construction of digital library management is a complex project, including how to implement the generic cabling system, what network technology to choose, and how to

W. Fu and L. Yun (Eds.): ADHIP 2022, LNICST 469, pp. 620–631, 2023.
https://doi.org/10.1007/978-3-031-28867-8_45

realize the construction of hardware projects such as computer network communication. However, to realize the functions of digital library, the corresponding software system is also indispensable. More importantly, it is necessary to realize the functions of collection and classification, management and storage, reservation query, access rights and so on of digital resources in the library, Only in this way can we provide users with convenient, fast and comprehensive information services. This is a two-sided problem. On the one hand, it manages the user's management needs for the system, that is, the requirements for the collection and processing of information resources, the storage and release authority management, and on the other hand, it is the reader user's functional requirements when using information. To sum up, the key core and lifeline of digital library construction is to build a digital resource system, and the establishment of the digital resource system is the precondition for the library to realize digitalization. Because the construction of digital library involves the cooperation and assembly of multiple modules, the data volume of the overall system is large, and the response is often slow when users use it, so it is necessary to design a new digital library management system.

Nowadays, the university network has been popularized. Transplanting the library to the network can not only solve the problems existing in the traditional library, but also transfer the record link data of library books in the past to the database, and complete a large number of data processing operations with computers, simple and convenient operation, well-organized data classification and storage methods, which greatly reduce the workload of librarians and improve work efficiency. Although the digital electronic library is being built in an orderly way, there are still many digital libraries with constantly updated and changing user requirements, gathering and expanding network resources. For example, how to push users' corresponding information, how to quickly and accurately query users' demand information, how to reasonably mine users' information, how to provide diversified user services, how to communicate with users more personalized, and how to display the final execution results in a friendly interface are all issues that need to be seriously considered at this stage.

With the rapid development of Internet technology, Web 2.0 has played an important role in the network and is an indispensable part of people's information life. Web 2.0 pays attention to respect for all individuals, pays more attention to users' own knowledge and information, and at the same time, it is also an important core idea of Web 2.0 [4]. However, at present, the library management systems in different environments are different, especially how to share information in the library and how to interact effectively between readers and the library. WEB-based book management information system is a system that uses computer system to manage books ordering, borrowing, returning books and searching. Through this system, the efficiency of book management is greatly improved, and readers' borrowing, information query and other needs are facilitated, which has incomparable advantages over manual management. According to the above analysis, in order to meet the above requirements, aiming at the problems of imperfect functions, complicated operation, slow efficiency and large hardware resource occupation of the previous library management system, this paper will study and design a digital library book management system based on the web platform, and apply the system to the actual library book management work, so as to enhance the professionalism of university library data service, improve service quality and service response efficiency, and ensure the needs of university library construction.

2 Hardware Design of Library Books Digital Management System Based on Web Platform

When designing the hardware part of the digital library books management system based on the web platform, the hardware part should meet the requirements of this paper's embedded design scheme. The overall architecture of the hardware part of the system is shown in Fig. 1.

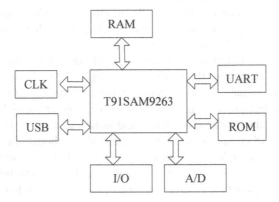

Fig. 1. Hardware architecture of library books digital management system

The core control of the hardware part of the library books digitization management system designed in this paper is the T91SAM9263 chip based on ARM core as the hardware control processor of the system. The chip has the highest clock frequency of 266 MHz, and has 2 USB OTG interfaces, extensible Ethernet interfaces, 4 UART, 2 CAN channels, 2 SSP controllers, 1 SPI interface, 3 12C interfaces, 2 input and 2 output 12S interfaces and 8 channels of 12-bit ADC. The ARM has two stack pointers, but only one of them can be used at a time. The main stack pointer (MSP) is the stack pointer used after reset, which is used for system kernel and exception handling. The process stack pointer (PSP) is used by application code written by users [5]. The hardware part of the system is externally connected with a 2 Mbits ferroelectric memory, the model is FM25CL64B-GTR, and the interface is standard SPI interface. Ferroelectric memory has the advantages of RAM and ROM, fast reading and writing speed and nonvolatile characteristics. In this design, there are two external memories, one external memory is used to load programs, and the system is started from this memory when it is reset [6]. Another piece of external memory saves the data backed up by hardware checkpoints, and restores the saved data to the system when the system is powered on next time.

In the main memory, this system chooses SDRAM as the main memory, which is mainly based on its advantages of fast processing speed and low price. At the same time, this system occupies a large amount of memory when running the GUI, so SDRAM with large capacity should be used as the main memory of the system. Based on this, this system selects the chip MT48LCl6M16A2 produced by MICRON Company, and this system uses two pieces of SDRAM with a total size of 64M as the main memory.

An idle register address in T91SAM9263 chip serves as the communication channel between SCORE and HCORE. SCORE compiles and generates the ".HEX" file in KEIL as the ROM initialization file of CPU. HCORE is designed by Verilog HDL, including command receiving and decoding module, task management module, delay management module, timer module, semaphore management module, mutually exclusive semaphore management module, edge detection module and task arbitration module. SFR bus has 8-bit address space, some of which are occupied by internal modules of T91SAM9263, such as timer and serial port, etc., which are defined and used as shown in Table 1.

Table 1. SFR address representation

Port				
P0	P1	P2	P3	P4
80h(1)	90h(1)	93h-97h	9Ah-9Fh	A0h(1)-A7h
9Ah-AFh	B0h(1)-B7h	B9h-BFh	C2h-C7h	C9h
CEh	CFH	D1h-D7h	D9h	DAh-DFh
E1h-E7h	E9h	EAh-EFh	F1h-F7h	F9h-FFh

In order to reduce the time of task switching and improve the way of interrupt triggering task switching, a multi-user register group working scheme is designed. The logic unit of FPGA is used to configure a set of RO-R7 registers, PSW (program status word), ACC (accumulator), B (one register), DPL (lower 8 bits of data pointer register), DPH (upper 8 bits of data pointer register) and stack space for each task. Each register group takes the highest priority of the task as chip selection, and the task switching can be realized without frequently moving data. In the auxiliary memory, the auxiliary memory is used to store a large amount of program code or information, so the size and price of the auxiliary memory become the decisive factors in choosing it. NANDFLASH has become the first choice for most embedded systems because of its large capacity and low price per unit capacity. In this design, the 8-8bit NANDFLASH memory K9F1208UOB is selected as the auxiliary memory of this system, and the size of K9F1208UOB is 64M, which can fully meet the requirements of this system.

In this design, the chip ISP1362 is selected as the USB host controller, and the interface between the ARM control chip and ISP1362 is designed. As the peripheral of this system, ISP1362 is mounted on the SFR system bus of T91SAM9263, and the peripheral address 9Ah is assigned to it. When reading the peripheral data with address 9A, first set the data on the sfr_addr address bus to 9A, so that the chip select signal sfr_cs_9A is valid. When the read control bus sfr_rd is valid, read the data into the data bus sfr_data_in, and then complete a data reading operation.

When the hardware part communicates with the upper computer of the software part of the library book digital management system, it communicates through SPI bus. The external RESET of the hardware part is to reset the chip by pulling the reset pin of the chip to low level. The reset signal can't be removed immediately after it appears. The reset signal can only be removed when the crystal oscillator runs stably and the XTAL1 pin of T91SAM9263 meets the specified clock signal. The RESET pin of T91SAM9263 is Schmidt trigger pin with an extra interference filter. The filter can filter out very short pulse signals, so that the processor will not be accidentally reset by interference pulses or reset many times by unstable reset signals.

Under the hardware framework of the library books digitization management system designed above, the effective management of university library books digitization is realized in the system software part combined with the web platform.

3 Software Design of Library Books Digital Management System Based on Web Platform

3.1 Building the Web Platform Architecture of Library Books Digital Management System

The core concepts of Web 2.0 are interaction, sharing and relationship. In this paper, B/S architecture is used to design the software structure of library books digital management system. In B/S architecture, the server is an important part, which integrates business logic. Users mainly use browsers to send requests to the server on the WEB side. The WEB transmits the requests to the server through the back-end modules and users' demands. The server processes the data accordingly and returns the final results to the users. The final results will be presented to the users in the form of hypertext markup language, which may contain some scripting languages. B/S architecture is mainly based on browser and server architecture. Users send their own requirements to the server through the browser, and the server returns the results to users through a series of processing. Compared with C/S, B/S does not need to maintain the client or update the client from time to time, which reduces the specificity of the client. Figure 2 is the functional structure diagram of the library book digital management system software based on web platform.

The system adopts a three-tier architecture model, including client node, server node and database node. The client is mainly IE browser, which requires IE 8.0 and above. It is mainly responsible for communicating with the server and realizing data interaction. In order to reduce the pressure and security of the server, this system adds a series of additional functions of authority logic judgment and data verification to the client. Tomcat is installed in the server, which deploys the main code of the system and interacts with the client and database nodes at the same time, mainly including system authority verification, data access, business logic processing, etc. The database node adopts SQL Server 2008, which mainly stores a large amount of data in the system and provides data support for the server. The running environment of the system is the local area network on campus and the external network of the school. Browser is responsible for the input, management and output of system data, and Server is responsible for the storage, access and processing of data.

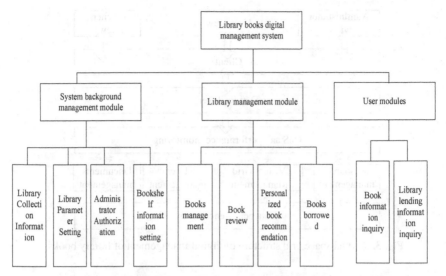

Fig. 2. Functional structure diagram of digital management of books

3.2 Grid Division of Virtual Space in Library Digital Management

With the expansion of systematic book resources, it is easy to cause confusion if all resources are managed together. At this time, by dividing the total resources into small subsets, the small subsets will not influence each other, and better results will be achieved. Virtual grid organically combines a series of basic functional modules to work together to ensure the efficient operation of the system and provide high-quality services to users. The functions required by the grid environment include management hierarchy, information service, communication service, security and authorization, distributed file system, system state and fault tolerance, resource management and scheduling, programming tools, user interface and so on. The basic functional modules for managing the virtual space grid of books are shown in Fig. 3.

In the same host, virtual machines and virtual machines are in a resource competition relationship. By setting the resource allocation of virtual machines, some virtual machines can be prevented from occupying most of the load while some virtual machines have no resources available, which affects the computing space in book management and scheduling. Resource allocation settings (share, reservation and limit) are used to determine the amount of CPU and memory resources provided to the virtual machine. In the grid, there are three entities: resource requester, resource intermediary and resource provider. The resource intermediary is the resource manager, and sometimes the resource provider is the resource itself. According to the different information flow paths in the process of book management, there are three forms of book management system. This design uses triangle structure to divide the virtual grid, that is, users make a request to the book resource manager, and the resource manager searches for suitable resources for users and drives resources to work, and at the same time, tells the resources in what form and address to return the service result to the requester, and tells the resource requester

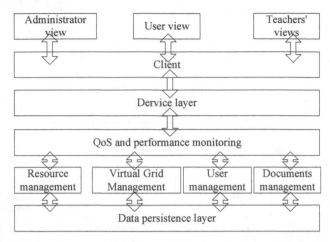

Fig. 3. Virtual space grid structure of digital management of library books

the service result according to the message provided by the manager after the service [7].

Grid users interact with the grid through user interface, such as resource request, result return and other functions. After selecting a resource, the grid provides the resource to users and manages the resource, which requires the function of resource management. In the process of providing services, the grid needs to transmit the grid data to the required nodes, and the job management module provides the report function of job operation. The cost and time for users to use resources are realized by the user accounting and management module. The whole process of grid users using the grid needs communication and security guarantee QoS guarantee.

Under the virtual grid, the system software can be divided into four layers: portal layer, service layer, logical function layer and data persistence layer. The portal layer is mainly responsible for the interaction with grid users, and provides users with various services related to virtual organization through the service layer providing clients, thus shielding the complexity of the underlying services for users. The service layer is mainly responsible for encapsulating the functions of each module in the logical layer according to the management policy, and providing services to grid service systems and grid users in the form of Web Service. The logic layer is mainly responsible for providing the upper layer with basic logic functions including users, resources, voucher management, etc. through the support of the data persistence layer. The data persistence layer communicates with the underlying database, providing data persistence support for the realization of upper logic functions and shielding the underlying complex data access operations.

3.3   Book Virtual Space Management Mode Setting

Virtualization technology can effectively reduce the waste of computing resources caused by deploying a large number of servers. In virtualization technology, multiple virtual machines can share a physical machine, but they are completely isolated from each other as if they are running multiple physical machines. Different operating systems and applications can be run on these virtual machines, and they do not interfere with each other. Moreover, the virtual machine is completely independent of the underlying physical hardware, which enables users to migrate the virtual machine from one physical machine to another without changing the device driver, operating system and application program. In the virtual grid, the concept of virtual organization is introduced to establish a flexible and safe resource sharing and management mode. At the same time, single sign-on and controllable cross-domain resource sharing are realized through certificate issuance and detection, so as to provide users with a unified and safe way to use grid resources and provide the system with simple and effective user and resource management services. In this paper, the book resource sharing structure of distributed library based on Web technology is the core part of digital management of university books. The servers are completely peer-to-peer P2P mode, but it is not required that the use of university book resources in grid system is also P2P mode. The users in the virtual organization of regional university book resources management are mainly university members, and share setting, reservation setting and restriction setting are required when corresponding users request book resources scheduling. The share values of virtual machines can be set to high, normal and low. When set to high, each CPU has 2000 shares, and the configured virtual memory has 20 shares per megabyte. When set to normal, each CPU has 1000 shares, and the configured virtual memory has 10 shares per megabyte; When set to low, each CPU has 500 shares, and the configured virtual memory has 5 shares per megabyte. In this system, the CPU resource reservation of the virtual machine is set to 1 GHz, and the memory resource reservation is 1 GB, which can be reserved when the virtual machine is created. After the virtual machine is created, the resource reservation of the virtual machine can also be reset by reconfiguration.

In the grid environment, any virtual organization can be regarded as an entity. According to the characteristics of resource organization and function, its functions are encapsulated into related services. For users, virtual organization members are not only a resource entity, but also a service entity. According to the commonness and difference of service functions of virtual organizations, the common services are integrated in the network, and the different services are interconnected, thus forming the resource sharing service model of virtual organizations. The resources of this system are all placed in the data center of the library, and the resources are divided by clusters, in which virtual machines can migrate in the hosts in the clusters. The problem to be solved in resource scheduling is to balance the load among hosts, that is, to transfer the load pressure of the host with heavy load to that with light load.

D is the wide-area dispatching service center, $D = \{d_1, d_2, \ldots, d_n\}$, d_i is the resource management and dispatching center. Following the grid resource allocation management (GRAM) of OGSA, D gathers all the basic service information of d_i, including d_i's address, service cost, service task application amount, etc., and is responsible for the dynamic grid service management and coordination of each d_i. The specific resource sharing and scheduling process is as follows:

Firstly, d_i inquire all the information of the books in the current collection from the local task management and dispatching service center of D Middle School or other places, and select the best book scheme according to the inquiry results, and then send a dispatching request directly to the local task management and dispatching service center to which the tasks will be transferred. d_i. After receiving the reply of consent, the task will be scheduled, otherwise, it will apply to other local task management and scheduling service centers again until the book borrowing or information update service operation is completed. In this way, through the coordination and sharing of resources among virtual organizations, the collaborative lending management service for library books is realized.

3.4 Implementation of Book Resource Management Recommendation

This design constructs the following recommendation model of university book resource scheduling, in which the book resource collection is defined as $R = \{r_1, r_2, \ldots, r_n\}$, assuming that there are $R = \{r_1, r_2, \ldots, r_n\}$ kinds of resources, r_i refers to the ith kind of resources, and its quantity is recorded as NR_i units. The total quantity of all resources is recorded as N, $N = \sum_{i=1}^{n} NR_i$. The user set is defined as $S = \{s_1, s_2, \ldots, s_m\}$, and s_j refers to the jth user. Assuming that there is a M number of users, M is always greater than N. When recommending student books, users' needs and interests should be taken into account. Therefore, this design uses the user interest model based on BTVSM to reflect users' interest in book resources more truly. Users' interests will change with the passage of time, some topics that users were originally interested in will be gradually forgotten, and new topics of interest will gradually emerge. In order to reflect users' interests more truly, the user interest model is represented by the vector space model (BTVSM) based on background and tense. That is to say, subjects and sections are introduced into the model as background constraints, and interest weight function $W_n(t_n)$ based on temporal changes is introduced into vector space to calculate the attenuation and update of user interest weight. At a certain time t, the user interest model is expressed as:

$$IS = \{A, G, K\} \tag{1}$$

$$K = \{(k_1, w(t_1)), (k_2, w(t_2)), \cdots, (k_n, w(t_n))\} \tag{2}$$

Among them, A represents the classified collection of books in university library; G indicates the grade of college students; K represent that keyword vector space of user's inter, and k_n is the n keyword describing interest; t_n represents the time set of each submission of keywords, and $w(t_n)$ is the weight function of keywords k_n with respect to time. For the attenuation and update of users' interests, the mechanism based on time window is used to calculate, that is, within a certain time window, if keywords are submitted, the weight will be increased; Otherwise, the weight is attenuated.

Assumptions:

1. Every time keyword k_n is submitted in each time window ΔT, the interest weight is increased by q units;
2. If keyword k_n is not submitted in each time window ΔT, the interest weight attenuation unit p; Then, at a certain time t, the interest weight function of keyword k_n is expressed as:

$$w(t_n) = \begin{cases} \sum \left[f((t_n + i\Delta T - \Delta T)) \cdot q - \xi p \right], & w(t_n) > 0 \\ 0, & else \end{cases} \tag{3}$$

where, f indicates the number of time windows in the time period. There may be multiple requesters for each resource. Suppose that for book resource R_i, if there are y units left unused, then an allocation queue is set up for resource R_i. When user s_j requests resource R_i, the user enters the allocation queue of resource R_i. It is easy for people to follow the principle of first come, first served. The user who requests earlier is at the top of the queue and has the right to use resource R_i first. In other words, users who wait longer have the right to use resources first. F_j indicates that the user j obtains the benefits generated by the resource combination R_i in the resource allocation. F represents the collection of book resources obtained by all users. Then the objective function of integer linear programming recommended by library books is as follows:

$$Q = \max \sum \sum F_j(R_i)x_{js} \tag{4}$$

$$st. \sum x_{js} \leq 1$$
$$R_i \in R \tag{5}$$

where, x_{js} indicates whether the user has been assigned books R_i, and if so, the value is 1, otherwise, it is 0. According to the actual storage situation of the corresponding books in the library, solve the above model and get the personalized recommendation scheme of books. When using the system, students can choose to borrow books according to the recommended results obtained by the system. Transplanting the software function part designed above to realize library book management to the system hardware part, that is, completing the design of library book digital management system based on web platform.

4 System Test

The function test of the system that has been designed above is the basis to ensure that the designed system can effectively and stably exert its functions in the actual use process. The following is the test content of the library books digital management system based on the web platform.

4.1 Test Content and Preparation

After the overall design of the system is completed, it is necessary to test the system to ensure that the online book management system can meet the design requirements. In this paper, the test of the library books digital management system based on web platform is to carry out various assembly tests and function confirmation tests on the whole system. After completing the test function of the system, it is necessary to test the single running effectiveness of the network, peripherals, software and hardware equipment of the system and the running effectiveness of this single element in the complex environment, which is the main test content of this test. The testing work of this design adheres to the principle of comprehensive testing, and it is planned to conduct an overall test on the whole product, which serves as the full coverage of the system functions in the test range, to check whether the design system can perfectly realize the designed functions of the design function design part, and to screen out some functions that are found to be inconsistent or contradictory in some tests, and make targeted revision and correction.

As an important part of software development, system testing can ensure that it has a positive impact on the effective development of the project only through repeated testing of the system. Based on this, this section tests the function and performance of the library books digital management system designed by this design, so as to analyze its feasibility in practical application, so as to avoid problems in its later investment and use. In this system test, the digital management system of library books based on web platform designed in this paper is compared with the library management system based on WeChat platform database and the library management system based on mobile AR. Evaluate the performance of the system by comparing the stress test of the system with the personalized recommendation effect of books.

4.2 Test Result

Firstly, the system hardware connection and basic operation function response designed in this paper are tested. After testing, the hardware part of the system designed in this paper is connected normally, and it can respond to the user's requests for books borrowing, administrators adding books information and other operations, and runs well as a whole.

The performance of the three systems is tested under different concurrency of book access requests, and the test results are shown in Table 2 below.

Table 2. System performance test data

Service quantity	Library management system based on web platform		Library management system based on mobile AR		Book management system based on WeChat platform database	
	Response time/ms	Error rate of book recommendation/%	Response time/ms	Error rate of book recommendation/%	Response time/ms	Error rate of book recommendation/%
100	197.3	2.8	219.4	6.6	313.4	**20.1**
200	223.9	3.5	235.3	11.3	357.3	19.8
300	239.4	4.2	269.8	7.4	371.8	8.3
500	242.7	**4.3**	282.9	**16.5**	427.6	14.8
800	250.6	3.8	331.7	12.1	463.3	10.4
1000	264.9	4.2	361.6	6.9	526.9	12.7
1500	**296.1**	3.6	**458.1**	7.8	**600.5**	16.4

By analyzing the data in Table 2, we can see that the system in this paper can still keep the response time less than 300 ms in the face of high concurrent requests. The response time of the system is obviously shorter than that of the other two comparison systems, which indicates that the system in this paper has a good response performance under pressure test. Judging from the error rate of book recommendation, the error rate of personalized recommendation of books to different users in this system is less than 5%, which indicates that it is more accurate and efficient to recommend books to system users. To sum up, the digital management system of library books based on web platform designed in this paper improves the efficiency and effect of book management, and can meet the demand of university teachers and students for personalized service of library. This is because the method in this paper uses the B/S structure to build the system architecture, so that the response between modules is faster, and sets the resource allocation of virtual machines to prevent some virtual machines from occupying most of the load when there are no available resources, configures the resources for book resource scheduling under the application of web technology, uses the vector space model based on background and temporal to determine user interest, and updates the attenuation value of interest in time, To ensure that the system can achieve rapid and accurate recommendation and management of book resources.

5 Concluding Remarks

This paper designs a digital library book management system based on web platform, which can use the web platform to improve the management efficiency of books and the user's satisfaction with the system. The B/S structure is used to establish the system architecture, which makes the response speed between modules faster, and can set the resource allocation of virtual machines, avoiding that some virtual machines occupy a large number of resources due to insufficient resources. The adjustment of book resources is configured using web technology, and the vector space model based on background and temporal is used to judge the user's interest, and the change of interest is updated in time, This ensures that the system can quickly and accurately recommend and manage book resources. The system test results show that the running performance and functions of the system meet the needs of digital management of books in university libraries, and the actual performance test results are better than the current book system.

References

1. He, Y.: Research on context-aware library management system based on mobile AR.Libr. J. **39**(08), 50–56+74 (2020)
2. Wang, S., Chen, W., Lu, K., et al.: University smart libraries 4.0: research on libraries in future based on industry 4.0 and web 4.0. Libr. Theory Pract. (01), 59–66 (2021)
3. Yu, J.: Multilevel force interactive simulation supporting embedded web server. Comput. Simul. **38**(07), 256–260 (2021)
4. Jiang, H.: Design and research of book lending information management system based on Wechat platform database. Mod. Electron. Tech. **43**(13), 134–137 (2020)
5. Wang, L.: The practice and enlightenment of data management service based on translational medicine——take University of Washington Health Sciences Library for example. Libr. Tribune **41**(02), 143–151 (2021)
6. Ding, Z., Wang, G.: Library reading seat management system based on Web technology. Jiangsu Sci. Technol. Inf. **37**(20), 7–11 (2020)
7. Xun, X., Cai, M., Yao, Z.: Student affairs management system based on information of book borrowing. J. North China Inst. Aerosp. Eng. **30**(01), 18–20 (2020)

Design of Numerical Control Machining Simulation Teaching System Based on Mobile Terminal

Liang Song[✉] and Juan Song

Sifang College, Shijiazhuang Tiedao University, Xuzhou 051132, China
s117751998160@163.com

Abstract. Today's manufacturing industries all over the world widely use CNC technology to improve the manufacturing capacity and level, and improve the adaptability and competitiveness of the dynamic and changeable market. The research and development of CNC technology and the promotion and application of CNC products require a large number of high-quality CNC professionals, and CNC teaching and training are therefore in a very important position. In order to improve the success rate of system requests, a numerical control machining simulation teaching system based on mobile terminals is designed. Using the combination of PC and motion control card, combined with the oscillation circuit inside the PIC16F877 microcontroller to form a complete oscillation circuit; using 3D graphics technology to simulate the CNC machining process, optimize the CNC machining process, build a tool database, and transfer these parameters to the In the simulation program, the cutting process is simplified as a one-dimensional Boolean operation along the line of sight, and the function of the simulation teaching system is designed by using the mobile terminal. Experimental results: The request success rate of the designed system is high, indicating that its use effect is better.

Keywords: Mobile terminal · CNC machining simulation · Teaching system · CNC machine tool · Virtual environment · Part program

1 Introduction

The machining process of CNC machine tools is controlled by the part program. The correctness of the part program directly determines the quality and efficiency of machining, and an incorrect machining program can also lead to accidents. Due to the complex and changeable part shape, and the specific machine tool structure and workpiece clamping method are generally not considered in the process of tool path generation, the generated part program may not be suitable for the actual processing situation. CNC technology is the basic technology for the realization of automation, flexibility and integrated production in the manufacturing industry. It is not only an indispensable material means to improve product quality and labor productivity, but also related industries based on

W. Fu and L. Yun (Eds.): ADHIP 2022, LNICST 469, pp. 632–646, 2023.
https://doi.org/10.1007/978-3-031-28867-8_46

it are related to the national strategic position and The technical level of an important basic industry that reflects the level of a country's comprehensive national strength has become an important symbol for measuring a country's industrial modernization.

Although the current CNC programming technology has made great progress in surface modeling, trajectory planning, tool position calculation, etc., it still does not ensure that the generated part processing program is completely correct and reliable. The main problems are: over-cut and under-cut during processing, collision between tool and machine tool components and fixtures, and cutting overload during processing. Especially in high-speed machining, these problems are often fatal. In severe cases, tools, workpieces, machine tools and even personal accidents will be damaged. Therefore, after the part program is generated, it needs to be checked for correctness and modified according to its existing problems until a qualified part program is formed. NC machining simulation is the mapping of index-controlled machine tools in a virtual environment, which integrates manufacturing technology, machine tool numerical control theory, computer-aided design (CAD), computer-aided manufacturing (CAM), and modeling and simulation technology [1–3]. In recent years, both at home and abroad have attached great importance to the teaching and training of numerical control technology, and strived to develop new training technologies and means. At present, the commonly used NC machining simulation teaching systems mainly include the NC machining simulation teaching system based on virtual reality and the NC machining simulation teaching system based on PC platform. The former specifically divides the functional modules of the virtual reality NC machining simulation system, designs and develops them module by module, and combines the virtual reality external hardware equipment to design and implement a variety of three-dimensional display modes for different applications of the system. Thus a virtual reality NC machining simulation system with high model authenticity, complete machining functions, and preliminary cutting simulation effect and immersion effect is developed. The latter focuses on the input, modification, display and access simulation of the NC program code, the translation of the NC program, and the simulation of the machining process of the NC machine tool. Although the above simulation teaching system meets the teaching needs of CNC machine tools to a certain extent, there is a problem that the success rate of system requests is low.

Numerical control teaching has its own unique characteristics, and the rapid development of numerical control technology constantly puts forward new requirements for numerical control teaching. How to adapt to the characteristics of numerical control teaching and keep up with the pace of numerical control technology development is an important topic in numerical control teaching. With the help of mobile terminal technology, the virtual environment of the three-dimensional simulation model generated by the computer can be intuitively perceived, and the CNC machining of parts can be carried out in the virtual environment before the real manufacturing, and the correctness and reasonableness of the CNC program can be checked before designing a new scheme or changing the scheme. The advantages and disadvantages of the processing plan are evaluated and optimized, so as to ultimately achieve the purpose of shortening the product development cycle, reducing production costs, and improving product quality and production efficiency.

Under the above background, this paper designs a numerical control machining simulation teaching system based on mobile terminal. In the system hardware design, the system adopts a combination of power-on reset and button reset, which improves the convenience of system operation; in the decoding process In the program segment, the syntax check is performed, and if syntax errors are found, an alarm will be issued immediately to avoid damage to tools, workpieces, and machine tools, and to reduce the probability of personal accidents.

2 Hardware Design of NC Machining Simulation Teaching System

Before the system software and hardware design, the overall architecture of the NC machining simulation teaching system is first given, as shown in Fig. 1.

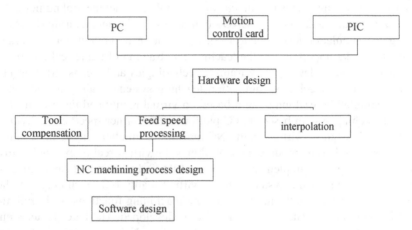

Fig. 1. Overall system architecture

When designing the hardware, try to use modular and standardized board-level modules, and use a combination of PC and motion control card in the system hardware. The motion control card is inserted into the PCI bus slot of the PC, and the A shared bus communication method is used between them. PIC is the prefix of the single-chip microcomputer series products produced by American Microchip Company. The hardware design of PIC series single-chip microcomputers is simple and the instruction system design is refined. It is easy to develop for beginners. It is one of the most widely used varieties of single-chip microcomputers. At the same time, efforts are made to reduce hardware costs, and economical stepping motors are used as driving components to form an open-loop feeding system with the stepping motor drivers. In this system, the PC system manages most of the control functions, including work mode scheduling, parameter and work state setting, emergency stop control, human-computer interaction, part program input, editing, translation, machine parameter interpolation, parts Graphic display, simulation processing display and dynamic display of tool coordinates etc.

At present, several well-known semiconductor companies in the world have developed a series of single-chip microcomputers that are compatible with pins by imitating the PIC series of single-chip microcomputers. It can be said that the PIC microcontroller series represents a new direction of microcontroller development. PIC 16F87X is a new product with distinctive features introduced by Microchip. Its most important advantage is that the PIC16F877 series of microcontrollers have FLASH program memory and built-in ICD function, which can realize online debugging and online programming. This is the MCS51 and MC68 series microcontrollers. Not available, this is exactly what the majority of microcontroller users need. The motion control card is the MPC02 motion control card of Step-servo Company. It transmits digital pulses to the X, Y, Z stepper motor drivers respectively through the I/O interface, and drives the stepper motor to drive the XY table and the Z spindle to respond accordingly. Exercise. The motion control devices of the numerical control system mainly include stepper motors and their drivers, spindle motors, numerical control actuators and operation panels. In the PIC 16F87X series, the PIC16F877 includes the functions of other types of microcontrollers, and other microcontrollers are partially simplified on the basis of their functions. The superiority of the PIC series of microcontrollers is mainly due to the unique design method adopted in the architecture of this series of microcontrollers. Separate from the command bus and use different widths.

Because this machine tool is a CNC milling machine for three-coordinate linkage teaching developed for CNC teaching and vocational technical training. That is to consider the function, performance requirements, but also take into account the needs of economic and practical. The X, Y, Z coordinates of the machine tool are all driven by linear rolling guides and ball screws, and are driven by stepper motors. This facilitates the realization of the "pipeline operation" of instruction fetching, that is, while executing one instruction, the instruction fetch operation is performed on the next instruction, which facilitates the realization of single-byte and single-cycle of all instructions, which is conducive to improving the execution of instructions by the CPU. Speed, so this system adopts PIC16F877 type single chip microcomputer. In order to meet the needs of different cutting speeds, the main shaft of the machine tool is rotated by a variable frequency speed regulation motor, so this machine tool has the advantages of stable movement, high movement precision, and stepless speed regulation. The lifting table is used to expand the range of workpieces to be processed, and all movements during processing are gearless, which greatly improves the transmission accuracy and helps reduce machine tool noise. The machine can also be equipped with automatic cooling, oil pump lubrication and lighting systems. In order to obtain a relatively stable oscillation timing signal, the clock circuit design adopts the XT method, selects a 4 MHz crystal oscillator and two 22PF capacitors, and connects the crystal oscillator and two capacitors to the OSC1 and OSC2 pins, combined with the internal PIC16F877 microcontroller. The oscillator circuit constitutes a complete oscillator circuit. The reset circuit of the hardware part of the CNC machining simulation teaching system is shown in Fig. 2:

As can be seen from Fig. 2, according to the requirements of PIC16F877 single-chip low-level reset and reset time of 5ms, the system adopts a combination of power-on reset and button reset. Although an all-round software control panel is designed in this numerical control system, for the convenience of operation and equipment safety, it is also equipped with an electrical operation panel of the machine tool, which is mainly used for the start and stop of the numerical control machine tool, and the control and

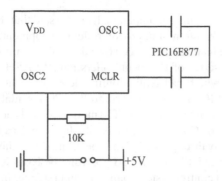

Fig. 2. System hardware reset circuit diagram

status display of the actuator. The machine tool is equipped with a special strong electric control cabinet, which is convenient for cognitive experiments of the control system. Such as the electrical control of spindle motor, stepper motor and cooling motor, as well as servo control, signal transmission, frequency conversion speed control and so on. The power-on reset is when the single-chip microcomputer is powered on, when the pin VDD rises to 1.6–1.8 V, the power-on reset circuit provides a reset pulse to reset directly. The key reset is to reset by the negative pulse generated by the key. The function of the peripheral output circuit is to indicate the current state of the machine tool by the light and shade of the indicator light on the machine tool operation panel. MPC02 control card as a platform for developing motion control system, its structure is open. The card is inserted into the PCI expansion slot of the PC, and the number of control cards and the number of control axes on each card can be easily configured. The MPC02 card provides a powerful motion control function library, and can make full use of the existing resources of the PC to develop a perfect motion control system. The MPC02 control card is divided into two types: A and B, and the number of control axes can be selected from 1 to 3 axes. Because the machine tool may be in a certain state for a long time, such as coolant startup, spindle forward travel, etc. Taking into account the characteristics of the status of the indicator light, a 8D latch tri-state unidirectional data buffer of a 74LS373 and a 8255A are used to expand, and the latch function of the 74LS373 is used to meet the requirements of the indicator light status.

3   Software Design of NC Machining Simulation Teaching System

3.1   Optimizing CNC Machining Process

The machining process of CNC machine tools is controlled by the part program. The correctness of the part program directly determines the quality and efficiency of machining, and an incorrect machining program can also lead to accidents. Input to the CNC controller usually includes part programs, machine parameters and compensation data [4, 5]. However, machine parameters and compensation data other than tool size are generally set during installation and debugging, so the main input is the part program and tool size compensation data. Due to the complex and changeable part shape, and the specific machine tool structure and workpiece clamping method are generally not considered in the process of tool path generation, the generated part program may not be suitable for the actual processing situation. The CNC input working modes include storage mode and NC mode. The so-called storage method is to input the entire part program into the CNC internal memory at one time, and then call out the blocks one by one from the memory during processing. The so-called NC mode means that the CNC performs processing while inputting, that is, when the previous program is processed, the content of the next block is input. Because the PC memory is relatively large now, this system adopts the storage method. Decoding processing is to process a block of the part program as a unit. Therefore, although the current CNC programming technology has made great progress in surface modeling, trajectory planning, tool position calculation, etc., it still cannot ensure that the generated part machining program is completely correct and reliable. The main problems are: over-cut and under-cut during processing, collision between tool and machine tool components and fixtures, and cutting overload during processing. That is, the part contour information (such as starting point, end point, straight line or arc, etc.), feed speed information (CF code) and other auxiliary information (M, S, T code, etc.) are interpreted into the computer according to certain grammar rules. A data form that can be recognized and stored in a designated memory dedicated area in a certain data format. The main steps of the CNC machining process are shown in Fig. 3:

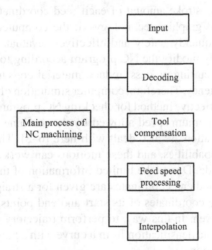

Fig. 3. The main process of CNC machining

As can be seen from Fig. 3, the main process of CNC machining mainly includes steps such as input, decoding, tool compensation, feed speed processing and interpolation. Especially in high-speed machining, these problems are often fatal. In severe cases, tools, workpieces, machine tools and even personal accidents will be damaged. Therefore, after the part program is generated, it needs to be checked for correctness and modified according to its existing problems until a qualified part program is formed. During the decoding process, the syntax check of the program segment should be completed, and if any syntax error is found, an alarm will be issued immediately. Tool compensation refers to the compensation of tool length and tool radius. In order to make the compiled part program independent of the tool size, that is, the compiled program will not be changed due to the change of the tool size, usually the part program is programmed with the part contour trajectory. The function of tool compensation is to automatically convert (or convert) the contour of the part into the tool center path according to the set tool size data.

There are several ways to check the part program. One of the methods is to let the machine tool "dry run" before the formal processing. The dry run can only make a rough estimate of whether the machine tool movement is correct and whether there is interference or collision. However, if the method of "trial cutting" is adopted, a more accurate judgment can be made on whether the processing process is normal and whether the processing results meet the requirements. However, trial cutting is a time-consuming and expensive work, its efficiency is very low, and it needs to increase production costs. In addition, the safety of the trial cutting process cannot be guaranteed. Tool compensation also includes automatic transfer between blocks and over-cutting judgment, which is the so-called C tool compensation function. The moving speed of the tool relative to the workpiece given by the CNC machining program is the speed in the direction of the composite movement of each coordinate, that is, the command value of the F code. The first work to be done in the speed processing is to calculate the sub-speed in the direction of each feed motion coordinate according to the composition, so as to prepare for the calculation of the stroke amount of each feed coordinate during interpolation. Using three-dimensional graphics technology on the computer to simulate the CNC machining process can quickly, safely and effectively evaluate the correctness of the NC program, and quickly modify the NC program according to the simulation results, avoiding repeated Trial cutting process, reduce material consumption and production cost, improve work efficiency. Therefore, computer simulation of NC machining process is an efficient, safe and effective method for checking NC programs. In addition, the limit of the minimum and maximum speed allowed by the machine tool and the automatic acceleration and deceleration are also dealt with here [6–8]. The MPC02 motion card has speed processing capabilities, and these motions can work in normal speed mode or trapezoidal speed mode. Due to the limited information of the command stroke, for example, only the start and end coordinates are given for a straight line. For a segment of arc, in addition to the coordinates of its start and end points, its center coordinates or arc radius are also given. In this way, to perform trajectory processing, CNC must automatically perform data densification from a curve with a known start and end point, which is called interpolation.

3.2 Build Tool Database

In the process of CNC machining, the selection of tools should be carried out according to different machining surfaces, the setting of blank size, the selection of fixtures, and the assembly of blanks and fixtures should be carried out according to different parts. For the management and application of these data, in CNC It is very important in the simulation system. Therefore, it is necessary to establish a database to complete the storage and management tasks of these data. By selecting the tool from the turning tool library, the main parameters of the turning tool are obtained, and these parameters are transferred to the simulation program, and the tool is drawn according to these parameters. Similarly, by setting the main parameters of the workpiece, these parameters are also passed to the simulation program, and the workpiece is drawn according to these parameters. In this paper, the main databases are: tool database, workpiece database, fixture database and assembly database. A database is the total container for data and its related objects. A table is a container for storing raw data. In order to reduce the memory footprint, a loop is used to read in a field—check and explain the field—record the field, and then read in a word—check and explain the field—record the field until the end of the NC code. In order to read in the NC code, a character array array with 10 elements is defined, which is used to temporarily store each character of a field of the NC code. As the user enters data, the table stores it in logical combinations of similar data. Tables organize information in rows and columns, where rows are called records and columns are called fields. The same column of data in the table has similar information, and the column entries of these data are fields.

When reading a file, first read a character from the NC code file, and then process it differently according to the character (this character can be letters, numbers, decimal points, spaces, line breaks, semicolons, Chinese characters, file terminators, etc.), for syntactic, semantic and error analysis. The first character in the string array gives the meaning of the character in the NC code, and the substring composed of the following characters gives the numerical information of the character. Each field is identified by a specific data type (e.g. text, number, date, etc.) and has a specific length. Each field also has a name that indicates its category of information. Rows of data in a table are records. Each piece of information is conceived as a separate entity that can be accessed or ordered as needed. The structure diagram of the tool definition in the database is shown in Fig. 4:

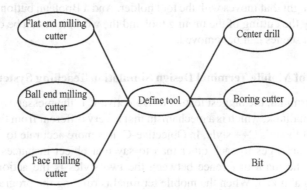

Fig. 4. Structure diagram of tool definition

As can be seen from Fig. 4, the structure diagram of the tool definition in the database includes six parts: flat end milling cutter, ball end milling cutter, face milling cutter, center drill, boring cutter and drill bit. There are three main purposes for checking and interpreting NC codes: checking the integrity and correctness of the entire field read and the correctness of each character in the field. If there is an error, even if an error message is given, the user is prompted to modify and terminate the program. Separates a substring of the first character of a string and subsequent numeric characters and converts the substring to a numeric value. The data management part of the tool library includes the input interface of the basic parameters of the tool and the data interface of the auxiliary parameters of the tool. According to the needs of the system, seven data tables are established. The base table is "Define Tool", which has six fields, namely: Tool No., Tool Name, Diameter, Tool Type, Corner Radius, and Tool Radius Form. These six fields represent the properties of a tool. Necessary information. The remaining six tables are face milling cutters, flat end milling cutters, ball end milling cutters, bell cutters, drill bits and center drills. The obtained data information is given to the tool motion variables in the program and subdivided to meet the real-time nature of dynamic simulation. The task of setting is to set the pixel format by calling the interface function of the CGL class, create and activate the rendering description RC (Rendering Content), and then set the position and size of the viewport, and set the shape of the viewfinder when creating a 3D view client area., size, position, and set the depth buffer, the clear value of the color buffer and the polygon drawing mode, etc., and finally set the material and lighting model. From this information, the geometric parameters of each tool in the tool library can be known. The operator of the CNC machine tool selects the appropriate tool from the tool library to process the workpiece according to the geometric shape and material properties of the workpiece to be processed.

When simulating machining, it is necessary to define, select or modify various parameters such as the shape, size, material and so on of the blank to be machined. Therefore, a database about the workpiece must be established. Here you can define a new blank, or select a previously defined blank, or modify the blank parameters defined in the past, so as to prepare for the subsequent simulation processing. In the turning simulation system, two light sources are used in the scene, one is a chandelier in a similar room, and the other is a work light that moves with the tool holder. And a Boolean button is used. After pressing it, only the cutting of the turning tool and the workpiece can be displayed, and other redundant scenes can be removed.

3.3 Function of Mobile Terminal Design Simulation Teaching System

In the mobile terminal, the biggest feature of Objective-C is the message delivery model derived from Smalltalk, which is a mechanism that is very different from the mainstream delivery style of today's C++ style. In Objective-C, it is more accurate to say that object instances pass messages to each other than to say that object instances call methods to each other. The main difference between the two styles is the action of calling a method/passing a model. When the mobile terminal is running the program, the classification code can access all the member variables, including the private class member variables of course. If methods in two categories have the same name, it is unpredictable at runtime which method is called.

Maintainability requires that the software system must have sufficient elasticity and scalability, so that the software system can cope with changes in customer needs for a long period of time, while maintaining the invariance of its overall structure, without the need for redesign and For development, it is only necessary to add a functional component in an appropriate place or replace an original functional module. In simulation technology, modeling is the key of simulation technology. It can be said that geometric simulation technology develops with the development of modeling technology. The description method of shape information by computer is referred to as modeling or modeling technology. The modeling technology is mainly composed of two parts: shape expression and shape operation. The task of shape expression is to simulate the structure of the shape with a data structure. This data structure describing the shape is called a model. Described in software engineering terms is to achieve "high cohesion, low coupling". For the numerical control teaching system, with the expansion of teaching breadth, more numerical control systems will be learned further, so how can many different types of numerical control systems be easily derived from one design structure, and how can one easily change a numerical control system? It is particularly important to reflect the shape characteristics of various CNC machine tools in order to reflect different types of machine tools and CNC systems.

The task of shape operation is to realize the operation of model generation, modification, synthesis, analysis, calculation, display, etc., so as to complete the modeling tasks in the design process. Solid modeling technology decomposes an object into a limited set of three-dimensional elements and a set of operations that can be applied to this set of elements. Depending on the different three-dimensional elements and their operations, solid modeling techniques can be divided into many types. In the design, it is more appropriate to design the part of the man-machine operation interface reflecting different types of CNC systems separately as a replaceable independent module. At the same time, as the process simulation part to deal with the corresponding type of CNC system processing, its core is the G/M code interpreter [9, 10]. Since different types of CNC systems use many common G/M codes, together with some exclusive special codes for special processing, they constitute all the processing codes of this type of CNC system. In addition, the solid modeling technology provides an accurate, complete and unambiguous description method for 3D solids, surfaces and curves. The direct Boolean operation algorithm based on the solid modeling technology is used for the geometric simulation of the machining process, and the cutting geometric information can be accurately obtained., not only for interference verification, but also for 3D dynamic simulation of multi-axis machining material removal process.

The task of the machining environment simulation module is to calculate and display the cutting condition of the workpiece in real time. Allows users to intuitively see the processing effect of each step. It directly accepts the tool path information sent from the machine state part, and then dynamically changes the corresponding position of the tool graphics in the graphics environment according to the information. The image space modeling method is to use the Z-Buffer idea similar to graphic blanking, to discretize the workpiece and the tool into a Z-Buffer structure according to the pixels of the screen, and the cutting process is simplified to a one-dimensional Boolean operation along the line of sight. This algorithm has a small amount of calculation, good real-time performance, and a better display effect in the simulation. Set (A_1, B_1) as the starting point of the tool movement and (A_2, B_2) as the end point of the tool movement, then the calculation formula of the unit vector of the movement direction of the tool center is:

$$A_1 = \frac{(A_2 - B_1)}{\left[(A_2 - B_1)^2 + (B_2 - B_1)^2\right]} \tag{1}$$

$$B_1 = \frac{(B_2 - B_1)}{\left[(A_2 - B_1)^2 + (B_2 - B_1)^2\right]} \tag{2}$$

Since the tool radius vector refers to the vector that is always perpendicular to the programmed path during the machining process, and whose size is equal to the tool radius value, the direction points to the tool center, which is represented by R_m. On the basis of formula (1) and formula (2), the expression formula of tool radius vector is obtained:

$$R_m = \frac{|m - n|}{A_1 - B_1} \tag{3}$$

In formula (3), m represents the transition vector at any point on the arc, and n represents the motion vector at any point on the arc. When the viewpoint is determined, the data structure of the algorithm is also determined. If you want to change the viewpoint to observe from another direction, you need to recalculate the data. The computer screen is used as the reference surface, so it will completely depend on the view. At the same time, it is judged whether the tool collides with the machine tool and whether it is in the state of cutting the workpiece, so as to decide whether to feedback error information to the machine interface, whether to call the real-time cutting calculation and display module to change the shape of the workpiece, and to convert the final The tool and workpiece status are displayed.

The discrete vector intersection method is mainly used in the estimation of machining errors. Error detection is accomplished by calculating the distance between the discrete point vector and the tool swept surface. The "point-vector" technology developed by Chappel has laid the foundation of this method. This method approximates the surface by selecting some points on the surface, selecting the normal vector direction of the point as the direction of the point vector, and extending the vector until it intersects with the blank body of the part or other surfaces. By calculating the intersection of the tool swept surface and the point vector, and calculating the distance between the starting point and the intersection of the point vector, the cutting process of the tool is simulated.

Because in the process of software design and development, it is considered that a part of the staff is transplanting the tasks and motion control parts of the real CNC system to the Windows platform, and it exists in the form of a library file. A segment is a part or whole of a graph. In the raster system, the bitmap segment (icon) can be obtained directly from the screen, and the acquired image segment is stored as a bitmap file and block on the disk or in the buffer, and then another image segment display operation is called., place them at the specified position of the screen, and by changing different output screen positions, the continuous movement of the segment graphics is generated, and the segment transformation animation is also generated. So compared with the processing simulation part developed by oneself, the transplanted library undoubtedly has more comprehensive, perfect and authoritative characteristics. In the simulation software, if this porting library file can be fully utilized, the simulation similarity will be stronger for the simulation software. Therefore, two interfaces are designed here, which can be switched between two different process processing modules as needed.

4   System Test

4.1   Test Preparation

This experiment chooses OpenGL as the main operating software. OpenGL is an open three-dimensional graphics software package, which is independent of the window system and operating system. The applications developed based on it can be easily transplanted among various platforms. In fact, in the CNC system, the conversion from NC code to tool path point is completed by continuously calling the NC code interpreter. OpenGL can be closely interfaced with Visual C++, which is convenient to realize the calculation and graphics algorithm of the manipulator, and can ensure the correctness and reliability of the algorithm. At the same time, OpenGL is easy to use and has high efficiency. In order to establish a realistic machining process simulation environment and improve the NC code processing process, a brand-new NC code interpreter is not directly implemented in the simulation system, but is further encapsulated on the basis of the existing SAItest version interpreter to make it Suitable for the current simulation environment.

4.2   Test Results

The CNC machining simulation teaching system based on virtual reality and the ARM-based CNC machining simulation teaching system are selected to compare and test with the CNC machining simulation teaching system designed this time. The request success rates of the three systems were tested under different system requests, and the experimental results are shown in Tables 1, 2 and 3:

Table 1. Number of requests 1000 system success rate (%)

Number of experiments	Numerical control machining simulation teaching system based on virtual reality	Simulation teaching system of CNC machining based on PC platform	The CNC machining simulation teaching system designed this time
1	86.915	89.645	95.001
2	87.104	85.117	93.678
3	88.699	86.302	92.515
4	86.815	87.944	93.008
5	87.649	88.964	92.548
6	86.774	889.311	94.316

(*continued*)

Table 1. (*continued*)

Number of experiments	Numerical control machining simulation teaching system based on virtual reality	Simulation teaching system of CNC machining based on PC platform	The CNC machining simulation teaching system designed this time
7	88.215	87.442	95.156
8	89.361	86.130	95.008
9	86.772	87.555	96.334
10	89.616	88.612	94.125
11	87.313	89.111	93.701
12	86.955	87.553	93.889
13	86.201	88.206	92.116
14	85.477	87.134	96.002
15	86.344	86.393	97.441

Table 2. The success rate of the system with 3000 requests (%)

Number of experiments	Numerical control machining simulation teaching system based on virtual reality	Simulation teaching system of CNC machining based on PC platform	The CNC machining simulation teaching system designed this time
1	65.312	65.205	73.644
2	63.014	66.449	74.812
3	65.802	65.813	76.915
4	64.812	64.919	77.388
5	59.133	63.485	76.512
6	61.225	63.145	78.454
7	62.948	66.312	75.911
8	63.462	62.588	76.021
9	64.815	64.051	77.535
10	63.775	63.974	78.446
11	65.004	64.802	75.904
12	63.815	63.114	74.119
13	62.901	64.819	73.464
14	64.228	63.550	75.008
15	63.455	64.122	76.499

Table 3. The number of requests 5000 system success rate (%)

Number of experiments	Numerical control machining simulation teaching system based on virtual reality	Simulation teaching system of CNC machining based on PC platform	The CNC machining simulation teaching system designed this time
1	43.615	44.815	62.145
2	42.779	43.712	59.887
3	41.080	46.330	63.001
4	43.664	45.877	58.132
5	44.812	44.212	59.877
6	45.903	45.913	65.447
7	46.774	46.822	66.115
8	45.812	47.114	65.923
9	46.337	45.993	66.748
10	45.208	44.877	65.264
11	46.977	43.606	64.717
12	46.288	44.555	65.808
13	45.317	45.182	64.144
14	46.211	46.907	65.312
15	45.667	45.316	66.715

It can be seen from Table 1 that the success rates of the CNC machining simulation teaching system designed this time and the other two CNC machining simulation teaching systems are: 94.323%, 87.347%, and 87.628% respectively; The success rates of the teaching system and the other two CNC machining simulation teaching systems are: 76.042%, 63.580%, and 64.423% respectively; from Table 3, it can be seen that the designed CNC machining simulation teaching system and the other two CNC machining simulation teaching systems have the same success rate. The success rates are: 63.949%, 45.096%, 45.415%.

5 Concluding Remarks

In order to improve the request success rate of the simulation teaching system, a numerical control machining simulation teaching system based on mobile terminal is designed. By combining power on reset and key reset, the convenience of system operation is improved; Build the cutter database, design the function of the mobile terminal design simulation teaching system, and complete the design of the simulation teaching system. The experimental results show that the request success rate of the system designed in this paper is significantly higher than that of the traditional system, which fully verifies its application value.

References

1. Zhou, M., Yin, J., Zhu, H., et al.: Error analysis and five axis NC machining method of precision casting blank of knotter bracket. Trans. Chin. Soc. Agric. Mach. **51**(12), 417–424 (2020)
2. Xiao, W., Cao, X., Zhao, G., et al.: Digital twin system for CNC machining. Aeronaut. Manuf. Technol. **63**(23), 46–55 (2020)
3. Wang, C., Shen, D., Bai, C., et al.: Simulation optimization on five-axis NC machining of shunt impeller. Tool Eng. **54**(11): 70–74 (2020)
4. Zuo, B., Huang, H., Tao, Y.: Innovative design of feed drive system in CNC machine tools based on TRIZ. Mach. Des. Res. **37**(1), 139–143, 155 (2021)
5. Su, C., Meng, X., Cui, J.: Development of remote operation monitoring and information management system for CNC machine tools. Comput. Eng. Des. **42**(12), 3576–3587 (2021)
6. Yang, Z.: Discussion on gauge parts in NC machining. Die Mould Ind. **46**(1), 62–65 (2020)
7. Zhang, S., Sun, J., Jiang, C.: Numerical control machining simulation of automobile crankshaft based on CAD/CAM. Manuf. Autom. **42**(2), 153–156 (2020)
8. Luo, Y., Hu, X.: Design of CNC machine tool automatic production system based on cloud manufacturing. Mod. Electron. Tech. **43**(22), 181–183, 186 (2020)
9. Sun, R.: Optimum design of NC machining process for plastic mold cavity. China Synth. Resin Plast. **37**(2), 54–58 (2020)
10. Li, L., Lian, Y.: Simulation of Information security transmission of large data mobile terminal network. Comput. Simul. **35**(6), 188–192 (2018)

Evaluation Method of Mobile Online Course Effect of Gender Psychological Education for Primary School Students

Yi Zhang[1][✉] and Tengfei Ma[2]

[1] Xi'an Siyuan University, Xi'an 710038, China
zytq13484541354@163.com

[2] College of Civil Engineering and Planning, Liupanshui Normal University, Liupanshui 55300, China

Abstract. The mobile Internet gender psychological education curriculum for primary school students takes positive development as the core idea, which has high value for improving the quality of primary school students' mental health education. The teaching evaluation of this course can improve the efficiency and effect of teaching, and provide reference and help for future teaching. However, the current effect evaluation methods have the problem of single evaluation. Therefore, this paper studies the evaluation method of mobile online course effect of gender psychological education for primary school students. This paper analyzes the content of gender psychology education for primary school students, and establishes a teaching model of gender psychology education on mobile networks by using the interactive relationship between learners and learning resources. Establish an evaluation index system according to the course content and teaching interaction mode. Based on the theory of fuzzy mathematics, the evaluation model of the effect of gender psychology education curriculum for primary school students is established. The corresponding weight values are obtained by using the fuzzy consistent judgment matrix, and the numerical results of comprehensive evaluation are obtained. The experimental results show that the designed course effect evaluation method can improve the accuracy of evaluation and the evaluation results are more convincing.

Keywords: Primary school students · Gender psychological education · Mobile online courses · Effect evaluation · Evaluation method · Mental health

1 Introduction

Positive mental health education inherits and integrates the theories and ideas of positive psychology, positive mental health, positive psychotherapy and positive education. It takes positive development as the core concept, human goodness as the value orientation, positive content, methods and means as the educational content, and positive development and cultivation of positive psychological quality, prevention and treatment

W. Fu and L. Yun (Eds.): ADHIP 2022, LNICST 469, pp. 647–661, 2023.
https://doi.org/10.1007/978-3-031-28867-8_47

of various psychological problems, and promoting the overall and harmonious development of individual body and mind as the educational goal. With the growing awareness of the importance of children's sex education and their parents' acceptance of sex education in advance. Parents believe that they have the responsibility to teach their children the methods of self-protection, but considering their lack of systematic professional knowledge, they are unable to comprehensively and scientifically implement sex education for their children, so schools need to provide students with comprehensive and scientific sex education guidance. As an integral part of the ideological and political education system, the quality of mental health education is obviously related to the overall quality of Ideological and political education. Complying with the educational requirements of the times and enriching the research results of gender mental health education evaluation of primary school students has both theoretical and practical value for the improvement of the quality of primary school students' mental health education and the development of Ideological and Political Education [1]. Teachers should use positive, sunny and positive contents and ways to cultivate and stimulate students' individual positive psychological quality, help students learn to experience and create happiness, experience and share happiness, maximize students' own positive potential, cultivate students' positive quality, promote students' self psychological self adjustment ability, and enrich and develop students' psychological potential and psychological quality in the best state. Schools bear the responsibility of teaching and educating people, and can use students' time in school to carry out systematic, scientific and comprehensive sex education for students [2]. However, many schools and teachers are often troubled by a series of problems, such as the lack of scientifically compiled sex education textbooks, the lack of paradigm in Teachers' teaching methods, and how to arrange classroom activities to be both scientific and clean. As a new type of education, online teaching is facing both opportunities and unprecedented problems and challenges. "Online teaching", "online course" and "online learning platform" have become hot topics in the development of education in recent years. The mobile online course of gender psychological education for primary school students makes full use of network information resources and modern information technology to make the teaching and learning methods of teachers and students more diversified. The research results of primary school students' mental health education evaluation, in turn, can promote and monitor educational practice. The core function of educational evaluation is to lead practical and specific practice and grasp the value direction of psychological education practice through objective scientific analysis. Exploring appropriate online teaching methods can not only enhance the efficiency and effectiveness of teachers' online teaching, but also enable students to effectively plan and manage their own effective time, and effectively integrate the corresponding teaching resources and existing theoretical achievements, so as to provide reference and help for future online teaching.

Related Work Introduction

Literature [3] proposes a method for evaluating the effect of mental health education courses for freshmen in Colleges and universities. This paper selects the mental health course of Freshmen in a university in Jiangsu Province as the research object. The Chinese college students' mental health screening scale was used to conduct a questionnaire survey before, during and after the mental health course. The development track of

Freshmen's psychological problems is divided into two types, which are "rising first and then falling" and "slowly falling", to achieve curriculum effect evaluation. Based on the authenticity evaluation theory, literature [4] constructs a curriculum teaching effect evaluation system from the five dimensions of evaluation objectives, real tasks, evaluation standards, evaluation gauges and evaluation feedback. In terms of evaluation objectives, the system takes "ability and accomplishment" as the benchmark; In terms of real tasks, the system is guided by "real situations"; In terms of evaluation standards, the system is based on the principle of "emphasizing participation, depth and attitude"; In the evaluation gauge, the system takes "transparent rules, clear grades and clear examples" as the standard; In terms of evaluation feedback, the system aims at "diligent thinking and dynamic adjustment" to achieve teaching effect evaluation.

However, the teaching effect evaluation of the above methods is not accurate enough, and the evaluation results are not convincing. This paper proposes a method to evaluate the effect of mobile online course of gender psychological education for primary school students. The research structure of this paper is as follows: firstly, the content of gender psychological education courses for primary school students is analyzed, and the teaching mode of gender psychological education on mobile networks is established by using the interactive relationship between learners and learning resources. Secondly, according to the course content and teaching interaction mode, the evaluation index system is established. Thirdly, based on the theory of fuzzy mathematics, the evaluation model of gender psychology education curriculum effect for primary school students is established. Finally, the corresponding weight values are obtained by using the fuzzy consistent judgment matrix, and the numerical results of comprehensive evaluation are obtained. The experimental results show that this method can improve the quality of primary school students' mental health education. Only by realizing a more reasonable distribution of the proportion of practical elements and making the balance of practical scale more accurate, can we better guide the subject and object of primary school students' mental health education to enjoy the theoretical and practical results of psychological education, and further enhance the effectiveness of psychological education.

2 Evaluation Method of Mobile Online Course Effect of Gender Psychological Education for Primary School Students

2.1 Curriculum Content Analysis of Gender Psychological Education for Primary School Students

According to the entities associated with curriculum and teaching, namely students, teachers, educational administrators, curriculum plans, curriculum standards, teaching materials, teaching plans, teaching processes and relevant institutions, the evaluation objects can be divided into: curriculum and teaching design, curriculum implementation and teaching activities, students' academic performance, curriculum and teaching system, and curriculum and teaching evaluation. The evaluation objects of gender education curriculum for primary school students involve students, teachers and the curriculum itself. The content of courses of different ages not only extends vertically, but also connects with the learning content of different sections of the course content horizontally

[5]. When implementing the sex education curriculum, teachers can also create the curriculum according to the actual situation and the needs of students, and do not have to follow the original curriculum plan. The evaluation of students' development examines the knowledge and skills, processes and methods, emotional attitudes and values of gender education obtained by students. The curriculum content is not invariable and is always developing. The existing curriculum structure and content only provide some support and reference [6]. The evaluation of teachers' development mainly involves teachers' professional knowledge on gender education, teaching activity design ability, implementation ability, reflection and improvement ability, etc. The evaluation of curriculum development includes whether the orientation of curriculum objectives is scientific and appropriate, whether the content organization is scientific and humanized, whether the process design pays attention to experience and perception, and whether various educational resources are effectively used. Traditional education is mostly used to the effect evaluation, which focuses on the achievement of the goal, so it can reflect the achievement effect of education to a certain extent.

At present, the school-based curriculum of sex education in primary schools is mainly organized and managed by and another full-time psychological teacher. In terms of curriculum construction, organization managers mainly grasp the orientation and development direction of the curriculum; At the level of curriculum development, set curriculum objectives, arrange curriculum content and Curriculum textbooks; In the process of curriculum implementation, organize teachers to discuss and help teachers improve teaching design. Do a good job in the management and supervision of curriculum implementation. According to the content of evaluation, whether students, teachers or the curriculum itself, primary schools pay more attention to development rather than the final clear results. The essence of evaluation is concerned with students' experience and growth, the progress and promotion space of teachers in professionalism, and the expectation that gender education as a developing curriculum can be continuously revised and improved [7]. The proposal of process evaluation is a great progress in the history of educational evaluation, which means that in the process of College Students' mental health education evaluation, we should pay more attention to the growth and development of educational

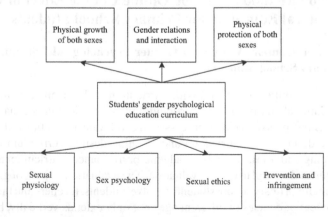

Fig. 1. Curriculum content of gender psychological education for primary school students

objects in the whole process of primary school students' gender mental health education, carry out educational evaluation according to the changes of students in the process of growth, timely adjust educational methods and improve educational mechanism. The gender psychological education course for primary school students is mainly divided into seven main contents, as shown in Fig. 1.

Among them, each content module is divided into different content modules according to the lower grade, middle grade and higher grade. In each content section, each grade section will accept two different content modules. The content of the lower grade curriculum requires children to know different animals and plants, including where they come from, and then lead to the combination of their parents' sperm and eggs; In the middle grade, through the theme of birthday, we will lead to the process of breeding and growing children in the mother's body, understand the hardships of parents raising children, and experience the hard won life; In the senior grade, it will be further expanded to let children understand the different growth stages and different characteristics of people's life. The development of school-based curriculum of gender psychological education for primary school students should not only avoid the influence of traditional culture and the resistance of social ideas, but also consider their own ideas and skills. Due to the particularity of gender education, the selection of content requires teachers to use their rich practical experience and theoretical experience to grasp the appropriate degree.

2.2 Establish a Mobile Online Teaching Model for Gender Psychological Education

Online teaching is a teaching activity carried out by using modern communication technology. The connotation of online teaching includes two aspects: first, it is a teaching activity. Second, online teaching is very different from the usual traditional teaching activities. Its educational approach is carried out through online education software. This kind of teaching activity reflects the main status of the corresponding learners, so as to explore the main teaching activities of the corresponding learners. Network technology supports online teaching. Online course interaction is the interaction between online learners and all online teaching resources. It includes the interaction between learners and learning auxiliary facilities (including learning materials, learning resources, learning hardware facilities, etc.), learners and learning organizations (learners and teachers, learners and learners). Mobile online teaching of gender psychological education should provide support for communication and interaction to realize the generation and creation of knowledge. In the learning process of mobile online teaching of gender psychological education in primary school, students not only passively accept the old knowledge, but also create new knowledge in the existing cognitive experience of students, and carry out the integration, innovation and application of knowledge independently. On the one hand, the huge network information database can provide teaching resources for teachers and students. On the other hand, network technology has the characteristics of strong adaptability, many exchanges, personalization and convenient learning records in teaching, so it is more suitable for the corresponding traditional teaching. And it has a series of advantages such as open cooperation and convenience [8]. The development of information technology enriches the forms of online teaching resources. This paper holds that online course interaction should include two aspects: online course itself and auxiliary teaching means. Each aspect specifically includes operation interaction, information interaction and concept interaction. The interactive mode of gender psychological education mobile online teaching established in this paper is shown in Fig. 2.

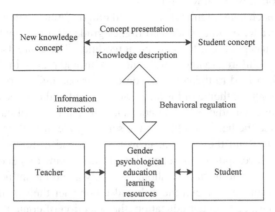

Fig. 2. Interactive mode of mobile online teaching

In this teaching relationship, the information interaction of each type of learning content is based on the operation interaction of each type of learning content, and the corresponding learning content has become a concept accepted by learners after information interaction. Course resources improve the effect of online teaching from two aspects. Schools must strengthen the management of online course content. The administrative department of education should clearly explain the learning connotation of online teaching and pay attention to students' physical and mental health, so as to help students develop good learning habits and healthy living habits. In fact, the process of transforming the concept in students' mind into the expected concept in teaching is the process of transforming the concept in students' mind, and then forming the expected effect. Teachers and administrators should choose diverse learning materials in the process of pushing course content, and try their best to meet the development and innovation of learning resources. During the communication in the Q & a interactive area, teachers and administrators should also make full use of the difficult problems raised by students, re integrate some effective problems and knowledge points into new curriculum learning resources and publish them in the curriculum, deepen students' impression and develop the potential value realized in students' learning process. The online teaching course guidance document has released the operability, effectiveness and general implementation plan. For example, give teachers full professional autonomy and encourage districts, counties, schools and teachers to conduct online teaching in a way suitable for students according to the differences between urban and rural areas, schools, students, grades and teachers. In the teaching interaction model, concept interaction is the top layer, and its presentation effect can be used as the adjustment basis of the whole teaching interaction process. In turn, in the whole teaching interaction process, teachers and students can improve the final concept interaction effect through reflection on the whole interaction behavior. In this way, it can better realize the integration and innovation of difficult problems in the learning process and curriculum push, promote students to give full play to their subjective initiative, and improve students' enthusiasm to participate in gender psychological education and mobile online learning.

2.3 Select the Evaluation Indicators of Primary School Students' Gender Psychological Education Curriculum

The standard of index evaluation is the evaluation standard of primary school students' gender mental health education. It is a yardstick that directly reflects the results of mental health education. The determination of evaluation criteria is one of the core elements of the construction of the evaluation system of gender mental health education for primary school students in the new era. At the same time, it is also an important basis for judging the quantity and quality of mental health education. The establishment of the index system needs to comply with the systematic principle, which means that the indicators of the evaluation index system of primary school students' gender mental health education can comprehensively reflect the evaluation object, summarize the characteristics of all aspects of primary school students' gender mental health education, and reproduce and reflect the objectives of school mental health education comprehensively and without omission. Due to the particularity and complexity of gender mental health education, a single evaluation standard can not meet the complex and huge evaluation system, and personalized evaluation standards should also be used and adopted.

The systematic principle of the index system requires that the indicators of the evaluation index system of primary school students' gender mental health education can evaluate the joint effect of all aspects of school mental health education, paying attention to both the overall coordination and the importance of individual. The change of teaching form is the key content of the evaluation of primary school students' gender mental health education in the new era. Online teaching is not only an innovative means of mental health teaching in the new era, but also an important and difficult part of the construction of evaluation indicators of mental health education in the new era. The principle of mutual independence of the indicators in the index system means that the indicators at the same level in the evaluation index system of positive mental health education in primary and secondary schools must not overlap and have no causality. The principle of mutual independence of various indicators in the index system is conducive to avoid increasing the workload and repeated scoring of evaluation due to repeated and lengthy indicators, and ensure the feasibility of evaluation and the scientificity of evaluation results. As a part of school mental health activity curriculum, gender education is essentially "human" education. Curriculum design not only includes relevant knowledge and skills, but also pays more attention to the cultivation of students' humanistic quality and the development of students. Online courses do not have mandatory teaching tasks, but are more aimed at assisting offline teaching and promoting each other. To strengthen the indicator construction of online teaching, we need to focus on the learning indicators of online app and online psychology courses to realize the suspension of classes without suspension. The evaluation index system of primary school students' gender psychological education curriculum constructed in this paper is shown in Table 1.

With the progress of science and technology, there are various forms of online teaching of mental health, and online courses have been launched one after another. The

Table 1. Evaluation index system of gender psychological education curriculum for primary school students

Primary index	Secondary index	Primary index	Secondary index
Content of courses	Clear teaching objectives	Teaching method	Guided teaching
	The teaching content is novel		Teaching with practice
	The teaching focus is prominent		High-quality teaching philosophy
	The teaching content is rich		Innovation in teaching methods
	The content is easy to understand		Flexible teaching methods
Teaching attitude	Full of teaching expression	Teaching efficiency	The realization of teaching objectives
	Serious teaching attitude		The classroom teaching atmosphere is active
	Enhance positive emotional experiences		Intelligent learning path
	Shape a positive mental health personality		Effective study and supervision
	Build a harmonious psychological environment		Parent feedback

combination of offline teaching and online teaching can promote the quality of gender mental health teaching for primary school students. In the specific implementation process, educators should use a positive attitude to understand and analyze students' psychological phenomena, guide students to actively understand and solve the social and psychological problems they face, promote students' positive experience, construct a positive environmental support system, and stimulate students' internal positive psychological potential and quality. Primary school teachers focus on students' inner world in their teaching activities, hoping to explore themselves with students through any teaching interaction process, verify the questions in the growth process, and meet the needs of students' life growth. All these are aimed at the comprehensive and healthy development of students' personality.

Through the teaching of gender psychological education mobile online course, students' psychological immunity and resistance can be greatly improved, and students' physical and mental harmony, health and sustainable development can be promoted, so as to create a beautiful spirit full of optimism, hope and positive progress matching with a harmonious society. It is the goal of positive mental health education in primary and secondary schools. Primary school teachers insist that students' development should be diverse, personalized and have their own characteristics. They hope that through their own efforts, school education can meet the needs of each student, and expect students' personality development: build the school into a learning paradise for students' happy growth.

2.4 Design Mobile Online Course Effect Evaluation Model

The application of qualitative standards in the evaluation system can make the provisions of evaluation standards macroscopically and guide the general direction of the improvement of mental health education activities. The quantitative standard can be a single value or a range. At the same time, it also has the division of grades, but it must have a quantitative number. According to the established evaluation index system of primary school students' gender psychological education curriculum, this paper uses fuzzy mathematics theory to establish the effect evaluation model of primary school students' gender psychological education curriculum. It can be used in the rigorous and scientific evaluation of primary school students' psychological health. Both qualitative standards and quantitative standards have their advantages and disadvantages. The organic combination of the two methods can be realized in the process of educational evaluation [9]. The evaluation element set is a set composed of factors affecting the evaluation objectives, expressed as:

$$R = \{R_1, R_2, R_3, R_4\} \tag{1}$$

In formula (1), R represents the set of evaluation objectives; R_1, R_2, R_3, R_4 corresponds to teaching content, teaching method, teaching attitude and teaching effect. Each primary indicator is divided into several secondary indicators that affect the superior indicators. Taking the teaching content as an example, the set of secondary index evaluation objectives can be expressed as:

$$R_1 = \{S_1, S_2, S_3, S_4, S_5\} \tag{2}$$

In formula (2), S_1, S_2, S_3, S_4, S_5 corresponds to clear teaching objectives, novel teaching content, prominent teaching focus, rich teaching content and easy to understand.

The weight of the evaluation index of the mobile online course effect evaluation system of gender psychological education for primary school students is an important part of the index system and a quantitative sign of the importance of an index in this evaluation index system. In order to reflect this different status and importance, we need to set the weight for each evaluation index, so as to meet the requirements of objectivity and comparability. In order to scientifically reflect the importance of evaluation elements, the weight set of evaluation elements is constructed accordingly. In order to avoid artificial subjectivity in the setting of weight value, the corresponding weight value is obtained by using fuzzy consistent judgment matrix. In the effect evaluation index system of mobile online course of gender psychological education for primary school students, the status and importance of various evaluation indexes in the whole evaluation index system are different. According to the constructed secondary factor set, leaders and experts are invited to use the quantitative scale of 0.1–0.9 to make fuzzy description judgment on the factor set and construct the fuzzy consistency judgment matrix. The judgment matrix can be expressed as:

$$
P = \begin{bmatrix} p_{11} & p_{12} & p_{13} & p_{14} \\ p_{21} & p_{22} & p_{23} & p_{24} \\ p_{31} & p_{32} & p_{33} & p_{34} \\ p_{41} & p_{42} & p_{43} & p_{44} \end{bmatrix} \tag{3}
$$

In formula (3), P represents the judgment matrix; p represents the subordinate relationship of the importance degree of factor S_x compared with the fuzzy relationship of S_y. The weight value is normalized and non negative, and the calculation formula is as follows:

$$
\vartheta_x = \frac{1}{u} - \frac{1}{2\lambda} + \frac{1}{u\lambda} \sum_{y=1}^{u} p \tag{4}
$$

In formula (4), ϑ_x represents the weight value of factor S_x; u represents the number of evaluation elements; x, y represents the serial number of the evaluation element; λ indicates the adjustment parameter.

The qualitative standard describes the significance of the evaluation standard through words, and the quantitative standard uses specific values to realize the scientific rationality of the evaluation standard. While ensuring the scientific and rigorous evaluation standard of primary school students' gender mental health education, it gives the significance of the evaluation standard data, making the evaluation standard closer to the purpose and significance of mental health education in the new era in the output process. The larger the value of adjustment parameter λ, the more decision-makers pay attention to the difference of element importance. The calculation formula is as follows:

$$
\lambda = \frac{u - 1}{2} \tag{5}
$$

Fuzzy comprehensive evaluation considers the influencing factors of the evaluation objectives comprehensively, and can be shown in the results with the characteristics most in line with the actual situation. From the fuzzy vector element set of secondary evaluation, the judgment matrix of primary evaluation can be obtained, and finally the numerical results of comprehensive evaluation can be obtained. So far, the design of the effect evaluation method of the mobile online course of gender psychological education for primary school students has been completed.

3   Experimental Study

3.1   Experimental Preparation

In order to verify the application effect of the mobile online course effect evaluation method of gender psychological education for primary school students proposed in this paper, an experimental test is carried out. The experiment takes a primary school as the research object. The gender curriculum of the primary school is taught in classes, facing six grades of the whole school. Infiltrating gender mental health education into subject teaching can exert a subtle influence, get twice the result with half the effort, face the whole and provide positive guidance for the knowledge and cultural education and mental health education of primary school students. The whole school students were randomly divided into five groups. The number of the first group was 100, and the other four groups increased by 100 in turn. Using the relevant data of gender psychological education mobile online courses collected on the online teaching platform, establish the index system and calculate the weight according to the method in this paper, as shown in Table 2.

Table 2. Weight of gender psychological education curriculum evaluation indicators for primary school students

Primary index	Weight	Secondary index	Weight
Content of courses	0.2843	Clear teaching objectives	0.0432
		The teaching content is novel	0.0501
		The teaching focus is prominent	0.0697
		The teaching content is rich	0.0428
		The content is easy to understand	0.0439
Teaching method	0.3061	Guided teaching	0.0682
		Teaching with practice	0.0603
		High-quality teaching philosophy	0.0869
		Innovation in teaching methods	0.0792
		Flexible teaching methods	0.0465
Teaching attitude	0.2534	Full of teaching expression	0.0498
		Serious teaching attitude	0.0535
		Enhance positive emotional experiences	0.0423
		Shape a positive mental health personality	0.0576
		Build a harmonious psychological environment	0.0358
Teaching efficiency	0.1562	The realization of teaching objectives	0.0272
		The classroom teaching atmosphere is active	0.0331
		Intelligent learning path	0.0357
		Effective study and supervision	0.0249
		Parent feedback	0.0493

The above evaluation index system and weights are used to obtain the effect evaluation results of mobile online courses of primary school students.

3.2 Results and Analysis

This experiment uses the evaluation accuracy index to measure the effect of the effect evaluation method of gender psychological education.The effect evaluation method of mobile online course based on hierarchical analysis (AHP) and BP neural network was selected as the control scheme to verify the effectiveness of this method.The results of evaluation accuracy of different student size tests are shown in Table 3, Table 4, Table 5, Table 6 and Table 7.

Table 3. Assessment accuracy for the 100 students

Test times	The effect evaluation method of gender psychological education in elementary paper	Evaluation method of mobile online curriculum based on AHP	Effect evaluation method of mobile online course based on BP neural network
1	0.9109	0.8845	0.8675
2	0.9248	0.8786	0.8446
3	0.9454	0.8864	0.8583
4	0.9361	0.8730	0.8760
5	0.9235	0.8852	0.8632
6	0.9426	0.8528	0.8555
7	0.9283	0.8616	0.8627
8	0.9144	0.8843	0.8618
9	0.9317	0.8775	0.8546
10	0.9252	0.8584	0.8522

In the test of 100 students, the accuracy of the primary school students was 0.9283, which improved 0.0540 and 0.0687 than the AHP-based and BP neural network-based evaluation methods.

In the test of 200 students, the accuracy of 0.8985 improved the accuracy of 0.0525 and 0.0712 over the AHP-based and BP-based neural network evaluation methods.

In the test of 300 students, the accuracy of the effectiveness of gender psychological education was 0.8570, with the accuracy of 0.0346 and 0.0565 over the AHP-based and BP neural network-based evaluation methods.

In the test of 400 students, the accuracy of the gender psychological education was 0.8291, which improved the accuracy of 0.0374 and 0.0549 over the AHP-based and BP neural network-based evaluation methods.

In the test with 500 students, the accuracy of the mobile online course effect evaluation method of gender psychological education for primary school students designed in this paper is 0.7969, which is 0.0414 and 0.0597 higher than that based on AHP and BP neural network. Therefore, through the comprehensive evaluation of the effect of mobile online course of gender psychological education for primary school students, the evaluation accuracy is improved. The design method of this paper retains the subjective and

Table 4. Assessment accuracy for the total number of 200 students

Test times	The effect evaluation method of gender psychological education in elementary paper	Evaluation method of mobile online curriculum based on AHP	Effect evaluation method of mobile online course based on BP neural network
1	0.8945	0.8346	0.8266
2	0.9088	0.8483	0.8195
3	0.8967	0.8365	0.8357
4	0.8834	0.8537	0.8434
5	0.8956	0.8424	0.8261
6	0.9013	0.8552	0.8325
7	0.9112	0.8318	0.8243
8	0.8945	0.8544	0.8212
9	0.9024	0.8481	0.8181
10	0.8967	0.8551	0.8254

Table 5. Assessment accuracy for the total number of 300 students

Test times	The effect evaluation method of gender psychological education in elementary paper	Evaluation method of mobile online curriculum based on AHP	Effect evaluation method of mobile online course based on BP neural network
1	0.8545	0.8167	0.8002
2	0.8687	0.8235	0.8178
3	0.8564	0.8354	0.8029
4	0.8431	0.8228	0.8056
5	0.8552	0.8286	0.7943
6	0.8525	0.8143	0.7985
7	0.8619	0.8212	0.7917
8	0.8546	0.8225	0.7844
9	0.8583	0.8261	0.8062
10	0.8652	0.8130	0.8033

objective consciousness of the evaluator in the evaluation results to the greatest extent, so as to make the evaluation results more accurate and persuasive.

Table 6. Assessment accuracy for the 400 students

Test times	The effect evaluation method of gender psychological education in elementary paper	Evaluation method of mobile online curriculum based on AHP	Effect evaluation method of mobile online course based on BP neural network
1	0.8344	0.8048	0.7847
2	0.8258	0.7886	0.7788
3	0.8163	0.7963	0.7896
4	0.8256	0.7835	0.7765
5	0.8385	0.7958	0.7624
6	0.8220	0.7822	0.7857
7	0.8434	0.8014	0.7702
8	0.8268	0.7947	0.7539
9	0.8362	0.7872	0.7663
10	0.8223	0.7823	0.7741

Table 7. Assessment accuracy for the total number of 500 students

Test times	The effect evaluation method of gender psychological education in elementary paper	Evaluation method of mobile online curriculum based on AHP	Effect evaluation method of mobile online course based on BP neural network
1	0.8048	0.7506	0.7464
2	0.8082	0.7655	0.7538
3	0.7960	0.7528	0.7326
4	0.7833	0.7536	0.7253
5	0.8056	0.7503	0.7475
6	0.8027	0.7612	0.7242
7	0.7915	0.7545	0.7311
8	0.7849	0.7484	0.7204
9	0.7973	0.7562	0.7487
10	0.7951	0.7621	0.7422

4 Conclusion

The key step of curriculum reform lies in curriculum implementation, and the perception at the practical level is the necessary condition for the formation of research. However, in order to make the effect of gender psychological education curriculum for primary school students more convincing, the more scientific research on the effectiveness of sex education curriculum development and implementation should be the better. This paper designs a mobile online course effect evaluation method of gender psychological education for primary school students. This method has high efficiency and feasibility, and can improve the accuracy of evaluation. Testing the evaluation system of primary school students' mental health education through scientific theory can improve the objectivity and authority of the evaluation conclusion, guide the practice of educational evaluation, and realize the scientific test of educational evaluation on the basis of conforming to the objective law of the development of educational evaluation.

References

1. Chen, W., Xiao, B., Liu, J., et al.: Validity and reliability of the possibility/cost of negative social events questionnaire among Chinese pupils. Chin. J. Clin. Psychol. **29**(5), 948–951 (2021)
2. Wang, X., Liu, Y., Lin, J., et al.: Analysis of aggression in primary and middle school students during COVID-19 pandemic and its influencing factors. J. Southwest Univ. (Nat. Sci. **43**(1), 12–21 (2021)
3. Zhang, J., Chen, J.: Effectiveness of the mental health education courses for freshmen in Jiangsu colleges. Chin. J. Sch. Health **42**(8), 1198–1200, 1205 (2021)
4. Guo, F.: Construction of evaluation system of ideological and political education effect of vocational education curriculum based on authenticity evaluation theory. Vocat. Tech. Educ. **43**(11), 62–68 (2022)
5. Zhang, J., Fu, M., Xin, Y., et al.: The development of creativity in senior primary school students: gender differences and the role of school support. Acta Psychol. Sin. **52**(9), 1057–1070 (2020)
6. Sun, Y., Wang, M.: Parental physical discipline and children's teacher-student relationships: the roles of children's self-control and gender. Chin. J. Clin. Psychol. **28**(1), 145–151 (2020)
7. Ma, Z., Su, Z.: Characteristics of longitudinal development of pupils' loneliness: a follow-up study. Chin. J. Sch. Health **41**(10), 1533–1535 (2020)
8. Chen, J., Li, D., Wang, J., et al.: Crowd behavior simulation and control optimization technology for large campus. Comput. Simul. **38**(7), 198–202, 409 (2021)
9. Jin, J., Yu, G.: The consideration on the setting of "mental health education" in the revision of compulsory education curriculum standards of morality and the rule of law. Curriculum Teach. Mater. Method **42**(1), 71–77 (2022)

Evaluation Method of Teaching Quality of Adolescent Health Physical Education Based on Mobile Education Technology

Yanfeng Wu[1](✉) and Xu Liu[2]

[1] School of Physical Education, Changchun University of Finance and Economics, Changchun 130122, China
wuyanfeng115@163.com
[2] School of Media and Communication, Changchun Humanities and Sciences College, Changchun 130117, China

Abstract. At present, when the teaching quality evaluation methods of adolescent health physical education courses encounter large-scale data, the evaluation speed is slow, In order to solve this problem, a teaching quality evaluation method of adolescent health physical education based on mobile education technology is proposed. Vector, evaluation of potential user's demand, evaluation of student's physical activity quality, vector of potential user's demand, evaluation of student's physical activity quality, estimation of potential user's demand, evaluation of student's physical activity quality, evaluation of potential user's demand, evaluation of student's physical activity quality. Experimental results: the average evaluation speed of the adolescent health physical education teaching quality evaluation method in this paper and the other two evaluation methods are 9.884 s, 16.184 s and 16.489 s respectively, which shows that the performance of the adolescent health physical education teaching quality evaluation method is more perfect after fully combining the mobile education technology.

Keywords: Mobile education · Teenagers · Health physical education curriculum · Teaching practice · Physical activity · Teaching quality evaluation

1 Introduction

Teaching quality evaluation is directly related to the quality of talent training [1, 2]. The traditional teaching quality evaluation in China mainly includes expert evaluation, peer evaluation and leadership evaluation. The teaching quality of adolescent physical education is a multidimensional and dynamic concept, which has the characteristics of society and the times. There are different evaluation standards for teaching quality according to the surrounding humanistic environment, facilities and equipment conditions of sports venues, teachers and teaching subjects. With the development of higher education in China, the main role of teenagers has become increasingly prominent. As

W. Fu and L. Yun (Eds.): ADHIP 2022, LNICST 469, pp. 662–675, 2023.
https://doi.org/10.1007/978-3-031-28867-8_48

a new evaluation method, teenagers' evaluation has attracted more and more attention. Therefore, the evaluation of physical education teaching quality cannot be judged from the traditional perspective, but from the perspective of development according to the needs of society [3, 4]. From the main research results, teaching quality evaluation plays a very important role in teaching theory research. Many scholars have done a lot of research work on teaching quality evaluation in Colleges and universities and achieved a series of research results. The teaching quality evaluation system should be the unity of adaptability, multidimensional and development. At present, most schools do not have a teaching quality evaluation form specifically for physical education courses. There may be physical education teaching quality evaluation form, but the satisfaction evaluation of teenagers is ignored, resulting in the evaluation results can not truly reflect the learning experience of teenagers. At the same time, due to the large amount of data involved in the evaluation, many research methods have a slow evaluation speed in the process of evaluation, which affects the efficiency of evaluation. Combined with the current situation of social development and the characteristics of teenagers' behavior, previous relevant research results, as well as the opinions and suggestions of experts, scholars and front-line teachers, this paper makes an all-round discussion on the connotation and standards of teaching quality. This study intends to evaluate the quality of physical education teaching from the aspect of teenagers' learning satisfaction. Studying the teaching quality from the aspect of teenagers' learning satisfaction can not only enable teenagers to express the physical education learning experience of the whole learning process, reflect the degree of fit between learning expectations and learning results, but also explore the loyalty of teenagers' physical education learning. The purpose of reconstructing the scientific evaluation system of teaching quality of youth physical education curriculum is to put forward the evaluation standards of physical education teaching suitable for social development from the perspectives of objectivity, comprehensiveness, reality, simplicity and operability. Overcome the problems existing in the previous expert evaluation, such as the obvious improvement of teachers' teaching quality when experts listen to classes, and the non objective evaluation factors such as human evaluation caused by peer evaluation. This study has certain reference significance for promoting the scientific development of physical education teaching quality evaluation. It has certain theoretical value for enriching and perfecting the evaluation system of college physical education teaching quality, effectively guiding physical education teaching practice, improving and improving physical education teaching quality, and finally realizing the training goal of physical education curriculum. The main aspects of the research are: first, the main body of teaching quality evaluation; Second, the content of teaching quality evaluation; The third is the method of teaching quality evaluation; Fourth, the system of teaching quality evaluation.

2   Evaluation Method of Teaching Quality of Adolescent Health Physical Education Based on Mobile Education Technology

In order to enhance the educational effect of adolescent health physical education curriculum and promote the improvement of teachers' teaching quality, this paper designs a teaching quality evaluation method of adolescent health physical education curriculum based on mobile education technology. The following is the flow of the teaching quality evaluation method of adolescent health physical education (Fig. 1).

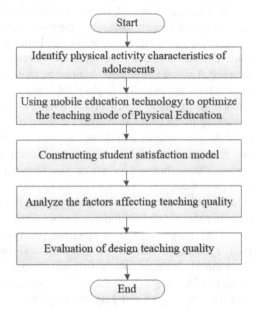

Fig. 1. Flow of teaching quality evaluation method of adolescent health physical education

2.1 Identify Physical Activity Characteristics of Adolescents

The definition of physical activity must be distinguished from physical exercise. Health education is a public health strategy and a professional field of public medicine. With the progress of society and the development of medical research and practice, it has been given new connotation. Physical exercise is the lower concept of physical activity, which refers to planned, structured and repetitive physical activities, in order to improve or maintain one or more physical abilities. Adolescent healthy physical activities cover five levels, as shown in Fig. 2.

As can be seen from Fig. 2, the main constituent elements of teenagers' healthy physical activities include: health policy, health education (health curriculum teaching, health activities, health consultation), physical exercise habits, school health services and so on. The World Health Organization believes that health education refers to making people pay attention to their health and know how to maintain their health and seek appropriate help when necessary. Motion sensor is a kind of mechanical or electronic device, which can specify the sensor in a certain part of the body and measure physical activity by sensing the movement or acceleration of limbs or trunk [5]. Health education is a purposeful, planned, targeted and evaluated educational activity through the school according to certain social requirements, conditions and norms. Common motion sensors are pedometer and accelerometer. The core of the accelerometer is the

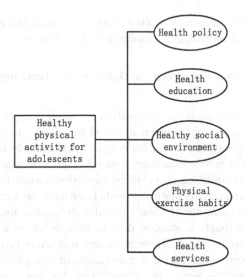

Fig. 2. Main components of healthy physical activities of adolescents

piezoelectric sensor composed of piezoelectric elements and vibrating body, which can sense the acceleration in motion. Health education is to disseminate health knowledge to individuals or groups through planned and organized systematic education activities, to establish healthy awareness and promote people's resources to adopt healthy and healthy behaviors and healthy lifestyles, so as to eliminate or reduce the risk factors of health, reduce incidence rate, mortality rate, promote health and improve the quality of life. The different understanding of the concept of health education makes the practice of health education in China reflected in two levels: one is the public health education carried out by public medical professionals, and the other is the student health education from the perspective of school. Then it acts on the piezoelectric element to generate an electrical signal. After computer processing, the acceleration count is obtained, and then the energy consumption is calculated. Based on the concept of health promotion, the concept of school health promotion focusing on student groups came into being. Adhering to the concept of health promotion refers to the education, policy environment and practice of promoting health. School health promotion originates from school health education, which emphasizes providing students with complete and positive experience and knowledge structure, creating a safe and healthy learning environment, providing appropriate health services, seeking extensive participation of families and communities, and jointly promoting students' health. In addition, the principle of pedometer is that the vertical acceleration generated during walking deflects the horizontal spring lever inside the pedometer, connects the closed circuit, measures the steps, calculates the walking distance according to the stride, and then calculates the energy consumption value. Pedometer method can effectively evaluate children's physical activity level, which is not affected by age, gender, height, waist circumference and body mass index. It is simple to operate and cheap. It is suitable for large sample size population survey. Adolescent health education and health promotion cannot replace each other. Health promotion pays more attention to seeking cooperation from the aspects of education, social and cultural background, environment, policy and so on.

2.2 Mobile Educational Technology to Optimize the Teaching Mode of Physical Education

Mobile education technology is the combination of mobile computing and digital learning. It includes learning resources at any time and anywhere, strong search ability, rich interaction, strong support for effective learning and performance-based evaluation. The key technology used in mobile education is mobile communication technology, which makes it different from digital learning. Mobile education is a new type of interactive digital learning method and activity that can provide learners at any time and anywhere with the help of mobile communication technology and mobile education terminal equipment. It is digital learning through information devices such as handheld computer, personal digital assistant or mobile phone. There are many restrictive factors in the setting of physical education curriculum, such as the limitation of discipline progression and students' acceptance ability, as well as the allocation of class hours in the overall teaching plan and so on. As we all know, in the teaching system, the factors that have the greatest impact on the teaching quality are teachers, students and managers. Therefore, the participants should also include these three parts. Different evaluators have different positions and different degrees of contact and observation. However, the main purpose of physical education curriculum is to develop students' sports skills and cultivate basic sports ability. First, more teaching objectives can be achieved on the premise of ensuring the development of students' full sports ability, We should not ignore the essential purpose in order to simply pursue extensive exploration. Such abandoning the basics and discarding the details is of no benefit to the improvement of teaching quality. The reference value of the evaluation results is also different. The evaluation of teaching quality by school expert group often focuses on the good classroom teaching order, the organization of teaching content and the advantages and disadvantages of teaching methods. This evaluation plays a direct role in promoting teachers to improve the quality of lesson preparation and lectures. Mobile education means that relying on the relatively mature wireless mobile network, Internet and multimedia technology, students and teachers can realize interactive teaching activities more conveniently and flexibly by using mobile devices (such as mobile phones). Mobile education features are shown in Fig. 3.

As can be seen from Fig. 3, mobile education is characterized by mobility, real-time, contextual relevance, interactivity and personalization. Mobile education depends on the development of digital learning. It is a special way of digital learning. As a new concept, it must be different from the traditional way of learning, otherwise it will lose its own significance. From the perspective of educational technology development, the definition of mobile education can take into account both broad and narrow orientations. Secondly, mobile education is digital learning based on portable mobile education terminal equipment. The transmission of learning content and knowledge depends on mobile communication technology, which can meet the needs of people to learn anytime and anywhere. Then, mobile education should realize the two-way effective interaction between teaching and learning, and carry out teaching and learning conveniently and flexibly. Therefore, on the premise of ensuring the full development of sports technology,

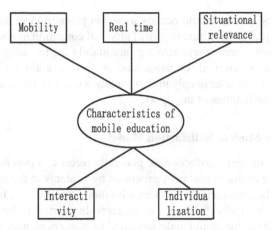

Fig. 3. Characteristics of mobile education

we should pursue diversified development. The setting of physical education teaching subjects is an important factor to improve the teaching quality. The scientific and reasonable setting of subjects can not only effectively develop students' sports skills, but also teach a variety of sports projects, broaden students' vision and develop sports thinking, which is the continuous pursuit of physical education curriculum system. Based on their own teaching experience and knowledge of the subject, we have a rational analysis of the teaching methods and means used by the teachers, the teaching attitude, the scientific nature and the advanced nature of the teaching contents. The goal of physical education teaching fully shows the curriculum designer's understanding of the value and orientation of physical education curriculum and the understanding of physical education teaching. It is the concretization of the purpose of school education in curriculum teaching. Moreover, the evaluation of teaching quality among peers is conducive to mutual learning between teachers and the construction of a good teaching style. Therefore, it is conducive to improving teaching quality. The broad concept of mobile education can take the literal meaning, that is, learners' learning activities in non-fixed places and non-fixed time with some media tools and equipment. In the narrow sense, mobile education refers to learners' learning activities in non-fixed places and non-fixed time with digital media tools and equipment based on mobile computing, mobile communication and mobile Internet technology. Finally, it needs special attention that mobile education occurs anytime, anywhere. It is an informal way of learning. The learning time is intermittent and the learning content is "fragmented". In other words, understanding mobile education from the literal meaning of mobile education is a broad understanding. Understanding mobile education from the perspective of the emergence and development of mobile education terms is a narrow understanding. Due to the new curriculum reform, most physical education colleges and universities begin to pay attention to the cultivation of innovative spirit and practical ability. Student evaluation should be an important part of the whole teaching quality evaluation system. They are the direct feelings of teachers' teaching, have the most comprehensive contact and understanding of teachers' teaching status and effect, and can reflect students' basic needs and teachers' teaching situation.

Therefore, students' scores should occupy a certain proportion in the teaching quality evaluation system [6, 7]. The ultimate goal of physical education major is to guide school sports activities, and school sports activities are mainly the teaching of sports technology. The main task of physical education teachers is to explain the sports technology of normal schools. Only after deeply understanding a sport can teachers realize the key parts and learning difficulties of this sport.

2.3 Constructing Student Satisfaction Model

This paper defines student satisfaction as: generally refers to a psychological feeling of happiness, pleasure or disappointment generated by students in the process of comparing the results of their educational services with their expectations. In terms of content setting, try to meet the individual needs of students. In short, it is the activities that students are interested in. We should make students' interests dominant and make physical education teaching universal. Only when a student's desire or desire has been fulfilled can he feel satisfied or satisfied. In the process of model construction, matrix decomposition technology is introduced, which is a technology for dimension reduction. This technology can decompose the original matrix into the product of two or more matrices, and is often used to fill some sparse matrices. Finally, the user potential feature vector and the project potential feature vector are used for inner product, so as to combine the user and the project. Then the expression formula of user vector is:

$$H = \beta \left(\frac{\alpha^F}{F} \right) \tag{1}$$

In formula (1), β represents the user's potential feature vector, F represents the project's potential feature vector, and α represents the relevance between the user and the project, which is generally expressed in the user's score on the project or the compliance of the project with the user's preference. Through the physical education teaching of students' interested projects, teachers can skillfully master one or two favorite sports skills and form sports behaviors that are not easy to change. At the same time, the user's score on the unknown item is also expressed by the inner product of the user's potential feature vector and the item's potential feature vector, so as to fill the sparse matrix:

$$D = \sum \left(\frac{1 - \varepsilon}{E} \right)^\delta \tag{2}$$

In formula (2), E represents the regular term coefficient, ε represents the activation function of the output layer, and δ represents the connection weight. Student satisfaction refers to the degree of satisfaction of students, that is, the degree to which students' actual feelings of receiving educational services are compared with their expectations. For example, why should we teach badminton, why should we teach roller skating, how much positive effect and value it has on them, and whether students have such needs? Therefore, we should consider the teaching contents and corresponding teaching methods from the interests, specialties and needs of students, which cannot be imposed on students. The application of customer satisfaction theory in the construction of educational service quality is the inevitable choice for colleges and universities to achieve student satisfaction and improve social reputation and recognition under the condition of market economy. The fierce market competition requires colleges and universities to use customer satisfaction theory to analyze and judge the satisfaction of educational service and enhance the recognition of students. Taking the student satisfaction of teenagers as the input sample, the corresponding hidden layer and output layer are defined as:

$$L_{in} = \frac{\sum |h - r|}{K} \tag{3}$$

$$L_{out} = \frac{\sum |r - 1|}{K} \times \sqrt{h} \tag{4}$$

In formula (3–4), h represents the weight, r represents the probability that the output category is 1, and K represents the weight of the full connection layer. In schools, teachers' teaching methods greatly affect students' learning methods and enthusiasm, but only teachers' continuous improvement in teaching methods and teaching methods can fully mobilize students' enthusiasm for learning and enable students to study independently and freely instead of passively. The connotation and significance of customer satisfaction has an important reference for student satisfaction, and provides a strong theoretical support for the research on practical teaching satisfaction of master of physical education. Teaching is composed of two parts, one is teaching and the other is learning. It is a two-way activity. Teachers play the main role of "teaching" in the process of teaching and learning, while students play the main role of "learning" in the process of teaching and learning. The two sides complement and cooperate with each other, so that the activities of "teaching" and "learning" can be carried out orderly and smoothly, and a good teaching environment can be formed in the teaching process. On the basis of formulas (1)–(4), the expression formula of students' interest is obtained:

$$T = h\left(K \begin{bmatrix} M^\eta \\ N^\gamma \end{bmatrix} \right)^{-1} \tag{5}$$

In formula (5), M represents the user word vector, N represents the course word vector, η represents the length of potential feature vector, and γ represents the length of input information. Practical teaching satisfaction evaluation expresses the students' evaluation of the service quality of practical teaching in higher education. It not only helps colleges and universities find problems, but also has the function of problem diagnosis, helps colleges and universities reasonably allocate educational resources and improve the service quality of practical teaching. Moreover, good and beautiful action demonstration can not only enable students to correctly understand the action itself, but also indirectly stimulate students' interest in learning, create a good learning atmosphere and increase students' learning motivation. Therefore, the basic technology of physical education is not only the direct manifestation of physical education teaching quality, but also directly related to the teaching level of school physical education and the realization of physical education teaching objectives. Teachers' teaching skills directly affect students' learning attitude. Teaching teachers can make appropriate adjustments according to the situation of teaching students, so as to teach students according to their aptitude, so that every student can learn what they have learned from teachers' teaching skills and make a breakthrough in students' majors.

2.4 Design Teaching Quality Evaluation Method

Teaching quality is the fundamental guarantee to achieve teaching objectives and complete teaching tasks. As the logical starting point of this study, teaching quality is also the basic premise of constructing the teaching quality evaluation of public physical education courses [8]. As an internal psychological component and a long-term intermediate variable between social stimulation and individual behavior, teaching attitude plays a preparatory role in teaching behavior, which affects the quality of teaching. According to the definition of the concept of teaching quality, the teaching process is the main factor affecting the quality of teaching, because the consistency between students' learning results and teaching objectives is directly related to the teaching process. Although teaching attitude can not completely determine the quality of teaching behavior and teaching quality, it has a great impact. Moreover, teachers' teaching attitude also affects the level of teaching efficiency. Every component and link of the teaching process may become relevant factors affecting the teaching quality. Objective factors may become factors affecting teaching quality, but subjective factors really play a key role. Teachers with real ability and level will not be affected by the objective environment and conditions of teaching. If teachers have a positive teaching attitude, they will take the initiative to increase the time for lesson preparation, widely involve all kinds of knowledge, and form a relatively complete knowledge structure. Try to meet the requirements of students with strong thirst for knowledge and wide interests, so as to win the trust of students, establish the prestige of physical education teachers, enhance the attraction and influence on students, and improve students' interest in learning. Therefore, from the perspective of subjective factors affecting teaching quality and taking each component of the teaching process as the object, the elements affecting teaching quality can be divided into the following points, as shown in Fig. 4.

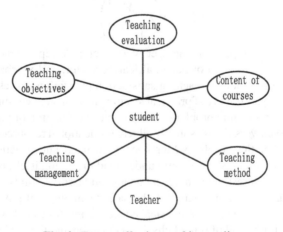

Fig. 4. Factors affecting teaching quality

As can be seen from Fig. 4, the factors affecting teaching quality include: teaching objectives, teaching contents, teaching evaluation, teaching management, teaching methods, teachers and students. These seven elements basically cover all the links that affect the subjective factors of teaching quality. Teaching objectives are the fundamental elements that affect the teaching quality. In terms of skill objectives, if the teaching objectives are set too high, which are inconsistent with the physical education basis of the teaching subject, and the students can't achieve it with any effort, the gap between the teaching effect and the teaching objectives will be insurmountable, the teachers will feel frustrated, the students may lose their confidence in learning, and the teaching quality will never reach the preset standard. So as to improve the teaching quality and strengthen teachers' student-centered awareness in the teaching process. The teaching content is the basis of the course, which determines the knowledge, skills and abilities of the talents trained by the specialty, and is also the premise and basis of teaching quality evaluation [9–11]. The choice of teaching content directly affects the initiative of College Students' sports skill learning. College students have active thinking and are easy to accept new things. They are the vanguard of leading social fashion. Conversely, social fashion sports are also the object they pursue. Teaching methods and means are one of the factors affecting teaching quality. Physical education teaching content refers to the knowledge of physical health care and various sports actions selected to achieve the goal of physical education teaching. It is the fundamental guarantee to achieve the goal of physical education teaching. Teaching methods and means are an important part of the teaching process. Appropriate teaching methods and means can improve teaching efficiency. Inappropriate teaching methods and means may get twice the result with half the effort. The organization and management of teaching is the most important guarantee to improve the quality of teaching. There is no doubt that the main body of the teaching content of the technical course of physical education is always the study of sports technology. As graduates of technical courses of physical education, they should not only have excellent physical education technology, but also have excellent physical education teaching ability. In other words, we should not only know technology, but also teach technology.

3   Experimental Study

3.1   Experimental Preparation

Resource Description Framework RDF, which is a recommendation standard based on XML syntax representation, is now used by more and more researchers. Because the server-side code of this platform is developed in Java, JDK should be used to provide developers with runtime and public basic class library. Its purpose is to establish a general resource description framework for a variety of metadata and promote the sharing, exchange and reuse of various resources. In addition, because PC terminals need to interact through a web browser, they need to provide a stable web server. Using Tomcat server to assume the role of web server can well meet the deployment of web applications based on Java. It is an open source and flexible plug-in integrated development environment. RDF uses XML to describe resources and express the content of learning resources and the relationship between resources. It is a real universal resource description technology. By adding corresponding web development plug-ins and Android application development plug-ins, the selection of platform tools is greatly simplified, so as to improve the efficiency of platform coding development. Because red5 server is based on RTMP protocol, it can well support the research scheme of mobile platform in this subject. The media elements contained in mobile learning resources mainly include text, graphics and images, audio, video and animation. According to the different combinations and presentation forms of these media elements, common mobile learning resources can be divided into SMS, MMS, e-book, web page, wechat, online course, educational game and other types. Nginx server has the advantages of lightweight, good concurrency and less system resources. It can support the response of up to 50000 concurrent connections, and can well support the server of web applications.

3.2 Experimental Result

In order to verify the effectiveness of the teaching quality evaluation method of the designed adolescent health physical education curriculum, an experimental test was carried out. The adolescent health physical education course of a junior high school in one semester was selected as the evaluation object. Among them, there were 6 teachers who taught the adolescent health physical education course. A total of 36 classes participated in the adolescent health physical education course in the whole semester. On average, each class received 32 adolescent health physical education courses in this semester. After statistics, 32000 pieces of data are obtained, 20000 pieces of effective data are retained as test data after removing redundant data, and they are divided into three data scales, namely 5000, 1000 and 2000. The teaching quality evaluation method of adolescent health physical education based on audio synchronization technology, the teaching quality evaluation method of adolescent health physical education based on data mining and the teaching quality evaluation method of adolescent health physical education in this paper are selected for experimental comparison. The evaluation speeds of the three methods are tested under different data scales. The experimental results are shown in Tables 1, 2 and 3.

From Table 1, it can be concluded that the average evaluation speed of the teaching quality evaluation method of adolescent health physical education in this paper is 3.157; The average evaluation speed of the teaching quality evaluation method of adolescent health physical education based on audio synchronization technology is 6.089; The average evaluation speed of the teaching quality evaluation method of adolescent health physical education based on audio synchronization technology is 5.760.

From Table 2, it can be concluded that the average evaluation speed of the teaching quality evaluation method of adolescent health physical education in this paper is 9.422; The average evaluation speed of the teaching quality evaluation method of adolescent health physical education based on audio synchronization technology is 14.628; The average evaluation speed of the teaching quality evaluation method of adolescent health physical education based on audio synchronization technology is 14.928.

From Table 3, it can be concluded that the average evaluation speed of the teaching quality evaluation method of adolescent health physical education in this paper is 15.050; The average evaluation speed of the teaching quality evaluation method of adolescent health physical education based on audio synchronization technology is 24.769; The average evaluation speed of the teaching quality evaluation method of adolescent health physical education based on audio synchronization technology is 25.576.

Table 1. Data scale 500 evaluation speed (s)

Number of experiments	Evaluation method of teaching quality of adolescent health physical education based on audio synchronization technology	Evaluation method of teaching quality of adolescent health physical education based on Data Mining	The evaluation method of teaching quality of adolescent health physical education course in this paper
1	5.615	5.366	3.221
2	6.315	6.293	3.154
3	5.948	5.784	3.058
4	6.215	5.884	3.569
5	5.976	6.546	3.211
6	6.355	4.917	3.025
7	5.941	5.628	2.978
8	6.258	4.996	2.996
9	5.949	5.312	3.205
10	6.313	6.878	3.154

Table 2. Data scale 1000 evaluation speed (s)

Number of experiments	Evaluation method of teaching quality of adolescent health physical education based on audio synchronization technology	Evaluation method of teaching quality of adolescent health physical education based on Data Mining	The evaluation method of teaching quality of adolescent health physical education course in this paper
1	15.612	15.206	9.161
2	14.347	14.978	8.557
3	14.619	14.696	9.253
4	13.548	15.313	9.467
5	15.098	14.899	8.949
6	13.647	13.262	8.649
7	14.220	14.649	10.202
8	15.361	15.212	9.667
9	14.515	16.955	10.154
10	15.315	14.113	10.161

Table 3. Data scale 2000 evaluation speed (s)

Number of experiments	Evaluation method of teaching quality of adolescent health physical education based on audio synchronization technology	Evaluation method of teaching quality of adolescent health physical education based on Data Mining	The evaluation method of teaching quality of adolescent health physical education course in this paper
1	26.155	26.164	15.669
2	23.610	25.484	14.677
3	22.994	26.913	15.293
4	23.157	26.556	15.204
5	24.055	25.377	16.117
6	23.649	26.199	14.323
7	25.691	26.184	15.266
8	26.384	25.109	14.311
9	25.677	24.616	15.262
10	26.315	23.154	14.379

4 Concluding Remarks

Physical education teaching assessment, as a test measure of technical teaching quality, is an effective way to ensure the smooth completion of teaching objectives, and can also play a positive role in the construction of a good style of study. In the environment of college enrollment expansion and lower assessment standards, the effectiveness of assessment is getting lower and lower. Strengthening the construction of curriculum assessment system plays an important role in improving the teaching quality of technical courses. In the future, we need to constantly expand the application field of the teaching quality evaluation method of adolescent health physical education in this paper.

Funding Project. "Fourteenth Five-Year" Social Science Project of Jilin Provincial Education Department: Research on the model of adolescent health promotion in the post-epidemic era between home, school and social sports (JJKH20211388SK).

Higher Education Teaching Reform Research Project of Jilin Province: Construction and Practice of Hybrid Teaching Mode of College Physical Education under MOOC Environment (20224BR6215009Z).

References

1. Gaixiao, Z., Enshan, L.: The pursuit of effective teaching from American ACOP classroom teaching quality evaluation system. Stud. Foreign Educ. **47**(5), 103–118 (2020)
2. Li, K.: Exploration of the problems and countermeasures of teaching quality evaluation of general undergraduate. Road Success 31, 10–11 (2020)
3. Xu, X.: Research on teaching quality evaluation of online courses in higher vocational colleges based on fuzzy analytic hierarchy process. J. Wuxi Inst. Commer. **20**(4), 90–93,112 (2020)
4. Jia, J.: Evaluation of mathematics teaching quality in higher vocational colleges based on cognitive load theory. Bull. Sci. Technol. **36**(8), 115–119 (2020)
5. Su, H.: Passive location simulation of multi-sensor and multi-target based on artificial intelligence. Comput. Simul. **37**(9), 399–403 (2020)
6. Longying, Z.: Research and implementation of higher vocational teaching quality evaluation based on AHP mathematical model. Sci. Technol. Innov. Herald **17**(26), 193–195 (2020)
7. Hong, R., Chang, S.: Research and practice on the construction of teaching quality evaluation system for clinical teaching bases of local medical colleges. Guide Sci. Educ. 24, 5–6 (2020)
8. Yanli, W., Chuangju, W., Yujiao, Y.: Internet teaching quality evaluation model based on evidence theory and neural network. Mod. Electron. Tech. **43**(19), 175–178 (2020)
9. Gan, T.: Research on the evaluation of distance teaching quality in colleges and universities based on decision tree classification algorithm. Mod. Electron. Technol. **44**(09), 171–175 (2021)
10. Chen, L.: The application of mobile teaching based educational APP in the teaching of asset evaluation. J. Changchun Instit. Eng. (Soc. Sci. Ed.) **22**(02), 112–116 (2021)
11. Wang, X.: Research on the effectiveness of online and offline hybrid teaching mode based on mobile education platform in higher vocational English teaching. Grade · Classics (17), 152–155+159 (2021)

Evaluation Method of Physical Education Teaching Quality in Higher Vocational Colleges Using Mobile Teaching Terminal

Wei He[1]([⊠]) and Daoyong Pan[2]

[1] Hunan Vocational Institute of Technology, Xiangtan 411100, China
hewei19830825@163.com
[2] Sports Institute, Kashi University, Kashgar 844006, China

Abstract. At present, there is a problem of single subject and mode in the evaluation of physical education teaching quality in higher vocational colleges, which affects the accuracy of the evaluation results. In order to improve the accuracy of teaching quality evaluation results, this paper proposes an evaluation method of physical education teaching quality in Higer Vocational Colleges by using mobile teaching terminal. Analyze the application strategy of mobile teaching terminal in physical education teaching; Collect physical education teaching data based on mobile teaching terminal; Design the teaching quality evaluation index system; Determine the index weight and build the quality evaluation model; The comprehensive evaluation of physical education teaching quality in higher vocational colleges is realized through the qualitative index fuzzy quantitative method. The test results show that the higher vocational physical education teaching quality evaluation method using mobile teaching terminal proposed in this paper has high detection rate and accuracy, so the evaluation results are more effective.

Keywords: Mobile teaching terminal · Higher vocational colleges · Physical education · Teaching quality · Evaluation method · Index system

1 Introduction

Teaching is the starting point and central link of all school education activities. As main implementers for school physical education, physical education teachers are responsible for imparting students' physical education knowledge and skills, cultivating students' good quality and improving the quality of school education. The core sports literacy of young students has risen to the height of national strategy. In order to develop teenagers' core literacy, the state endowed the potential and innovative teaching mode. Physical evaluation is the physical education teaching process. Therefore, through the evaluation system, we can put forward feedback about teachers' teaching situation, so as to improve teachers' teaching ability and promote the improvement of school education quality. This not only provides policy guarantee for the smooth development of specialized teaching reform. Teachers' professional development has always been the focus of educational

W. Fu and L. Yun (Eds.): ADHIP 2022, LNICST 469, pp. 676–687, 2023.
https://doi.org/10.1007/978-3-031-28867-8_49

reform. It is a process from immature to mature for teachers as professionals to develop from novice teachers to expert teachers [1]. There are serious deviations in this model, which is one of the factors lagging behind the reform. The evaluation for teachers' teaching ability is very cumbersome, and considering the particularity, the evaluation work for it is very difficult, so build a relatively scientific and system. By constructing ability, we diagnose teachers' teaching ability scientifically and reasonably, improve teaching ability and promote the specialization [2]. The teaching ability is the focus of development and the core ability of effective teaching. If you want to improve professional ability, you must have higher teaching ability as support. Driven by mobile teaching terminal, the evaluation results can reflect the real teaching resultsphysical education. This can not only promote relevant teachers to adopt correct values when engaging in teaching activities, but also urge relevant teachers to continuously improve their quality, so as to judge their teaching activities relatively fairly, and finally help the overall physical education management and fundamentally improve teachers' teaching ability. In order to improve the evaluation effect of teaching and assist the physical education teaching, this paper puts forward a method of higher vocational physical education teaching quality evaluation by using mobile teaching terminal, so as to promote the implementation of higher vocational physical education teaching evaluation, and also hopes to provide reference for relevant research.

2   Evaluation Method of Higher Vocational Physical Education Teaching Quality Using Mobile Teaching Terminal

2.1   Application Analysis of Mobile Teaching Terminal in Physical Education

School education cultivate people with all-round development. Based on practical level, the ability of such teachers should be defined. The ability can be divided into two parts, mainly aiming at the cognitive and operational abilities produced in the specific teaching. Digital learning resources rely on various storage technologies and database technologies, which can realize the storage of massive learning resources, so that students can choose suitable learning resources from massive learning resources. Cognitive ability means that teachers of related disciplines need to make a basic judgment on the abilities of students from their own understanding of the established teaching outline and plan, and then have an overall judgment. Teachers' cognitive ability is helpful to help students master the required sports skills and sports knowledge, and shape students' good ideology and morality. On the one hand, it has greater autonomy in choosing learning resources; On the other hand, for students, they can understand the knowledge points more thoroughly through sufficient data. Operational ability refers to the ability of teachers of this kind of discipline to constantly deal with various problems in their specific teaching work. The sports learning based on mobile terminal frees teachers from the busy demonstration actions, and can interact with students on some key knowledge of sports learning at any time, and discuss some more profound sports theories and tactics with students. Clarifying the characteristics and connotation of sports specialized teaching can not only make more people know and contact specialized teaching, but also form a joint force to promote the development of specialized teaching [3]. Teachers' operational ability is helpful to develop students' physical fitness and cultivate students' healthy sports behavior. Teachers pass on their training experience to the students through the course. Students receive these teachings and build their own knowledge system, so as to avoid injury during training and to enhance their physical fitness. The physical training system based on mobile terminals is shown in Fig. 1.

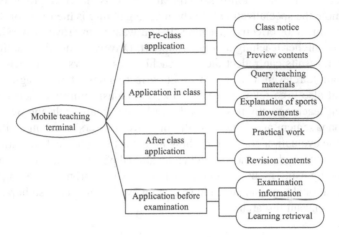

Fig. 1. The physical training system

In new and old curriculum standards, although there are four different characteristics in the physical education curriculum, there are some changes in the specific content, changing the fitness of the former to the selectivity of the latter. Teachers can observe their own defects in action through video, correct them in time, and avoid inaccurate actions in the courseware as much as possible. Students can grasp the essentials of an action accurately through action freeze frame, and draw inferences from one instance through careful analysis. The practical content in the new version covers the practicality and fitness of the old version, and complements the selectivity. The selectivity mainly focuses on students to cultivate different sports specialties based on their own physical quality, and form the habit of regular exercise at the same time. Students' learning experience can also be spread among students through data sharing to help other students in the group learn.

2.2 Collect Physical Education Teaching Data Based on Mobile Teaching Terminal

Mobile terminal equipment has strong flexibility and can be carried around at any time, so mobile terminal has become a necessary basic tool for mobile learning. The mobile teaching terminals have function of student measurement of learning situation. The data stored of the student evaluation is as few as millions of thousands of pieces, as many as tens of millions of pieces. These data are still rising sharply, and the amount of data is unimaginable. No matter where learners do, they can choose and obtain the knowledge content they need through the equipment they carry. Mobile teaching terminal can help the rapid improvement of physical education teachers' teaching ability, so as to achieve the most fundamental purpose to continuously improve the horizontal of talent training. In process for mobile learning, users can realize rapid and efficient communication, synchronous sharing of resources and information, and even face-to-face interaction through various networked ways. In addition, the communication, assessment and evaluation between teachers and students can also be more diversified.

There are inevitably some abnormal data in the sports teaching data collected by the mobile teaching terminal. Through the analysis of the initial data, it is found that there are many noise data in the data, which are mainly divided into noise caused by subjective factors and noise caused by objective factors. If not handled, it will affect the judgment of teachers' current teaching situation. Clean physical education teaching information and remove the strange information from the sample set, so as to ensure the effectiveness and authenticity of the evaluation. The subjective factors mainly include: students have a certain perfunctory psychology towards the teaching evaluation activity itself; Each student has different evaluation criteria for teachers; Students may maliciously slander individual teachers in the process of teaching evaluation. Objective factors include: different course assessment methods, students' different attitudes towards their teaching evaluation and so on. The information was classified based on the similarity, and then the abnormal elements in the teaching evaluation were eliminated. The similarity of all attribute dimensions in the sample set was calculated, and the samples were divided according to the similarity calculation results for each attribute value. The equation for the similarity of the sample is as follows:

$$\gamma(\varphi, \vartheta) = \frac{\sum\limits_{x=1}^{a} [(\varphi_x - u_1)(\vartheta_x - u_2)]}{\sqrt{\sum\limits_{x=1}^{a} (\varphi_x - u_1)^2} \sqrt{\sum\limits_{x=1}^{a} (\vartheta_x - u_2)^2}} \tag{1}$$

In formula (1), φ, ϑ represent two dimension attributes; γ represents similarity; x, a represent the serial number and total number of sample data respectively; φ_x and ϑ_x represent the sample data of two dimension attributes respectively; u_1 and u_2 represent the average value of the value range of the two dimensions[4, 5].

Select the nearest point as the category of the data. After this visit, all samples are divided into the category. In order to eliminate the difference of evaluation scale of different individuals, the data are standardized, and the data are scaled and mapped to the [0,1] interval. The mapping method is as follows:

$$\varphi_x' = \frac{\varphi_x - m_1}{m_2 - m_1} \tag{2}$$

In formula (2), φ_x' represents the corresponding value of the standardized sample data; m_1 and m_2 are the minimum and maximum values of sample data respectively.

According to the standardized data, the data are merged according to the method of calculating the average value of each column, so that the teaching evaluation data of each PE teacher becomes a piece of data [6, 7]. Teachers can identify, develop, accumulate and use physical education curriculum resources. Similarly, it itself is the most basic resource condition. The professional teachers themselves is an important physical education curriculum resource. Continuously surpass in development and utilization of physical education curriculum resources, and have achieved regeneration and created curriculum resources beyond their own value. The real significance and value of these data stored in mobile teaching terminals lies in timely understanding and mastering the overall state of teachers' teaching work through its analysis and mining, and giving supervising and guiding teaching, enhancing teachers' internal driving force of teaching, purposeful training, targeted management and so on.

2.3 Design the Evaluation Index System of Physical Education Teaching Quality in Higher Vocational Colleges

Teachers' ability should be improved, and the evaluation can truly reflect the realization of students' learning objectives of the curriculum. The focus of evaluating the learning of the curriculum is the formation of the core quality of the discipline, the related problems in real life, whether students develop healthy habits in real life, and the attitude and ability[8, 9]. When choosing the evaluation content, we should refer to the new curriculum standard, evaluate PE teachers reasonably and scientifically, and clarify the direction of teachers' professional development. When evaluating PE teachers, we should consider their past and future, so as to fully mobilize the enthusiasm of teachers.

At present, concomitant evaluation is an evaluation system that combines summative evaluation with process evaluation, and pays equal attention to both result and process. Higher vocational physical education takes initiative, teachers' will, teachers' thinking and teachers' professional ethics are all important factors that restrict students' learning. Let physical education teachers actively participate in the evaluation work from inside to outside, so that they can continuously improve their teaching ability according to the results of evaluation [10, 11]. Whether teachers pay close attention to students' personality development, whether students master effective learning methods, whether students acquire knowledge, develop ability and have positive emotional experience are the standards to measure the quality of teaching effect. The primary index are shown in Table 1.

The selected indicators and formulated standards should have scientific basis, and the technical evaluation method of evaluation should also be scientific. The design scheme, collected information, results and treatment of related matters should not lack reliability, so as to ensure that it is effective [12]. The diversity of evaluation content is the diversity of content. For example, sports theoretical knowledge and physical fitness, but also evaluate learning attitude, learning ability, progress of sports skills and so on.

2.4 Construction of Higher Vocational Physical Education Teaching Quality Evaluation Model

The evaluation system is composed of multiple complex indexes, so the evaluation index system is more complex. This can make the evaluation index more objective and accurate. Weight refers to comparing and measuring the importance in overall things in a certain quantitative form. The change of weight value is directly related to the evaluation results. Before starting the construction, we need to calculate the sum and average of the

Table 1. Evaluation Index System

Secondary index	Symbol	Tertiary indicators	Symbol
Teaching preparation	A1	Reasonable preparation of teaching documents	B1
		Site equipment is fully prepared	B2
		Choose teaching methods	B3
		Classroom content design	B4
Teaching implementation	A2	Create a teaching environment	B5
		The course progress is reasonable	B6
		The teaching content is novel	B7
		Action demonstration is accurate	B8
Teaching guidance	A3	Teaching concept	B9
		Teaching attitude	B10
		Teaching coordination ability	B11
		Teaching strain ability	B12
Teaching result	A4	Students master sports skills	B13
		Good communication between teachers and students	B14
		Improve students' physical literacy	B15
		Spread the scientific fitness knowledge	B16

importance scores of various experts on the indicators at all levels, and then compare the scores of the two indicators, so as to find out the difference in the importance of each indicator. The judgment matrix is:

$$W = \left\{ \begin{matrix} 1*p*q \\ \frac{1}{p*1*r} \\ \frac{r*p}{q*1} \end{matrix} \right\} \qquad (3)$$

In formula (3), W represents the judgment matrix; p, q and r are the importance degree of each element in the same level with respect to a certain criterion in the previous level. Calculate the n-th root of the product of each row, normalize the vector, and obtain the characteristic vector, that is, the weight of each index.

$$F = \frac{\sqrt[n]{W}}{\sum_n \sqrt[n]{W}} \qquad (4)$$

In formula (4), F represents the eigenvector; n represents the number of rows of the judgment matrix.

Carry out a sequence test and basic satisfaction consistency test on the feature vector, and the test shows that the weight is acceptable. When CR < 0.1, it can be considered that the evaluation is completely consistent, and calculated coefficient can more perfect mapping the relative importance index, otherwise the matrix must be readjusted to make it completely consistent [13]. According to the value of the expert opinions and the indicators, the weight sum of the overall evaluation system is calculated, and each weight index is ranked. The higher the ranking is, the more important it is, and the score increases accordingly. Then, the teaching level is vaguely and comprehensively evaluated, and the model is shown in Fig. 2.

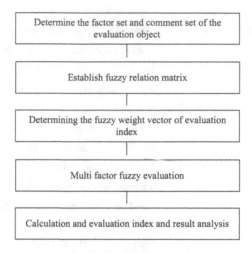

Fig. 2. Evaluation Model

For each student, the evaluation system users can get enough scoring elements and scoring attributes to make a comprehensive evaluation of the teaching level. The evaluator may make a collection of various general evaluation result elements for the evaluation object, that is, the comprehensive evaluation result of physical education teachers' teaching quality in higher vocational colleges. Then, through the unified quantitative processing of each factor, the evaluation set of each index constitutes a general evaluation matrix. The fuzzy evaluation comprehensive result vector can be obtained by calculating in the way of fuzzy operation and normalizing the results. So far, the design of higher vocational physical education teaching quality evaluation method using mobile teaching terminal has been completed.

3 Experimental Study

3.1 Experimental Preparation

Collect the PE teaching data in this paper from mobile teaching terminal. Using the method of the above parameters, the weight of the scoring indicators at all levels is calculated. The preparation results are detailed in Table 2.

It can be seen from Table 2 that the "Action demonstration is accurate" weight index is the highest. The data of Table 2 was inserted into the higher vocational physical education quality were analyzed.

Table 2. Index weight coefficient

Secondary index symbol	Weight	Tertiary indicators symbol	Weight
A1	0.1563	B1	0.0527
		B2	0.0423
		B3	0.0751
		B4	0.0872
A2	0.3121	B5	0.0815
		B6	0.0818
		B7	0.0864
		B8	0.0995
A3	0.2275	B9	0.0527
		B10	0.0536
		B11	0.0492
		B12	0.0473
A4	0.3041	B13	0.0517
		B14	0.0528
		B15	0.0415
		B16	0.0447

3.2 Results and Analysis

Verifing the higher vocational method using teaching terminal, the detection rate and accuracy ability of evaluation method. The test samples are divided into three levels: 1000, 5000 and 10000. The teaching quality is evaluated, and evaluation method are calculated. The test results are also compared with the two conventional methods. The comparison results are shown in Tables 3, 4, 5.

In the test with a sample number of 1000, the detection rate of the higher vocational physical education quality evaluation method using the mobile teaching terminal was 0.7753, which was improved by 0.1120 and 0.1349 compared with the neural network-based and SVM-based evaluation method.

In the test with a sample number of 5000, the detection rate of higher vocational physical education quality evaluation method using mobile teaching terminal was 0.7254, which improved 0.0989 and 0.1302 compared with the neural network based and SVM based evaluation method.

In the test with 10000 samples, the detection rate of higher vocational physical education method using teaching terminal is 0.6858, which is 0.1048 and 0.1191 higher than other methods. From the index of detection rate, the performance of this paper is better than the comparative physical education teaching evaluation method. Then, three method was evaluated as the precision index. The precision comparison results of the evaluation methods are shown in Tables 6, 7, 8.

Table 3. Comparison results of detection rate of 1000 samples

Test times	The method of the article	Traditional method 1	Traditional method 2
1	0.7954	0.6849	0.6244
2	0.7847	0.6713	0.6477
3	0.7781	0.6804	0.6716
4	0.7528	0.6588	0.6183
5	0.7612	0.6496	0.6555
6	0.7856	0.6761	0.6268
7	0.7965	0.6854	0.6037
8	0.7530	0.6670	0.6621
9	0.7825	0.6212	0.6552
10	0.7628	0.6383	0.6385

Table 4. Comparof detection rates for 5000 samples

Test times	The method of the article	Traditional method 1	Traditional method 2
1	0.7104	0.6552	0.6021
2	0.7247	0.6384	0.5983
3	0.7411	0.6148	0.5954
4	0.7188	0.6211	0.5883
5	0.7055	0.6025	0.5863
6	0.6926	0.6502	0.5972
7	0.7463	0.6266	0.6044
8	0.7389	0.6333	0.5997
9	0.7322	0.6157	0.5926
10	0.7434	0.6074	0.5872

In the test with a sample number of 1000, the precision rate for the evaluation method using the teaching terminal was 0.8206, which was improved by 0.0570 and 0.0852 compared with the neural network-based and SVM-based evaluation method.

In the test with a sample number of 5000, the precision rate for the method using the teaching terminal was 0.7505, which was improved by 0.0677 and 0.1032 compared with the neural network-based and SVM-based evaluation method.

In the test with 10000 samples, the precision rate for method using teaching terminal is 0.7021, which is 0.0900 and 0.1296 higher than other methods. However, the paper method is basically not affected by the increase of samples, and still maintains a high precision, while the accuracy of other methods is reduced.

Table 5. Comparison of detection rates for the 10000 samples

Test times	The method of the article	Traditional method 1	Traditional method 2
1	0.6649	0.5842	0.5569
2	0.6786	0.5786	0.5448
3	0.6852	0.5958	0.5687
4	0.6964	0.5667	0.5824
5	0.6952	0.5833	0.5855
6	0.6957	0.5820	0.5636
7	0.6975	0.5954	0.5773
8	0.6712	0.5777	0.5615
9	0.6823	0.5816	0.5608
10	0.6908	0.5642	0.5651

Table 6. Precision rates of the 1,000 samples

Test times	The method of the article	Traditional method 1	Traditional method 2
1	0.8143	0.7466	0.7262
2	0.8210	0.7505	0.7384
3	0.8054	0.7678	0.7457
4	0.8285	0.7844	0.7548
5	0.8368	0.7717	0.7676
6	0.8132	0.7621	0.7123
7	0.8256	0.7465	0.7235
8	0.8027	0.7532	0.7352
9	0.8372	0.7856	0.7044
10	0.8214	0.7678	0.7461

Based on comprehensive detection rate and accuracy rate indicators, method using mobile teaching terminal proposed in this paper has better evaluation effect and has positive significance in theory and practice. Through this evaluation, strengthen the construction of physical education teachers, strengthen management, ensure the quality of physical education and teaching, and promote the professional development of physical education teachers.

Table 7. Precision rates of the 5,000 samples

Test times	The method of the article	Traditional method 1	Traditional method 2
1	0.7543	0.6926	0.6453
2	0.7782	0.6848	0.6506
3	0.7664	0.6873	0.6643
4	0.7437	0.6694	0.6782
5	0.7258	0.6715	0.6223
6	0.7776	0.6757	0.6303
7	0.7518	0.6921	0.6657
8	0.7352	0.6984	0.6572
9	0.7423	0.6861	0.6414
10	0.7293	0.6701	0.6181

Table 8. Precision rates of the 10000 samples

Test times	The method of the article	Traditional method 1	Traditional method 2
1	0.6842	0.6327	0.5762
2	0.6974	0.6151	0.5805
3	0.6888	0.6048	0.5647
4	0.7156	0.5976	0.5738
5	0.7003	0.6259	0.5652
6	0.7225	0.5914	0.5719
7	0.7161	0.5804	0.5827
8	0.7337	0.6309	0.5595
9	0.6852	0.6258	0.5673
10	0.6773	0.6165	0.5832

4 Conclusion

This paper puts forward an evaluation method of physical education teaching quality in Higher Vocational Colleges by using mobile teaching terminal. This method innovatively collects data of mobile terminals; Design the evaluation system; Build the quality evaluation model; The comprehensive evaluation of teaching quality is realized through the qualitative index fuzzy quantitative method. The test results illustrate that this method can basically reflect all elements. When using this system to evaluate, we can not only consider the specific situation of the evaluator, such as the identity of the evaluator and the resources that the evaluator can control, but also consider the specific situation of the evaluated party, such as the scale of the evaluated party and the number of students.

According to the above situation, we should reasonably decompose the evaluation system and use the corresponding evaluation forms to make the evaluation activities faster, objective and effective.

References

1. Xue-xia, Z.: Study on the Indexes of the scale of students' evaluation of physical education in colleges and universities. J. Pingxiang Univ. 37(1), 103–106 (2020)
2. Hang, G.: Analysis on the diversification model of college physical education evaluation. Bull. Sport Sci. Technol. 28(4), 82,84 (2020)
3. Kun, Y., Ling-li, L.: Mobile terminal network survivable database security anti-tampering simulation. Comput. Simul. 37(1), 456–459,483 (2020)
4. Youmei, Z.: Design of physical education teaching quality evaluation system based on BS framework. China Comput. Commun. 33(1), 237–239 (2021)
5. Tang Like, Xu Ying. Evaluation of Physical Education Teaching Quality Based on Back Propagation Neural Network Optimized By Golden Sine Algorithm[J]. Modern Scientific Instruments, 2021, 38(5):260-264.
6. Qian, H.U., Shaona, L.I.N.: Construction of teaching quality evaluation system for college physical education teachers based on competency model. Contemp. Sports Technol. 11(3), 247–250 (2021)
7. Tong-qing, Z., Jing-jing, G., Jian-guo, S.: Research on the evaluation system of college pe classroom teaching quality from the perspective of sports core quality——taking Hexi university as an example. J. Hexi Univ. 37(2), 72–76 (2021)
8. Wang, Q.: Research on teaching quality evaluation of college English based on the CODAS method under interval-valued intuitionistic fuzzy information. J. Intell. Fuzzy Syst. 41(1), 1–10 (2021)
9. Zhang, X., Wei, Z., Han, T.: PHP-based undergraduate data reporting and teaching quality evaluation information system. J. Phys: Conf. Ser. 1827(1), 121–124 (2021)
10. Yuan, T.: Algorithm of classroom teaching quality evaluation based on Markov chain. Complexity 2021(21), 1–12 (2021)
11. Chinkina, M., Ruiz, S., Meurers, D.: Crowdsourcing evaluation of the quality of automatically generated questions for supporting computer-assisted language teaching. ReCALL 12(32), 14–19 (2020)
12. Johnson, J.K., Paul, B., Tina, F., et al.: A starter's guide to learning and teaching how to coproduce healthcare services. Int. J. Qual. Health Care 33(2), 55–62 (2021)
13. González, C.S.G., Infante, A., Moro, J.: Implementation of e-proctoring in online teaching: a study about motivational factors. Sustainability 12(8), 3390–3404 (2020)

Simulation of Impact Force Absorption of Synthetic Rubber Materials in Volleyball Court Based on Discrete Element Method

Jun Xie[1]([✉]) and Gang Qiu[2]

[1] School of Physical Education and Health, A'ba Teachers' University, Wenchuan 623002, China
xiejun1456@163.com
[2] School of Information Engineering, Changji University, Changji 831100, China

Abstract. In order to accurately simulate the impact force absorption of synthetic rubber materials in volleyball courts, a simulation method of impact force absorption of synthetic rubber materials in volleyball courts based on discrete element method was proposed. According to the related theory of discrete element method, the mathematical equation is established to solve the maximum impact load that can be absorbed by the synthetic rubber material in the volleyball field in the process of human jumping and falling; According to the model results and the stress characteristics of human lower limb bones, the threshold of impact load borne by synthetic rubber materials in volleyball courts is obtained, and the impact absorption rate and deformation characteristics of volleyball courts under safe conditions are determined, so as to provide a simulation basis for the optimization of shock absorption structural parameters of volleyball courts. Finally, the experiment proves that the simulation method in this paper has high practicability in the practical application process and fully meets the research requirements.

Keywords: Discrete element method · Volleyball court · Synthetic rubber · Impact force

1 Introduction

With the increase in the number of large-scale sports events held at home and abroad, the improvement of people's quality of life and the enhancement of national sports fitness awareness, the construction area of sports facilities is increasing. Volleyball field is the basis of indoor sports facilities. Under the background of the increasing demand and development space of volleyball field and the lack of domestic volleyball field production and testing standards and methods [1]. Domestic and foreign scholars have gradually improved their research on the structure and surface mechanical properties of volleyball court, mainly focusing on the optimization of volleyball court structure and the improvement of volleyball court surface properties. Among them, the performance index of volleyball court structure optimization generally takes the impact load as the research object, and the surface performance improvement takes the rolling load performance index as the analysis goal. Volleyball court is a special floor with the function of

© ICST Institute for Computer Sciences, Social Informatics and Telecommunications Engineering 2023
Published by Springer Nature Switzerland AG 2023. All Rights Reserved
W. Fu and L. Yun (Eds.): ADHIP 2022, LNICST 469, pp. 688–702, 2023.
https://doi.org/10.1007/978-3-031-28867-8_50

bearing people's high-intensity sports and providing sports protection. It is an important part of the ground pavement facilities of gymnasium [2]. The research on volleyball field in China started late and the foundation is weak. The research on mechanical properties, testing methods and testing instruments of volleyball field is relatively scarce. Based on the theory of discrete element method and the methods of human motion simulation analysis, algorithm prediction and experimental verification, this paper studies the influence of impact load on the impact force absorption of synthetic rubber materials in volleyball field; According to the impact load experiment and simulation method, the structural parameters of damping pad in volleyball court are optimized to improve the impact absorption performance of volleyball court; Based on the principle of laser displacement measurement, the rolling load detection method of volleyball field is proposed; Combined with the discrete element method and the design requirements of detection standards, a comprehensive load detection simulation method of volleyball court is designed.

2   Simulation of Impact Force Absorption of Synthetic Rubber Materials in Volleyball Field

2.1   Impact of Human Sports on Synthetic Rubber Materials in Volleyball Field

During the jumping movement of sports athletes on the volleyball field, jumping and falling will produce impact load on the floor. According to the relationship between force and reaction force, the impact load acts on the human body in the form of impact rebound. If the impact rebound is too large, it will damage the human sports joints [3]. Based on the mathematical model and simulation solution process established by the discrete element method, the load borne by the human body in the jumping process is obtained: the minimum absorption rate of the impact load of the volleyball court is determined according to the load threshold of the human femoral head joint, which provides a theoretical basis for the subsequent structural optimization of the volleyball court. In the current sports competition, the physical stress status, physical and chemical index changes, psychological and emotional changes and other factors of athletes in the process of sports will have an important impact on the competition results and athletes' physical and mental health [4]. Therefore, protecting athletes' physical and mental health and making athletes have a good competitive state has developed into a multifaceted and comprehensive interdisciplinary subject. Among them, sports medicine, sports physiology, sports biochemistry, discrete element method, sports psychology, sports nutrition, training equipment and other technologies constitute an important scientific and technological support system for sports training [5]. From the perspective of analyzing the force on the human body, the discrete element method plays a positive guiding role in solving the impact load on the human leg in the process of jumping. In order to more clearly describe the impact load and load absorption process during human jumping, the impact load absorption flow chart shown in Fig. 1 is drawn.

The specific process is: the human body generates an impact force F → the volleyball court absorbs part of the impact load and the human body's own legs bend to absorb the impact load b → the human body actually bears the load F1 → according to the

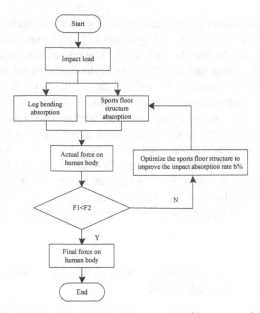

Fig. 1. Shock load absorption process of human motion

human body discrete element method to determine whether the impact load is less than the human bone joint force threshold F2 → If it matches (ie F1 < F2), solve the final impact absorption load F3 of the floor; if it does not meet (ie F1 ≥ F2), change the structure of the volleyball court to improve the absorption performance of impact load a. Among them, the calculation formula of the relationship between the impact load F of the human body and the actual force F1 of the human body is:

$$F_1 = F \times s(1 - a) \times (1 - b) \tag{1}$$

where, F is the impact load generated by human motion, s is the actual force of human motion, a is the bending impact absorption rate of human legs, and b is the impact absorption rate of volleyball court structure. m represents the internal force between the hinges of the knee; g represents the internal force distance, which is the control torque of the human body and is generally a function of $m\ddot{y}$. The dynamic equation of the system can be listed by Newton mechanics, momentum theorem and momentum moment theorem as follows:

$$\begin{cases} m\ddot{y} = F_y - mg \\ (m\ddot{y} + mg)l\sin\theta = M(\theta) \end{cases} \tag{2}$$

where, l is the body mass, F_y is the ground rebound force, and θ is the angle between the lower leg and the vertical direction; $M(\theta)$ is the moment of human leg, ω is the length of human leg, d is the second derivative of the height of the leg contacting the ground from the ground, which describes the constraint equation between the height t between the body and the ground and the included angle θ between the lower leg and the vertical direction when the human foot contacts the ground:

$$\begin{cases} y = 2l\cos\theta \\ \dot{y} = 2l\sin\theta, \omega = -\frac{d\theta}{dt} \\ \ddot{y} = -2l\omega\left(\frac{d\omega}{d\theta}\sin\theta + \omega\cos\theta\right) \end{cases} \tag{3}$$

You can get:

$$M(\theta) = 2F_y l\sin\theta \tag{4}$$

Let $x = 2$, the first-order linear differential equation can be obtained:

$$\frac{dx}{d\theta} + (2\cot\theta)x = \frac{1}{l\sin\theta}\left[g - \frac{M(\theta)}{ml\sin\theta}\right] \tag{5}$$

The expression of the analytical solution of the differential equation is:

$$x = \frac{1}{\sin^2\theta}\left[\int \frac{\sin\theta}{l}\left[g - \frac{M(\theta)}{ml\sin\theta}\right]d\theta + C\right] \tag{6}$$

In order to further explore the motion and stress process of various parts of human lower limbs and analyze the stress of important joints of human lower limbs, the method of human simulation model analysis is adopted. This method is the key technology of the cross scientific research of computer and modern discrete element method, and it is also one of the frontier directions in the field of virtual reality and computer vision [6]. Adams dynamic modeling and simulation software, which has the advantages of fast calculation speed, high simulation accuracy and two-way solution of kinematic and dynamic parameters, is selected to analyze the stress status of each joint of the leg in the process of human jumping, and the three-dimensional modeling software is used to call different human body size parameters to draw the leg model of the human lower limb. The size and mass of each part of human lower limb are shown in Tables 1 and 2.

Table 1. Size and mass of human lower limbs

Project	Company	Average value	Standard deviation
Thigh length	mm	502.758	38.582
Calf length	mm	376.252	21.695
Foot length	mm	249.256	11.165
Thigh weight	kg	8.265	0.865
Calf weight	kg	2.365	0.305
Full weight	kg	0.832	0.065

Taking volleyball as an example, the maximum reaction force F generated by the ground to the human body in the take-off process is 6500N, the standard adult male

Table 2. Hourglass energy of impact deformation of material structure

Cross section	Ehg(J)	Eint(J)	RHI(%)
SQUARE	6.325	652.962	0.98
NCMC12	21.358	1429.658	1.51
NCMC20	43.925	2350.200	1.89

weight is 60kg, the length of thigh and calf is 400m, and the maximum bending angle of leg buffer jump is calculated according to the actual jumping movement law of human body θ The variation range is 0° ~ 30° [7]. Due to the complexity of the functional relationship, in order to clearly and intuitively express the parameter relationship in the functional relationship, the functional relationship between the human leg bending offset force F and the leg bending angle is constructed, programmed with MATLAB software, and the functional relationship image between the human leg bending offset force and the leg bending angle is drawn, as shown in Fig. 2.

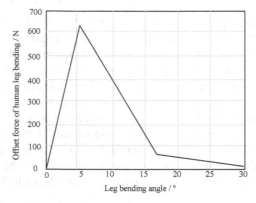

Fig. 2. Relationship between human leg bending offset force and leg bending angle

It can be seen from Fig. 2 that the angle between the resistance generated by the human leg and the vertical direction increases rapidly and then decreases gradually. When the bending angle is 5°, the maximum offset force generated by the human body is 650N, when the bending angle is 18°, the offset force is reduced to 50N, and then the offset force continues to decrease. Therefore, it can be obtained that in the process from the human body's take-off contact with the ground to the bending of the legs, taking the maximum impact force of 6500N as an example, the body itself can absorb 650N impact force by relying on the leg movement mode, that is, 10% of the total impact force received by the body. This data acquisition lays a foundation for subsequent simulation analysis and the design of floor impact absorption and impact instrument.

2.2 Impact Deformation Characteristics of Synthetic Rubber Materials

The synthetic rubber surface sports ground is different from the hard pavement. The impact absorption test must be carried out with a special synthetic material surface sports ground impact absorption tester based on the impact absorption test specification in the national standard GBT36246–2018 to determine FR. FR is the percentage value of the impact absorption performance of the synthetic material surface layer relative to the rubber material base layer. Therefore, before calculating FR, not only the synthetic material surface layer needs to be tested, but also the peak impact force of the hard rubber material base layer needs to be measured [8]. Finally, the peak impact force of synthetic material surface and hard rubber base is calculated, and H is obtained. The energy absorption of each corner part in a folding cycle is:

$$E = M(\theta)\left(16IF_1/x + 2\pi y + 4CIH^2/b\right) \tag{7}$$

where, C is the partial plate length of each corner. Therefore, the energy absorption of a non convex section thin-walled material with I corners in one cross section is:

$$\frac{P_m}{M_0}2H\chi = E\left(16I_1\frac{b}{t}H + 2\pi C + 4I_3\frac{H^2}{b}\right) \tag{8}$$

where, χ represents the effective folding coefficient and N is the perimeter of the material cross section, then:

$$\frac{P_m}{M_0} = \frac{1}{\chi}\left[N\left(8I_1\frac{b}{t} + 2I_3\frac{H}{b}\right) + \pi\frac{L_c}{H}\right] \tag{9}$$

Hourglass deformation may lead to invalid numerical simulation results. Therefore, hourglass deformation must be restrained as much as possible. From the deformation after compression, the deformation of each rubber material is very smooth, and no obvious hourglass deformation is found. The hourglass deformation energy of each material when the impact deformation is 0.73 times the initial material length is listed in the table.

It can be seen from the table that the hourglass deformation energy of the material is lower than 2% of its internal energy. Therefore, the hourglass deformation is well restrained, and the calculation results are effective. The design of specific performance structure is a very common and important problem in engineering application. The energy absorption properties of square materials, multi cell square materials and non convex section materials will be compared and analyzed from the three aspects of the same material dosage, equal energy absorption and equal maximum peak load [9]. The weight of energy absorbing structure is not only directly related to its manufacturing cost, but also very important to the cost of the structure in the whole life cycle. The energy absorption performance of non convex cross-section discrete element is studied by the combination of theoretical analysis based on discrete element method and explicit nonlinear finite element analysis. Based on the discrete element method, it is extended to the discrete element including the cross part, and the calculation formula of the average force of the nonconvex discrete element is deduced. Secondly, the axial deformation of the nonconvex discrete element is analyzed by using the explicit nonlinear finite element analysis software ANSYS /LS-DYNA, and the detailed energy absorption performance parameters of this kind of structure are obtained. Finally, the theoretical prediction and numerical simulation results are compared and discussed, and the advantages of non convex section discrete element are revealed. The schematic cross-section of non convex material is shown in Fig. 3.

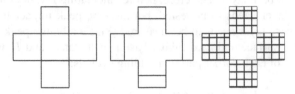

Fig. 3. Cross section of non convex material

Figure 3 shows the geometric composition and energy absorption of the T-shaped part composed of two plates and three plates. Based on the observation of the deformation pattern of the non convex discrete element in the cross part, the composition and energy dissipation of the cross part deformation mechanism are given, and the energy absorption of the deformation mode of the cross part is the same whether at the concave corner of the outer profile or in the middle of the discrete element [10]. According to the impact load in the functional index of the floor used in the standard gymnasium, the impact absorption rate values in the purposes of competition, training, teaching and fitness are shown in Tables 3 and 4.

Table 3. Impact absorption rate of volleyball courts for different purposes

Floor use	Impact absorptivity(%)
competition	≥ 55
Training and teaching	≥ 36
Bodybuilding	≥ 36

Table 4. Constraint setting of sports field on synthetic material surface

Parts	Upper surface	Bottom surface	Front	After	Left	Right
Synthetic surface	Contact constraint	Contact constraint	X = 0;Y = 0	X = 0;Y = 0	X = 0;Y = 0	X = 0;Y = 0
Concrete base	Contact constraint	Fix suport	Fix suport	Fix suport	Fix suport	Fix suport

The impact absorption rate f of volleyball court is 41%, which is the value that can withstand the impact in the range of 35% ~ 53%.

2.3 Simulation of Impact Force Absorption of Synthetic Rubber Materials

As a deformable body, rubber base is usually represented by volume element in the numerical model of impact absorption test, so there are only translational degrees of freedom in three directions, so only translational degrees of freedom constraints are considered in the numerical model. Since the upper and lower surfaces of the synthetic material surface are in contact with the steel dynamometer and the rubber base respectively, it is only necessary to set displacement constraints on the front, rear, left and right surfaces of the synthetic material surface, that is, fully constrain in the XY direction and release only the Z axis. Based on the actual situation, except the contact surface between the rubber material base course and the synthetic material surface course, all other surfaces are added with fixed constraints to suppress all displacement degrees of freedom. The specific settings are shown in the table.

In ANSYSWorkbench, contact description can be applied to the connection relationship between objects. By setting different contact relations, the force and energy transfer algorithm of contact coincidence area between different objects can be defined. The contact types given in ANSYSWorkbench include binding, non separation, no friction, roughness and friction, among which binding and friction are the most representative. Binding contact is used to describe the connection behavior that two contact objects do not separate in the normal direction of the contact surface and can not slide relative in the tangent direction. If there is friction contact, it describes the connection relationship that the normal direction can be separated and the tangent direction can not slide relative. The contact setting needs to set the target surface and contact surface for the contacted object. Generally, the object with fine grid, small surface area and low material hardness is the contact body, on the contrary, it is the target body. The contact surfaces on the contact body and the target body are the contact surface and the target surface respectively. From the complete three-dimensional model of impact absorption test, it can be seen that there are four groups of contact, Among them, normal separation can occur between the heavy object and the anvil, and relative sliding cannot occur in the tangential direction, so it is defined as frictional contact. The other three groups of contacts cannot be separated in the normal direction and relative sliding cannot occur in the tangential direction, so it is set as binding contact. To sum up, the contact definition in the numerical model is shown in Table 5.

In the process of impact force absorption simulation, when the initial displacement and velocity conditions and all time step loads need to be given, the displacement of the next time step can be obtained according to the initial acceleration, and then the velocity and acceleration can be obtained respectively. The displacement, velocity and acceleration of all time steps can be obtained by calculating in turn.

In the pre-processing stage, it is necessary to compare the real physical system and establish a scientific and reasonable numerical model from the aspects of research object model, material model and boundary conditions. The approximation between the numerical model and the real physical system plays a decisive role in the accuracy of

Table 5. Material contact settings

Contact serial number	Contact type	Interface	Target surface
A	Friction	Lower surface of weight M	Upper surface of anvil T
B	Binding	Anvil T lower surface	Upper surface of anvil T
C	Binding	Lower surface of steel side force platform G	Upper surface of moving material layer
D	Binding	Lower surface of moving material surface	Upper surface of concrete base

simulation results. In the calculation stage, select appropriate solution items for calculation according to the research purpose. After the solution is completed, enter the post-processing stage to process the data, evaluate and analyze the calculation results. When it is determined that the numerical model can meet the analysis requirements, output charts, analysis reports, etc. The rebound impact load results obtained are consistent with the results in sports structural mechanics that the first peak range of the vertical component of the ground reaction force of volleyball players after taking off and landing is 1000 ~ 2000N, while the second peak range is 1000 ~ 6500n, which can ensure the correctness of the simulation.

3   Analysis of Experimental Results

Volleyball courts can be divided into prefabricated rubber courts, mixed rubber courts and composite rubber courts according to different materials and pavement structures.

Prefabricated rubber track: it is composed of anti-skid layer and bottom rubber layer, all of which are made of PU and EPDM elastomer. The thickness of Pu is generally 9-13mm, with high resilience. It is a professional volleyball stadium.

Mixed rubber field: the rubber surface of this rubber field has a double-layer structure, with an anti-skid layer and a bottom layer. The bottom layer contains 15% - 25% of waste tire particles, and the thickness is generally 9-13mm. A layer of PU material is poured as the friction surface layer.

Compound rubber field: there is a rubber primer layer (40% - 60%) of rubber bonded waste tire rubber particles, with a general thickness of 9-25mm.

When the impact absorption test of rubber material base is carried out, the test steps are consistent with the impact absorption test of synthetic material surface sample. The main difference is that 11 impact tests are carried out on rubber material base, the results of the second to eleventh tests are valid data, and the final impact force peak F of rubber material base is the average value of the last 10 test results. The calculation method is shown in formula (10):

$$F_{conceere} = \frac{F_2' + F_3' + \cdots + F_{10}' + F_{11}'}{10} \qquad (10)$$

where: $F_2', F_3', \cdots, F_{10}', F_{11}'$ represents the 2nd to 11th times on the hard rubber material. Prepare one composite and one mixed composite surface sample, numbered respectively

Test_A and Test_B. The test scheme is designed according to the impact absorption test specification, as shown in Tables 6, 7, 8 and 9.

Table 6. Design of impact absorption test scheme

Test piece No	Test No	Weight mass(kg)	Test interval(s)	Length of test piece(mm)	Specimen width(mm)	Specimen height(mm)
Test_A	A	20	60	400	300	14
	B	20	60	400	300	14
	C	20	60	400	300	14
Test_B	A	20	60	400	300	14
	B	20	60	400	300	14
	C	20	60	400	300	14

Table 7. Damping pad parameters.

Serial number	Length (mm)	Width (mm)	Thickness (mm)	Hardness (HA)	Quality (g)	Material properties	Elongation (%)	Proportion(g/cm-3)
A	62	51	20	46	30	Natural rubber	300	1.5
B	77	50	20	46	69	Natural rubber	300	1.5
C	66	66	14	47	22	Natural rubber	300	1.5
D	78	50	10	48	13	Natural rubber	300	1.5
E	80	56	18	47	68	Natural rubber	300	1.5

The sample type is used as the control variable, and the peak impact force acting on the specimen surface during the test is used as the dependent variable. During the test, conduct three impact absorption tests on each sample at an interval of 60s, record the peak impact force measured in each test, and calculate the average of the peak impact force measured in the second and third tests of the sample. The calculation results are used as the final results of the peak impact force of the sample for FR calculation of the sample. The impact test takes the single-layer keel volleyball field as the research object. The test floor is composed of: the surface floor is made of hard miscellaneous wood such as maple; The wool floor adopts waterproof elastic multilayer board; The keel adopts solid wood keel; The experimental assembly size is 3600mm × 3600mm; It is produced by a domestic sports floor Co., Ltd. The impact test uses the five most common damping pads in the market as the experimental materials, and the parameters are shown in the table.

Table 8. Structural parameters of damping pad and experimental results of back impact force

Serial number	Damping pad parameters							Experimental result
	L(mm)	W(mm)	T(mm)	H(HA)	M(g)	A(%)	V(%)	F(kN)
A	61	51	20	46	30	77	92	1.53
B	77	51	20	46	69	42	68	1.99
C	66	66	14	47	22	22	60	1.84
D	78	51	10	48	13	46	83	1.53
E	75	56	18	47	68	69	90	1.65

Table 9. Parameter standardization coefficient

Model parameter	Standardization coefficient
L	−3.652
W	−4.989
T	−3.128
H	−2.658
M	−2.908
A	3.658
V	−4.256

The single-layer keel assembled volleyball field produced by a special floor Co., Ltd. is used as the experimental specimen in the impact test. Combined with the layout requirements of the test points of the impact load test and the floor assembly size used in the test, the center of the sports point is taken as the starting point, five points are selected as the actual test points, and marked on the floor surface. Based on the stress-strain data of composite and mixed composite surface samples obtained from uniaxial tensile test, in ANSYSWorkbench/EngineeringData, Mooney rivilin model is used to fit the Mooney-Rivilin test data of two surface materials. The fitting is shown in the Fig. 4.

According to the structural parameter requirements of the shock pad installed in the actual single-layer keel volleyball court, the parameters affecting the impact performance of the shock pad in the single-layer keel volleyball court include: length (L), width (W), thickness (T), hardness (H), mass (m), area proportion (A), volume proportion (V), etc. In order to determine the reasonable structural parameters of the damping pad, five kinds of damping pads with different structural parameters are selected for impact test to obtain the anti impact force F of the volleyball court under this condition. The structural parameters of the damping pad and the experimental results of the anti impact force are shown in the table.

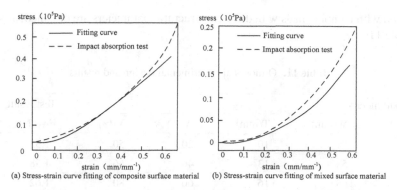

Fig. 4. Curve fitting of stress-strain data of synthetic material surface

The structural parameters of the damping pad and the experimental results of the back impact force of the single-layer keel volleyball field in the table are analyzed and calculated by using the fitting algorithm of SPSS multiple linear regression equation, and the standardized coefficient values corresponding to the 7 model parameters (structural parameters of the damping pad) can be obtained, as shown in the table.

The greater the absolute value of the standardization coefficient corresponding to each model parameter in the table, the greater the weight coefficient of the parameter, that is, the greater the impact on the back impact force, and the absolute value of 3 ~ 5 is the best. The absolute values of L, W, T, A, V standardization coefficients in the table are greater than 3. Therefore, five parameters L, W, T, A, V are selected as the structural parameters affecting the back impact force of the damping pad. By determining the five main influencing parameters of the damping pad, further explore the relationship between the parameters and the back impact force, and construct an orthogonal test factor level table of 5 factors and 4 levels according to the experimental requirements, as shown in Table 10.

Table 10. Factor level of orthogonal test

Level	Factor				
	L(mm)	W(mm)	T(mm)	A(%)	V(%)
5	61	51	10	20	51
10	68	58	14	40	66
15	75	65	16	60	80
20	82	72	19	80	96

According to the orthogonal test factor level table, the L(4) orthogonal test design scheme with the number of experiments of 16 can be designed. Through the impact test of single-layer keel sports floor, the test results of back impact force of sports floor installed with damping pads with different structural parameters are obtained, as shown in Table 11.

Table 11. Orthogonal experimental design and results

Test parameters					Test result
L(mm)	W(mm)	T(mm)	A(%)	V(%)	Fw(kN)
60	51	10	20	50	1.55
60	58	13	40	66	1.58
60	65	16	60	80	1.62
60	72	19	80	96	1.64
68	51	13	60	96	1.72
68	58	10	80	80	1.71
68	65	19	20	66	1.69
68	72	16	40	50	1.67
75	50	14	60	66	1.75
75	58	10	80	50	1.72
75	65	19	20	95	1.78
75	72	16	40	80	1.85
82	50	19	40	80	1.99
82	58	16	20	95	20.22
82	65	13	80	50	1.89
82	72	10	60	66	1.80

The primary and secondary order of the test parameters (influencing factors) in the table has been determined and screened by SPSS multiple linear regression equation fitting algorithm. In order to avoid repetition, the significance analysis of parameter influence on the test results will not be carried out. The impact absorption curves of two combined material surface samples are shown in the Fig. 5.

The abscissa represents the number of pressure data acquisition points, and the ordinate f represents the untreated impact force value. By comparing and analyzing the test curves, it can be seen that during the impact absorption test, the impact force borne by the sample increases first and then decreases, and the impact force peak is obtained at a certain time in the impact process, and then the untreated impact force peak is determined through the test curve. The establishment of the test table and the acquisition of the experimental results (anti impact force), It can be preliminarily concluded that under the action of a variety of structural parameters of damping pad through orthogonal test, the value of back impact force of damping pad of sports floor changes obviously, indicating that the selection of structural parameters of damping pad is more reasonable.

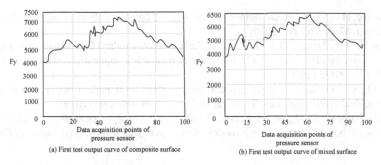

Fig. 5. Impact absorption curve of synthetic material surface sample

4 Conclusion

Combined with the discrete element method and the design requirements of the test standard, and according to the research on the test standard method of impact absorption load in volleyball court, the overall structure design and problem solution of the comprehensive load tester in volleyball court are put forward; The structural design and finite element static simulation analysis of the impact power control mechanism and rolling load operation mechanism of the comprehensive tester are carried out to verify the rationality, feasibility and safety of the design of the comprehensive tester for volleyball field. The research provides a new idea for solving the method of impact load on human lower limbs, provides a theoretical basis for improving the sports protection function of volleyball court by optimizing the damping structural parameters, provides a new method for improving the detection accuracy of impact force absorption mold of synthetic rubber materials in volleyball court, and promotes the development of volleyball court production and detection technology in China.

Fund Projects. This work was supported by Aba Normal University, 2021 research projects focus: New Material Technology for the development of volleyball.(NO. ASA21–15).

References

1. Jiangyang, Z., Defu, Z.: Simulation of human motion information capture in time-space domain based on virtual reality. Comput. Simul. **38**(08), 391–395 (2021)
2. Graziosi, P., Kumarasinghe, C., Neophytou, N.: Impact of the scattering physics on the power factor of complex thermoelectric materials. J. Appl. Phys. **126**(15), 155701 (2019)
3. You Liming, D., Wei, H.X., et al.: Simulation of stress-strain of chloroprene rubber hose before and after hot oil aging. China Syn. Rub. Ind. **43**(06), 458–462 (2020)
4. Saputra, D.A., Husin, S., Gumelar, M.D., et al.: Preparation and characterization of hard rubber and soft rubber for marine rubber fender. Macromol. Symposia **391**(1), 1900189 (2020)
5. Banlusan, K., Amornkitbamrung, V.: Effects of free volume on shock-wave energy absorption in a metal-organic framework: a molecular dynamics investigation. J. Phys. Chem. C **124**(31), 17027–17038 (2020)
6. Jiang, H., Li, B., Zhao, B., et al.: Aging characterization of 500-kV field-serviced silicone rubber composite insulators with self-normalized photothermal radiometry. Infrared Phys. Technol. **23**(3), 103763 (2021)

7. Haiming, H., Qin, W.: Finite element simulation and fatigue analysis of tire mold base. China Rub. Ind. **67**(01), 65–68 (2020)
8. Hidayat, A.S., Jayatin, D.K.A., et al.: Optimization of standard Indonesian rubber and ethylene propylene diene monomer blending for ship launcher application. Macromol. Symposia **391**(1), 1900135 (2020)
9. Lu, Q., Qi, D., Li, Y., et al.: Impact energy absorption performances of ordinary and hierarchical chiral structures. Thin-Walled Struct. **140**, 495–505 (2019)
10. Qi, Z., Jianwen, S., Shuangfu, S., et al.: Finite element analysis of rubber materials based on Mooney-Rivlin models and Yeoh models. China Syn. Rub. Ind. **43**(06), 468–471 (2020)

Research on False Public Opinion Recognition Method of Social Network Based on Fuzzy Cluster Analysis

Gang Qiu[1]([✉]) and Jun Xie[2]

[1] School of Information Engineering, Changji University, Changji 831100, China
coral1001@163.com
[2] School of Physical Education and Health, A'ba Teachers' University, Wenchuan 623002, China

Abstract. In order to solve the problem that the classification of network pseudo public opinion events is too subjective, a social network pseudo public opinion recognition method based on fuzzy cluster analysis is proposed. Combined with the principle of fuzzy cluster analysis, a new index system for identifying false and emotional events is established. On this basis, the relevant index data of network false and emotional events are collected, and the classical fuzzy cluster analysis algorithm is used to cluster and analyze the network false and emotional events, so as to obtain different types of network false and emotional event sets, analyze and summarize the characteristics of all kinds of false and emotional events, and finally confirm through experiments, The social network false public opinion identification method based on fuzzy cluster analysis has high practicability, provides a new method for the identification and classification of network false and sentiment, and provides a reference for relevant departments to accurately control all kinds of false and sentiment by using network big data.

Keywords: Fuzzy cluster analysis · Social networks · Identification of false public opinion

1 Introduction

With the continuous development of computer technology and artificial intelligence, network public opinion analysis in the era of big data has attracted much attention. At present, the number of Internet users in China has exceeded 900 million, and the Internet has gradually become a distribution center of Ideological and cultural information and an amplifier of public opinion. With the gradual integration of network public opinion into people's life, some fictional, untrue and misguided remarks or events, that is, network pseudo public opinion, are disturbing people's life, enterprise operation and government decision-making. Therefore, how to use big data technology to deeply mine historical data, clarify network pseudo public opinion events, improve the pertinence of pseudo public opinion management, and provide reference basis for relevant departments to

W. Fu and L. Yun (Eds.): ADHIP 2022, LNICST 469, pp. 703–719, 2023.
https://doi.org/10.1007/978-3-031-28867-8_51

accurately control network pseudo public opinion [1]. The decision tree method of network pseudo public opinion collects and preprocesses network public opinion, classifies the processed data through the decision tree method, and finally completes the identification of social network pseudo public opinion on the classified data. However, this method is relatively simple and has the problem of low classification accuracy. In this regard, this paper proposes a method for identifying network pseudo public opinion based on the combination of fuzzy clustering analysis and k-nearest neighbor. To sum up, the existing research on the classification of network pseudo public opinion is only limited to text analysis and manual classification of features, and most of them refer to the existing index system of network public opinion classification. In view of the particularity and complexity of pseudo public opinion, it has many features different from real public opinion [2]. This paper will comprehensively consider the characteristics of network pseudo public opinion, build a brand-new pseudo public opinion index system, use fuzzy clustering analysis algorithm to classify and identify network pseudo public opinion, and combine network big data to provide a new idea for the classification of pseudo public opinion.

2   Identification Method of False Public Opinion in Social Network

2.1   Collection of Pseudo Public Opinion Characteristics of Social Networks

Fuzzy clustering analysis algorithms aim to divide existing datasets into multiple mutually exclusive groups or clusters, so that objects in a cluster are as similar as possible and different from objects in other clusters. Fuzzy cluster analysis algorithm is by far the most widely used and mature data mining cluster analysis algorithm. It has the advantages of being simple, fast and suitable for processing large-scale data. Its basic idea is to randomly select from a dataset containing a large number of data objects. Select k data objects as the initial centroids, calculate the distance between each data object and d centroids, and divide all sample data into the clusters represented by the centroids closest to it. The mean value updates the C_i centroids until the x centroids no longer change in the iterative process, then the clustering result is output [3]. The traditional fuzzy clustering analysis algorithm is adopted, and the specific steps are as follows: select the appropriate k value, and use the sum of squares of errors as the objective function to measure the clustering quality, as shown in the formula:

$$\text{SSE} = g \sum_{i=1}^{K} k \sum_{x \in C_i} d(C_i, x)^2 \tag{1}$$

where g is the standard Euclidean distance between two objects in Euclidean space, select the most suitable k value according to the fuzzy cluster analysis rule, and randomly select S centroids from the data set, usually the selected centroids are The n actual data points already in the sample dataset. Calculate the distance between each data in the data set and S centroids respectively, and divide the data objects into the classes represented by the nearest centroids. This article will use Euclidean distance to measure and calculate the distance between two data points. The centroid is updated with the mean of all data objects contained in each cluster, and the centroid of the i cluster is defined, where e is the object cluster, and m is the number of objects in the i cluster:

$$|S| = \frac{1}{m_i} \sum_{X \in c_i} \text{SSE} - Sen \tag{2}$$

The discretization algorithm based on normal gain generally calculates the p_i of each attribute, and then selects the attribute value corresponding to the extreme value of NG(A, S) as the discrete segmentation point. The normal gain calculation formula is as follows:

$$\text{NG}(A, S) = \frac{\sum\limits_{j \in value(A)} \frac{1}{|S|} \sum\limits_{i=1}^{k} -p_i \log_2 p_i}{\log_2 kn} \tag{3}$$

where p_i is the proportion of examples belonging to category i in training set S, and k is the total number of categories. $x(i, k)$ is an attribute, and k is the set of values. $y(i)$ is the total number of samples, and $y(j)$ is the number of samples whose attribute y is j. The normalized input-output correlation value of each sample attribute is calculated as an index for evaluating the importance of the attribute. This paper proposes to use the following formula to calculate the normalized input-output correlation value of the attribute:

$$IOC(k) = \text{NG}(A, S) \sum |x(i, k) - x(j, k)| \times \text{sign}|y(i) - y(j)| \tag{4}$$

Confirm whether the iterative process is terminated according to whether the centroids of various types have changed. If the centroids have not changed, the clustering results will be output. The main flow of the fuzzy clustering analysis algorithm in this paper is shown in Fig. 1.

First, input the data into the data set, randomly select k data objects as the initial centroids, calculate the distance between each data object and the S centroids, divide all sample data into the clusters represented by the nearest centroid, update the centroids according to the average value of all data objects in the newly generated clusters, confirm whether the iterative process ends according to whether the centroids change, and return to recalculate the new centroids if centroids change, If there is no transformation, each point in the data set is assigned to the nearest centroid to form a cluster, and the cluster result is output.

Online public opinion includes real online public opinion and online fake public opinion. The essence of identifying online fake public opinion is to classify the real online public opinion and online fake public opinion[4]. The network pseudo-public opinion recognition model based on fuzzy clustering analysis is shown in Fig. 2.

Combined with the principle of fuzzy clustering analysis, a brand-new pseudo intelligence identification index system is established. On this basis, the relevant index data of network pseudo intelligence events are collected, and the collected data are preprocessed. Randomly generate a training set and a test set from the processed data. If the proportion of the training set data to the support vector machine is $> 80\%$, the parameter

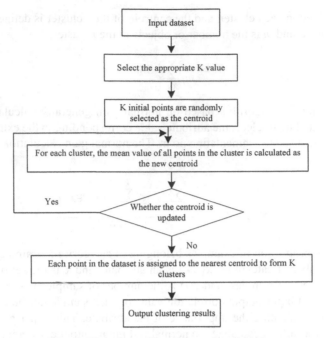

Fig. 1. Flowchart of fuzzy clustering algorithm

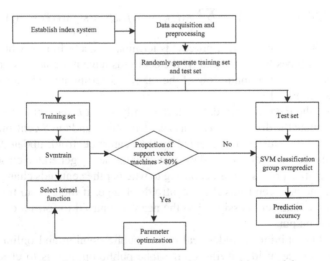

Fig. 2. Network fake and emotion recognition model

optimization will continue, and if it is not > 80%, the iterative process will be terminated. Then the accuracy is tested by SVM prediction of SVM classification group. And the test set is directly tested by SVM prediction of SVM classification group.

Establish a discriminant index for online pseudo-public opinion. Network public opinion can be described by some main index data. There is a relationship between entities and attributes between these evaluation indicators. The selection of reasonable and effective evaluation indicators is the basis and premise of effectively identifying online pseudo-public opinion. Secondly, Collect data based on public opinion indicators [5]. The collected public opinion data includes both real public opinion and false public opinion, and at the same time pay attention to the quantitative balance of the two types of data. The collected data needs to be preprocessed, mainly including the determination of the weight of the information source data, the calculation of the direct information coefficient of the public opinion information from different channels, the supplementation of missing data, and the normalization of data of different dimensions, etc. [6].

2.2 Construction of the Evaluation Index System of Pseudo-public Opinion Features

An important purpose of false public opinion features is to provide a theoretical basis for identifying and monitoring false public opinion. However, not all pseudo-public opinions are both false and artificial [7]. Modeling is carried out according to partial differential equations. First, the relevant symbols are given as shown in Table 1.

Table 1. Description of symbols of pseudo-public opinion dissemination model

Symbol	Description (probability)	Symbol	Description (probability)	Symbol	Description (probability)
a1	$P \to F$	a5	$F \to R$	a9	$B \to R$
a2	$R \to F$	a6	$AS \to R$	a10	Equilibrium value
a3	$F \to AS$	a7	$IS \to R$	a11	Node transition state time
a4	$F \to IS$	a8	$B \to AS$		

The conversion rate in the social network public opinion crisis information dissemination model can be determined by the multi-subject attribute. The subject attribute is usually statistically analyzed by the expert scoring method and the survey method. Because the subject attribute statistics are carried out by means of dimensionless quantity statistics or dimensionless expert scoring values. Given, the conversion rate and the influence rate should be controlled within the range of [0,1], so all conversion rate formulas calculated by using the main attributes are expressed as ratios. The main attributes are shown in Table 2.

The three types of pseudo-public opinions A, B, and C are defined as follows: if some public opinions are false, that is, the events or opinions that caused the public opinions are not in line with the facts, then no matter whether there is human intervention or whether the publishers have a tendency to the opinions, The resulting public opinion discussions are also false public opinion. Such pseudo-public opinions are category A pseudo-public opinions, such as false information such as "radish can cure cancer" [8].

Table 2. Attributes of pseudo-public opinion subjects

Subject	Main attributes
government	Authority, credibility, processing speed, information disclosure and transparency
netizen	Participation, attention, conformity and polarization
media	Attitude, media influence, communication intensity
Internet media	Credibility, activity and reporting frequency
event	Hazard, fuzziness, explosiveness and sensitivity
Mobile social platform	Social influence, communication speed, opinion leader force

Falsehood is a necessary condition for Class A false public opinion. If the pseudo-public opinion itself cites real events or a certain personal opinion, likes and dislikes, but uses human intervention to create and guide public opinion, the final public opinion that violates the objective development law of online public opinion is also a kind of pseudo-public opinion, which is called Type B pseudo public opinion, that is, the opinion or event of public opinion is not false or does not involve authenticity but the process of formation is false. The creation of such pseudo-public opinion often has a clear purpose, and the views held have a clear tendency, otherwise it will be meaningless to artificially intervene in the dissemination of topical events or speeches. Fundamentally speaking, the reason why the B type pseudo-public opinion is called pseudo-public opinion is that it adopts the means of artificially intervening in the formation and development of public opinion, that is, the artificial characteristic is the necessary condition for the B type pseudo-public opinion. A considerable part of pseudo-public opinion may have both false and artificial characteristics, which is called C type pseudo-public opinion [9]. The use of false, forged, tampered, and exaggerated events and the use of various means to induce or mislead netizens in order to achieve their own goals is often the most serious harm of such false public opinion, and it is also the focus of online public opinion supervision. Falsehood and artificiality are the necessary and sufficient conditions for Class C false public opinion. Therefore, the relationship between pseudo-public opinion features and types is shown in Fig. 3.

In order to effectively warn the public opinion crisis of mobile social networks, this paper uses the fuzzy analytic hierarchy process to construct the crisis early warning index system, and uses the triangular fuzzy number to determine the index weight, so as to determine the risk index of the network public opinion crisis early warning index. The network public opinion crisis early warning index system is shown in Fig. 4.

From the perspective of six main factors, the popularity of online public opinion is mainly affected by social networks and public opinion; for online public opinion attitude, it is mainly affected by the government and netizens; for online public opinion behavior, it is mainly affected by events, media, online media and social influence, therefore, for different indicators, they correspond to different main influencing factors, this study will give each main factor and its internal influencing factors in the simulation of crisis early warning effect on the network public opinion crisis early warning. Overall effect [10].

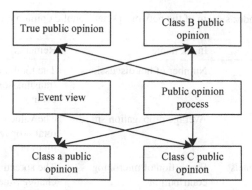

Fig. 3. Relationship between characteristics and types of pseudo public opinion

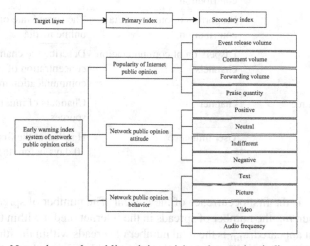

Fig. 4. Network pseudo-public opinion crisis early warning indicator system

For the subjective influence level, the triangular fuzzy comment value given by experts is used to determine the importance of the index, and the evaluation judgment matrix and the fuzzy positive and negative value matrix are determined. Combined with the above-mentioned indicators obtained by intuitive observation, some indicators that have no hysteresis defects in the past network pseudo-public opinion research are used, and the final indicator system is constructed as shown in Table 3.

For each event, the propagation trend data of three channels are collected, and the raw data of the kth network public opinion event is

$$X_k = \left\{ x_{t_{k1}}, x_{t_{k2}}, x_{t_{k3}}, \cdots, x_{t_{ki}}, y_k \right\} \tag{5}$$

Among them: x is the propagation times of the k event t_k time, and its expression is

$$\overrightarrow{x_{t_{kl}}} = \left(t_{ki}, w_{t_{ki}i}, e_{t_{ki}i}, n_{t_{ki}}, s_{t_{ki}i} \right) \tag{6}$$

Among them: t_{ki} is the i hour after the first occurrence of the k event, in hours, $w_{t_{ki}i}$ is the number of social communication within the ki hour after the first occurrence of

Table 3. The index system of the network pseudo-public opinion recognition model

Category	Index name	Remarks
degree of heat	Number of hot discussions	The total number of peaks in the transmission trend after the event
	Average propagation speed	The value of the ratio of the total propagation to the duration up to now
Communication channels	Proportion of microblog contribution	The spread rate of microblog to change time
	Proportion of wechat contribution	Wechat's spread rate of the event
	Proportion of online media contribution	The spread rate of the event by online media
	Dispersion of communication channels	Describe the channel concentration of events in the communication trend
information source	channel	Channels of initial information sources
	User influence	H index of the first user within 30 days before the event

the t event, $e_{t_{ki}i}$ is the first occurrence of the k event. The number of spreads in WeChat within the i hour, $n_{t_{ki}}$ the number of spreads in the Internet media within the i hour after the k event first appeared, $s_{t_{ki}i}$ is the total number of spreads within the ith hour after the k event first appeared y_k is The authenticity of the artificially marked k public opinion event, 0 represents the real public opinion, and 1 represents the network fake public opinion. The degree of early warning of online public opinion crisis is analyzed through the fuzzy comprehensive evaluation results. The delineation of the police limit can fully describe the effect of early warning of online public opinion crisis, provide a scientific and accurate judgment basis for the dynamic simulation of the online public opinion crisis system, and ensure the early warning of social network public opinion crisis. The sensitivity of the model is ensured, and the government and relevant departments can timely detect changes in the police situation and make corresponding responses. The delineation of the police limit is shown in Table 4.

In order to effectively control the occurrence and development of the crisis of network public opinion, it is necessary to enhance the awareness of the crisis of network public opinion in relevant government departments, and the managers of network public opinion should unify the opinions raised by netizens when solving emergencies and network public opinion problems before making decisions., Comprehensively consider the real situation of the incident and compare it with the public feedback information, so as to avoid the occurrence of mass incidents. After emergencies occur, the government

Table 4. Delineation of early warning limits for online public opinion crisis

Indicator color	Alarm degree	Warning limit
Yellow	D	[0,0.2]
Orange	C	[0.2,0.3]
Powder	B	[0.3,0.7]
Purple	A	[0.7,0.9]
Crimson	S	[0.9,1.0]

should publish authoritative and authoritative information in a timely manner, and open communication channels between the public and the government, so that the public's right to speak is not limited to social networking, so as to avoid online public opinion caused by the spread of rumors deterioration occurs. The media must be guided by the mainstream opinions of the real-time information released by the media, and explain in a rational and objective way. In the process of information dissemination of online public opinion crisis, the media mainly takes the role of "gatekeeper" to screen and transmit information. Effective control, and the authenticity and reliability of media information dissemination determine the success of social network public opinion crisis early warning. Combined with the multi-agent interactive analysis of the public opinion crisis information dissemination mode, and based on the SIR model to explain the network public opinion crisis information dissemination, the network public opinion crisis information dissemination mode is shown in Fig. 5.

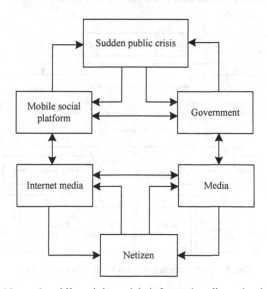

Fig. 5. Network public opinion crisis information dissemination mode

The development law of network public opinion is contained in various public opinion index data, and collecting effective indicators can improve the rationality and accuracy of pseudo-public opinion classification. At present, the research on fake public opinion on the Internet mainly focuses on improving the efficiency and accuracy of identifying fake public opinion. Therefore, the public opinion indicators selected by these studies focus on the characteristics that distinguish fake public opinion on the Internet from real public opinion. In order to achieve a reasonable classification of fake public opinion on the Internet On the basis of relevant research, this paper adds relevant indicators for the overall monitoring of public opinion, and on the basis of following the principles of scientificity, practicability, flexibility and clarity of division, the following 9 main indicators are selected as the identification of false public opinion on the Internet Indicator: Influence Index. Based on the social media and online media data of the whole network, weighted summation is made according to the dissemination effect of public opinion events on social media (mainly social media, WeChat) and online media, and the event influence after the summation Then, the event influence index ranging from 0 to 100 is obtained through normalization operation, which is an authoritative index used to reflect the effect of a public opinion event on the Internet.

2.3   Realization of Network Fake Public Opinion Recognition

The connotation of network public opinion Since it was put forward, public opinion usually refers to a collection of attitudes and opinions expressed by a certain number of people towards a certain event. The Internet public opinion originates from life but is formed on the Internet, and it is spread through the Internet. The relevant analysis of network public opinion is given below, as shown in Fig. 6.

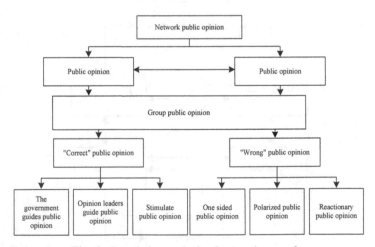

Fig. 6. Correlation analysis of network examples

Public opinion in a broad sense mainly includes three types, one is social public opinion, which can reflect public opinion at different levels from the national level, social level and class level, and highlights the social problems existing at different levels. The second is public opinion, also known as public opinion, which can reflect the perceptions, views, wishes and demands of people of different classes and strata towards the real society. The third is Internet public opinion. People use the Internet as a public platform to express their opinions and opinions, and use Internet language, video, pictures, comments, likes, forwarding and other forms to evaluate public events. The determination of the overall risk index of the social network public opinion crisis warning is also calculated. Through the calculation, the risk index of the social network public opinion crisis warning is equal to 0.612. The state of the crisis warning in this period is a severe level, and the indicator color is orange. At the same time, the government and relevant public opinion management departments should be reminded to increase the management of online public opinion Table 5. The crisis in which the risk index of all levels of indicators is in:

Table 5. Hazard index of indicators at each level

Index hierarchy	Variable	Hazard index	Index hierarchy	variable	Hazard index
Primary index	S1	0.358	Secondary index	S11	0.591
				S12	0.716
				S13	0.429
				S14	0.218
	S2	0.462		S21	0.0258
				S22	0.148
				S23	0.0035
				S24	0.796
	S3	0.629		S31	0.832
				S32	0.593
				S33	0.133
				S34	0.723
				S35	0.599

Therefore, from the perspective of practical analysis, it can be seen that the essence of network public opinion in a broad sense is still a form of public opinion that condemns and resists social events in the real world. Network public opinion in a narrow sense refers to an act of using network media to discuss emotions and attitudes. The study uses the broad form of network public opinion to extract social hot events, so as to carry out statistics of relevant information and data, and analyze the narrow network public opinion from different angles, in order to determine the development trend of network public opinion in emergencies, so as to correctly guide the development of public opinion. Provide theoretical support. Constant parameters are given as Table 6.

Table 6. Constant parameters of pseudo public opinion recognition model

Variable name	Value	Variable name	Value
Event publicity degree	36.85	Platform audience	1000
Event sensitivity	74.65	Platform activity	68.15
Hazard degree of the event	79.33	Media Authority	42.15
Emergency degree	46.85	Media activity	46.98
Polarization degree of Internet users	66.85	Online entertainment authority	46.86
Internet users' psychological intensity of curiosity	43.19	Online entertainment activity	52.28
Internet users' emotional intensity	75.26	Information disclosure and transparency	72.6
Role of opinion leaders	45.83	Authority of press spokesman	62.6
Internet users' forwarding intention	58.65	Government response speed	35.32
Content audit degree	33.65	Satisfaction with event handling	21.36

Based on social public opinion activities, from the perspective of multiple subjects, it determines the relevant factors that affect the formation of online public opinion crisis, and identifies the role of public opinion content in netizens, the government and society. It fully describes the size and judgment process of the crisis that may be caused by the public opinion formed by the event, and clarifies the judgment basis and process of the early warning signal. Social Network Public Opinion Crisis Early Warning is a universal public opinion crisis early warning system that can strengthen the protection of online public opinion. The specific framework is described in Fig. 7.

The network public opinion crisis early warning system used in social networking can accept comments from netizens at any time, and can timely, refine and deeply understand the trend of group intentions, and conduct targeted online education for netizens to avoid netizens because of group conformity. Psychology tends to polarize public opinion; scandal revelations, etc., in order to prevent public opinion from being pushed to the brink of group polarization and let the irrational behavior of netizens prevail, it is possible to effectively and timely respond to the crisis that may be about to arise in network public opinion. Forecast, minimize the degree of harm suffered by netizens, and correct the role of the media in responding to the crisis of online public opinion. The media, as the main medium for information dissemination, should supervise the positive development of online public opinion with a benign orientation in the development process of the crisis of online public opinion. The media has a certain "information buffer" effect on the network public opinion formed by sudden events. The media has different influences on online public opinion in the early, middle and late stages of the incident. In the early stage, because netizens reported the emergency through WeChat, social networking and other social networks, the public had an intuitive view of the malignant impact of the

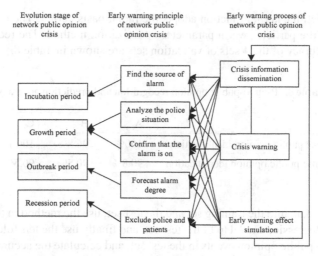

Fig. 7. Designing the network as a framework for public opinion identification

emergency. The feeling of the Internet public opinion has transformed from the gestation period to the growth period, which shows the backwardness of traditional media and Internet media and the mobile social network platform. In the middle and later stages of emergencies, the real-time reports on the extent of the impact of emergencies on the country and society, the extent of emergency handling, and the government's handling of such emergencies are unclear. Increased guesswork.

3   Analysis of Results

Using the scikit-learn library in Python to realize the construction of Logistic Regression and the recognition model in this paper, in order to verify the application effect of using the propagation trend data of public opinion events in identifying fake public opinion on the Internet, according to the authenticity of public opinion events and the prediction of the model As a result, the classification results can be obtained as shown in Table 7.

Table 7. Identification result confusion matrix

		Forecast category	
Real category	yes	yes	no
		YP (real)	WN(false negative)
	no	WP (false positive)	YN (true negative)

In order to further verify the effectiveness of the method in this paper, the four types of public opinion A, B, C, and D are unified into pseudo public opinion, and the four sets of experimental data are respectively applied to the proposed fuzzy clustering analysis model, in which the kernel function adopts the radial basis kernel. The parameters are optimized by the particle swarm parameter optimization method. The recognition and prediction accuracy of the 4 sets of validation sets are shown in Table 8.

Table 8. Pseudo-public opinion recognition and prediction accuracy

Group	A	B	C	D
Accuracy of true public opinion identification	100%	96.8%	100%	96.5%
Accuracy of false public opinion identification	90.8%	90.5%	90.5%	91.3%

Divide the data into the training set and the test set, use the method in this paper and the Logistic Regression model to fit the test set, and finally use the ten-fold cross-check to identify the public opinion events in the test set, and calculate the accuracy, precision and recall of the two models respectively. The average of the rate scores, as shown in Table 9.

Table 9. SVM and Logistic Regression model multi-index evaluation table

	Accuracy	Accuracy rate	recall	F1 score
Paper method	0.8335	0.8223	0.8335	0.7667
Logistic Regression	0.6811	0.6	0.69	0.6155

Compared with the data in the table, it is found that the prediction accuracy of the method in this paper is higher than that of the fuzzy cluster analysis method in the two models. The final cluster centers of the three categories obtained by mean clustering are shown in Table 10.

It is worth noting that the model only divides the authenticity of public opinion, while the method proposed in this paper needs to further divide the specific categories of false public opinion. The specific impact of online false public opinion events is shown in Fig. 8.

The impact on the formation degree of online public opinion crisis is shown in the curve in the figure above. The crisis formation degree shows decreasing changes, especially the decrease in the sensitivity of the event leads to a greater reduction in the formation degree of online public opinion crisis. For any pseudo-public opinion example, If it is only classified as false public opinion, the judgment is correct. Because the training set is randomly selected, there are differences in the dividing points and the number of intervals after discretization, but they are roughly stable at 4 or 5 intervals from high to low. The rules for each experiment were not exactly the same. Based on this comparison of the actual dissemination of fake public opinion, the public opinion tracking and identification of the method in this paper and the traditional method are compared and recorded, as shown in Fig. 9.

Table 10. Cluster center result table

Category	Influence index	Total number of participating media	Central level media participation	Proportion of public opinion field	Peak event heat
A	55.9755	28.9509	0.2425	0.0425	316.8565
B	59.377	32.9	0.2758	0.0723	1099.856
C	62.5	41.42859	0.385	0.1358	2392.652
Category	Peak propagation velocity	Duration	Number of communication channels	information source	Number of events
A	49.4985	182.6528	96.785	1.985	62
B	113.9	129.6	149.9856	1.8668	16
C	160.2859	195.1652	232.7658	1.7256	8

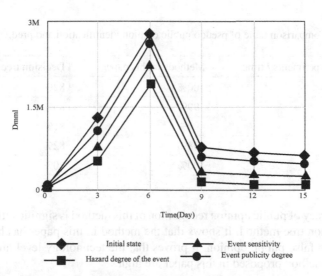

Fig. 8. Influence of online pseudo-public opinion events

In the process of establishing the fuzzy clustering identification model, the number of posts per hour, the duration of the first media, and the proportion of the largest opinion are the most important attributes of the root node, which are basically consistent with the identification of actual public opinion. The accuracy is relatively high.

In order to verify the effectiveness of the method in this paper, taking the accuracy of pseudo public opinion recognition as the experimental index, the method in this paper and the decision tree method are used for experimental testing. The test results are shown in Table 11.

It can be seen from the results shown in Table 11 that the accuracy of the method in this paper is up to 100%, and that of the decision tree method is up to 82%. It can be seen

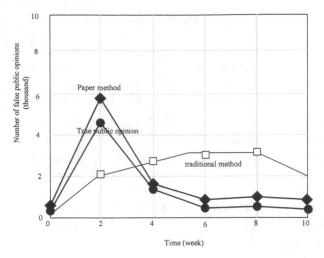

Fig. 9. Shows the situation of public opinion recognition

Table 11. Comparison table of pseudo public opinion identification and prediction accuracy

Number of experiments / time	Methods in this paper	Decision tree method
10	100%	82%
20	99%	80%
30	99%	81%
40	98%	82%
50	100%	80%

that the accuracy of public opinion recognition of this method is significantly higher than that of decision tree method. It shows that the method in this paper has high accuracy in identifying false public opinion. It proves that the technology level and application value of the method proposed in this paper are high.

4 Concluding Remarks

A set of new index system is established for the identification of online pseudo-public opinion, which covers the macro-index data characteristics of most pseudo-public opinion events, and further uses the classical fuzzy clustering analysis algorithm to divide the network pseudo-public opinion events into three categories, and analyzes them in detail. The characteristics of various kinds of pseudo-public opinion events are enriched, and the classification research of online pseudo-public opinion is enriched, which provides a reference for the relevant departments to accurately control and guide all kinds of pseudo-public opinion events, and contributes to the harmonious and stable development of the network society.

5 Fund Project

The work was supported by the 2021 Autonomous Region Innovation Environment (Talents, Bases) Construction Special-Natural Science Program (Natural Science Foundation) Joint Fund Project(2021D01C004) and 2020 Changji University Teaching and Research Project (20jyyb002).

References

1. Li, A., Fu, J., Shen, H., et al.: Indoor positioning algorithm based on fuzzy clustering and cat swarm optimization. Chinese J. Sci. Instr. **41**(01), 185–194 (2020)
2. Huang, L., Yao, B., Chen, P., et al.: Superpixel segmentation method of high-resolution remote sensing image based on fuzzy clustering. Acta Geodaetica et Cartographica Sinica **49**(05), 589–597 (2020)
3. Qz, A., Zw, A., Aha, C., et al.: A two-step communication opinion dynamics model with self-persistence and influence index for social networks based on the DeGroot model. Inf. Sci. **519**, 363–381 (2020)
4. Yin, J., Chen, H., Wang, J.: Negative public opinion propagation mechanism and evolutionary game analysis in online social networks. Inf. Sci. **38**(04), 153–162 (2020)
5. Ma, H., Wang, J., Lin, H., et al.: A multi-view network for real-time emotion recognition in conversations. Knowl.-Based Syst. 107751.1–107751.10 (2022)
6. Han, P., Gao, S., Yu, Z., et al.: Case-involved public opinion news summarization with case elements guidance. J. Chinese Inf. Process. **34**(05), 56–63+73 (2020)
7. Chen, L., Zheng, J., Cao, J., et al.: Intelligent traffic sign recognition method based on capsule network. J. Comput. Appl. **40**(4), 1045–1049 (2020)
8. Wang, L., Gan, S., Lin, N., et al.: Construction of collective system and analysis framework for china-related public opinion. J. Intell. **39**(06), 135–142 (2020)
9. Zhang, X., Wang, Y.: Information dissemination efficiency evaluation of network public opinion of the government microblogs based on input-output analysis. Inf. Sci. **38**(05), 43–48 (2020)
10. Liu, X., Nan, Y., Xie, R., et al.: DDPG optimization based on dynamic inverse of aircraft attitde control. Comput. Simul. **37**(07), 37–43 (2020)

Research on Abnormal Target Recognition of Full Information Mobile Monitoring Based on Machine Vision

Yudong Wei[✉] and Yuhong Xia

Chengdu College of University of Electronic Science and Technology of China,
Chengdu 611731, China
weiyudong@cduestc.edu.cn

Abstract. When the color of moving object is close to the background, the accuracy of moving object recognition is affected. So the method of moving object recognition based on machine vision is designed. In order to reduce the distortion of image edge position, the moving object is calibrated and corrected by vision. In order to reduce the influence of noise to a controllable range, the full information mobile monitoring image is enhanced to preserve the image details. The edge features obtained from view and template are calculated by moment, and the similarity is obtained. Then the contour feature of moving monitoring target is extracted based on machine vision. Segmentation of the background region, according to the moving object trajectory center point information such as speed, direction and so on to determine whether the trajectory is abnormal events. The proposed method is tested on INRIA dataset and Vehicle Reld dataset, and the results show that the proposed method can improve the accuracy and recall rate and has good detection performance.

Keywords: Machine vision · Full information · Mobile monitoring · Abnormal target · Target identification · Monitoring objectives

1 Introduction

Full-information mobile monitoring can be networked transmission of information, and has the ability to remote real-time video viewing. Full-information mobile surveillance anomaly target recognition can analyze surveillance video streams, detect anomalous behavior and alert in a timely manner at less cost [1]. At present, most of the video surveillance is still a simple function that only provides the collection, storage and playback of video. If there is an abnormality, we need to rely on the monitors to view and analyze the video scene through human operation and human eyes detection. For the massive video data, data analysis is difficult, the workload is very large, and a lot of human resources are needed. How to get the information which we are interested in from the massive surveillance video data, this is the question which the video surveillance technology develops to the intelligent direction faces. Traditional video surveillance

© ICST Institute for Computer Sciences, Social Informatics and Telecommunications Engineering 2023
Published by Springer Nature Switzerland AG 2023. All Rights Reserved
W. Fu and L. Yun (Eds.): ADHIP 2022, LNICST 469, pp. 720–733, 2023.
https://doi.org/10.1007/978-3-031-28867-8_52

can not meet the needs of practical application, so the research of intelligent video surveillance system based on computer vision has been paid more and more attention.

Chen et al. [2] proposed the research of full information alarm and monitoring system based on mobile network. Through the location positioning class, the A-GPS module is used to obtain location information, store the mobile multimedia information center and automatically forward it to the preset number monitoring terminal, so as to realize the recognition and alarm of abnormalities in full information monitoring. Ma [3] proposed the design of video image tracking and monitoring system based on target detection, which uses DSP + FPGA to form the core part of hardware, converts video signals into recognizable TTL level signals, and applies drive servo technology to achieve stable target tracking. Hu et al. [4] proposed an end-to-end SSD real-time video monitoring abnormal target detection and localization algorithm. By setting a target preselection box in the convolutional neural network, the abnormal classification and abnormal target boundary box are obtained to complete the abnormal detection. However, the above three methods do not consider the problems of image distortion and noise, resulting in low detection accuracy.

Therefore, an abnormal target recognition method based on machine vision for full information mobile monitoring is proposed. Through the visual calibration and correction of the moving monitoring target, the distortion of the image edge position is reduced. The method of moments is used to calculate the edge features of the view and the template, and the similarity is obtained. According to machine vision, the contour features of moving objects are extracted to detect abnormal objects. This method has certain theoretical and practical significance for improving intelligence, real-time and efficiency.

2 Abnormal Target Recognition Method of Full Information Mobile Monitoring Based on Machine Vision

2.1 Visual Calibration and Correction of Moving Monitoring Targets

The object in the real environment is projected onto a plane by the imaging system to form an image. The color of each pixel in the picture shows the "reaction" of the real environment after receiving the light at the same position. The confirmation of coordinate system is the key to locate objects. Vision systems often contain worlds, pixels, images, and camera coordinates. For the determination of the placement of the camera and the object, there must be a reference to describe their corresponding positions. The position of the pixel in the image is geometrically related to the corresponding position in the real environment. The projection model of an imaging system concretizes and digitizes this geometric connection. The most useful optical imaging model is the central projection model, i. e. the pinhole imaging model. A coordinate system is established in space according to the datum, so that any object in the same space can describe its position in the coordinate system. Before using machine vision for target recognition, the first choice is to calibrate and correct the moving monitoring target visually. The camera coordinate system is generally based on the camera, and its origin is the center of plane projection. The focal length of the camera is the distance between the origin of its coordinate system and the origin of projection plane coordinate system. The pixel coordinate system is unique in the view and can not be set at will, so it is difficult to calculate the conversion relationship between the pixel coordinate system and the camera position. The focal length varies with the position of the camera. The basic unit is the millimeter. The camera coordinate system coincides with the optical axis, and the X and Y axes are parallel to each other, which are similar to the projection plane coordinate system. The camera is calibrated to obtain its internal and external parameters, which represent the change of the azimuth of the view and the camera as well as the position of the camera in the world coordinate system. The center projection can be used between the camera coordinate system and the image coordinate system, and the conversion relationship is as follows:

$$r \begin{bmatrix} \alpha \\ \beta \\ 1 \end{bmatrix} = \begin{bmatrix} z\ 0\ 0\ 0 \\ 0\ z\ 0\ 0 \\ 0\ 0\ 1\ 0 \end{bmatrix} \begin{bmatrix} p \\ q \\ r \\ 1 \end{bmatrix} \tag{1}$$

In formula (1), z represents the camera focal length, (p, q, r) represents the camera coordinate system, and (α, β) represents the image coordinate system. Camera parameters are determined, not to be changed, is the camera's basic characteristics, scale factor, the coordinates of the pixel coordinates are included. The internal parameters will deviate from the values in the specification, so it is necessary to re-calibrate them in the application of machine vision. Camera calibration is the link between 2D image information and 3D information in reality. Zhang Zhengyou calibration method can keep high accuracy and robustness while considering lens distortion, and only need to print a piece of standard checkerboard as calibration object in practice. Zhang Zhengyou's calibration method has been widely used in the field of computer vision since its inception because of its high accuracy and flexibility. In this paper, Zhang Zhengyou calibration method, calibration template selected standard checkerboard. The homogeneous coordinates of corresponding points are obtained by projecting the points on the calibration template plane onto the image plane, and the mapping matrix can be expressed as follows:

$$W = \vartheta s \begin{bmatrix} \varphi_1 & \varphi_2 & \lambda \end{bmatrix} \tag{2}$$

In formula (2), W represents the mapping matrix, ϑ is the camera internal parameter matrix, s represents any standard vector, φ_1, φ_2 and λ are the rotation matrix and translation vector of the camera coordinate system relative to the world coordinate system. In the process of calibration, the image will be distorted or distorted, and the position of the real object in the picture will be different from the expected position. This is called distortion. The distortion parameters are calculated by maximum likelihood estimation. Because the location of distortion is usually the edge of the image, the selection region should be as close to the edge as possible, and the average value should be obtained after calculating multiple selection points. Based on this, the visual calibration and correction of moving monitoring target are completed.

2.2 Full Information Mobile Monitoring Image Enhancement Processing

In the process of image acquisition, the image quality is often damaged by the interference of external factors. In addition, the result of image post-processing is not ideal. Eliminating these disturbances is a necessary condition for the next operation. The image processing process is shown in Fig. 1.

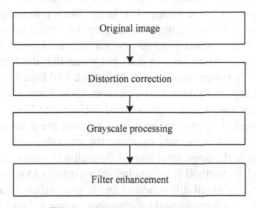

Fig. 1. Image processing process

Because the image captured by the camera is described and stored based on RGB color model, the color image is gray-processed. The key of image preprocessing is to reduce the influence of noise to a controllable range while preserving image detail features. Aiming at the characteristics of noise gray value, the filter is used to suppress the influence of noise and keep the useful and reliable information is the common denoising method. Because the gray image contains only a single channel array of pixels, the amount of data is small, which is conducive to improve the processing speed of image video, and can meet the basic requirements of video applications. Therefore, when processing and monitoring video images, color images in RGB format are usually converted to gray images [5]. The mean values of R, G and B components are taken as the gray values of pixels after gray-leveling, and the calculation formula is as follows:

$$G(\alpha, \beta) = \frac{X_1(\alpha, \beta) + X_2(\alpha, \beta) + X_3(\alpha, \beta)}{3} \qquad (3)$$

In formula (3), $G(\alpha, \beta)$ is the gray value of grayed pixels, $X_1(\alpha, \beta), X_2(\alpha, \beta), X_3(\alpha, \beta)$ respectively correspond to the R, G and B components of the pixel at position (α, β). Because of the aberration of optical system, the diffraction of optical imaging, the nonlinear distortion of imaging system, the random noise of image capturing device and other factors, the image quality will be degraded in the process of capturing information moving monitoring image. The Gaussian filter takes into account the influence of the near distance, so its ability to reduce the influence of noise to a controllable range is better than that of the neighborhood mean filter. In this paper, Gaussian low-pass filtering is used to enhance the image. This process can be expressed as follows:

$$F(\alpha, \beta) = e^{\frac{-h^2(\alpha,\beta)}{2\varepsilon^2}} \tag{4}$$

In formula (4), $F(\alpha, \beta)$ represents the enhanced image, e represents the natural constant, h represents the distance between the pixels at position (α, β) and the central pixel, and ε represents the standard deviation. The adjustment of ε can change the smoothness of the preprocessed image. The larger the ε parameter, the smoother the image. Image enhancement can highlight or suppress the image information of a specific part according to the specific needs to achieve rapid extraction of feature information. But the quality of image preprocessing does not mean that the higher the smoothness is, the more blurred the image features will be, which will lead to errors in the process of object extraction, especially in texture detection. Gray linear transform is a common method to enlarge the range of image contrast difference. This method enlarges the range of image contrast difference in the way of keeping the pixel value and gray value unchanged. Therefore, it is necessary to adjust the smoothness degree of ε parameter reasonably according to the noise level and the detection of image target features. Image enhancement is usually a method of enhancing image quality to solve the problems such as blurring caused by the small difference of image gray value, small range of contrast difference and so on, which are caused by improper parameter setting [6]. The image is converted to HSV space, the lightness value of video image is calculated and counted, and then the lightness graph is formed. According to the global distribution of V-value, V-value can be divided into different ranges, and the number of ranges can be selected according to the needs of feature detection and calculation. The contrast difference range is enlarged to improve the edge contour feature of the processed image and increase the success rate and accuracy for the subsequent edge contour extraction.

2.3 Extraction of Contour Features of Full Information Mobile Monitoring Targets Based on Machine Vision

When the color of a moving object in an image is the same or similar to that of a moving object in the background, the correctness of the tracking may be affected by the number of iterations. Therefore, combining the hue and contour features of the target, the local texture of the background object with the same color as the target in the window can be further judged, thus improving the accuracy of target tracking under complex background. The convex polygon is the convex hull of an object, which consists of all the points in the set of points. A useful method of understanding the shape or contour of an object is to calculate the convex hull of the object, and judge the degree of defect by the size of the convex hull area and the contour area. Edge extraction is to set the region with obvious difference between the gray values of each pixel in the image as the boundary line. Because the foreground and background of the binary image can be distinguished by image segmentation, the difference between the gray value of the boundary region and the gray value of its adjacent region can be distinguished obviously, and the edge extraction becomes very easy. Object contour detection is such a process, specifically refers to a digital image containing the target and background, ignoring the impact of the target interior texture and background, using a specific algorithm to extract the target peripheral point set. Object recognition, object detection, object tracking and shape analysis are all inseparable from the basic contour detection. As a feature based on the surface of an object, which is not dependent on the change of color and illumination intensity, the contour features reflect the rough and smooth degree of the object. At the same time, in the image, the contour mainly reflects the local structural features of the image, specifically, some changes of pixel gray level or color within a certain region [7]. First, the gray difference of pixels is calculated according to the convolution template in the x and y directions, and then the gradient in the x and y directions is calculated according to the formula (5).

$$
\begin{cases}
W(\alpha, \beta) = \sqrt{G_1(\alpha, \beta)^2 + G_2(\alpha, \beta)^2} \\
\tau = -\arctan\left(\dfrac{G_1(\alpha, \beta)}{G_2(\alpha, \beta)}\right)
\end{cases}
\tag{5}
$$

In formula (5), $W(\alpha, \beta)$ represents the gradient value, τ represents the gradient direction, $G_1(\alpha, \beta)$ and $G_2(\alpha, \beta)$ represent the gray difference in X and Y directions. Contour features are invariant in rotation, scale and noise, so the statistical or structural relationship between pixel values of local images is needed to be calculated for extracting contour features. The scale-invariant feature method decomposes the target image into squares. Finally, the feature points of the whole image are composed of the directions of these squares. Then the Hu moment feature matching is performed on the contour of the moving monitoring target. The essence of Hu moment is to construct several moment characteristics of normalized center moment by linear change, and it has good invariance of affine, translation and image transformation. The description effect of the same object in different directions is consistent [8]. If an image is described as a large rectangular box, it contains sixteen small rectangular boxes, or subregions. There are twenty-five characteristic points in each small rectangular box, through which the wavelet eigenvalues of small rectangular boxes can be obtained. Calculate the moments of the edge features obtained from the view and the template map, take their similarity, and then identify them. The specific process is as follows:

$$
\begin{cases}
\varpi = \sum_{7} \left| \dfrac{u_1 - u_2}{u_1} \right| \\
u_1 = sign(d_1)\, \lg|d_1| \\
u_2 = sign(d_2)\, \lg|d_2|
\end{cases}
\tag{6}
$$

In formula (6), ϖ represents similarity, u_1 and u_2 are Hu moment parameters, d_1 represents template contour, and d_2 represents matching contour. For the center pixel, the positions of the other 8 peripheral pixels are uniformly distributed around the center pixel at the same distance, and their effects are uniform and equal. But when the center pixel is given the weight, if the starting point is at a different pixel position, then it will be encoded differently, and these different encodings have exactly the same meanings and effects. According to the above method, the target contour features of full information moving monitoring are identified, and the flow chart is shown in Fig. 2.

The target contour extracted from the template image is basically the same as the actual contour of the target object. This contour is obtained by manually simulating

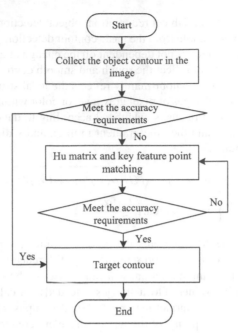

Fig. 2. Contour feature recognition process of full information mobile monitoring target

the similar environment. It can be compared with the object contour in the image to be recognized.

2.4 Establish the Abnormal Target Recognition Model of Mobile Monitoring

In video surveillance applications, surveillance cameras are usually placed in places where they can monitor important information, such as entrances, corridors, roads, etc., to monitor possible targets. However, the scene of video surveillance usually represents the moving range of the moving object in the surveillance area, and it is also the "visual" range of surveillance. Therefore, it is of great significance to analyze and study the monitoring scene area for the acquisition and analysis of moving objects. The normal video stream is divided into many segments in P frames, and then the convolution neural network self-encoder is trained with these segments. For the video stream waiting to be detected, the segment is divided into P frames, and the reconstruction error of P frames is used to measure the degree of anomaly. When the reconstruction error is large, the probability of the detected P frame containing abnormal behavior is considered to be high. In the full information mobile surveillance, it is difficult to analyze the scene and its content in the compressed domain, so in this paper, we first need to decode a small number of I frames, i. e. obtain the monitored scene model. Computing the background information of video requires decoding a small amount of I-frame weighting to obtain a relatively clean background information, and the background is processed accordingly. This process needs to continuously update the background as the video frame plays. Weighted gray histogram or color histogram is used to model the target, so that the pixels near the center of the target have greater weight, so that the edge noise and partial occlusion can be overcome. The specific background acquisition process is as follows:

$$V_t = \frac{Z_{t+c} + Z_{t+2c} + \cdots + Z_{t+mc}}{m} \tag{7}$$

In formula (7), V_t represents the background image after I frame superposition, Z represents the pixel value, t represents the time, c represents the interval between I frames in the video sequence, m represents the number of I frames, and the number of I frames required to obtain a background model does not exceed three. The behavioral trajectory of moving object is analyzed by segmenting the background region. Firstly, the histogram of luminance component Y is selected as the feature input for clustering segmentation. If the H value of the tracking target background is similar to the target, it is easy to drift.

Considering that the brightness is easily affected by illumination, while the saturation is not sensitive to illumination and is easy to distinguish between background and target, saturation S and hue H are introduced here to jointly count the color histogram of the target area. The number of clusters is set, and then the final clustering segmentation result is obtained by iterative operation according to the clustering rules of color histogram, and the classification result is used as the basis to distinguish active region and inactive region. The color distribution histograms of H component and s component of HSV color space are counted respectively, and then the back projection calculation is carried out to obtain the probability projection map. In the compressed domain, particle filter is used to track the target. For the tracked target, four tracking points are set, that is, the coordinates of the four corners of the tracking block diagram. The coordinates of the center point of the target can be calculated from the coordinates of the four points of the tracking frame. Images containing specific targets can produce very sparse and stable features through the deep convolution network with excellent classification performance. In video surveillance, since most pedestrians or other targets walk in a normal way or pass through the monitoring area, whether the track is an abnormal event is determined according to the information carried by the track center point, such as speed and direction. When judging the speed, use the following formula:

$$\eta = \sqrt{(B_N - B_{N-1})^2 + (A_N - A_{N-1})^2} \tag{8}$$

In formula (8), η represents the speed of tracking the target, (A_N, B_N) and (A_{N-1}, B_{N-1}) respectively represent the tracking point coordinates of the tracking target in frame i, N represents the number of frames the target spans. In this feature space, the changes of a small number of pixels and other ordinary changes in the input image will be ignored. Due to the pre training in the target classification training set, these convolution networks only respond to the structures that can represent the type of target and the structures of human interest.In order to detect the abnormal trajectory, this paper uses the velocity and direction information of the moving target as the input of the generalized regression neural network. For each moving target trajectory input, predict whether the current input trajectory is an abnormal event according to the historical information. If the trajectory is abnormal, it can be considered that its point velocity and direction change greatly, and then extract the abnormal target. So far, the design of abnormal target recognition method for full information mobile monitoring based on machine vision has been completed.

3 Experimental Study

3.1 Experimental Preparation

The moving targets in full information mobile surveillance video are mainly vehicles and pedestrians. In order to facilitate the experiment, this paper takes vehicles and pedestrians as the main objects. The algorithm framework in the experiment is implemented by tensorflow. The hardware environment is i7-5960x CPU, 64GB memory and NVIDIA Titan x graphics card. The running speed when detecting the test video is 20fpsa. This paper uses INRIA public pedestrian data set and vehiclereld vehicle data set as the training sample set. INRIA Pedestrian dataset (https://zhuanlan.zhihu.com/p/106216763)It is a video data containing pedestrians. The training set contains 614 positive samples (including 2416 pedestrians) and 1218 negative samples; The test set has 288 positive samples (including 1126 pedestrians) and 453 negative samples, which can be used for machine vision tasks such as pedestrian detection and recognition. Vehiclereld vehicle data set (https://zhuanlan.zhihu.com/p/106216763)It contains more than 50000 images of 776 vehicles, which were captured by 20 cameras and covered an area of 1.0 square kilometers in 24 h, which makes the dataset scalable enough for vehicle re ID and other related studies. The images are captured in the real-world unconstrained monitoring scene, and are marked with different attributes. The sample size is adjusted according to the scale characteristics of vehicles and pedestrians. Adjust the pedestrian sample size to 128x256 and the vehicle sample size to 256x256. The specific settings of the experimental data set in this paper are shown in Table 1.

Table 1. Experimental data set

Parameter	INRIA data set	VehicleReld data set
Number of positive samples	2358	2652
Number of negative samples	1164	1026
Sample size	128×256	256×256
Feature dimension	3685	1758

Adjust the size of the training negative samples of the two data sets for random cutting, and each negative sample generates five sample images with the same size as the pixels of the positive sample. The training sample set is used to train the model. The prediction network used is full convolution. The spatial resolution of the training data can be different from the test data. Therefore, the trained network can be applied to any resolution video.

This experiment measures the detection effect by calculating the accuracy and recall. The larger the ratio of accuracy and recall, the better the recognition effect of the proposed method.

Accuracy indicates the proportion of the detected abnormal targets in the total number, where TP indicates the actual abnormal data set, TN indicates the predicted abnormal data set, FP indicates the actual normal data set, and FN indicates the predicted normal data set:

$$Accutacy = \frac{TP + TN}{TP + TN + FP + FN} \tag{9}$$

Recall rate refers to the proportion of the number and total number of abnormal targets predicted and actually:

$$Recall = \frac{TP}{TP + FN} \tag{10}$$

3.2 Results and Analysis

After the training of target recognition network, the performance of the designed full information mobile monitoring abnormal target recognition method based on machine vision is tested on the test data set. In order to measure the performance of the abnormal target recognition method designed in this paper, target recognition is carried out on the test samples, and the detection effect is measured by accuracy and recall. The test results of this method are compared with the abnormal target recognition methods based on inter frame difference and motion vector. In the test of INRIA data set and vehiclereld data set, the accuracy comparison results of different recognition methods are shown in Table 2–Table 3.

In the test of INRIA data set, the accuracy of the full information mobile monitoring abnormal target recognition method based on machine vision is 96.26%, which is 9.09% and 10.44% higher than the other two methods respectively.

In the test of vehiclereld data set, the accuracy of the full information mobile monitoring abnormal target recognition method based on machine vision is 92.83%, which is 9.26% and 10.04% higher than the abnormal target recognition methods based on inter frame difference and motion vector, respectively. From the perspective of detection accuracy, the recognition effect of abnormal targets in this paper is better than the two comparison schemes. In the tests of INRIA dataset and vehiclereld dataset, the comparison results of recall rates of different recognition methods are shown in Table 4–Table 5.

In the test of INRIA data set, the recall rate of the full information mobile monitoring abnormal target recognition method based on machine vision is 96.00%, which is 9.84% and 12.72% higher than the abnormal target recognition methods based on inter frame difference and motion vector, respectively.

In the test of vehiclereld data set, the recall rate of the full information mobile monitoring abnormal target recognition method based on machine vision is 93.29%, which is 10.92% and 10.19% higher than the abnormal target recognition methods based on inter frame difference and motion vctor, respectively. Based on the test results of accuracy and recall, the proposed method has good performance and can meet the recognition requirements of abnormal targets in full information mobile monitoring.

Table 2. Comparison of accuracy of INRIA data set (%)

Number of tests	Abnormal target recognition method of full information mobile monitoring based on machine vision	Abnormal target recognition method of full information mobile monitoring based on inter frame difference	Abnormal target recognition method of full information mobile monitoring based on motion vector
1	96.17	88.42	84.46
2	95.04	87.08	85.87
3	97.51	88.27	86.64
4	96.22	85.34	88.58
5	95.66	86.61	84.21
6	96.81	87.55	87.45
7	95.52	85.26	85.69
8	96.54	86.93	86.33
9	97.36	87.65	84.22
10	95.76	88.32	86.07
11	96.88	87.09	85.52
12	95.64	87.55	84.84

Table 3. Comparison of accuracy of vehiclereld data set (%)

Number of tests	Abnormal target recognition method of full information mobile monitoring based on machine vision	Abnormal target recognition method of full information mobile monitoring based on inter frame difference	Abnormal target recognition method of full information mobile monitoring based on motion vector
1	92.48	82.63	81.79
2	93.57	83.17	82.81
3	92.81	84.74	82.47
4	91.65	85.39	82.54
5	92.36	82.62	83.19
6	93.58	82.84	82.68
7	92.22	83.57	82.43
8	94.13	83.22	84.25

(*continued*)

Table 3. (*continued*)

Number of tests	Abnormal target recognition method of full information mobile monitoring based on machine vision	Abnormal target recognition method of full information mobile monitoring based on inter frame difference	Abnormal target recognition method of full information mobile monitoring based on motion vector
9	92.52	84.58	82.71
10	93.20	82.47	83.08
11	92.64	83.20	82.62
12	92.81	84.46	82.96

Table 4. Comparison of recall rate of INRIA dataset (%)

Number of tests	Abnormal target recognition method of full information mobile monitoring based on machine vision	Abnormal target recognition method of full information mobile monitoring based on inter frame difference	Abnormal target recognition method of full information mobile monitoring based on motion vector
1	94.49	86.17	82.51
2	95.54	85.59	82.06
3	96.62	84.85	82.59
4	97.38	86.66	83.38
5	96.16	87.32	84.51
6	94.17	88.24	82.14
7	95.02	85.57	82.27
8	96.58	86.83	83.63
9	97.54	85.64	84.05
10	97.96	86.37	84.48
11	95.37	85.26	84.62
12	95.22	85.45	83.11

Table 5. Comparison of recall rate of vehiclereld dataset (%)

Number of tests	Abnormal target recognition method of full information mobile monitoring based on machine vision	Abnormal target recognition method of full information mobile monitoring based on inter frame difference	Abnormal target recognition method of full information mobile monitoring based on motion vector
1	92.49	83.47	83.97
2	96.18	80.74	82.56
3	90.56	81.58	84.41
4	96.22	82.16	82.18
5	90.60	83.03	82.43
6	95.85	80.25	83.74
7	93.24	82.31	83.86
8	92.37	82.64	84.58
9	92.61	83.88	84.05
10	92.52	81.26	81.32
11	94.73	84.50	81.61
12	92.16	82.65	82.48

4   Conclusion

Video surveillance technology is widely used to maintain the security of public places and homes. The video surveillance system that relies on manpower has been increasingly unable to adapt to this era of big data. Therefore, integrated artificial intelligence and intelligent computer vision technology become the inevitable development direction of video surveillance in the future. Based on machine vision, this paper designs an abnormal target recognition method for full information mobile monitoring, which can improve the accuracy and recall rate, and has good detection performance. In this study, only THE HOG feature is used to describe the target, and then the target representation with multi-feature fusion can be considered to describe the target more fully to improve the target recognition effect. In this study, the research on target recognition is still insufficient, and the research on the recognition of human target pose is insufficient. In view of such problems, in-depth research is needed in future research.

References

1. Zhao, X., Ding, B., Xi, Y.: Weak supervised abnormal behavior detection using improved YOLOv3 under video surveillance. Opt. Technique **47**(1), 120–128 (2021)
2. Chen, Z., Yan, B., Zhang, S., et al.: Research of whole information alaeming and monitoring system based on mobile network. Laser J. **41**(10), 144–148 (2020)
3. Ma, Y.: Video moving target design of video image tracking and monitoring system based on target detection. Modern Electron. Technique **43**(11), 47–50+54 (2020)

4. Hu, Z., Zhang, Y., Li, S., et al.: End-to-end SSD real-time video surveillance anomaly detection and location. J. Yanshan Univ. **44**(05), 493–501 (2020)
5. Mo, H., Gan, J.: Network digital video image anomaly target detection area recommendation simulation. Comput. Simul. **37**(6), 396–400 (2020)
6. Hu, Z., Zhang, Y., Li, S., et al.: End-to-end SSD real-time video surveillance anomaly detection and location. J. Yanshan Univ. **44**(5), 493–501 (2020)
7. Dong, Y., Hu, G.: Detection and identification of the abnormal behavior in video surveillance systems. Mach. Des. Manuf. Eng. **49**(3), 66–70 (2020)
8. Ji, G., Xu, Z., Li, X., et al.: Progress on abnormal event detection technology in video surveillance. J. Nanjing Univ. Aeronaut. Astronaut. **52**(5), 685–694 (2020)

Author Index

Printed in the United States
by Baker & Taylor Publisher Services